Reviews in Frontiers of Modern Astrophysics

Petr Kabáth • David Jones • Marek Skarka
Editors

# Reviews in Frontiers of Modern Astrophysics

## From Space Debris to Cosmology

Springer

*Editors*
Petr Kabáth
Astronomical Institute of the Czech
Academy of Sciences
Ondrejov, Czech Republic

David Jones
Instituto de Astrofisica de Canarias
La Laguna, Spain

Marek Skarka
Department of Theoretical Physics
and Astrophysics
Masaryk University
Brno, Czech Republic

ISBN 978-3-030-38508-8 ISBN 978-3-030-38509-5 (eBook)
https://doi.org/10.1007/978-3-030-38509-5



This Springer imprint is published by the registered company Springer Nature Switzerland AG.
The registered company address is: Gewerbestrasse 11, 6330 Cham, Switzerland

# Foreword of the Director of the PI Institute

Centuries ago, astronomy was the first of classical sciences that had developed an observational basis to verify and, eventually, falsify our theories about the processes shaping the cosmic bodies and the Universe as a whole. To this end, the development of highly sophisticated instrumentation has been crucial, as this allows astronomers to carry out precise observations and quantify and record results. However, along with the empirical approach, equally important has been the tight relationship with mathematics that provides the basis for deeper understanding of the origin and evolution of cosmic systems. Astronomy was the first to develop and employ procedures that are now routinely accepted by other branches of science as well as in engineering and technology. Nowadays, modern astronomy combines the cutting-edge technology of innovative engineering solutions for highly complex optical systems and space devices with the popular culture and citizen science, where the general public is engaged in research. This is also reflected by chapters of "Reviews in Frontiers of Modern Astrophysics: From Space Debris to Cosmology."

Does the basic research in astronomy provide also some kind of direct benefit to the society? Numerous practical applications of the "blue-sky" astronomy research are known very well: Communication methods of today owe a lot to radio astronomy research that had started several decades ago and helped to develop techniques of transmitting, compressing, detecting, and decoding weak signals. Besides other applications, this laid basis to the modern wireless technology. Construction of ever bigger telescopes and ever more sensitive detectors is motivated by an unending hunt for photons from the most distant galaxies, quasars, and other faint objects. Clearly, this is useful and relevant in numerous other areas of technology development far away from astronomy. Astronomers have realized the importance of charge-coupled devices for imaging, and these detectors are now present in all digital cameras. Likewise, active optics has been greatly improved in recent years in order to achieve amazingly sharp images and removing the degrading effect of atmosphere; the same technique is now employed in other areas of image processing. Indeed, the basic research in astronomy creates valuable knowledge and experience.

Students of astronomy are trained in mathematical methods, theoretical and experimental physics, computing and informatics, as well as modern technology.

They have to be; otherwise, a continuous flood of discoveries would quickly fade away. However, discoveries in astronomy cannot be planned. In fact, nowhere in science any attempt of perfect planning of next steps has ever worked. Instead, serendipity is what defines future directions for those who are prepared to recognize newly emerging opportunities. The discovery of cosmic microwave background radiation is a clear-cut example of an incidental finding which shaped our view of the cosmic history by supporting the hypothesis of the Big Bang. Understanding how Universe functions and evolves, what is our role in it, what will be the future of stars, of our Sun, and of the Solar System is clearly the main motivation for the endless effort of astronomers. It also attracts an enormous interest of general public and motivates new generations to learn. Exchange visits for young scientists, summer schools, workshops, research stays as well as participation in observations are essential ingredients of future astronomer's education.

Astronomers at several European research institutes have joined their forces in professional education and training within the ERASMUS+ scheme (Fig. 1). The five collaborating institutions are: Astronomical Institute of the Czech Academy of Sciences (ASU, the coordinating institution) in Ondřejov, Czech Republic; Instituto de Astrofísica de Canarias (IAC) in La Laguna and Breña Baja, Canary Islands; Masaryk University (MU) in Brno, Czech Republic; Comenius University (UK) in

**Fig. 1** Scientific research in astronomy has been for centuries traditionally concentrated around major observatories. Because of the evolving global environment—the gradual climate change, the population growth, and the technological advances—many of the historical sites struggle to perform their original purpose of cutting-edge observations. Still the persisting atmosphere of the scientific institutions, their existing infrastructure and culture offer an ideal setup for a new role contributing to challenges of the present-day society; in particular, teaching students, educating the young generation of scientists, promoting diversity and equity in research. ERASMUS+ program contributes to this effort in a very significant way. Ondřejov Observatory with its upgraded 2 m optical telescope is no exception in the global trend. Photo credit: Jaroslav Horák

Bratislava, Slovakia; and Astronomical Institute of Slovak Academy of Sciences (AI SAV) in Tatranská Lomnica, Slovakia. Under the active supervision by the project lead Dr. Petr Kabáth, this book is the lasting outcome of the partnership which had started before and will extend beyond the funding period of the program (2017–2020).

Individual chapters were prepared by senior researchers active in diverse topics of current expertise and research carried out at the participating institutes, where the students have been trained during extended internships. The focus has been concentrated on close environment of the Earth and the Solar System bodies, solar physics and stellar objects, exoplanets, as well as extragalactic astrophysics.

Director Vladimír Karas
Astronomical Institute of the Czech Academy of Sciences
Prague, Czech Republic

# Acknowledgement

This work was supported by the ERASMUS+ project 2017-1-CZ01-KA203-035562

# Contents

# Space Debris: Optical Measurements

**Jiří Šilha**

## Contents



**Abstract** Space debris is a major threat to the satellite infrastructure. A collision with even small particle, e.g. 1 cm of size, can cause a catastrophic event when the parent body, spacecraft or upper stage, will break up into hundreds of trackable fragments. Space debris research helps to discover, monitor and characterize these objects, identify their origin and support their active removal. Surveys with optical telescopes aim to discover new objects for cataloguing and to increase the accuracy of space debris population models. The follow-up observations are performed to improve their orbits or to investigate their physical characteristics. We will present the space debris population, its orbital and physical characteristics and we will discuss the role which the optical telescopes play in space debris research. We will

---

J. Šilha (✉)
Comenius University, Faculty of Mathematics, Physics and Informatics, Bratislava, Slovakia
e-mail: jiri.silha@fmph.uniba.sk



P. Kabáth et al. (eds.), *Reviews in Frontiers of Modern Astrophysics*,
https://doi.org/10.1007/978-3-030-38509-5_1

also discuss the adopted astronomical techniques like astrometry, photometry and spectroscopy used in the space debris domain.

# 1 Space Debris

Space debris, also known as orbital debris, can be defined as man-made object which is situated on geocentric orbit and have no longer any purpose. There are many sources and types of space debris with different origins, trajectories and physical parameters.

## *1.1 Satellite Infrastructure*

Since the first launch of *Sputnik 1* in 1957 thousands of satellites have been put on orbit around the Earth. They fulfill various tasks, from telecommunication, trough scientific and meteorological missions, and from military support, to broadband internet coverage. Additionally, the continuous satellite launches, the long exposure of hardware to the harsh space environment, and fragmentations of satellites and upper stages led to an unwanted population of debris objects orbiting the earth.

## *1.2 Spatial Distribution*

The orbital distribution of catalogued debris is directly associated with the operational orbits of satellites. A snapshot of the space debris spatial distribution as of January 2019 can be seen in Fig. 1, rendered by using data from the US public catalogue [1]. As seen in Fig. 1, there are several types of geocentric orbits. They can be classified according to their orbital elements.

There are several slightly different definitions of orbital regimes in literature. The following definitions should be considered as more generic, but they follow the general consensus. The most populated regime is the low earth orbit (LEO) with mean altitude above the earth surface lower than 2000 km, which corresponds to orbital periods of $P < 2.2$ h. In general, LEOs have small orbit eccentricities ($e < 0.1$) and inclinations of $<100°$. According to the public catalogue, almost 80% of all catalogued objects are located on LEO, with most of them fragments. In Fig. 1 the LEO population is represented by the dense region closely surrounding the earth.

A very unique type of orbit is the geosynchronous earth orbit (GEO). Its orbits have periods close to 24 h, inclinations between 0° and 15° and eccentricities close

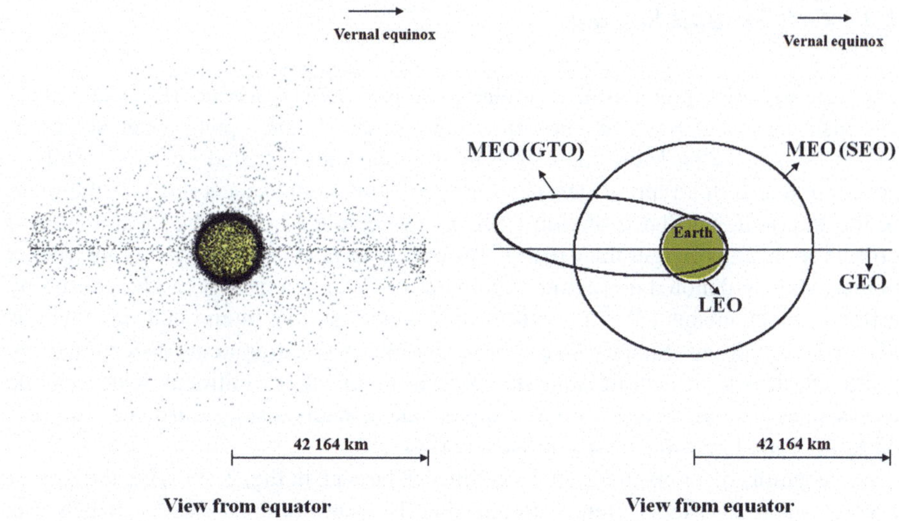


**Fig. 1** Space debris situation as of January 2019. The line of sight is in the equator and perpendicular to the vernal equinox direction. Data were obtained from [1]

to 0. Specifically, truly geostationary orbits are near-circular with inclinations close to 0° and orbital periods of one sidereal day (23 h, 56 min and 4.1 s), with mean altitudes above the earth surface of 35,786 km. Spacecraft located on this type of orbit remain above the same longitude in the equatorial plane. According to the public catalogue about 7% of objects are in or on GEO. They are mostly payloads and rocket bodies. In Fig. 1 the GEO population is represented by the slightly dispersed ring with orbit radii around 42,000 km and inclinations up to 15°.

Objects on medium earth orbits (MEO) have periods between 2.2 and 24 h, and eccentricities covering wide ranges. Part of MEO are semi-synchronous orbits (SEO). These are used for navigation systems such as the US Global Position System (GPS), the Russian Globalnaya navigatsionnaya sputnikovaya sistema (GLONASS), and the European Galileo navigation system. These are sometimes referred to as Global Navigation Satellite Systems (GNSS). SEOs have periods around 12 h and are near-circular ($e$ close to 0). Very common are eccentric MEO orbits known as geosynchronous transfer orbits (GTO) ($i$ mostly below 30°) and Molniya orbits ($i$ between 60° and 70°). About 8% of all catalogued objects are located in the MEO regime. It is populated by satellites, rocket bodies, fragmentation debris and mission-related objects. In Fig. 1 the MEO population can be seen between the LEO and GEO regions.

Objects orbiting earth above GEO altitudes make only a small fraction of all catalogued objects. They are usually rocket bodies and science missions. Their orbits can be denoted as high elliptic orbits (HEO), or super GEO.

## 1.3 Origins and Sources

The catalog orbit population is mainly composed of space debris. The largest and also easiest to track are non-functional payloads and spent upper stages of rocket bodies (R/B). More than 97% of the total mass located on earth orbits is concentrated in this type of debris, along with functional spacecraft [2] (the mass of the International Space Station (ISS) is not included here). The most abundant objects are roughly larger than 10 cm. They are fragments from payloads and rocket bodies, shortly denoted as fragmentation debris. Fragmentations can be caused by different mechanisms [3], e.g., explosions, collisions [4], intentional destructions [5], or malfunctions. Shapes, sizes and material types of fragments differ for every piece. During the spacecraft launching process many other additional objects can be released. Protective covers, launch adapters and objects lost by astronauts are part of debris called mission-related objects (MRO) [6].

An example of a non-functional satellite can be seen in Fig. 2, showing the former ESA mission ENVISAT (international identification no. 2002-009A), which was successfully operating for about 10 years. In April 2012, however, ESA suddenly lost contact with ENVISAT and was not able to recover the connection. A few weeks later ENVISAT was declared as non-operational and the mission was terminated [7]. Currently, ENVISAT is the heaviest civilian LEO satellite which is not operational and therefore it is a likely target for future active debris removal (ADR) missions [2].

Since the 1960s micrometer (dust) to centimeter (slag) sized particles were created during the burning process of solid rocket motor (SRM) in space. Such residues are mostly composed of aluminium oxide, mixed with SRM liner material [8]. All space objects, such as satellites, rocket bodies and fragments are exposed

**Fig. 2** Scientific ESA satellite ENVISAT (COSPAR no. 2002-009A). ESA lost contact with ENVISAT in April 2012. It is the largest civilian non-functional satellite on low earth orbit and possible target of a future active debris removal mission. Image credit: ESA

to the rough space environment. Spaceweather effects, such as extreme ultraviolet (UV) radiation and interaction with atomic oxygen, cause erosion on the satellite surface which slowly ages and releases small particles into the space environment. Erosion has a strong influence mostly on painted surfaces and thermal protection materials. Because of its most common source this type of debris is denoted as *paint flakes*. The large amount of small debris particles and meteoroids can create another population type of debris, called *ejecta* [9]. After small particle impacts on painted surfaces, solar arrays or other type of spacecraft surface materials, impact craters are formed and some material is ejected into the space environment. Paint flakes and ejecta are usually sub-millimeter particles which do not pose a major risk to space missions, but they can cause degradation of surfaces and optical instruments.

A specific type of debris are particles released from spacecraft with small additional velocities. Often they are caused by unknown mechanisms and referred to as *anomalous debris*. One of the candidates for anomalous debris are multilayer insulations (MLI). MLI is material used as a thermal protection for sensitive systems of spacecraft (see Fig. 2 for highly reflective gold-colored MLI). During breakup events, or under the influence of the space environment (impacts from small particles, extreme ultraviolet radiation) the MLI parts can be detached from the parent body. Typically, these objects have very high area-to-mass ratios A/M (HAMR) and highly reflective surfaces [10]. Because of solar radiation (and the atmospheric drag for LEO passes), high A/M values have a strong influence on the dynamics of anomalous debris and may cause a drift away from the satellites' operational orbits [11].

A unique population from a dynamic point of view are Temporarily Captured Orbiters (TCO) with an artificial origin. Those can be stages from former lunar missions which were gravitationally ejected into heliocentric orbits. Such objects have been observed several times in the last two decades once they were re-captured in the gravitational field of the earth and consequently have been mistaken for a Near Earth Asteroids (NEA) [12, 13]. These objects are on HEOs.

Parts of the space debris population do not have reproducing sources anymore, and therefore their populations are decreasing over time. Such population is cooling liquid droplets released during the Russian missions called Radar Ocean Reconnaissance Satellites (RORSAT). These droplets are consisting of a *sodium–potassium alloy* (NaK) [14]. Clusters of small, micrometer size needles created during the *Westford Needles* project in 1960s are another example a non-reproducing debris sources [15].

A closer look at the LEO population from the perspective of object type versus orbit is plotted in Fig. 3. Shown is the orbital inclination as a function of mean altitude above the earth's surface for different types of debris populations. The data were plotted by using the public catalogue.

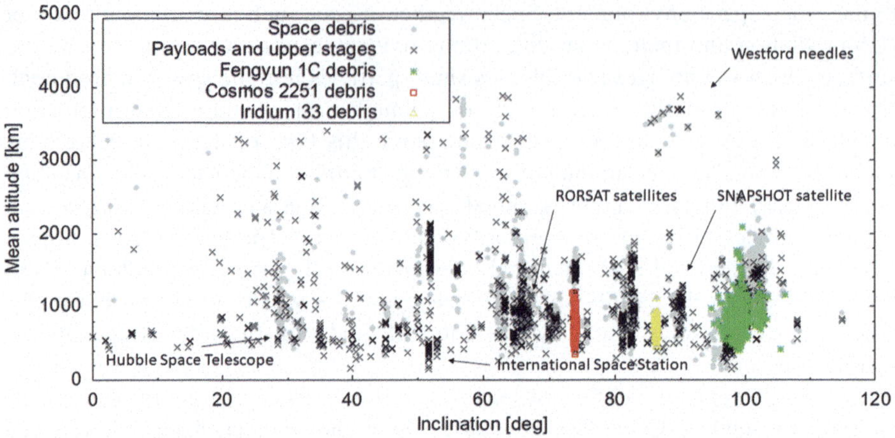


**Fig. 3** Inclination vs mean altitude of catalogued object orbits with altitudes less than 4000 km. It is possible to distinguish intact objects, such as payloads and rocket bodies, fragments from satellites Fengyun-1C (mean inclination of 99.0°), Cosmos 2251 (mean inclination of 74.0°), Iridium 33 (mean inclination of 86.4°) and other catalogued debris. Marked are also orbits of the International Space Station (*ISS*), Hubble Space Telescope (*HST*), Russian *RORSAT* satellites, US *SNAPSHOT* satellites and Westford Needles clusters. Data were obtained from the public catalogue [1] (September 2019)

### *1.4 Population Growth*

The future state of the space debris environment depends on several factors. The crucial one is the ADR technology [16, 17] which can help to remove potentially hazardous objects with high risk of fragmentation which eventually can cause a catastrophic cascade effect also known as Kessler syndrome [18, 19]. Additionally, cheaper space technologies and new participants involved in space industry such as commercial companies and academic institutions created a new form of satellite infrastructures like cubesats [20] and broadband world-wide coverage mega-constellations [21, 22].

## 2 Measurement Techniques

There are several ways how to measure the space debris population, in terms of its size, mass, composition, and spatial distribution. Using optical telescopes help us to track objects of lower sizes from 10 to 50 cm on LEO and GEO orbits, respectively, and study their physical properties such as brightness and rotation. Thanks to radar measurements we are able to track and monitor LEO objects down to 10 cm in size. In-situ detectors and returned spacecraft hardware allow us to measure micrometer-sized debris and micrometeoroids, in terms of their flux, size and mass distributions.

## 2.1 Passive Optical

The principle of passive optical space debris observations is based on collecting the sun light reflected from the object's surface. Therefore, to acquire such data several conditions need to be fulfilled: the object must be illuminated by the sun; it must be observed during the night; and the sky needs to be transparent, hence the weather conditions must be optimal.

Passive optical measurements provide so-called angle measurements and astrometric data, which are then used for orbit determination and object cataloguing. Additionally, surveys are performed to discover new objects and to gather new sampling data for statistical purposes. Light curve acquisitions help to monitor the rotational behaviour of objects, while multi-band photometry along with spectroscopy help to characterize the object and its surface properties. More about optical passive measurements can be found in Sect. 3.

## 2.2 Optical Active, Satellite Laser Ranging

Satellite Laser Ranging (SLR) systems, which are active optical systems, acquiring range and angular measurements. These observations are based on the photon reflection from the target. Contrary to passive optical measurements, a photon is emitted by the SLR system toward the object and reflected back toward the observer, where it is detected. The obtained information is the photon's duration of flight which can be recalculated to the range between the SLR system and the object. SLR measurements can be conducted during the day and night but they require good weather conditions.

SLR systems are primarily used to accurately measure an object's range with accuracy up to few millimeters [23]. Two types of SLR targets can be distinguished. (a) The object is equipped with retroreflector(s) (RR), a highly reflective device which reflects the light back to the source. These are cooperating targets which are used for scientific applications such as geodesy [24]. Cooperating targets are observable by any SLR station but their total number is limited (a few dozen worldwide) [25]; (b) The object is not equipped with retro-reflector. Such non-cooperative objects, are the vast majority, including all space debris. To observe these targets it is necessary to use much stronger lasers to get a statistically meaningful number of photons reflected back to the system. Those are usually experimental set-ups such as in [26] or [27]. The final product of SLR measurements are range measurements which, apart from position information, can also provide information about the target's attitude state [28, 29].

## 2.3 Radar

Radar measurements, in contrast with optical measurements, not limited by meteorological conditions and are executable during the whole day. Most radars are able to measure 2-way signal travel time, the Doppler shift between received and transmitted frequency, the azimuth angle $A$ and elevation angle $h$, and the received power and polarization changes in the radar pulse [15]. The parameter which describes the target's ability to reflect the radar energy is denoted as radar cross section (RCS). The RCS depends on the target's size, the ratio of its characteristic length $Lc$ to the radar wavelength, and the target shape, orientation and material properties. The RCS is expressed in decibel square meters (dBsm) and can be understood as equivalent to the visual magnitude used during the optical measurements [15]. The RCS values for some debris objects can be retrieved from the public catalogue [1].

There are several research radar systems around the world participating in space surveillance and tracking. These are for example the US Goldstone and Haystack radars [30], and European EISCAT [31] and TIRA radars [32].

## 2.4 In-Situ

In-situ measurements mainly help to model the debris population, smaller than 1 mm. There are several models which use this information, for example ESA's MASTER model (Meteoroid and Space Debris Terrestrial Environment Reference) and NASA ORDEM model [33]. There were several in-situ experiments such as the Long Duration Exposure Facility (LDEF) [34] or Space Shuttles surfaces returned for impacts analyses [35].

A good example of in-situ measurement is the LDEF mission. The mission lasted from April 1984 till January 1990, when the satellite was exposed to the space environment in LEO. The purpose of the LDEF mission was mapping the micrometeoroid and space debris (MMOD) environment, testing various materials (such as plastics and glass) and their suitability for space missions, and study effects of radiation and other space hazard. The satellite was located in LEO (mean altitude about 458 km and inclination around 28.5°) and exposed to the space environment for more than 5.75 years [36]. There were 86 trays installed on board of LDEF with 57 different experiments and a total of exposed area of 130 $m^2$. The post mission examination of the thick and thin targets was performed to get the size, mass and flux distribution of the MMOD population [37].

## 3 Research with Passive Optical Telescopes

This section discusses different research performed with passive optical telescopes. It covers the survey and cataloguing, light curve processing, color photometry and spectroscopy.

### 3.1 Survey, Astrometry and Cataloguing

Survey observations serve to discover new objects, either for statistical population modeling, or for tracking and cataloguing [38, 39]. Space agencies such as NASA and ESA are monitoring the MEO and GEO regions to obtain statistical information about faint non-catalogued objects [40–42].

Since 2001 NASA uses its Michigan Orbital Debris Telescope (MODEST) to scan the GEO region. MODEST is a 0.6-m Curtis-Schmidt telescope equipped with the charged coupled device (CCD) camera with a 1.3 × 1.3 degrees field of view [42, 43]. The telescope is located at the Cerro Tololo Inter-American Observatory (CTIO) in Chile. The acquired observations are processed by internal Image Reduction and Analysis Facility (IRAF) routines for photometry and astrometry, and the obtained results are compared with the public catalogue [1]. This helps to distinguish correlated targets (CT, i.e. objects have been correlated with the public catalogue) from un-correlated targets (UCT). An additional information from the data is the brightness (apparent magnitude) $m_{\mathrm{app}}$ of the object, its size $L_{\mathrm{c}}$ by using some assumptions on the shape and albedo, and a preliminary subset of orbital elements, namely the orbital inclination $i$, right ascension of ascending node $\Omega$ and mean motion $n$.

There were usually from 20 to 40 nights of observations per year, each split into three observations runs. The basic observation strategy was to choose equatorial coordinates for the field of view, with right ascension $\alpha$ and declination $\delta$ close to the Earth's shadow (or anti-solar point) to maximize the detection rate. Once the object is observed under small phase angles ($\varphi$ close to 0°) it maximizes its reflection of the sunlight toward the observer. $\varphi$ is defined as the angle between the Sun-object and object-observer directions. The chosen field of view with the given $\alpha$ and $\delta$ was tracked during the whole night using a broad R filter centered at 630 nm, with a the full width at half maximum of 200 nm. The exposure time was set to 5 s, which led to a signal-to-noise ratio of 10 for objects of 18th magnitude. During the observations the time delay integration (TDI) method was used, where the charge on the CCD is shifted opposite to the sidereal rate. Hence, the stars on the exposure images are displayed as streaks and objects located on GEO are displayed as points or small streaks.

Once the apparent magnitude $m_{\mathrm{app}}$ is obtained (see Sect. 3.3) it is possible to get a characteristic length $L_{\mathrm{c}}$ of the object, which characterizes the size of asymmetric debris fragments. It is defined as the average of three orthogonal dimensions X, Y and Z, where X is the longest dimension of the object, Y is the longest dimension perpendicular to the X axis, and Z is the longest dimension perpendicular to the other axes [44]:

$$L_c = \frac{X + Y + Z}{3}. \tag{1}$$

To get $L_c$ from $m_{app}$ we use following formula assuming the object is a Lambertian sphere [45]:

$$L_c = 10^{(m_{sun} - m_{app})/5} \frac{R}{\sqrt{A}} \sqrt{\frac{6\pi}{\sin\phi + (\pi - \phi)\cos\phi}}, \tag{2}$$

or

$$L_c = 10^{(m_{sun} - m_{abs})/5} \frac{R}{\sqrt{A}} \sqrt{6}, \tag{3}$$

where the variable $L_c$ [km] is the diameter of a diffusely reflecting Lambertian sphere, $R$ is the observer-satellite distance [km], $m_{sun}$ is the apparent magnitude of the sun, $\varphi$ is the solar phase angle [rad], $A$ is the Bond albedo, and $m_{abs}$ is the absolute magnitude of object. $m_{abs}$ is defined as the magnitude of the object corrected for the phase angle ($\varphi = 0°$). The magnitude $m_{abs}$ can be determined as follows:

$$m_{abs} = m_{app} - 5\log\left(\frac{\pi}{\sin\phi + (\pi - \phi)\cos\phi}\right). \tag{4}$$

According to [45] the value A = 0.175 can be adopted as a transformation albedo to calculate the object size from Eqs. 2 and 3. The absolute magnitude distribution of 3143 objects which were observed during MODEST observation runs between 2007 to 2009 can be seen in Fig. 4 [43]. Objects were on GEO and are separated into the functional CTs, UCTs and non-functional CTs, which are objects that are freely drifting in the north-south direction and east-west direction. The peak for the functional CTs is at $m_{abs}$ = 10.5 mag which under previous assumptions (see previous paragraph) can be interpreted as an object size of $L_c$ = 6.3 m. For non-functional objects the peak is at $m_{abs}$ = 12.5 mag, with corresponding $L_c$ = 2.5 m. The peak of the UCTs is an artificial one due to the limited detection capabilities of the telescope, and it can be assumed that the population continues rising with decreasing object size. The faintest object detected in 2007–2009 had a magnitude of $m_{abs}$ = 18.3 mag which corresponds to $L_c$ = 17 cm.

Once the object is discovered and sufficient follow-up observations are performed (at least two), its orbit can be determined and the object can be catalogued. There are several different perturbation forces affecting the dynamics of objects orbiting the earth. They need to be considered during the orbit determination and propagation process. These are gravitational effects such as earth's spherical

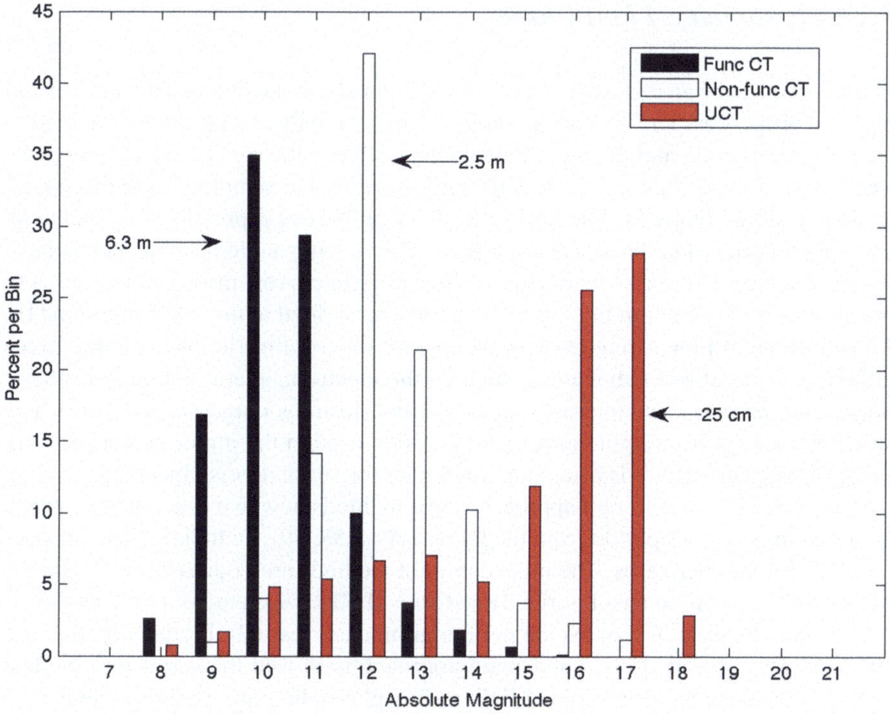


**Fig. 4** The absolute magnitude distribution of 3143 objects observed by the MODEST telescope during the years 2007, 2008 and 2009. Objects are separated into functional CTs, non-functional CTs and UCTs. Corresponding object sizes were calculated assuming a diffuse Lambertian reflection with albedo 0.175. Figure taken from [43]

harmonics and perturbations from the sun and moon, and solar radiation pressure and atmospheric drag. Thrust forces can also play a role. They are originating from the object but are not usually accounted for during routine orbit determination/prediction processes. There are several dynamical models available which deal with the satellite's orbital dynamics, e.g., described in [46, 47].

To get accurate orbits, highly accurate astrometric data need to be acquired. This specially goes for space debris objects which can reach apparent angular velocities up to few degrees per second. In this case even a small time error at the level of milliseconds can lead to an astrometric error of a few arc-seconds. For that reason optical sensors need to be monitored for possible time biases and inaccuracies. There are methods to perform such analysis, e.g., described in [48].

## 3.2 Photometry, Light Curves

Light curves carry an extensive set of information about the object's dynamical and physical properties. The brightness variation present in the data is directly related to the object's rotation and the mutual geometry between the sun, target and observer. The shape of the light curve is directly influenced by the sampling used during the data acquisition. Object's shape, its reflectivity properties expressed via the albedo, surface properties and the aspect angles, and the viewing angle from the perspective of the observer. There are many applications of light curves. Information related to the attitude determination [49, 50] can directly be applied to the ADR problem [16, 17, 51]. Regular monitoring of the rotation change over time helps to model forces influencing the object's dynamics, such as the electromagnetic field, atmospheric drag, solar radiation, and internal processes (fuel sloshing, outgassing) [52, 53]. The shape estimation from light curves is quite often used in the minor planets domain [54, 55], and currently it is also being exploited for space debris objects [56, 57].

There are several different approaches and methods how to extract the frequency or apparent (synodic) period from the light curves [58, 59], including visual inspection [53] for simpler cases. The most common methods are Fourier-based methods. These are for example Fast Fourier Transforms (FFT), Discrete Fourier Transforms [60], Lomb–Scargle [61] or Welch periodogram analyses [62]. Frequently used are string-length methods [59]. Their main approach is to fold the series into the test period, where the resulting folded light curve, phase diagram, is further analyzed. Those methods are more robust, and applicable to non-equally sampled data. Just to mention a few, there is epoch folding [63], wavelet analysis [64], the Lafler–Kinman method [65] and Phase Dispersion Minimization [66].

Between the years 1987 and 2004 the authors of [67] acquired photometric data of 20 GEO satellites, where the majority were satellites of type Gorizont, Raduga, Ekran and Geizer. The authors used a two-mirror Cassegrain telescope with 0.5 m aperture equipped with a photoelectrical photometer setup, with a fast photometry mode, operated at the Sayan observatory, Russia. This system was able to acquire data up to 1 kHz with a time accuracy of 1 ms. To extract the rotation periods the authors used FFT, the Lafler–Kinman method, and wavelet analysis. An example of an acquired light curve and reconstructed phase diagram constructed by using the Lafler–Kinman algorithm is plotted in Figs. 5 and 6, which show data for the non-functional spacecraft Raduga 14 (1984-016A). The light curve, as well the phase diagram contains several sharp peaks which are typical for rotating, box-wing-type spacecraft. For these specific data the obtained apparent rotation period was 53 s.

Authors of [67] monitored several objects over long periods of time. They provided information about the change of rotation as a function of time. According to the variation types the authors distinguished three types of behaviour for rotating objects: long-term systematic increase/decrease of rotation; step-like variations; and sporadic, anomalous increase in period over several days. The latter variation type has been observed for Raduga 14, for which data are plotted in Figs. 5 and 6.

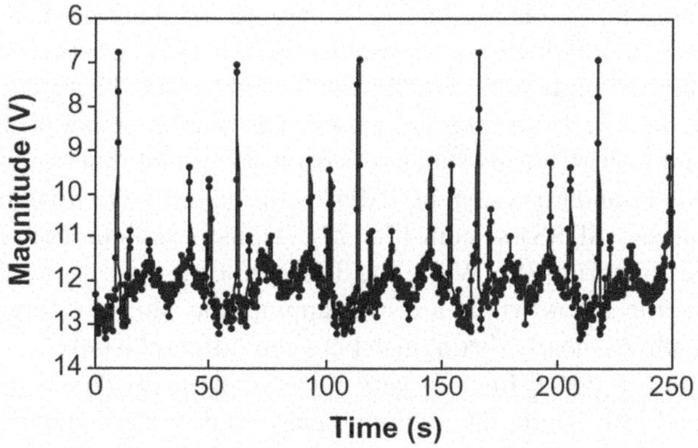


**Fig. 5** The original light curve of the spacecraft *Raduga 14* (1984-016A). Figure taken from [67]

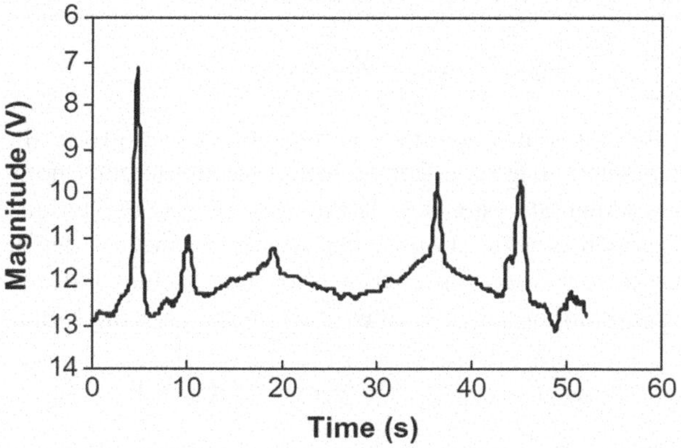


**Fig. 6** The averaged light curve of the spacecraft *Raduga 14* (1984-016A) developed by the Lafler–Kinman algorithm. Figure taken from [67]

Research published in [67], as well as [68–70] showed that debris, including upper stages and spacecraft, cover wide range of rotation rates, up to few revolutions per second.

## *3.3 Color Photometry*

Color photometry is based on multi-band photometric measurements. The difference between captured brightness in different bands is depended on the object's surface properties such as material color, roughness, albedo, etc. Color photometry can help to distinguish material types of an object's surface by comparing the results with laboratory experiments [71, 72] or at least it can help us to categorize an object according to its color [73, 74]. Additionally, one can use photometric data to monitor the aging of the material due to space weather effects [75] (see Sect. 1.3).

The photometric bands are defined trough their filters. There are several filter types in astronomy which are commonly used. Those are for example the $UBVR_{\mathrm{c}}I_{\mathrm{c}}$ filters for the Johnson–Cousins standard system [73], and more modern $u'g'r'i'z'$ filters of the Sloan standard system used during the extensive campaigns of the Sloan Digital Sky Survey (SDSS) system [76, 77]. Transformations between those two systems can be found in [78]. While the Johnson–Cousins system allows broader interval for specific filters which are overlapping, the Sloan system has a narrow interval which allows clearly distinguish between different bands.

Working with $UBVR_{\mathrm{c}}I_{\mathrm{c}}$ filters is very convenient mostly thanks to the extensive work of [79] and [80]. Here, the authors measured dozens of stars located around the equator by using $UBVR_{\mathrm{c}}I_{\mathrm{c}}$ filters. These so-called Landolt standard stars are used till today for the transformation from instrumental magnitudes to the Johnson–Cousins standard system. The conversion has a following form [81]:

$$m_{\mathrm{app}} = m_{\mathrm{ins}} - Z - \kappa X - color, \tag{5}$$

where $m_{\mathrm{app}}$ is the apparent magnitude of the object in a given filter, $m_{\mathrm{ins}}$ is the measured instrumental magnitude, $Z$ is the photometric zero point between the standard and instrumental systems, $\kappa$ is the atmospheric extinction coefficient, $X$ is the airmass (which is a function of the zenith distance $z$ [deg]), and $color$ is the color term transformation for a given filter, hence color index in instrumental system. For a specific filter, e.g. a $V$ filter, the equation has the following form:

$$m_{\mathrm{app,V}} = m_{\mathrm{ins,V}} - Z - \kappa X - C(B - V), \tag{6}$$

where $C$ is the color coefficient and $(B - V)$ is the color index. By observing the Landolt standard fields during photometric nights, the parameters $color/C$, $Z$ and $\kappa$ can be determined for a given optical system.

In [73] the authors observed thirteen GEO objects in total, including seven upper stages, one non-operational spacecraft, two fragments and three operational spacecraft. The authors used Johnson–Cousins $BVR_{\mathrm{c}}I_{\mathrm{c}}$ filters to acquire photometric data. The Landolt stars were used for the transformation from the instrumental to the standard system. The acquired light curves and obtained values for color indices can be seen in Fig. 7, showing measured color index $B - R$ versus color index $V - I$, along with the color indices of the sun. The numbers in Fig. 7 are the North American Aerospace Defense Command (NORAD) numbers which uniquely identify the object in the public catalogue. The results revealed that all objects (except one debris object) are redder than the sun in both indices. The authors also plotted color indices as a function of the object's launch date to investigate possible spaceweather effects. No visible trends have been observed.

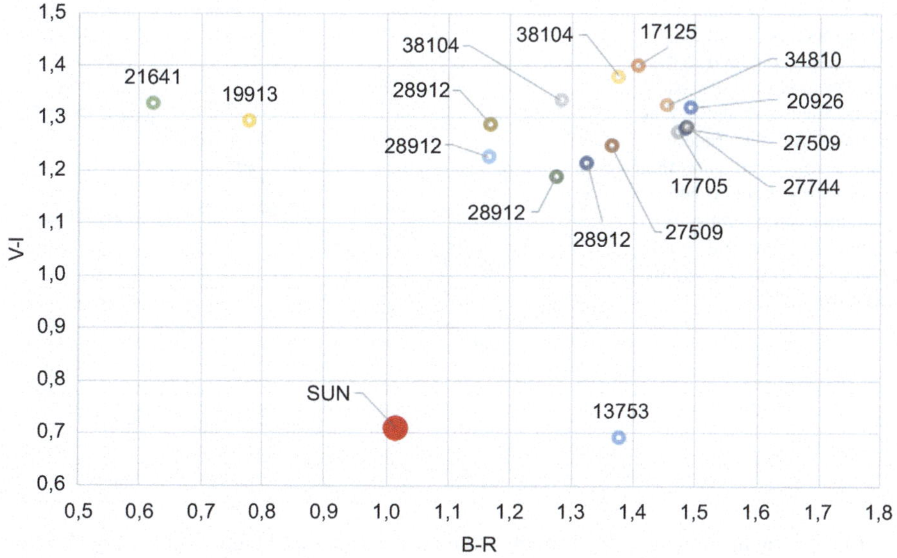


**Fig. 7** Color indices $B - R$ vs $V - I$ obtained for thirteen GEO objects to investigate the surface properties and spaceweather effects. Figure taken from [73]

The study [73] showed that the standard deviations of measurement points for one specific object are rather high. This is usually problematic for space debris objects where rotation rates are quite high (see Sect. 3.2), and therefore the acquired photometric points can be covering large or relative large fractions of one rotation or even several rotations of the object. Therefore, the knowledge of the rotation properties of debris objects is essential once such type of measurements is acquired.

### *3.4 Reflectance Spectroscopy*

The reflectance spectroscopy is measuring solar light reflected from the target. Its final output is a light spectrum, which is the reflectance as a function of wavelength. Reflectance spectra can provide, similar to color photometry (see Sect. 3.3), information about the material composition. It can be used to characterize the object [75, 82, 83] or it can reveal whether the object is artificial or not [12]. Reflectance spectroscopy has been adapted to the space debris domain decades ago, and its results are often used for comparisons with the spectra obtained in laboratory experiments for commonly used space materials such as solar panels, MLI, aluminium alloys, white paint, etc. [84, 85].

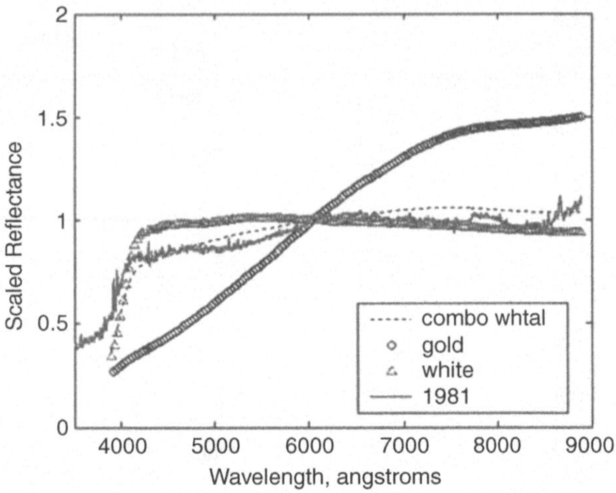


**Fig. 8** Normalized reflectance spectra of an upper stage launched in the year 1981. Plotted are also spectra for materials such as "white paint", "degraded white paint, "gold", and their combination "combo whtal" (with an assumed ratio between "white" and "gold" paint to be 90–10%). Figure taken from [75]

In [75] the authors acquired spectral data for various LEO and GEO rocket bodies and spacecraft within two campaigns: one concluded in 1999, and one performed in 2001. These data were acquired by the 1.6-m telescope located on at Maui, Hawaii, which is equipped with a spectrometer of three selectable gratings. The authors focused on three primary goals. The first goal was to compare the measured spectra with laboratory experiments to determine surface compositions. In Fig. 8 reflectance spectrum is plotted for an upper stage launched in 1981. For comparison, laboratory spectra for "white paint" and "white paint after degradation" and "gold" are provided. Additionally, a combination of two materials, "combo whtal", is shown, with assumed a ratio between "white paint" and "gold" of 90–10%. This is the combination for which the authors got a very good match with the measured spectra.

As a secondly goal, the authors focused on an investigation of aging of the surface material. This could be achieved by observing the same type of upper stage with different launching dates. The results are plotted in Fig. 9, where reflectance spectra for four different upper stages of the same type (shape and surface material) are provided. The observed spectra showed an inversely proportional effect to the increase in age. As a third goal the authors also compared spectra between different types of objects, e.g., spacecraft versus upper stage. Spectra revealed that the most of the difference is in the blue part of the spectra, where satellites can reach much higher reflectance due to the presence of solar panels.

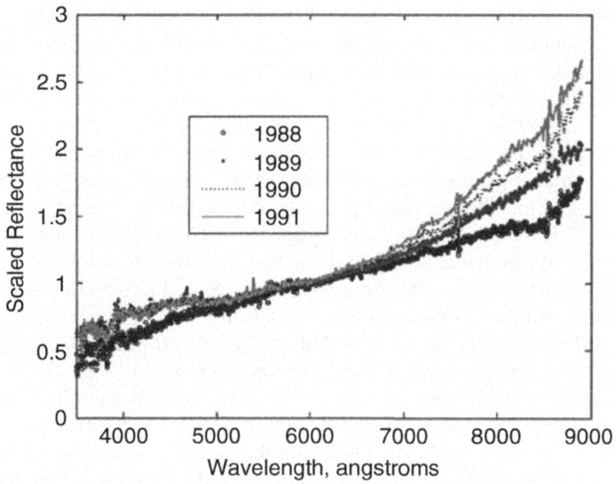


**Fig. 9** Normalized reflectance spectra acquired for four different upper stages with similar properties, shape and surface material, but with different launching dates. This specific type of upper stage type spectra shows an inversely proportional effect for the increase with age. Figure taken from [75]

## 4 Conclusions

Optical measurements play a strong role in space debris research. They are essential and irreplaceable for the tracking and cataloguing of objects on higher orbits such as geosynchronous and highly elliptical orbits. Typical astronomical methods such as sky surveys, photometry, or spectroscopy provide information about debris origins and creation mechanisms. The space debris population is constantly increasing and continuous improvements in optical data acquisition and processing can help to effectively deal with this problem.

## References


1. USSTRATCOM, Space-track.org
2. J.-C. Liou, An active debris removal parametric study for LEO environment remediation. Adv. Space Res. **47**(11), 1865–1876 (2011)
3. V. Braun, S. Lemmens, B. Reihs, H. Krag, A. Horstmann, Analysis of breakup events, in *Proceedings of the 7th European Conference on Space Debris*, Darmstadt, 18–21 April 2017
4. C. Pardini, L. Anselmo, Physical properties and long-term evolution of the debris clouds produced by two catastrophic collisions in earth orbit. Adv. Space Res. **48**(3), 557–569 (2011)
5. J.-C. Liou, N. Johnson, Characterization of the cataloged Fengyun-1C fragments and their long-term effect on the LEO environment. Adv. Space Res. **43**(9), 1407–1415 (2009)
6. N.L. Johnson, E. Stansbery, D.O. Whitlock, K. Abercromby, D. Shoots, History of on-orbit satellite fragmentations, 14th edn. NASA Technical Report, NASA TM-2008-214779
7. 'ESA', ESA declares end of mission for Envisat. Space Res. Today **184**, 13–14 (2012)
8. M.F. Horstman, M. Mulrooney, An analysis of the orbital distribution of solid rocket motor slag. Acta Astronaut. **64**, 230–235 (2009)
9. W.P. Schonberg, Characterizing secondary debris impact ejecta. Int. J. Impact Eng. **26**(1), 713–724 (2001)

10. T. Schildknecht, R. Musci, T. Flohrer, Properties of the high area-to-mass ratio space debris population at high altitudes. Adv. Space Res. **41**(7), 1039–1045 (2008)
11. J.-C. Liou, J.K. Weaver, Orbital dynamics of high area-to ratio debris and their distribution in the geosynchronous region, in *Proceedings of the 4th European Conference on Space Debris ESA SP587 18–20 April 2005 ESA/ESOC Darmstadt* (2005)
12. A. Buzzoni, G. Altavilla, S. Fan, C. Frueh, I. Foppiani, M. Micheli, J. Nomen, N. Sánchez-Ortiz, Physical characterization of the deep-space debris WT1190F: a testbed for advanced SSA techniques. Adv. Space Res. **63**(1), 371–393 (2019)
13. E. ESA, Space oddity: the mystery of 2013 qw1. ESA Publications Division, Noordwijk (2013)
14. C. Wiedemann, S. Flegel, J. Gelhaus, H. Krag, H. Klinkrad, P. Voersmann, NaK release model for MASTER-2009. Acta Astronaut. **68**(7), 1325–1333 (2011)
15. H. Klinkrad, *Space Debris - Models and Risk Analysis* (Springer, New York, 2006)
16. J.-C. Liou, N. Johnson, N. Hill, Controlling the growth of future LEO debris populations with active debris removal. Acta Astronaut. **66**(5), 648–653 (2010)
17. W. Wang, X. Song, K. Li, L. Chen, A novel guidance scheme for close range operation in active debris removal. J. Space Saf. Eng. **5**(1), 22–33 (2018)
18. D. Kessler, K. Jarvis, Obtaining the properly weighted average albedo of orbital debris from optical and radar data. Adv. Space Res. **34**(5), 1006–1012 (2004). Space Debris.
19. J.D. Kessler, N. Johnson, J.-C. Liou, M. Matney, The Kessler syndrome: implications to future space operations. Adv. Astronaut. Sci. **137**, 47–62 (2010)
20. M. Matney, A. Vavrin, A. Manis, Effects of cubesat deployments in low-earth orbit, in *Proceedings of the 7th European Conference on Space Debris ESA, ESA/ESOC Darmstadt*, 18–21 April 2017
21. I. del Portillo, B.G. Cameron, E.F. Crawley, A technical comparison of three low earth orbit satellite constellation systems to provide global broadband. Acta Astronaut. **159**, 123–135 (2019)
22. S.L. May, S. Gehly, B. Carter, S. Flegel, Space debris collision probability analysis for proposed global broadband constellations. Acta Astronaut. **151**, 445–455 (2018)
23. L. Combrinck, Satellite laser ranging, in *Sciences of Geodesy-I* (Springer, Berlin, 2010), pp. 301–338
24. D. Strugarek, K. Sosnica, A. Jaeggi, Characteristics of GOCE orbits based on satellite laser ranging. Adv. Space Res. **63**(1), 417–431 (2019)
25. M. Pearlman, J. Degnan, J. Bosworth, The international laser ranging service. Adv. Space Res. **30**(2), 135–143 (2002)
26. G. Kirchner, F. Koidl, F. Friederich, I. Buske, U. Volker, W. Riede, Laser measurements to space debris from Graz SLR station. Adv. Space Res. **51**(1), 21–24 (2013)
27. Z. Zhang, H. Zhang, M. Long, H. Deng, Z. Wu, W. Meng, High precision space debris laser ranging with 4.2 W double-pulse picosecond laser at 1 kHz in 532 nm. Optik **179**, 691–699 (2019)
28. D. Kucharski, G. Kirchner, H.-C. Lim, F. Koidl, Spin parameters of high earth orbiting satellites Etalon-1 and Etalon-2 determined from kHz satellite laser ranging data. Adv. Space Res. **54**(11), 2309–2317 (2014)
29. J.-N. Pittet, J. Šilha, T. Schildknecht, Spin motion determination of the Envisat satellite through laser ranging measurements from a single pass measured by a single station. Adv. Space Res. **61**(4), 1121–1131 (2018)
30. C. Stokely, E. Stansbery, R. Goldstein, Debris flux comparisons from the Goldstone Radar, Haystack radar, and HAX radar prior, during, and after the last solar maximum. Adv. Space Res. **44**(3), 364–370 (2009)
31. J. Markkanen, M. Lehtinen, M. Landgraf, Real-time space debris monitoring with EISCAT. Adv. Space Res. **35**(7), 1197–1209 (2005). Space Debris
32. D. Mehrholz, Radar observations in low earth orbit. Adv. Space Res. **19**(2), 203–212 (1997). Space Debris
33. P. Krisko, S. Flegel, M. Matney, D. Jarkey, V. Braun, ORDEM 3.0 and MASTER-2009 modeled debris population comparison. Acta Astronaut. **113**, 204–211 (2015)

34. J. Mandeville, L. Berthoud, From LDEF to Eureca: orbital debris and meteoroids in low earth orbit. Adv. Space Res. 16(11), 67–72 (1995). Space Debris
35. J. Hyde, E. Christiansen, D. Lear, Shuttle MMOD impact database. Proc. Eng. **103**, 246–253 (2015)
36. N. McBride, A. Taylor, S. Green, J. McDonnell, Asymmetries in the natural meteoroid population as sampled by LDEF. Planet. Space Sci. **43**(6), 757–764 (1995). Asteroids, Comets and Meteors 1993-V
37. J. McDonnell, D. Gardner, Meteoroid morphology and densities: decoding satellite impact data. Icarus **133**(1), 25–35 (1998)
38. I. Molotov, V. Agapov, V. Titenko, Z. Khutorovsky, Y. Burtsev, I. Guseva, V. Rumyantsev, M. Ibrahimov, G. Kornienko, A. Erofeeva, V. Biryukov, V. Vlasjuk, R. Kiladze, R. Zalles, P. Sukhov, R. Inasaridze, G. Abdullaeva, V. Rychalsky, V. Kouprianov, O. Rusakov, E. Litvinenko, E. Filippov, International scientific optical network for space debris research. Adv. Space Res. **41**(7), 1022–1028 (2008)
39. M. Weigel, H. Fiedler, T. Schildknecht, Scoring sensor observations to facilitate the exchange of space surveillance data. Adv. Space Res. **60**(3), 531–542 (2017)
40. T. Schildknecht, T. Flohrer, R. Musci, R. Jehn, Statistical analysis of the ESA optical space debris surveys. Acta Astronaut. 63(1), 119–127 (2008). Touching Humanity - Space for Improving Quality of Life. Selected Proceedings of the 58th International Astronautical Federation Congress, Hyderabad, India, 24–28 September 2007
41. J. Silha, T. Schildknecht, A. Hinze, T. Flohrer, A. Vananti, An optical survey for space debris on highly eccentric and inclined MEO orbits. Adv. Space Res. **59**(1), 181–192 (2017)
42. K.J. Abercromby, P. Seitzer, H.M. Rodriguez, E.S. Barker, M.J. Matney, Survey and chase: a new method of observations for the Michigan orbital debris survey telescope (MODEST). Acta Astronaut. **65**(1), 103–111 (2009)
43. P. Seitzer, S. Lederer, E.B.H. Cowardin, K. Abercromby, J. Silha, A. Burkhardt, A search for optically faint GEO debris, in *Proceedings of the Advanced Maui Optical and Space Surveillance Technologies Conference, held in Wailea*, Maui, Hawaii, September 13–16, 2011
44. T. Hanada, J.-C. Liou, T. Nakajima, E. Stansbery, Outcome of recent satellite impact experiments. Adv. Space Res. **44**(5), 558–567 (2009)
45. M.K. Mulrooney, M.J. Matney, A new bond albedo for performing orbital debris brightness to size transformations, in *Proceedings of the 59th International Astronautical Congress, England*, 29 Sept–3 Oct 2008
46. G. Beutler, *Methods of Celestial Mechanics, Volume II: Application to Planetary System, Geodynamics and Satellite Geodesy* (Springer, Berlin, 2005)
47. O. Montenbruck, E. Gill, *Satellite Orbits: Models, Methods, and Applications* (Springer, Berlin, 2000)
48. B. Jilete, T. Flohrer, A. Mancas, J. Castro, J. Siminski, Acquiring observations for test and validation in the space surveillance and tracking segment of ESA's SSA programme. J. Space Saf. Eng. **6**, 197–200 (2019)
49. V. Williams, Location of the rotation axis of a tumbling cylindrical earth satellite by using visual observations: part i: theory. Planet. Space Sci. **27**(6), 885–890 (1979)
50. F. Santoni, E. Cordelli, F. Piergentili, Determination of disposed-upper-stage attitude motion by ground-based optical observations. J. Spacecr. Rocket. **50**(3), 701–708 (2013)
51. J.L. Forshaw, G.S. Aglietti, T. Salmon, I. Retat, M. Roe, C. Burgess, T. Chabot, A. Pisseloup, A. Phipps, C. Bernal, F. Chaumette, A. Pollini, W. H. Steyn, Final payload test results for the RemoveDebris active debris removal mission. Acta Astronaut. **138**, 326–342 (2017). The Fifth International Conference on Tethers in Space
52. J.R. Wertz, *Spacecraft Attitude Determination and Control* (Kluwer, Dordrecht, 1978)
53. M.A. Earl, G.A. Wade, Observation and analysis of the apparent spin period variations of inactive box-wing geosynchronous resident space objects, in *Proceedings of the 65th International Astronautical Congress*, Toronto, 29 Sept–03 Oct 2019 (2014)
54. M. Kaasalainen, J. Torppa, Optimization methods for asteroid lightcurve inversion: I. Shape determination. Icarus **153**(1), 24–36 (2001)

55. J.L. Crowell, E.S. Howell, C. Magri, M.C. Nolan, Y.R. Fernandez, J.E. Richardson, B.D. Warner, S.E. Marshall, A. Springmann, R.J. Vervack, Radar and lightcurve shape model of near-earth asteroid (1627) Ivar. Icarus **291**, 254–267 (2017)
56. R. Linares, M. Jah, J. Crassidis, Inactive space object shape estimation via astrometric and photometric data fusion. Adv. Astronaut. Sci. **143**, 217–232 (2012)
57. B.K. Bradley, P. Axelrad, Lightcurve inversion for shape estimation of GEO objects from space-based sensors, in *Proceedings of the 24th International Symposium on Space Flight Dynamics*, Laurel, MD, 5–9 May, 2014
58. M. Graham, A. Drake, G. Djorgovski, A. Mahabal, C. Donalek, V. Duan, A. Maher, A comparison of period finding algorithms. Mon. Not. R. Astron. Soc. **434**, 3423 (2013)
59. E. Paunzen, T. Vanmunster, Peranso - light curve and period analysis software. Astron. Nachr. **337**, 239–245 (2016)
60. T.J. Deeming, Fourier analysis with unequally-spaced data. Astrophys. Space Sci. **36**, 137–158 (1975)
61. J.D. Scargle, Studies in astronomical time series analysis. ii - statistical aspects of spectral analysis of unevenly spaced data. Astrophys. J. **263**, 835–853 (1982)
62. P. Welch, The use of fast Fourier transform for the estimation of power spectra: a method based on time averaging over short, modified periodograms. IEEE Trans. Audio Electroacoust. **15**, 70–73 (1967)
63. S. Larsson, Parameter estimation in epoch folding analysis. Astron. Astrophys. Suppl. **117**, 197–201 (1996). http://dx.doi.org/10.1051/aas:1996150
64. N. Astafyeva, Wavelet analysis: basic theory and some applications. Phys. Usp. **39**, 1085 (1996)
65. J. Lafler, T.D. Kinman, An RR lyrae star survey with the lick 20-inch astrograph ii. The calculation of RR lyrae periods by electronic computer. Astrophys. J. Suppl. **11**, 216 (1965)
66. R.F. Stellingwerf, Period determination using phase dispersion minimization. Astrophys. J. **224**, 953–960 (1978)
67. P. Papushev, Y. Karavaev, M. Mishina, Investigations of the evolution of optical characteristics and dynamics of proper rotation of uncontrolled geostationary artificial satellites. Adv. Space Res. **43**(9), 1416–1422 (2009)
68. B.D. Pontieu, Database of photometric periods of artificial satellites. Adv. Space Res. **19**(2), 229–232 (1997). Space Debris
69. C.R. Binz, M.A. Davis, B.E. Kelm, C.I. Moore, Optical survey of the tumble rates of retired GEO satellites, in *Proceedings of Advanced Maui Optical and Space Surveillance Technologies Conference (AMOS), Conference 2014*, Maui, Hawaii (2014)
70. J. Silha, J.-N. Pittet, M. Hamara, T. Schildknecht, Apparent rotation properties of space debris extracted from photometric measurements. Adv. Space Res. **61**(3), 844–861 (2018)
71. H. Cowardin, P. Seitzer, K. Abercromby, E. Barker, T. Schildknecht, Characterization of orbital debris photometric properties derived from laboratory-based measurements, in *Proceedings of Advanced Maui Optical and Space Surveillance Technologies Cenference (AMOS), Conference 2014*, Maui, Hawaii (2010)
72. D. Bedard, P. Seitzer, A. Willison, P. Somers, In-situ VIS/NIR measurements of space environment effects on spacecraft surfaces, in *Proceedings of Advanced Maui Optical and Space Surveillance Technologies Cenference (AMOS), Conference 2014*, Maui, Hawaii (2016)
73. T. Cardona, P. Seitzer, A. Rossi, F. Piergentili, F. Santoni, BVRI photometric observations and light-curve analysis of GEO objects. Adv. Space Res. **58**(4), 514–527 (2016)
74. X.-F. Zhao, H.-Y. Zhang, Y. Yu, Y.-D. Mao, Multicolor photometry of geosynchronous satellites and application on feature recognition. Adv. Space Res. **58**(11), 2269–2279 (2016)
75. K. Jorgensen, J. Africano, K. Hamada, E. Stansbery, P. Sydney, P. Kervin, Physical properties of orbital debris from spectroscopic observations. Adv. Space Res. **34**(5), 1021–1025 (2004). Space Debris
76. M. Fukugita, T. Ichikawa, J.E. Gunn, M. Doi, K. Shimasaku, D.P. Schneider, The Sloan digital sky survey photometric system. Astron. J. **111**, 1748 (1996)

77. Y. Lu, C. Zhang, R. Sun, C. Zhao, J.-N. Xiong, Investigations of associated multi-band observations for GEO space debris. Adv. Space Res. **59**(10), 2501–2511 (2017)
78. P. Castro, T. Payne, J. Moody, S. Gregory, P. Dao, R. Acosta, Transformation between the Johnson-Cousins and sloan photometric systems for SSA, in *The Advanced Maui Optical and Space Surveillance Technologies Conference* (2018), p. 36
79. A.U. Landolt, UBVRI photometric standard stars in the magnitude range 11.5–16.0 around the celestial equator. Astron. J. **104**, 340–371 (1992)
80. A.U. Landolt, UBVRI photometric standard stars around the celestial equator: updates and additions. Astron. J. **137**(5), 4186 (2009)
81. P.J. Benson, Transformation coefficients for differential photometry. Int. Amateur-Professional Photoelectric Photometry Commun. **72**, 42–52 (1998)
82. A. Vananti, T. Schildknecht, H. Krag, Reflectance spectroscopy characterization of space debris. Adv. Space Res. **59**(10), 2488–2500 (2017)
83. A. DeMeulenaere, E. Harmon, F. Chun, Simultaneous glint spectral signatures of geosynchronous satellites from multiple telescopes, in *Proceedings of Advanced Maui Optical and Space Surveillance Technologies Cenference (AMOS), Conference 2018*, Maui, Hawaii (2018)
84. D.P. Engelhart, R. Cooper, H. Cowardin, J. Maxwell, E. Plis, D. Ferguson, D. Barton, S. Schiefer, R. Hoffmann, Space weathering experiments on spacecraft materials, in *Advanced Maui Optical and Space Surveillance (AMOS) Technologies Conference* (2017), p. 21
85. M. Bengtson, J. Maxwell, R. Hoffmann, R. Cooper, S. Schieffer, D. Ferguson, W.R. Johnston, H. Cowardin, E. Plis, D. Engelhart, Optical characterization of commonly used thermal control paints in a simulated GEO environment, in *Proceedings of Advanced Maui Optical and Space Surveillance Technologies Cenference (AMOS), Conference 2018*, Maui, Hawaii (2018)

# Meteors: Light from Comets and Asteroids

**Pavol Matlovič and Juraj Tóth**

## Contents



**Abstract** In studies of the oldest solar system bodies—comets and asteroids—it is their fragments—meteoroids—that provide the most accessible planetary material for detailed laboratory analysis in the form of dust particles or meteorites. Some asteroids and comets were visited by spacecrafts and returned interplanetary samples to Earth, while missions *Hayabusa 2* and *OSIRIX-REx* visiting asteroids Ryugu and Bennu are ongoing. However, the lack of representative samples of comets and asteroids opens the space to gain more knowledge from direct observations of meteoroids. At collision with the Earth's atmosphere, meteoroids produce light phenomena known as meteors. Different methods can be used to observe meteors, allowing us to study small interplanetary fragments, which would otherwise remain undetected. Numerous impressive meteor showers, storms and meteorite impacts have occurred throughout the recorded history and can now be predicted and analyzed in much more detail. By understanding the dynamics, composition and physical properties of meteoroids, we are able to study the formation history and dynamical evolution of the solar system. This work presents an introduction to meteor astronomy, its fundamental processes and examples of current research topics.

P. Matlovič (✉) · J. Tóth
Comenius University in Bratislava, Bratislava, Slovakia
e-mail: matlovic@fmph.uniba.sk; toth@fmph.uniba.sk



P. Kabáth et al. (eds.), *Reviews in Frontiers of Modern Astrophysics*,
https://doi.org/10.1007/978-3-030-38509-5_2

# 1 Introduction: The Space of Meteoroids

## *1.1 Overview*

Studying the solar system, as our home planetary neighborhood, the outlook for the first space travels, and the only recognized source of life has always been one of the main interests of astronomy and science in general. Despite the significant progress in our understanding of how the solar system was formed, what bodies constitute it, and what mechanisms influence their motion, there are still numerous unanswered questions regarding the complex nature of our planetary system.

Many of the key information we have about the solar system come from the studies of the oldest remnants left over from the planetary formation in the early protosolar disk—asteroids, comets, and meteoroids. On one side, these bodies can reveal the processes and conditions occurring in the early stages of the solar system. On the other, they can give rise to the dangers of catastrophic impacts, which have subjected our planet numerous times in the history. In each case, they are the topic of high scientific interest. The studies of asteroids and comets are however often complicated by the small size and low albedo of these bodies. Usually only limited information can be obtained from direct observations. Meteor observations during the interaction of meteoroids with the Earth's atmosphere allow us to study small solar system bodies, which would otherwise remain undetected. The importance of meteoroid studies was eloquently stressed by Ceplecha et al. [1], by simply plotting the mass versus size diagram of objects in the observable universe (Fig. 1). This plot demonstrates the multitude of meteoroids in the observable universe.

Recently, meteor astronomy has been gaining popularity among professional and amateur astronomers, due to the arising possibilities of using inexpensive technologies, particularly sensitive CCD cameras to effectively observe meteors and provide valuable scientific data. Most of these efforts are focused on studying the identification methods and activities of meteor showers [2, 3], determining original heliocentric meteoroid orbits or detecting potential meteorite impacts from the brightest fireballs [4–7]. Furthermore, the research focused on physical and compositional properties of meteoroids is progressing by applying photographic and video spectrographs to study emission spectra of meteors [8–10]. Physical parameters such us meteoroid masses, strengths and densities can be determined by studying their atmoshperic ablation and deceleration [11–13]. Example of a meteor shower captured by a photographic system and a meteor spectrum observed by an all-sky video spectrograph is on Fig. 2.

The aim of this work is to provide basic introduction to meteor astronomy, its fundamental processes, observational techniques and relevant examples of specific research topics. A more comprehensive reviews of meteor studies beyond the presented individual examples can be found in [1, 14–17] and the references therein.

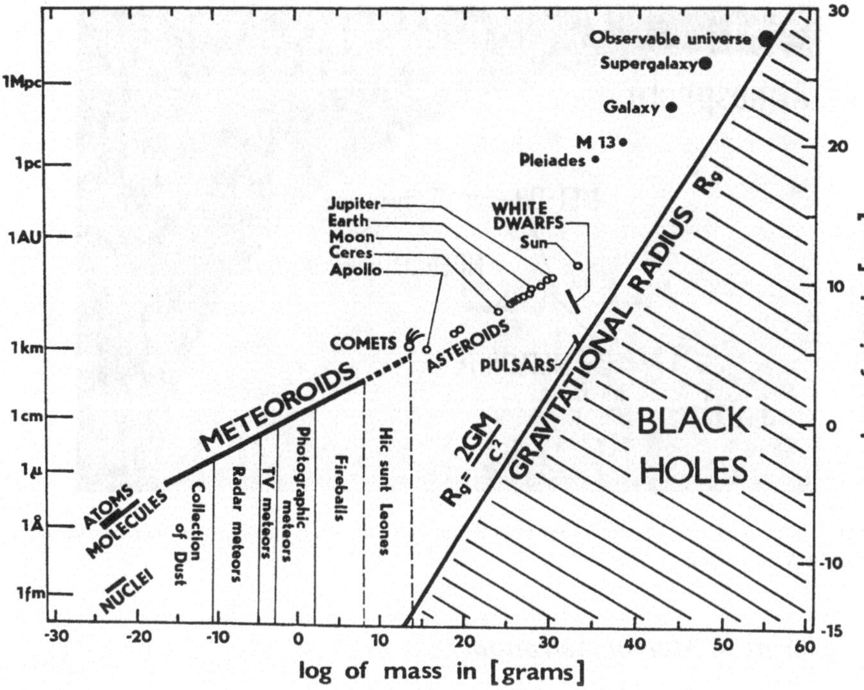


**Fig. 1** Mass versus size diagram containing some known objects of the observable universe and showing the significance of the meteoroid complex (from [1])

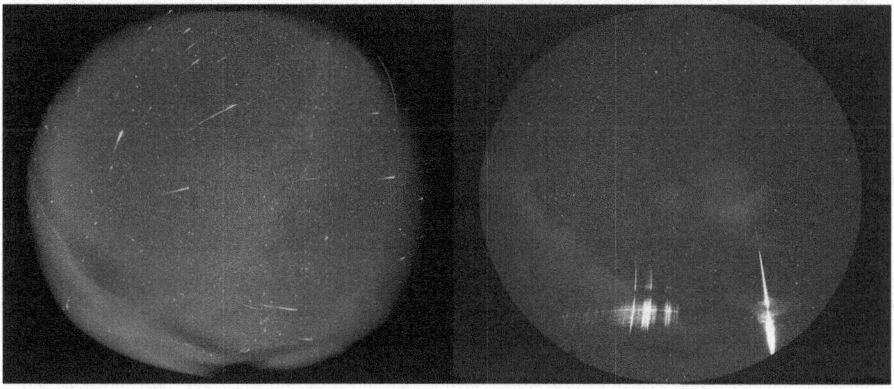

**Fig. 2** Left: Leonid meteor shower observed during the 1998 outburst by a photographic camera at the Modra Observatory. Right: a fireball captured by All-sky Meteor Orbit System (AMOS) spectrograph, along with the first order emission spectrum (images by J. Tóth and P. Matlovič)

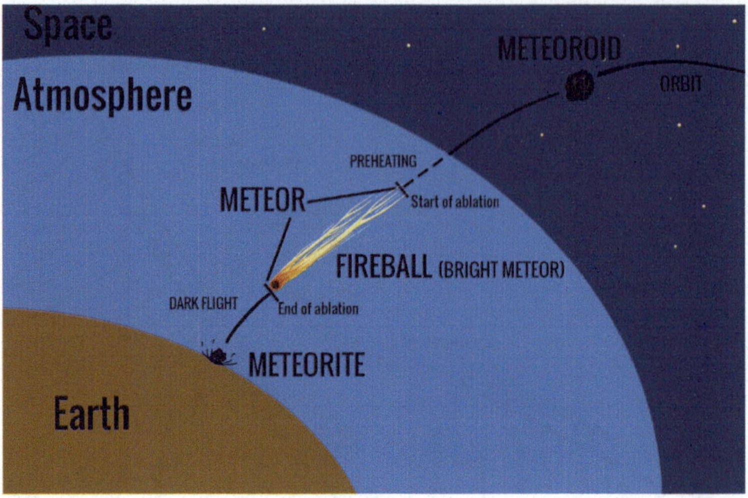


**Fig. 3** Different phases of meteoroid interaction and characteristic terms in meteor astronomy (image by V. Vojáček)

## 1.2 *Terms in Meteor Astronomy*

*Meteoroid* is currently defined as a solid natural object of a size roughly between 30 μm and 1 m moving in, or coming from, interplanetary space.[1] The size limits have been set by agreement and do not represent a physical boundary. In the context of meteor observations, any object causing a meteor can be termed a meteoroid, irrespective of its size. Bodies smaller than 30 μm tend to radiate heat away more efficiently and not to vaporize during the atmospheric entry. These smaller bodies are known as *interplanetary dust* particles. *Meteorite* is any natural solid object that survived the meteor phase in a gaseous atmosphere without being completely vaporized. Meteorites smaller than 1 mm in size are also called micrometeorites. Depending on their speed, these may be too small to experience ablation in an atmosphere. Graphical interpretation of the meteor terminology is given in Fig. 3.

Meteoroids are dominantly produced as the decay products of comets and asteroids. Only a minority of meteoroids come from the surfaces of planets (e.g. Mars) and planetary satellites (e.g. the Moon) or from interstellar space. Though meteoroids are conglomerates of materials formed in primordial solar nebula, the dynamical lifetime of objects in near-Earth space is assumed to be of the order of 10 Myr [18, 19]. This means that no meteoroids could have stayed on current orbits from the beginning of the solar system.

We consider three main processes which lead to the separation of meteoroids from their parent bodies. Comets produce meteoroids through the process of

[1]Defined in 2017 by the IAU Commision F1 on Meteors, Meteorites and Interplanetary Dust.

sublimation and following gas drag [20]. During the cometary activity near its perihelion, the drag of vapors from evaporating ices also releases solid particles, dust and meteoroids. Secondly, catastrophic disruption of comets can produce secondary nuclei and numerous dust particles and meteoroids (see Chapter 21 in [3]). The third process is related to direct collisions of solar system bodies, particularly among the main belt asteroids, which produce many collisional fragments [21]. The separation velocities of these fragments are always much smaller than the original orbital velocity. Therefore, young meteoroids follow very similar orbits as their parent body. In this early stage, it is relatively easy to link the *meteoroid stream* with its parent body. With time, gravitational perturbations from planets and various non-gravitational forces such as the Poynting–Robertson effect [22] cause the dispersion of the stream and separation from its parent orbit [23].

In this respect we can distinguish stream meteoroids, which are usually just recently separated from their parent object (typically up to few thousand years) and *sporadic meteoroids* on seemingly random orbits, significantly altered from the parent orbit. We assume that sporadic meteoroids were freed from their parent object thousands to millions of years ago. Compilation of known parent objects of meteoroid streams can be found e.g. in [24] or [3]. The sporadic complex was described e.g. by Wiegert et al. [25], Nesvorný et al. [26], and Levasseur-Regourd et al. [27]. The topics of meteor studies will be further discussed in Sect. 4.

Meteoroids may dynamically evolve to pass near Earth and interact with our atmosphere. The interaction may, for certain meteoroid sizes and velocities, result in a luminous phenomenon known as *meteor*. Meteoroid streams entering the Earth's atmosphere generate so-called *meteor showers*. For observers, meteor showers appear to originate from the same point and direction in the sky known as the *radiant*. The name of a meteor shower is derived from the apparent position of the radiant in the night sky. For example, Leonid meteor shower has a radiant in the Leo constellation.

Owing to the interaction with air molecules, a meteoroid entering the Earth's atmoshpere heats up to high temperatures and starts to melt and vaporize. A column of ionized and excited plasma is produced along the meteoroid trail, producing light, ionization, and for larger particles, shock waves [28]. The meteor phenomenon can exhibit several phases. The brightest part is called the meteor *head*. Ionization of air along the meteor path forms an ion *train* which reflects radio waves in the decameter range. The ion trains of bright meteors can be visible even to the naked eye and those of particularly brilliant meteors may persist for seconds or even minutes (also known as persistent trains). *Wake* of the meteor is the luminosity extending directly behind the meteor head and forming comet-like appearance of bright meteors. Meteor wake has different spectral features compared to the spectrum of meteor head. At a given position, the wake duration is only fraction of a second.

During the atmospheric flight, the meteoroid loses mass through processes of vaporization, fusion (melting), and fragmentation. Generally, the process of mass loss by a meteoroid is known as *ablation*. The resistance of the atmosphere causes meteoroid to decelerate. Ablation and deceleration affect one another, since ablation depends on the meteoroid velocity and deceleration on its mass. Therefore, the equations describing deceleration and mass loss (see Eqs. 3 and 5 in the next section) of the body must be solved simultaneously.

The collisions with air molecules and dynamical load generally cause the fragmentation of meteoroids. There are several ways in which meteoroids break up. The most significant fragmentation processes include progressive fragmentation [29, 30], in which the meteoroid fragments into parts which continue to crumble; (quasi-)continuous fragmentation [31, 32] describing continuous detachment of small particles; and gross (sudden) fragmentation [12, 33] characteristic during brilliant bursts, in which meteoroid suddenly disrupts into large number of fragments.

Recent studies suggest that most meteoroids undergo some form of fragmentation during the meteor phase. The most successfully applied ablation and fragmentation models follow the concept of the dustball meteoroid [11, 12]. Still, many of the methods we use to study meteoroid properties assume meteoroid interaction as a single non-fragmenting body (next section). While this approximation is sufficient for most of our estimates, precise description of meteor deceleration and mass loss must be based on a model of effective fragmentation. Neglecting meteoroid fragmentation was one of the main reasons behind the discrepancies of determined masses in past analyses (e.g. the discussion in [34, 35].

## 2 Atmospheric Interaction: Basics of Meteor Physics

The motion and ablation of a meteoroid in the Earth's atmosphere is most often described by the single body theory (Chapter 3 in [1]). Single body theory refers to the mass loss, deceleration, luminosity and ionization related to the motion of a single non-fragmenting body. The theory assumes that the heat transfer, ionization, luminosity and drag coefficient are during this path constant [28]. The presented standard equations of single body theory (following the formalization used by Weryk [36]) are used to describe meteors both before and after their disruption.

Collisions of atmospheric molecules with meteoroid can either liberate atoms from the surface of a meteoroid directly (also known as sputtering), or heat the material to its boiling point of approximately 2000 K. At these temperatures, material starts to evaporate from the surface of the meteoroid in a phenomenon known as thermal ablation. It is assumed that thermal ablation is the dominant process of mass loss (see [37] and the discussion therin). Excess of thermal stress or stagnation pressure can cause meteoroid to fragment into numerous smaller pieces. These fragments continue to collide with atmospheric molecules and ablate on their own as single bodies. The ablated material colliding with atmospheric molecules produces a trail of ionized and excited plasma. Rather than individual atomic emission, ablation may also take the form of a dust emission/removal, releasing a trail of heated small particles which cause the meteor wake [1]. The ablation behavior differs significantly among meteoroids. Large variations are observed in the beginning heights and light curve shapes among smaller meteoroids, and in end heights among larger meteoroids.

The basis of the mathematical form describing the motion of a meteoroid in atmosphere was first presented by Whipple [38] who used Hoppe's solution with constant coefficients. More elaborate solutions to these differential solutions were later presented by Levin [39] and Bronshten [28].

Let us assume meteoroid passing through a distance $v\Delta \mathrm{t}$ in time period $\Delta t$. A meteoroid with a cross-section area $S$ will encounter atmospheric mass (with atmospheric density $\rho_{\mathrm{a}}$) of $m_{\mathrm{a}} = \rho_{\mathrm{a}} S v \Delta t$. The cross-section area can be rewritten by introducing the dimensionless shape factor $A = S/V^{2/3}$. Meteoroid volume V is related to meteoroid bulk density and meteoroid mass as $V = m/\rho_{\mathrm{m}}$. We will assume that $A$ is constant and for simplicity usually corresponds to a sphere ($A = 1.21$). The rate of air mass encountering the meteoroid is defined as:

$$\frac{dm_{\mathrm{a}}}{dt} = \frac{\Delta m_{\mathrm{a}}}{\Delta t} = \frac{A v \rho_{\mathrm{a}} m^{\frac{2}{3}}}{\rho_{\mathrm{m}}^{\frac{2}{3}}} \tag{1}$$

We can express the momentum transfer from the atmosphere to the meteoroid by:

$$\frac{d\,(mv)}{dt} = \frac{dm}{dt} v + \frac{dv}{dt} m = \Gamma v \frac{dm_{\mathrm{a}}}{dt} \tag{2}$$

The drag coefficient $\Gamma$ in Eq. 2 is defined as the fraction of momentum transferred to the body from the ongoing molecules of air. The drag coefficient can vary between 0 (for no transfer of momentum) and 2 (perfect reflection of air molecules). For small meteoroids, the term $dm/dt$ can be neglected [1]. Next, by substituting Eq. 1 into Eq. 2, we get the drag equation:

$$\frac{dv}{dt} = -\frac{\Gamma A \rho_{\mathrm{a}} v^2}{\rho_{\mathrm{m}}^{\frac{2}{3}} m^{\frac{1}{3}}} \tag{3}$$

The drag equation is the first fundamental equation in meteor physics. It describes the deceleration of a meteoroid during its flight in the atmosphere (the deceleration is emphasized by the negative sign at the right-hand side of the equation).

The second fundamental equation is called the mass-loss equation. The mass loss rate is determined by the kinetic energy transferred from the atmosphere to the meteoroid. We assume that a certain fraction of the kinetic energy of the oncoming air molecules is expended on ablation of mass (vaporization or fusion and spraying) of the meteoroid [28]. The loss of mass during ablation can be expressed as:

$$\frac{dm}{dt} = -\frac{\Lambda E_{\mathrm{a}}}{\xi \Delta t} = -\frac{\Lambda v^2}{2\xi} \frac{dm_{\mathrm{a}}}{dt} \tag{4}$$

where $E_a$ is the kinetic energy of interacting air molecules and $\xi$ is the heat of ablation, representing the energy required to melt/vaporize one unit of meteoroid mass $dm$. Substituting Eq. 1 into Eq. 4 gives the conventional form of the mass-loss equation:

$$\frac{dm}{dt} = -\frac{\Lambda A \rho_{\mathrm{a}} v^3 m^{\frac{2}{3}}}{2\xi \rho_{\mathrm{m}}^{\frac{2}{3}}} \tag{5}$$

The heat-transfer coefficient $\Lambda$ is valued between zero and unity, since the energy used on ablation cannot exceed the total kinetic energy of interacting air molecules. Part of this kinetic energy will be used to heat up the body of a meteoroid, part will be re-radiated, and part will be expended for excitation and ionization of the meteoroid and surrounding air molecules. If fragmentation takes place, some of this energy is also expended to break the mechanical bonds between meteoroid grains.

It is assumed that the amount of light produced during this process is also related to the mass-loss rate, as it is proportional to the kinetic energy lost by the meteoroid [1]. The energy released by the meteoroid in the form of radiation, typically in the visible spectrum is described in the luminosity equation:

$$I = -\tau \frac{dE_{\mathrm{m}}}{dt} = -\tau \left( \frac{v^2}{2} \frac{dm}{dt} + \frac{dv}{dt} m v \right) \tag{6}$$

Here, $I$ is the radiative power (bolometric or in specific band pass) and $\tau$ is the luminous efficiency, which is defined as the fraction of the kinetic energy loss of a meteoroid transformed into radiation. Generally, the luminous efficiency is dependent on the wavelength of radiation, the chemical composition of the meteoroid body and the atmosphere, on the meteoroid velocity and possibly on the meteoroid mass. The deceleration term $(dv/dt)$ can be neglected for fast and faint meteors [1].

Most of the meteor radiation comes from line emissions in evaporated meteoroid atoms [28]. Clearly, the chemical composition of the meteoroid plays significant role in the nature of the produced emission, since different chemical elements are represented by different line strengths in the visible spectrum. The ionization produced during the interaction with atmosphere can be described using the ionization equation:

$$q = -\frac{\beta}{\mu v} \frac{dm}{dt} \tag{7}$$

In this equation, $q$ is the electron line density, which represents the number of electrons per unit trail length. The ionization coefficient $\beta$ describes the average number of electrons produced per ablated atom, while the atomic mass of standard meteoroid atom is labeled as $\mu$. The amount of ionization is again dependent on the mass-loss rate.

To determine meteoroid mass from the observed photon and electron count, we need the values of $\tau$ and $\beta$, which is often problematic. Usual process of obtaining the ionization mass is based on measuring the electron line density $q$, assuming and ionization curve and integrating Eq. 7. Similarly, the photometric mass is determined by measuring $I$ along the meteor trail and integrating Eq. 6.

Equations 3, 5, 6 and 7 contain several variables that need to be determined in order to solve the fundamental equations of meteor physics. Some of them are given by the physical properties of the meteoroid, some are determined from observations and others can be estimated theoretically. The atmospheric density $\rho_a$ is usually defined by a model of the atmosphere for different heights of meteoroid-atmosphere interaction. Currently the most commonly used video observations allow us to determine the velocity of a meteor as a function of time and luminosity. The heat of ablation is defined by the character of ablation (vaporizing or spraying) and is thus dependent on the composition of a meteoroid. Lastly, the luminous efficiency, heat transfer coefficient and drag coefficient must be obtained from theoretical models, experiments, or by estimating based on observational data.

## 3 The Nature of Meteor Radiation: Spectra and Meteoroid Composition

Meteor radiation is produced mainly by the excitation of atoms, due to the mutual collisions with atmospheric atoms and molecules, and due to the recombination of free electrons in the surrounding ionized gas and subsequent cascade transitions [28]. The radiation of a meteor originates in the plasma envelope of air and meteoric vapor surrounding the meteoroid. Meteor spectra consist primarily of atomic emission lines and molecular bands. Spectral analyses indicate that it is mainly the atoms and ions of the meteoroid vapor which radiate. Although meteor spectra have been observed since 1864, the knowledge gained from these observations is rather scarce. Most early studies were focused on the description of the spectra and identification of lines. The most extensive identifications are given by Halliday [40] for Perseid meteors (velocity of app. 60 km s$^{-1}$), by Ceplecha [41] for a 32 km s$^{-1}$ meteor and by Borovička [42] for a 19 km s$^{-1}$ meteor. Overview of reliably identified atoms and ions is presented e.g. in Section 3.3 of [1].

Not all lines in meteor spectra can be explained by a single temperature. Borovička [43] revealed that meteor spectra are composed of two distinct components with different characteristic temperatures. The lower temperature component is called the main spectrum and its origin is in the radiating gas of meteoroid and atmospheric vapors. The temperature lies usually in the range 3500–5000 K and does not generally depend on meteor velocity. The main spectrum consists of several hundreds of lines, mostly neutral lines of atoms of meteoric origin. The most notable lines present in the main spectrum are of Fe I, Mg I, Na I, Ca I, Cr I, Mn I and Ca II. Thermal equilibrium is nearly satisfied, although some lines may deviate. It was shown that chemical composition of the radiating plasma can be computed from the main spectrum.

The high temperature component is also known as the second spectrum and has characteristic temperature of nearly 10,000 K. The high temperature region is probably formed in the front of the meteoroid, near the shock wave of the meteor [8]. The chemical composition cannot be derived exactly, nevertheless, the determined elemental abundances from fireball spectra were found to be consistent with common meteorite composition. The typical lines for the second spectrum are the high excitation lines of Mg II, Si II, N I and O I. The low excitation transitions in singly ionized atoms can be present in both spectral components. This is most notably the case of Ca II, which is bright in both spectra, but also fainter lines of Ti II and Sr II can be present in both components. The second spectrum is strong in fast meteors while it can be absent in slow meteors with velocity of about 15 km $s^{-1}$ [1]. The ratio of gas masses involved in the production of both components was found to be a steep function of velocity.

The ratio of meteoric vapors to the atmospheric species in both components shows interesting disparities. Nitrogen and oxygen lines are not present in the main component. This is caused by the absence of allowed low excitation transitions in these atoms. The atoms simply do not radiate at 5000 K. Based on the pressure balance with surrounding atmosphere, [8] concluded that about 95% of atoms in low temperature gas were the invisible atmospheric species. The second spectrum demonstrates both meteoritic and atmospheric emissions. Their ratio varies widely. In faint meteors of medium and high velocity the meteoric emissions in the second spectrum are often absent, while O I and N I lines and $N_2$ bands are still present. The $N_2$ bands sometimes appear very early on the trajectory [44]. Other works [45] described cases in which meteoritic emission invisible at the start of the trajectory burst out later, while the atmospheric lines brighten only moderately. Based on these effects, it appears that atmospheric emissions are less dependent on the ablation rate, which is rather expected.

The previously described spectral lines are characteristic for meteor head—the brightest part of a meteor. Other phases of the meteor phenomenon present specific spectral features. The spectrum of the meteor wake consists chiefly from low excitation lines. Typical wake lines belong to Na I, Fe I, Mg I and Ca I. The short-duration trains are formed by only one spectral line, the forbidden green auroral line of neutral atomic oxygen at 557.7 nm. The luminosity is probably produced by the atmospheric oxygen. Persistent trains are still not well understood phenomena. Several spectra have been taken in the recent years, which show different features from case to case. The spectra show both continuous or quasi-continuous radiation and atomic lines. The most important and most persistent line, common for all spectra, is the sodium doublet near 589.2 nm. This suggests that the long-living luminosity is due to similar mechanism which produces the sodium airglow—a luminous layer in the Earth's mesosphere (80–105 km) of characteristic yellow color.

While meteor spectroscopy presents the most efficient way to study of meteoroid composition from unbiased sources, the composition of the radiating plasma during meteoric interaction does not fully reflects the composition of the original meteoroid [8]. Meteoroid composition must cover a wide range of material types including asteroidal, lunar and martian samples known from meteorite laboratory studies (for mineralogical overview, see [46]) and the lesser known cometary materials, so far only covered by few spacecraft probes such as the Stardust mission to comet 81P/Wild 2 [47] and Rosetta mission to comet 67P/Churyumov-Gerasimenko [48].

Most of the known meteorite samples come from primitive rocky asteroidal bodies formed in the early solar system which never reached the melting limit temperature in their interiors. This primitive material presents valuable chemical clues preserved from the early stages of the solar system formation. Most of the known rocky meteorites are chondritic, meaning that they contain chondrules, 0.1–1-mm sized objects of glass and crystalline silicates formed by melting followed by rapid cooling (Fig. 4). They also contain Ca- and Al-rich inclusions (CAIs) composed of refractory oxides and silicates and formed by condensation of high-temperature nebular gases. Besides the primordial material from the early solar system, it was detected that chondrites also contain particles from other stars [50]. Primitive chondritic composition has been also detected in samples of cometary dust (Fig. 5).

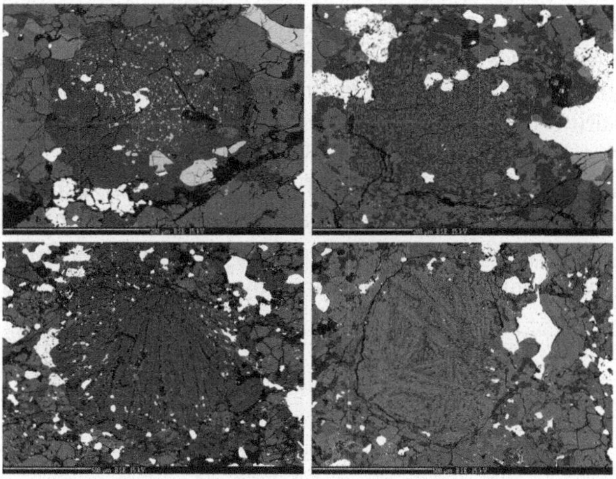

**Fig. 4** Different types of chondrules (large circular shapes in the middle of the pictures) in the Košice ordinary chondrite meteorite (image by D. Ozdin, further description is provided in [49])

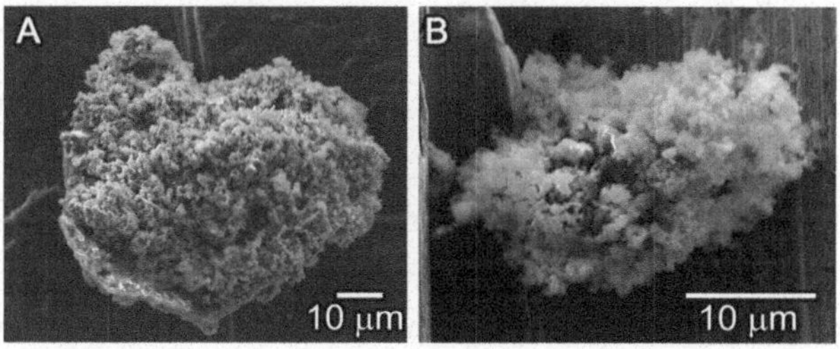


**Fig. 5** Electron images of chondritic porous micrometeorites of cometary dust (from [51])

Not all interplanetary bodies are chondritic. Variations of stony achondrites, iron and stony-iron materials have been identified from meteorite samples on Earth. They point towards more complicated formation processes and material differentiation in larger bodies. Still, together they only constitute approximately 7% of all known meteorite samples. The ratios of identified meteorite materials is however biased by the fact that only stronger asteroidal samples can withstand the ablation in the atmosphere and reach the Earth's surface. Given the dust production rates of comets and the observed activities of major meteor showers, it is likely that majority of smaller meteoroids in interplanetary space are of cometary origin. Precise orbital data determined from multi-station meteor observations along with the compositional information from emission spectroscopy allow us to study the real distribution of materials in the solar system.

To achieve this goal, larger surveys of meteor spectra based sensitive video spectrographs were initiated [10, 52, 53]. Generally, these instruments yield low-resolution spectra from which elemental abundances cannot be reliably determined. To reveal variations of meteoroid composition, the method of spectral classification was established. The method is based on the relative intensities of the three main emission multiplets Na I, Mg I and Fe I representative of the different components of meteoroid composition (volatile, silicate and metallic respectively). Figure 6 shows the results of such studies for different size populations of meteoroids.

While this method can be used to distinguish distinct meteoroid types from common chondritic bodies, the distinction between specific material types (e.g. specific chondritic classes) is difficult [55]. For this purpose, detailed model of meteor spectrum based on the solution of radiative transfer needs to be applied to records captured in high-resolution. First such model assuming thermal equilibrium and self-absorption in the radiating plasma was developed by Borovička [8] and applied to an excellent spectrum captured by a photographic system. Relative

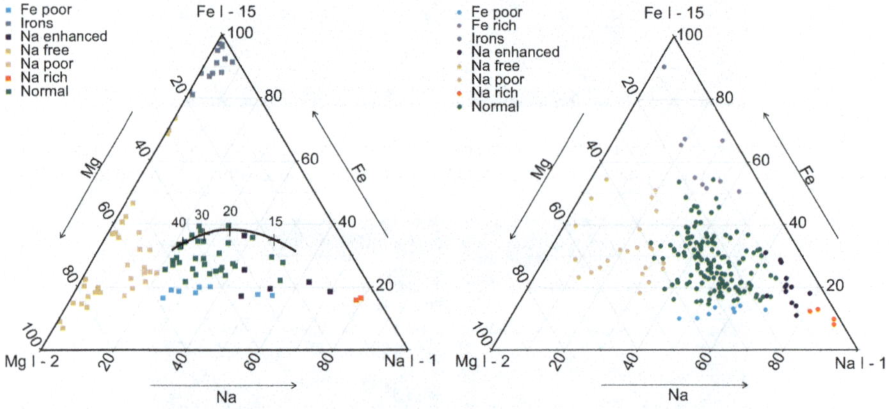


**Fig. 6** Spectral classification of mm-sized and mm-m-sized meteoroids showing range of material types detected from meteor observations (from [10] and [54])

**Fig. 7** The simulated ablation of a chondritic meteorite at the probe of the plasma wind tunnel (image by the High Enthalpy Flow Diagnostics Group, IRS)

abundances for nine elements were determined (Fe, Na, Mg, Ca, Ti, Cr, Mn, Ni, and Al) in agreement with laboratory meteorite measurements. Similar method has been since used for individual fireball spectra by few different authors [9, 56]. Similarly detailed analyses of meteor spectra for larger quantities of meteoroids samples are still missing.

Recent experiments suggest that our abilities to study meteoroid composition and ablation from ground-based observations can be also improved by laboratory analyses. Several teams have focused on using laser-induced breakdown spectroscopy of meteorite samples to study meteoroid composition [57, 58]. Alternatively, the simulated ablation of meteorites in plasma wind tunnels has been used to successfully reproduce the atmospheric meteoroid interaction [56, 59, 60] (Fig. 7). Besides high-resolution Echelle spectra, these experiments provide valuable data on the ablation processes for different meteoroid types. The next step is to quantitatively link the phenomena observed in the laboratory with real meteor observations in the atmosphere.

## 4 Meteor Observations and Meteoroid Population Studies

For meteor scientists, Earth's atmosphere serves as a large detector of meteoroid inflow in the region of 1 au from the Sun. Various observational methods can be used to observe meteors and provide different information about meteoroids. The light from meteor ablation can be observed visually by naked eye or telescopes, by photographic and video cameras, radar systems, seismic and infrasonic detectors.

Simple visual observations of meteors, though the least comprehensive, have historically been crucial in gaining basic knowledge of the inflow of meteors and describing shower activities. The earliest records of individual meteor sightings and probable meteor showers were made in ancient Mesopotamia and China, at the beginning of second millennium BC [61, 62]. Still today, visual reports are useful for constraining information on potential meteorite impacts, or calibration of instrumental efficiencies. Reports from visual observations can be submitted to the International Meteor Organization database.[2]

Optical video and photographic observations are currently the most utilized by astronomers, due to their ability to precisely determine meteor fluxes and constrain atmospheric trajectories and speeds of meteoroids. By triangulation of meteor trajectories observed from multiple stations and backwards propagation of the directional and velocity information, original heliocentric orbit of a meteoroid can be determined (e.g. [63]). Numerous groups have initiated development of video or photographic networks to provide large sky coverage with multi-station observations. These networks include the European Fireball Network [64] in Central Europe, Cameras for Allsky Meteor Surveillance (CAMS) network [65] in Northern America, Desert Fireball Network in Australia [66], global All-sky Meteor Orbit System (AMOS) network [67] based in Slovakia, Fireball Recovery and InterPlanetary Observation Network (FRIPON) in France [68], and many others. Orbital data from these networks is collected in databases such as the SonotaCo [69], EDMOND [70] and CAMS database [2].

Trajectories and speeds of meteors can be also determined by radar systems operating in the 15–500 MHz frequency range. Radars can be used to study inflow of even the faintest meteors (caused by roughly micrometer particles) and can operate even during the daytime, allowing to identify activities of daytime showers. Some of the most renowned radar surveys providing crucial meteor data include the Advanced Meteor Orbit Radar Facility (AMOR) in New Zealand [71], the Canadian Meteor Orbit Radar (CMOR) [72] or the Southern Argentina Agile MEteor Radar (SAAMER) [73]. Radar observations have been also used to constrain the mass-dependent flux of interplanetary particles on Earth [74, 75]. The standard mass distribution was previously determined from lunar crater counts and taking into account meteoroid and interplanetary dust measurements [76].

During observations, meteor showers caused by Earth intersecting meteoroid streams appear to originate from the same point and direction in the sky, known as the radiant. These meteoroids were ejected from their parent comet or asteroid relatively recently and can be used to probe the properties of larger parent bodies (e.g. [77–80] or [3] for overview). For some major streams, the association to parent object is well known (e.g. Perseids and comet 109P/Swift-Tuttle, Orionids and comet 1P/Halley or Draconids and 21P/Giacobini-Zinner). However, there are numerous minor streams and newly detected streams (for overview, see the IAU Meteor Data Center[3] [81]), which are yet to be confirmed in activity and linked to their parent object. A new review of the current state of minor streams a sporadic background research is presented in [82].

[2]http://www.imo.net.

[3]https://www.ta3.sk/IAUC22DB/MDC2007/.

In some cases, the link to a meteoroid stream or a parent body can be established based on orbital similarity. Dissimilarity parameters such as the D-criterion based on orbital [83, 84] or geocentric parameters [85, 86] can be used. Recent works focus on applications or improvements of these methods to identify potential meteoroid associations in orbital databases [87–89]. Alternatively, wavelet transform methods can be used to search for meteoroid streams in radar data [90]. Furthermore, modeling of the meteoroid stream evolution and prediction of meteor shower activities has improved significantly with the development of more powerful computational facilities [91, 92].

Sporadic meteors are not associated with any meteoroid streams, but they dominate the meteoroid influx at Earth. They include particles from the interplanetary meteoroid cloud (from previous asteroid and comet dust production processes) and in small proportion also interstellar particles [27]. The identification of their origin is hampered by orbital alteration caused by gravitational, radiation and collision processes. While among brighter meteors, sporadic particles comprise comparable datasets as shower meteors, their proportion increases dramatically among fainter meteors corresponding to smaller meteoroids (e.g. 99% of the meteors in the AMOR radar database [93]). This effect is caused by different ejections velocities and radiation forces affecting the evolution of smaller particles. All of the instrumentally observed meteorite impacts have originated from sporadic meteors [55]. The improved observational techniques and larger sky coverage allows for higher efficiency in the location of meteorite impacts [4–7]. The most advanced techniques for trajectory and photometric measurements today allow for very precise prediction of the meteorite strewn field [94].

Large datasets of sporadic meteoroids from optical and radar detections have revealed six sporadic meteor sources at the Earth. These sources do not generally correspond to physical meteoroid structures, but rather describe observed concentrated regions in the radiant space. The detected meteor sources are affected by the motion of the Earth around the Sun [95]. The strongest sources are the helion and anti-helion source in the ecliptic plane at aproximately 70° from the apex direction. These sources are likely formed by particles produced by Jupiter-family comets [25, 96]. We also observe north and south apex sources centered towards the direction of Earth's movement and toroidal sources 60° north and south from the apex. The apex and toroidal sources seem to originate from Halley-type or long-period comets [25, 97]. Furthermore, radiant distribution of sporadic meteors reveals a ring depleted in meteor radiants at 55° from the apex [98]. The ring is attributed to high-inclination meteoroids undergoing Kozai oscillations [25]. Studies of sporadic meteors have also provided constrains for the speed distribution of incoming particles, showing that most meteoroids impact the Earth at low speeds ($\approx$11–20 km $s^{-1}$) [99, 100].

Meteor observations can be also used to study the presence of interstellar bodies in the solar system. Interstellar dust was first detected by the Ulysses spacecraft [101] during a flyby of Jupiter and since studied by multiple other missions. The measured interstellar particles are usually not larger than a few microns. The detection of larger interstellar meteoroids would provide significant implications for the dust-to-gas mass ratio in close interstellar medium. Generally, meteoroids with speed above approximately 72 km $s^{-1}$ (given by the sum of the escape speed from the solar system and orbital speed of Earth) are on orbits not bound to the Sun. Slower meteoroids can still originate from the interstellar space, but cannot be identified as such only based on their speed. Typically large uncertainties of the determined meteoroid velocities for very fast meteors must be taken into account. The first detection of interstellar meteoroids was reported from the AMOR radar data [102], though these results are still being debated. Analysis of CMOR radar observations showed that the interstellar meteoroids would present a fraction too small to be statistically meaningful [72]. A search in the video meteor databases has revealed that majority of identified hyperbolic meteoroids are caused by the velocity estimation error [103].

As briefly discussed in previous section, observations of meteor emission spectra can provide information on meteoroid composition. Generally, these studies are either focused on detailed analyses of exceptional fireballs and can provide relative element abundances [8, 9, 104], focus on characterization of meteoroid streams [105, 106] and their parent bodies, or utilize larger datasets of low-resolution spectra to study variations of meteoroid composition from different orbital sources in the solar system [10, 52, 54, 107]. Meteor spectra studies have revealed that the depletion of volatiles in meteoroids is mainly caused by solar radiation in close proximity of the Sun and partially by cosmic-ray irradiation in the Oort cloud [10]. Depletion of sodium was also shown to have implications for meteoroid structure and material strength [108]. Besides atomic emission lines, several molecular bands have been identified in fireball spectra [109]. One of the interesting goals of meteor spectroscopy is to detect organic matter in meteoroids. While the detection of organic carbon can be difficult in meteor spectra [110, 111], the commonly observed hydrogen $H_\alpha$ line can be used as a tracer for the presence of organics and water [112].

Meteor trajectories and their light curves can be used to study physical properties of meteoroids. Meteoroid masses can be estimated based on optical meteor brightness or electron line density in radar data. Unfortunately, the accuracy of mass estimation is still limited mainly by the uncertainties of luminous and ionization efficiency parameters [113]. The most consistent estimates of meteoroid masses were yielded by models combining meteor light curves and atmospheric deceleration [12, 114]. The beginning and terminal heights of meteor luminous trajectory can be used to infer the material strength of a meteoroid. The empirical classification of [115] differentiates between the most fragile cometary (Draconid-type), standard and dense cometary, carbonaceous and ordinary chondritic material strengths. Using the meteoroid dustball model [11], distribution of bulk densities of meteoroids was studied by Kikwaya et al. [13]. It was revealed that on average, meteoroids on asteroidal orbits have densities of 4200 kg $m^{-3}$, on Jupiter-family orbits 3100 kg $m^{-3}$ and between 260 and 1900 kg $m^{-3}$ on Halley-type orbits. Grain densities of meteoroids can also be estimated using the heat conductivity equation and combining meteor trajectory measurements with laboratory data for different rock and mineral types [116, and references therein]. The difference between the bulk and grain density can be used to infer porosities of meteoroids.

Recent works and high-definition meteor observations [117] suggest that majority of meteoroids undergo different forms of fragmentation in the atmosphere, which is not accounted for in the single body theory (Sect. 2). Furthermore, it has been shown that ablation of meteoroids composed of iron–nickel alloy differs significantly from standard chondritic meteoroids [118]. For accurate determination of meteoroid physical properties, more complex models that describe the ablation and fragmentation of meteoroids are required [12, 119, 120].

## 5 Summary

We have presented a brief introduction to meteor astronomy and the fundamental processes of meteor ablation. The approach to constrain meteoroid properties from various ground-based observation techniques was described. Meteoroids are small interplanetary bodies continuously produced by sublimation, outgassing and collisions of comets and asteroids. During their interaction with the Earth's atmosphere, they create luminous phenomena known as meteors. This process can be studied from various points of view and eventually allows us to study dynamical processes and distribution of interplanetary material in the solar system. The current meteor research topics focus on different aspects of the phenomenon. Due to improved and efficient video and photographic techniques, numerous observational networks were created around the world and produce large datasets of orbital data. These observations can be used to study activities of meteor showers and link meteoroid streams with parent comets and asteroids.

Due to the improved sky coverage, the number of instrumentally observed meteorite falls rises and allows higher efficiency in locating meteorite impacts. The sensitive radar observations enable studies of interplanetary dust inflow, sources of the sporadic background and activities of daytime showers. More complex models of meteoroid ablation and emission spectra can be used to determine physical properties and composition of meteoroids and provide implication for the processes of material transfer in the solar system. Future analyses of meteors can be improved by utilizing laboratory facilities such us wind tunnels and shock tubes to simulate meteor ablation in controlled environment. While we already know a lot about the processes of meteor interaction and meteoroid populations, there are still numerous open questions to be answered and more details be obtained to better understand our solar system and prepare for potential Earth impacts and spacecraft shielding.

**Acknowledgements** This work was supported by the ERASMUS+ project 2017-1-CZ01-KA203-035562, by the Slovak Research and Development Agency grant APVV-16-0148 and the Slovak Grant Agency for Science grant VEGA 01/0596/18.

## References


1. Z. Ceplecha, J. Borovička, W.G. Elford, D.O. Revelle, R.L. Hawkes, V. Porubčan, M. Šimek, Space Sci. Rev. **84**, 327 (1998). https://doi.org/10.1023/A:1005069928850
2. P. Jenniskens, Q. Nénon, J. Albers, P.S. Gural, B. Haberman, D. Holman, R. Morales, B.J. Grigsby, D. Samuels, C. Johannink, Icarus **266**, 331 (2016). https://doi.org/10.1016/j.icarus.2015.09.013
3. P. Jenniskens, *Meteor Showers and their Parent Comets* (Cambridge University Press, Cambridge, 2006)
4. P. Spurný, J. Oberst, D. Heinlein, Nature **423**, 151 (2003). https://doi.org/10.1038/nature01592
5. P. Jenniskens, M.H. Shaddad, D. Numan, S. Elsir, A.M. Kudoda, M.E. Zolensky, L. Le, G.A. Robinson, J.M. Friedrich, D. Rumble, Nature **458**(7237), 485 (2009). https://doi.org/10.1038/nature07920
6. P.G. Brown, J.D. Assink, L. Astiz, R. Blaauw, M.B. Boslough, J. Borovička, N. Brachet, D. Brown, M. Campbell-Brown, L. Ceranna, Nature **503**(7475), 238 (2013). https://doi.org/10.1038/nature12741
7. J. Borovička, J. Tóth, A. Igaz, P. Spurný, P. Kalenda, J. Haloda, J. Svoreå, L. Kornoš, E. Silber, P. Brown, M. HusáRik, Meteorit. Planet. Sci. **48**, 1757 (2013). https://doi.org/10.1111/maps.12078
8. J. Borovička, Astron. Astrophys. **279**, 627 (1993)
9. J.M. Trigo-Rodriguez, J. Llorca, J. Borovička, J. Fabregat, Meteorit. Planet. Sci. **38**, 1283 (2003). https://doi.org/10.1111/j.1945-5100.2003.tb00313.x
10. J. Borovička, P. Koten, P. Spurný, J. Boček, R. Štork, Icarus **174**, 15 (2005). https://doi.org/10.1016/j.icarus.2004.09.011
11. M.D. Campbell-Brown, D. Koschny, Astron. Astrophys. **418**, 751 (2004). https://doi.org/10.1051/0004-6361:20041001-1
12. J. Borovička, P. Spurný, P. Koten, Astron. Astrophys. **473**, 661 (2007). https://doi.org/10.1051/0004-6361:20078131
13. J.B. Kikwaya, M. Campbell-Brown, P.G. Brown, Astron. Astrophys. **530**, A113 (2011). https://doi.org/10.1051/0004-6361/201116431
14. D. Koschny, R.H. Soja, C. Engrand, G.J. Flynn, J. Lasue, A.C. Levasseur-Regourd, D. Malaspina, T. Nakamura, A.R. Poppe, V.J. Sterken, Space Sci. Rev. **215**(4), 34 (2019). https://doi.org/10.1007/s11214-019-0597-7
15. G.O. Ryabova, D.J. Asher, M.J. Campbell-Brown, *Meteoroids: Sources of Meteors on Earth and Beyond* (Cambridge University Press, Cambridge, 2019)
16. J.M.C. Plane, G.J. Flynn, A. Määttänen, J.E. Moores, A.R. Poppe, J.D. Carrillo-Sanchez, C. Listowski, Space Sci. Rev. **214**(1), 23 (2018). https://doi.org/10.1007/s11214-017-0458-1
17. P. Jenniskens, Planet. Space Sci. **143**, 116 (2017). https://doi.org/10.1016/j.pss.2017.01.008
18. B. Gladman, P. Michel, C. Froeschlé, Icarus **146**, 176 (2000). https://doi.org/10.1006/icar.2000.6391
19. L. Foschini, P. Farinella, C. Froeschlé, R. Gonczi, T.J. Jopek, P. Michel, Astron. Astrophys. **353**, 797 (2000)
20. F.L. Whipple, Astrophys. J. **113**, 464 (1951). https://doi.org/10.1086/145416
21. D. Nesvorný, W.F. Bottke, H.F. Levison, L. Dones, Astrophys. J. **591**, 486 (2003). https://doi.org/10.1086/374807
22. J.A. Burns, P.L. Lamy, S. Soter, Icarus **40**, 1 (1979). https://doi.org/10.1016/0019-1035(79)90050-2
23. J. Vaubaillon, P. Lamy, L. Jorda, Mon. Not. R. Astron. Soc. **370**, 1841 (2006). https://doi.org/10.1111/j.1365-2966.2006.10606.x
24. J. Borovička, in *Near Earth Objects, our Celestial Neighbors: Opportunity and Risk, IAU Symposium*, ed. by G.B. Valsecchi, D. Vokrouhlický, A. Milani, vol. 236, pp. 107–120 (2007). https://doi.org/10.1017/S1743921307003134

25. P. Wiegert, J. Vaubaillon, M. Campbell-Brown, Icarus **201**, 295 (2009). https://doi.org/10.1016/j.icarus.2008.12.030
26. D. Nesvorný, D. Janches, D. Vokrouhlický, P. Pokorný, W.F. Bottke, P. Jenniskens, Astrophys. J. **743**(2), 129 (2011). https://doi.org/10.1088/0004-637X/743/2/129
27. A.C. Levasseur-Regourd, J. Agarwal, H. Cottin, C. Engrand, G. Flynn, M. Fulle, T. Gombosi, Y. Langevin, J. Lasue, T. Mannel, Space Sci. Rev. **214**(3), 64 (2018). https://doi.org/10.1007/s11214-018-0496-3
28. V.A. Bronshten, *Physics of Meteoric Phenomena* (D. Reidel, Dordrecht, 1983)
29. L. Jacchia, Astron. J. **60**, 165 (1955). https://doi.org/10.1086/107133
30. F. Verniani, Space Sci. Rev. **10**, 230 (1969). https://doi.org/10.1007/BF00212686
31. R.L. Hawkes, J. Jones, Mon. Not. R. Astron. Soc. **173**, 339 (1975). https://doi.org/10.1093/mnras/173.2.339
32. P.B. Babadzhanov, Astron. Astrophys. **384**, 317 (2002). https://doi.org/10.1051/0004-6361:20020010
33. Z. Ceplecha, P. Spurny, J. Borovička, J. Keclikova, Astron. Astrophys. **279**, 615 (1993)
34. Z. Ceplecha, Smithsonian Contributions Astrophys. **11**, 35 (1967)
35. F. Verniani, Smithsonian Contributions Astrophys. **11**, 61 (1967)
36. R. Weryk, Simultaneous radar and video meteors. Ph.D. Thesis, The University of Western Ontario (2012)
37. L.A. Rogers, K.A. Hill, R.L. Hawkes, Planet. Space Sci. **53**(13), 1341 (2005). https://doi.org/10.1016/j.pss.2005.07.002
38. F.L. Whipple, Proc. Am. Philos. Soc. **79**, 499 (1938)
39. B.J. Levin, Bull. Astron. Inst. Czechoslovakia **7**, 58 (1956)
40. I. Halliday, Publ. Dominion Obs. Ottawa **25**, 3 (1961)
41. Z. Ceplecha, Bull. Astron. Inst. Czechoslovakia **22**, 219 (1971)
42. J. Borovička, Astron. Astrophys. **103**, 83 (1994)
43. J. Borovička, Planet. Space Sci. **42**, 145 (1994). https://doi.org/10.1016/0032-0633(94)90025-6
44. A.F. Cook, C.L. Hemenway, P.M. Millman, A. Swider, NASA Spec. Publ. **319**, 153 (1973)
45. J. Borovička, J. Boček, Earth Moon Planet. **71**, 237 (1995). https://doi.org/10.1007/BF00612965
46. J.J. Papike, *Planetary Materials*, vol. 36 (Mineralogical Society of America, Chantilly, 1998)
47. D. Brownlee, D. Joswiak, G. Matrajt, Meteorit. Planet. Sci. **47**(4), 453 (2012). https://doi.org/10.1111/j.1945-5100.2012.01339.x
48. Y. Langevin, M. Hilchenbach, N. Ligier, S. Merouane, K. Hornung, C. Engrand, R. Schulz, J. Kissel, J. Rynö, P. Eng, Icarus **271**, 76 (2016). https://doi.org/10.1016/j.icarus.2016.01.027
49. D. Ozdín, J. Plavčan, M. Horåáčková, P. Uher, V. Porubčan, P. Veis, J. Rakovský, J. Tóth, P. Konečný, J. Svoreå, Meteorit. Planet. Sci. **50**, 864 (2015). https://doi.org/10.1111/maps.12405
50. S. Messenger, L.P. Keller, F.J. Stadermann, R.M. Walker, E. Zinner, Science **300**(5616), 105 (2003). https://doi.org/10.1126/science.1080576
51. T. Noguchi, N. Ohashi, S. Tsujimoto, T. Mitsunari, J.P. Bradley, T. Nakamura, S. Toh, T. Stephan, N. Iwata, N. Imae, Earth Planet. Sci. Lett. **410**, 1 (2015). https://doi.org/10.1016/j.epsl.2014.11.012
52. V. Vojáček, J. Borovička, P. Koten, P. Spurný, R. Štork, Astron. Astrophys. **580**, A67 (2015). https://doi.org/10.1051/0004-6361/201425047
53. R. Rudawska, J. Tóth, D. Kalmančok, P. Zigo, P. Matlovič, Planet. Space Sci. **123**, 25 (2016). https://doi.org/10.1016/j.pss.2015.11.018
54. P. Matlovič, J. Tóth, R. Rudawska, L. Kornoš, A. Pisarčíková, Astron. Astrophys. **629**, A71 (2019). https://doi.org/10.1051/0004-6361/201936093
55. J. Borovička, P. Spurný, P. Brown, *Small Near-Earth Asteroids as a Source of Meteorites* (2015), pp. 257–280. https://doi.org/10.2458/azu_uapress_9780816532131-ch014
56. A. Drouard, P. Vernazza, S. Loehle, J. Gattacceca, J. Vaubaillon, B. Zanda, M. Birlan, S. Bouley, F. Colas, M. Eberhart, T. Hermann, L. Jorda, C. Marmo, A. Meindl, R. Oefele,

F. Zamkotsian, F. Zander, Astron. Astrophys. **613**, A54 (2018). https://doi.org/10.1051/0004-6361/201732225
57. M. Ferus, J. Koukal, L. Lenža, J. Srba, P. Kubelík, V. Laitl, E.M. Zanozina, P. Váňa, T. Kaiserová, A. Knížek, P. Rimmer, E. Chatzitheodoridis, S. Civiš, Astron. Astrophys. **610**, A73 (2018). https://doi.org/10.1051/0004-6361/201629950
58. M. Dell'Aglio, M. López-Claros, J.J. Laserna, S. Longo, A. De Giacomo, Spectrochimica Acta **147**, 87 (2018). https://doi.org/10.1016/j.sab.2018.05.024
59. S. Loehle, F. Zander, T. Hermann, M. Eberhart, A. Meindl, R. Oefele, J. Vaubaillon, F. Colas, P. Vernazza, A. Drouard, J. Gattacceca, Astrophys. J. **837**, 112 (2017). https://doi.org/10.3847/1538-4357/aa5cb5
60. B. Helber, B. Dias, F. Bariselli, L.F. Zavalan, L. Pittarello, S. Goderis, B. Soens, S.J. McKibbin, P. Claeys, T.E. Magin, Astrophys. J. **876**(2), 120 (2019). https://doi.org/10.3847/1538-4357/ab16f0
61. S. Imoto, I. Hasegawa, Smithsonian Contributions Astrophys. **2**, 131 (1958)
62. J.K. Bjorkman, *Meteors and meteorites in the ancient Near East* (Arizona State University, Tempe, 1973)
63. Z. Ceplecha, Bull. Astron. Inst. Czechoslovakia **38**, 222 (1987)
64. J. Oberst, S. Molau, D. Heinlein, C. Gritzner, M. Schindler, P. Spurny, Z. Ceplecha, J. Rendtel, H. Betlem, Meteorit. Planet. Sci. **33**(1), 49 (1998). https://doi.org/10.1111/j.1945-5100.1998.tb01606.x
65. P. Jenniskens, P.S. Gural, L. Dynneson, B.J. Grigsby, K.E. Newman, M. Borden, M. Koop, D. Holman, Icarus **216**(1), 40 (2011). https://doi.org/10.1016/j.icarus.2011.08.012
66. P.A. Bland, P. Spurný, A.W.R. Bevan, K.T. Howard, M.C. Towner, G.K. Benedix, R.C. Greenwood, L. Shrbený, I.A. Franchi, G. Deacon, Aust. J. Earth Sci. **59**(2), 177 (2012). https://doi.org/10.1080/08120099.2011.595428
67. J. Tóth, L. Kornoš, P. Zigo, Š. Gajdoš, D. Kalmančok, J. Világi, J. Šimon, P. Vereš, J. Šilha, M. Buček, A. Galád, P. Rusňák, P. Hrábek, F. Ďuriš, R. Rudawska, Planet. Space Sci. **118**, 102 (2015). https://doi.org/10.1016/j.pss.2015.07.007
68. F. Colas, B. Zanda, J. Vaubaillon, S. Bouley, C. Marmo, Y. Audureau, M.K. Kwon, J.L. Rault, S. Caminade, P. Vernazza, *International Meteor Conference Mistelbach* (2015), p. 37
69. P. Vereš, J. Toth, WGN, J. Int. Meteor Organ. **38**(2), 54 (2010)
70. L. Kornoš, P. Matlovič, R. Rudawska, J. Tóth, M. Hajduková Jr., J. Koukal, R. Piffl, *Meteoroids 2013* (2014), pp. 225–233
71. W.J. Baggaley, R.G.T. Bennett, D.I. Steel, A.D. Taylor, Q. J. R. Astron. Soc. **35**, 293 (1994)
72. R.J. Weryk, P. Brown, Earth Moon Planets **95**(1–4), 221 (2004). https://doi.org/10.1007/s11038-005-9034-x
73. D. Janches, S. Close, J.L. Hormaechea, N. Swarnalingam, A. Murphy, D. O'Connor, B. Vandepeer, B. Fuller, D.C. Fritts, C. Brunini, Astrophys. J. **809**(1), 36 (2015). https://doi.org/10.1088/0004-637X/809/1/36
74. P. Pokorný, P.G. Brown, Astron. Astrophys. **592**, A150 (2016). https://doi.org/10.1051/0004-6361/201628134
75. R.C. Blaauw, M.D. Campbell-Brown, R.J. Weryk, Mon. Not. R. Astron. Soc. **414**(4), 3322 (2011). https://doi.org/10.1111/j.1365-2966.2011.18633.x
76. E. Grun, H.A. Zook, H. Fechtig, R.H. Giese, Icarus **62**(2), 244 (1985). https://doi.org/10.1016/0019-1035(85)90121-6
77. V. Porubčan, L. Kornoš, I.P. Williams, Contributions Astron. Obs. Skalnate Pleso **36**, 103 (2006)
78. D.J. Asher, S.V.M. Clube, D.I. Steel, Mon. Not. R. Astron. Soc. **264**, 93 (1993). https://doi.org/10.1093/mnras/264.1.93
79. P. Spurný, J. Borovička, H. Mucke, J. Svoreň, Astron. Astrophys. **605**, A68 (2017). https://doi.org/10.1051/0004-6361/201730787
80. P. Jenniskens, J. Vaubaillon, Astron. J. **139**, 1822 (2010). https://doi.org/10.1088/0004-6256/139/5/1822

81. T.J. Jopek, Z. Kaňuchová, Planet. Space Sci. **143**, 3 (2017). https://doi.org/10.1016/j.pss.2016.11.003
82. I.P. Williams, T.J. Jopek, R. Rudawska, J. Tóth, L. Kornoš, *Minor Meteor Showers and the Sporadic Background* (Cambridge University Press, Cambridge, 2019), p. 210
83. R.B. Southworth, G.S. Hawkins, Smithsonian Contributions Astrophy. **7**, 261 (1963)
84. J.D. Drummond, Icarus **45**(3), 545 (1981). https://doi.org/10.1016/0019-1035(81)90020-8
85. G.B. Valsecchi, T.J. Jopek, C. Froeschle, Mon. Not. R. Astron. Soc. **304**(4), 743 (1999). https://doi.org/10.1046/j.1365-8711.1999.02264.x
86. T.J. Jopek, R. Rudawska, P. Bartczak, Earth Moon Planets **102**(1–4), 73 (2008). https://doi.org/s10.1007/s11038-007-9197-8
87. A.V. Moorhead, Mon. Not. R. Astron. Soc. **455**(4), 4329 (2016). https://doi.org/10.1093/mnras/stv2610
88. L. Neslušan, M. Hajduková, Astron. Astrophys. **598**, A40 (2017). https://doi.org/10.1051/0004-6361/201629659
89. R. Rudawska, P. Matlovič, J. Tóth, L. Kornoš, Planet. Space Sci. **118**, 38 (2015). https://doi.org/10.1016/j.pss.2015.07.011
90. P. Brown, R.J. Weryk, D.K. Wong, J. Jones, Icarus **195**(1), 317 (2008). https://doi.org/10.1016/j.icarus.2007.12.002
91. J. Vaubaillon, F. Colas, L. Jorda, Astron. Astrophys. **439**(2), 761 (2005). https://doi.org/10.1051/0004-6361:20042626
92. G.O. Ryabova, Solar Syst. Res. **47**(3), 219 (2013). https://doi.org/10.1134/S0038094613030052
93. D.P. Galligan, W.J. Baggaley, Mon. Not. R. Astron. Soc. **359**(2), 551 (2005). https://doi.org/10.1111/j.1365-2966.2005.08918.x
94. A. Bischoff, J.A. Barrat, K. Bauer, C. Burkhardt, H. Busemann, S. Ebert, M. Gonsior, J. Hakenmüller, J. Haloda, D. Harries, Meteorit. Planet. Sci. **52**(8), 1683 (2017). https://doi.org/10.1111/maps.12883
95. J. Jones, P. Brown, Mon. Not. R. Astron. Soc. **265**, 524 (1993). https://doi.org/10.1093/mnras/265.3.524
96. D. Nesvorný, P. Jenniskens, H.F. Levison, W.F. Bottke, D. Vokrouhlický, M. Gounelle, Astrophys. J. **713**(2), 816 (2010). https://doi.org/10.1088/0004-637X/713/2/816
97. P. Pokorný, D. Vokrouhlický, D. Nesvorný, M. Campbell-Brown, P. Brown, Astrophys. J. **789**(1), 25 (2014). https://doi.org/10.1088/0004-637X/789/1/25
98. M.D. Campbell-Brown, Icarus **196**(1), 144 (2008). https://doi.org/10.1016/j.icarus.2008.02.022
99. N. McBride, J.A.M. McDonnell, Planet. Space Sci. **47**(8–9), 1005 (1999). https://doi.org/10.1016/S0032-0633(99)00023-9
100. A.V. Moorhead, P.G. Brown, M.D. Campbell-Brown, D. Heynen, W.J. Cooke, Planet. Space Sci. **143**, 209 (2017). https://doi.org/10.1016/j.pss.2017.02.002
101. E. Grun, H.A. Zook, M. Baguhl, A. Balogh, S.J. Bame, H. Fechtig, R. Forsyth, M.S. Hanner, M. Horanyi, J. Kissel, Nature **362**(6419), 428 (1993). https://doi.org/10.1038/362428a0
102. A.D. Taylor, W.J. Baggaley, D.I. Steel, Nature **380**(6572), 323 (1996). https://doi.org/10.1038/380323a0
103. M. Hajduková, L. Kornoš, J. Tóth, Meteorit. Planet. Sci. **49**(1), 63 (2014). https://doi.org/10.1111/maps.12119
104. P. Jenniskens, Adv. Space Res. **39**, 491 (2007). https://doi.org/10.1016/j.asr.2007.03.040
105. P. Matlovič, J. Tóth, R. Rudawska, L. Kornoš, Planet. Space Sci. **143**, 104 (2017). https://doi.org/10.1016/j.pss.2017.02.007
106. R. Rudawska, J. Zender, P. Jenniskens, J. Vaubaillon, P. Koten, A. Margonis, J. Tóth, J. McAuliffe, D. Koschny, Earth Moon Planets **112**, 45 (2014). https://doi.org/10.1007/s11038-014-9436-8
107. J.M. Madiedo, Planet. Space Sci. **143**, 238 (2017). https://doi.org/10.1016/j.pss.2016.12.005
108. V. Vojáček, J. Borovička, P. Koten, P. Spurný, R. Štork, Astron. Astrophys. **621**, A68 (2019). https://doi.org/10.1051/0004-6361/201833289

109. J. Borovička, A.A. Berezhnoy, Icarus **278**, 248 (2016). https://doi.org/10.1016/j.icarus.2016.06.022
110. P. Jenniskens, E.L. Schaller, C.O. Laux, M.A. Wilson, G. Schmidt, R.L. Rairden, Astrobiology **4**(1), 67 (2004). https://doi.org/10.1089/153110704773600249
111. A.A. Berezhnoy, J. Borovička, Icarus **210**(1), 150 (2010). https://doi.org/10.1016/j.icarus.2010.06.036
112. P. Jenniskens, A.M. Mandell, Astrobiology **4**(1), 123 (2004). https://doi.org/10.1089/153110704773600285
113. M.D. Campbell-Brown, J. Kero, C. Szasz, A. Pellinen-Wannberg, R.J. Weryk, J. Geophys. Res. (Space Phys.) **117**(A9), A09323 (2012). https://doi.org/10.1029/2012JA017800
114. M. Gritsevich, D. Koschny, Icarus **212**(2), 877 (2011). https://doi.org/10.1016/j.icarus.2011.01.033
115. Z. Ceplecha, Bull. Astron. Inst. Czechoslovakia **39**, 221 (1988)
116. P.B. Babadzhanov, G.I. Kokhirova, Astron. Astrophys. **495**, 353 (2009). https://doi.org/10.1051/0004-6361:200810460
117. D. Vida, P. Brown, M. Campbell-Brown, in *AAS/Division for Planetary Sciences Meeting Abstracts #50*, (2018), p. 100.01
118. D. Čapek, P. Koten, J. Borovička, V. Vojáček, P. Spurný, R. Štork, Astron. Astrophys. **625**, A106 (2019). https://doi.org/10.1051/0004-6361/201935203
119. M.D. Campbell-Brown, J. Borovička, P.G. Brown, E. Stokan, Astron. Astrophys. **557**, A41 (2013). https://doi.org/10.1051/0004-6361/201322005
120. E. Stokan, M.D. Campbell-Brown, Mon. Not. R. Astron. Soc. **447**(2), 1580 (2015). https://doi.org/10.1093/mnras/stu2552

# Extrasolar Enigmas: From Disintegrating Exoplanets to Exoasteroids

**Jan Budaj, Petr Kabáth, and Enric Palle**

## Contents



---

J. Budaj (✉)
Astronomical Institute, Slovak Academy of Sciences, Tatranska Lomnica, Slovakia
e-mail: budaj@ta3.sk

P. Kabáth
Astronomical Institute of Czech Academy of Sciences, Ondřejov, Czech Republic
e-mail: petr.kabath@asu.cas.cz

E. Palle
Instituto de Astrofísica de Canarias, Santa Cruz de Tenerife, Spain
e-mail: epalle@iac.es



P. Kabáth et al. (eds.), *Reviews in Frontiers of Modern Astrophysics*,
https://doi.org/10.1007/978-3-030-38509-5_3

**Abstract** Thousands of transiting exoplanets have been discovered to date, thanks in great part to the *Kepler* space mission. As in all populations, and certainly in the case of exoplanets, one finds unique objects with distinct characteristics. Here we will describe the properties and behaviour of a small group of 'disintegrating' exoplanets discovered over the last few years (KIC 12557548b, K2-22b, and others). They evaporate, lose mass unraveling their naked cores, produce spectacular dusty comet-like tails, and feature highly variable asymmetric transits. Apart from these exoplanets, there is observational evidence for even smaller 'exo-'objects orbiting other stars: exoasteroids and exocomets. Most probably, such objects are also behind the mystery of Boyajian's star. Ongoing and upcoming space missions such as *TESS* and *PLATO* will hopefully discover more objects of this kind, and a new era of the exploration of small extrasolar systems bodies will be upon us.

# 1 Introduction

The exoplanet science discoveries kicked-in after 1992–1995 when the first exoplanets were discovered [1] first around a pulsar and then a hot Jupiter around a solar type star 51 Peg [2]. The exoplanet 51 Peg b was detected from observations of radial velocities ('RV') from the ground with a 1.92-m telescope located at Observatoire de Haute Provence.

Later hundreds of exoplanets were discovered using the radial velocity method. In 2000, the first *transiting* exoplanet HD 2019458b, again a Jupiter-sized planet in a close-in orbit, was detected [3]. New automated ground-based projects to detect transiting exoplanets were started in the first decade of twenty-first century. The most successful of such projects to date is the WASP survey[1] which has discovered about 200 transiting planets (April 2019), and there are a number of other successful ground-based exoplanet surveys as well, such as HAT [4] or KELT [5].

A real breakthrough came with the launch of the *CoRoT* space mission in 2006. The *CoRoT* satellite was a french-led ESA mission carrying a 27-cm aperture telescope equipped with 4 CCD detectors dedicated to asteroseismology and exoplanetary transit detections [6]. The *CoRoT* mission was terminated in 2013 and it reported 33 substellar objects which are all fully characterized and thus we know both their masses and radii.

In 2009 a very successful NASA space mission *Kepler* was launched carrying a telescope with a mirror of 1.4-m with a large array of CCD detectors [7]. *Kepler*, and later its continuation *K2* mission, discovered during their lifetimes

[1] http://www.superwasp.org.

from 2009 until 2018 about 4000 transiting exoplanets. *Kepler/K2* photometric data likely still contain many more new planetary candidates. However, only a few hundreds of the *Kepler/K2* planets have been fully characterized, so that we know their masses and radii. This fact is due to the relative faintness of the *Kepler/K2* targets and the difficulty of carrying out ground-based follow-up spectroscopic RV observations. However, despite these limitations, *Kepler/K2* was able to deliver extremely interesting candidates, among them low mass and rocky planets in the habitable zone such as Kepler-62f [8], ultra-short period planets such as Kepler-78b [9], and multiple planetary systems [10]. Also, new types of objects such as Boyajian's star [11] and 'disintegrating' planets [12] were found with *Kepler/K2*.

In the following text, we will focus on the physics behind the more recently discovered enigmatic objects such as disintegrating and evaporating planets. A significant number of such objects are also expected to be discovered with the most recent and upcoming missions like *TESS* and later *PLATO*. Before discussing the physics of the disintegrating objects, we briefly introduce the observing strategies which led to the discoveries of these interesting types of exo-objects.

In Sect. 2 we describe the methods and observing strategies used to discover or characterize these 'dusty objects'. Section 3 contains a crash course on the dust properties which are important to understand the content of this chapter. Sections 4 and 5 describe the most interesting disintegrating exoplanets and minor bodies in exoplanetary systems. The special case of Boyajian's star is discussed in Sect. 6. Finally, Sect. 7 deals with ongoing and future space missions which may bring new fascinating discoveries and open a new era in the study of these extrasolar objects. For a reference, another recent review of disintegrating exoplanets can be found in [13].

## 2 Observing Methods and Strategies

The most successful methods of exoplanet detection are the transit and radial velocity measurements. Both methods benefit from their combination, and, in general, all planets detected by the transit method need follow-up radial velocity measurements for mass determination. Therefore, all exoplanetary transit space missions try to ensure that the targets in their prime sample can be followed-up spectroscopically from the ground.

### *2.1 Radial Velocities (RV)*

The method of discovering and characterizing exoplanets by precise radial velocity measurements is based on Kepler's laws. If the system consists of a star and a planet, these orbit around their common center of mass causing the star to move toward and away from the observer with a given radial velocity that is a function of the mass of the planet. Detailed derivations of the expression for the semi-amplitude $K$ of the radial velocity curve can be found in numerous publications [14, 15]; therefore, we limit ourselves here to only presenting the final expression for the semi-amplitude of the radial velocity curve $K$:

$$K = \frac{1}{\sqrt{1-e^2}} \left(\frac{2\pi G}{P_{\rm orb}}\right)^{1/3} \frac{M_{\rm plan} \sin i}{(M_{\rm star} + M_{\rm plan})^{2/3}} \tag{1}$$

where $G$ is the gravitational constant, $P_{\rm orb}$ the orbital period, $M_{\rm star}$ the stellar mass, $M_{\rm plan}$ the planetary mass, $i$ planetary orbital inclination angle, and $e$ the eccentricity of the planetary orbit. As can be seen from the above equation, the resulting radial velocity and the corresponding semi-amplitude $K$ can be obtained from the observed spectroscopic time series that adequately samples the orbital phases. However, this method can not provide a determination of the inclination, $i$, of the planetary orbital plane. Therefore, the value of planetary mass $M_{\rm plan}$ obtained from the RV measurements is only a lower limit since the value of $i$ is unknown without making use of the photometric transit data. One example of an RV curve is illustrated in Fig. 1

The typical radial velocity semi-amplitude of a large gas planet is of order of tens to hundreds of m s$^{-1}$. On the other hand a typical radial velocity signature of an Earth-sized planet can be as low as few cm/s.

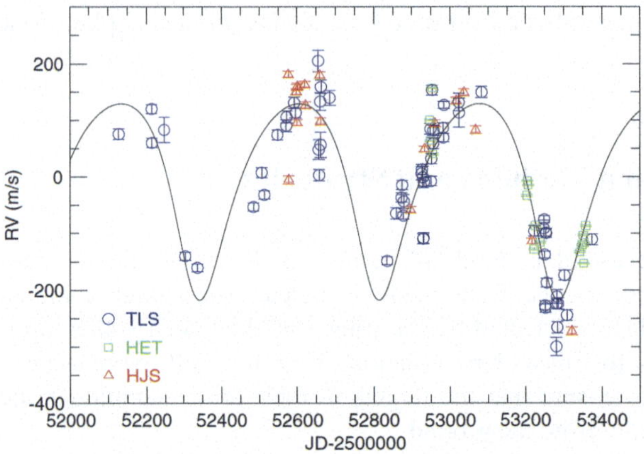


**Fig. 1** Figure shows a typical RV curve of a gas planet obtained with various telescopes around the globe. Figure Credit: [16] https://doi.org/10.1051/0004-6361:20052850, reproduced with permission © ESO

## 2.2 The Transit Method

If a planet passes in front of the stellar disc along observer's line of sight, then one can observe a periodic dimming of the stellar light, i.e., a transit. Typically a photometric time series with good sampling is obtained a few hours before, then during the transit, and finally a few hours after the transit ends. The basics of the method have been described in great detail elsewhere [17]. Here we limit ourselves to expressing the transit depth, $\delta$, as:

$$\delta \propto \frac{\Delta F}{F} = \frac{R_{\text{plan}}^2}{R_{\text{star}}^2} \tag{2}$$

where $\Delta$F is the observed change of flux during a transit, $F$ the flux of the star, $R_{\text{plan}}$ the planetary radius, and $R_{\text{star}}$ the stellar radius. An advantage of this method is that it can be used to determine the inclination of the planet's orbital plane if the stellar parameters of the host star are known. It is clear that the photometric transit method needs to be combined with spectroscopic observations of a given system in order to fully characterize the exoplanet.

The detection of hot-Jupiters can be accomplished even with small-aperture telescopes as the typical transit depth, $\delta$, due to a transit of a hot-Jupiter is a few percent of the stellar flux for a main sequence dwarf star. However, the detection of Earth-sized planets requires ultra-precise photometry, typically measured in parts per million ('ppm'). CoRoT-7b was the first example of a small rocky exoplanet showing a transit depth of only a few hundred ppm [18]. The smallest exoplanet currently known to orbit a solar-like star is Kepler-37b [19] and it was discovered by the transit method. Its light curve along with the light curves of two other larger planets in the system are shown in Fig. 2.

## 2.3 Transmission Spectroscopy and Exo-Atmospheres

Over the past decade, the characterization of exo-atmospheres has started to gain in importance. The first detection of sodium in the exo-atmosphere of a gas giant HD 2019458b was made from space with the Hubble Space Telescope (*HST*) [20], followed by the spectroscopic detection, also with *HST*, of an extended hydrogen atmosphere for the same planet [20, 21]. Ground based detection of exo-atmospheres with the transmission spectroscopy method using high spectral resolving power succeeded nearly 6 years later when sodium was detected in the atmosphere of HD 189733b [22].

Transmission spectroscopy uses the basic idea that, during a transit, the stellar light has to pass through the exo-planetary atmosphere which forms a thin annulus around the planet. If the atmosphere contains an absorber, such as sodium or any

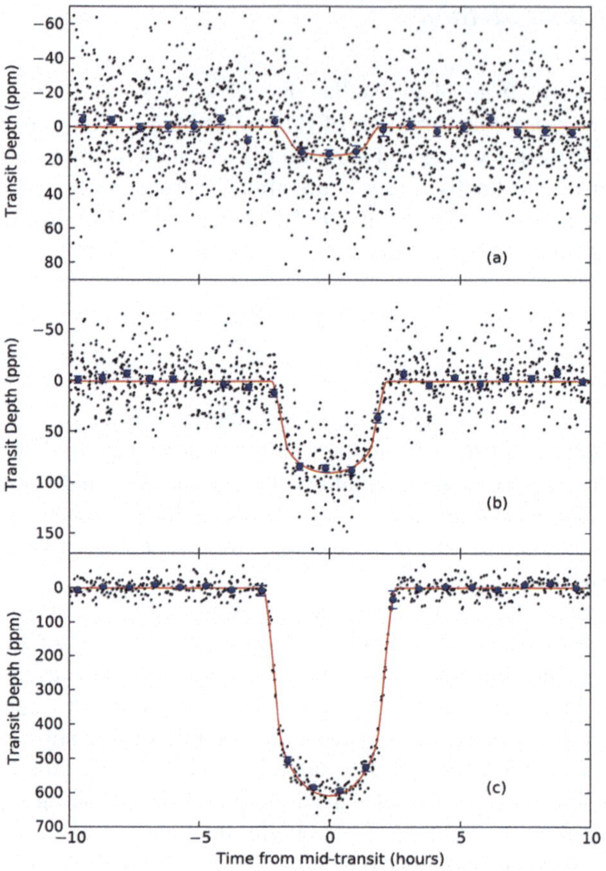


**Fig. 2** Figure shows a comparison of light curves obtained with *Kepler* for various sized exoplanets from the system Kepler-37, with the smallest being Kepler-37b (upper panel). Reprinted by permission from Springer Nature: Nature, [19] https://www.nature.com/articles/nature11914, ©2013

other species, the radius of the planet appears larger at the corresponding wavelength as the species blocks the stellar light.

When using transmission spectroscopy, typically, a time series of spectra with low spectral resolving power is recorded before, during, and after the transit. Each of the observed spectra from the time series is split into defined photometric bands and then the resulting spectrophotometric light curves are produced and evaluated. The variation in transit depth in the different spectral bands provides information on the absorbing species. This method has successfully confirmed atmospheres for a handful of planets. A metal rich atmosphere was confirmed for the Neptune-sized exoplanet GJ 1214b from the ground [23, 24], followed by many other detections for predominantly gas planets [25, 26]. Lately, reports of elements other than sodium and hydrogen have been reported, such as lithium and perhaps a first detection of

TiO features [26, 27]. However, this method also has potential application to rocky planets around late type dwarfs [28] that are recently discovered by *TESS* and will be found later by *PLATO* and ELT from the ground.

A slightly different approach is to use a spectrograph with high resolving power. A spectroscopic time series is again obtained on either side of, and during, the transit. In this case the actual spectra from in- and out-of-transit phase are directly compared. Before a search for planetary atmosphere signatures can start, a careful analysis of the telluric features in the spectra has to be performed and, if necessary, telluric features are removed [29]. Furthermore the Rossiter–McLaughlin effect which can affect the planetary signal needs to be taken into account [30]. Regions of prominent lines, such as the sodium doublet (NaD), or potassium region as well as hydrogen lines are typically investigated. The ratio of in- and out-of-transit spectra can reveal a planetary signature [22, 31–35] as the in-transit spectra also possess an excess signal from the planetary atmosphere.

## 2.4 Observing Strategies of Exoplanetary Space Missions

The detection of exoplanets is most efficient from space with the transit method. Therefore, we will introduce the observing strategies and principles of such missions. Different space missions dedicated to the search of planets via transit detection have followed different observing strategies. The first one, *CoRoT*, monitored several fields for a series of long (150 days) and short (30 day) periods. On the contrary, the space mission *Kepler* monitored a single field for 4 years. The selected field in the region of the Cygnus and Lyra constellations contained more than 150,000 stars [7] that were monitored. This part of the *Kepler* mission yielded about 2000 exoplanets and several thousand candidates. In 2013 the *Kepler* team needed to adopt a different observing strategy due to problems with the spacecraft gyroscopes. The mission was renamed *K2* and it observed one field for typically 70 days and then pointed towards a new field. Over the ensuing 4 years, the *K2* mission yielded about 1000 exoplanets and several hundred additional candidates [36]. There were numerous interesting discoveries among these missions, and many "firsts" reported, such as: the circumbinary planet Kepler-16b [37], the oldest known multiplanet system Kepler-444 [38], the first *Kepler* rocky planet Kepler-10b [39], and the first planet with a radius smaller than the Earth [40]. The *K2* mission was retired in late 2018 when the fuel was depleted.

However, *Kepler* also discovered a new class of 'disintegrating' planets. In the following text, we lay the theoretical ground for understanding these highly enigmatic planets among the known types of exoplanetary systems.

## 3 Dust Environment in Exoplanets

In this section we introduce the basic physical properties of astrophysical dust which will be important for understanding the subsequent sections. At sufficiently low temperatures and high density, grains of condensates can be formed out of a gas phase. Such grains are usually called "dust", although some authors use the more generic term "condensates". At the same time, the term "grain" often includes not only solid grains but also liquid droplets. Such condensates are usually confined to "clouds". These can not only be clouds in the atmospheres of cool objects but also vast interstellar dust clouds.

The reason why dust is so important for our objects will become obvious from the following everyday experience. Our atmosphere contains water. If this water is in the form of a gas one can easily see distant mountains which are 100 km away. However, once the water condenses and forms clouds or a fog, the visibility can drop to 10 m or even less. Thus the opacity, which is a measure of the non-transparency of the material (see Sect. 3.2), could be much higher if the material were in the form of dust rather than gas. Figure 3 illustrates the opacity of gas and dust in the visible and near infrared regions per gram of material. The opacity of the gas in this example is based on an assumed solar chemical composition and a density $\rho = 10^{-14}$ g cm$^{-3}$ [41]. For the dust opacity we used the illustrative mineral forsterite with a particle size of about 0.1 and 1 μm [42]. It should be mentioned that, as a rule, not all the gas can turn into a condensate. For solar composition material, dust can account for roughly 1% of the mass. Still, as can be seen from the figure, the dust opacity will easily overtake that of the gas.

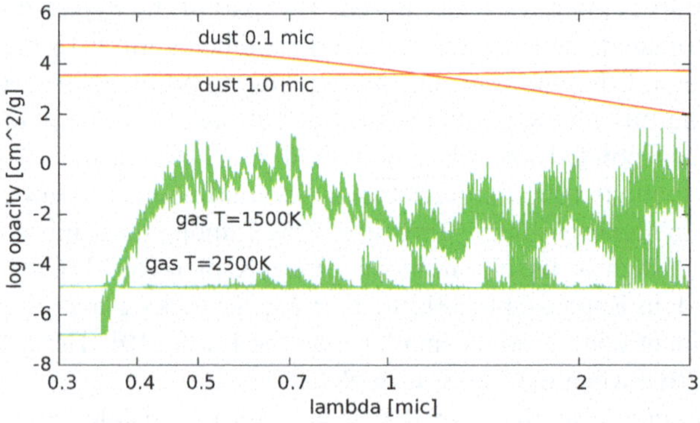


**Fig. 3** Comparison between the gas opacities at two temperatures and dust opacity of forsterite for two particle sizes

## 3.1 Absorption, Scattering, and Extinction

The optical properties of condensates may not only influence, but fully govern, the emerging spectrum and even the structure of a dusty object. Dust can absorb the impinging radiation and convert it directly into heating the grains. This process is called 'absorption' or 'true absorption' to emphasize that the photon is destroyed or thermalized. It is quantified by the absorption opacity.

Dust can also scatter the radiation in a process called 'scattering'. Scattering mainly changes the direction of the photon without significantly affecting its energy. So the scattered radiation is somewhat decoupled from the medium and flows through and around it without heating it. This process is characterized by the scattering opacity. Furthermore, scattering can be highly anisotropic, a property that is described by means of the phase function, which depends on the scattering angle (the deflection angle from the original direction of the impinging radiation). The most prominent feature is a strong forward scattering peak for large values of the so-called 'scaled particle size', $X = 2\pi a/\lambda$ where $a$ is the particle size (radius) and $\lambda$ is the wavelength of the radiation.

The combined effect of absorption and scattering is referred to as the 'extinction'. Finally, formation of dust can also affect the chemical composition of an object. It removes the condensed elements from the gas phase within the dust cloud. Subsequently, various processes and forces may decouple gas and dust, creating chemical inhomogeneities.

Absorption and scattering by large particles (relative to the wavelength, i.e., large $X$) is wavelength independent. However, *scattering* by small particles has a very strong, $\lambda^{-4}$, dependence (Rayleigh scattering) and *absorption* by small particles has a $\lambda^{-1}$ dependence. Blue light is scattered and attenuated more efficiently, and for this reason dust generally causes a reddening of the light passing through a dust cloud. The extinction at some wavelength (or filter) in magnitudes is the difference between the observed and intrinsic brightness: $A(V) = V_{\rm obs} - V_{\rm int}$. Reddening (selective extinction/color excess) is usually expressed as a difference between the observed and intrinsic color index:

$$E(B-V) = (B-V)_{\rm obs} - (B-V)_{\rm int} = A(B) - A(V) \tag{3}$$

A relative slope of the wavelength dependence of the extinction can be characterized by a single parameter $1/R(V)$ where R(V) is [43]:

$$R(V) = \frac{A(V)}{E(B-V)} = \frac{A(V)}{A(B)-A(V)} \tag{4}$$

$R(V)$ is sensitive to the particle size. The typical value of $R(V)$ for interstellar dust in our Galaxy is $3.1 \pm 0.2$. The absolute amount of the extinction as a function of $\lambda$ (the extinction curve) can be characterized by two parameters: $R(V)$ and $E(B-V)$.

### 3.2 Cross-Section, Opacity, and Phase Function

The optical properties of the dust are given by the complex index of refraction of the material it is made from (which is a function of wavelength), and further depends on the size and shape of the particles. These properties of the grains are often expressed in the form of cross-sections for absorption, scattering, and extinction $C_a$, $C_s$, $C_e$, respectively. The cross-sections are related to the projected area of the dust particles of radius $r$ via efficiency factors $Q_a$, $Q_s$, and $Q_e$, for absorption, scattering, and extinction, respectively:

$$C_a = Q_a \pi r^2, \quad C_s = Q_s \pi r^2, \quad Q_e = Q_a + Q_s. \tag{5}$$

The cross-sections are related to the absorption and scattering opacities, $\kappa_{\nu,a}$ and $\kappa_{\nu,s}$, of the condensates at radiation frequency, $\nu$, by:

$$\kappa_{\nu,\{a,s\}} \equiv C_{\nu,\{a,s\}}/m_g = \frac{3}{4}\frac{Q_{\nu,\{a,s\}}}{\rho_g r} \tag{6}$$

where $\kappa$ is the cross section per unit mass and has the dimensions of cm$^2$ g$^{-1}$, $m_g$ is the mass of a dust grain, $\rho_g$ is the bulk material density of the grain, and we have assumed spherical particles for simplicity. Note that $\kappa$ is nearly exclusively a property of the material and may not depend at all on the mass in dust grains per unit volume of the medium, $\bar{\rho}$. Finally, for completeness, we note that the quantity $\alpha \equiv \bar{\rho}\kappa$ is defined as the linear extinction coefficient (with units cm$^{-1}$), and may be useful in certain circumstances.

Using these opacities, the monochromatic optical depth due to scattering and absorption along the line of sight $z$ is then given by:

$$\tau_\nu = \int \bar{\rho}(z) \left[\kappa_{\nu,a}(z) + \kappa_{\nu,s}(z)\right] dz. \tag{7}$$

The sum of the absorption and scattering opacities is referred to as a total opacity. For the special idealized case of single-size dust particles, this reduces to

$$\tau_\nu = \frac{3}{4}\frac{\left(Q_{\nu,a} + Q_{\nu,s}\right)}{\rho_g r} \int \bar{\rho}(z) dz. \tag{8}$$

We can see from this, that for a fixed amount of dust mass per unit volume of the medium, i.e., $\bar{\rho}$ = constant, the optical depth would become monotonically larger with decreasing particle size, as $1/r$. However, in Mie scattering, once the particle size becomes substantially less than the radiation wavelength, $\lambda = c/\nu$, then the $Q$ factors for the cross section drop dramatically, and the optical depth stops rising with further decreases in particle size. This is the reason why, for observing wavelengths in the visible, it is often stated that particle sizes comparable to a micron are the most efficient at blocking light.

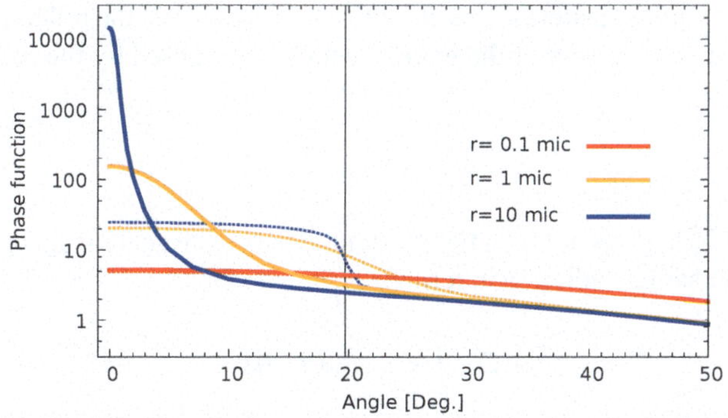


**Fig. 4** Phase functions assuming a point source of light (solid) versus the phase functions assuming the finite dimension of the stellar disc (dotted). Example is for enstatite at 600 nm for different dust particle radii. The vertical line illustrates the angular radius of the stellar disc of WASP-103 as seen from the planet WASP-103b

The angular distribution of the scattered light is described by the phase function, $p(\theta)$, where $\theta$ is the scattering angle which measures the deflection of the scattered photon from its original direction. The phase function is normalized such that its integral over all solid angles is $4\pi$. An example of the dust phase function at small angles is displayed in Fig. 4. One can see a strong increase towards zero phase angle which is called the forward scattering peak. The amplitude and width of this peak are quite sensitive to the particle size and wavelength. Calculations of phase functions usually assume an incident parallel beam of light. However, if the dust cloud were very close to the star, its angular dimension (as seen by the dust grain) could be comparable to, or wider than, the width of the forward scattering peak. This will be the case in our objects and one has to take that into account [44–46]. The same figure also illustrates this effect on dust particles located in the atmosphere of the exoplanet WASP-103b [47, 48].

It is sometimes useful to define a mean cosine of the scattering angle $g$, also known as the asymmetry parameter. It has values from $-1$ to 1 and is calculated from the phase function:

$$g = \int p(\theta)\cos(\theta)d\Omega\ /4\pi. \tag{9}$$

## 3.3 *Albedo, Equilibrium Temperature, and Radiative Acceleration*

Let's assume that a dust particle is irradiated by its host star with effective temperature $T_*$, solid angle $\Omega_*$, and intensity approximated by the Planck function $B_\nu(T_*)$. The particle can scatter some of the light from the star, and we define a quantity called single-scattering albedo, $\varpi$, which describes the reflecting properties of the grains. It is a fraction of the energy which is scattered by the particle:

$$\varpi_\nu = \frac{C_{\nu,\mathrm{s}}}{C_{\nu,\mathrm{a}} + C_{\nu,\mathrm{s}}} . \tag{10}$$

This scattered light does not heat the particle. Apart from scattering, the particle can also absorb the stellar radiation at a rate:

$$\Omega_* \int C_{\nu,\mathrm{a}} B_\nu(T_*) d\nu \tag{11}$$

This energy heats the particle to a temperature $T_\mathrm{g}$. Subsequently, the grain emits thermal radiation and cools at a rate:

$$4\pi \int C_{\nu,\mathrm{a}} B_\nu(T_\mathrm{g}) d\nu, \tag{12}$$

A balance between the absorbed and re-radiated energy sets the grain equilibrium temperature (provided that the grain is not also sublimating; but see Eq. 21). It can be obtained by solving the radiative equilibrium equation for $T_\mathrm{g}$:

$$\Omega_* \int C_{\nu,\mathrm{a}} B_\nu(T_*) d\nu = 4\pi \int C_{\nu,\mathrm{a}} B_\nu(T_\mathrm{g}) d\nu \tag{13}$$

Assuming that the opacities are grey (i.e., they do not depend of the frequency) the grain temperature is simple given by:

$$T_\mathrm{g}^\mathrm{grey} = T_* \left( \frac{\Omega_*}{4\pi} \right)^{1/4} . \tag{14}$$

A dust grain irradiated by a star with effective temperature $T_*$, mass $M_*$, radius $R_*$, and surface flux $F_\nu$ experiences a radiative acceleration $a_\mathrm{R}$. It is usually expressed as a parameter $\beta$ relative to the gravitational acceleration $a_\mathrm{G}$:

$$\beta = \frac{a_\mathrm{R}}{a_\mathrm{G}} = \frac{R_*^2}{G M_* c} \int \left[ \kappa_{\nu,\mathrm{a}} + (1-g)\kappa_{\nu,\mathrm{s}} \right] F_\nu(T_*) d\nu . \tag{15}$$

where $G$ is gravitational constant, $c$ is speed of light, and $g$ is the previously mentioned asymmetry parameter. Thus, in the two extreme cases of forward vs back scattering of the stellar radiation, the scattering adds either nothing to the radiative acceleration or has a factor of 2 enhancement relative to the absorption term.

Extensive online tables of such dust properties devoted mainly to exoplanets are publicly available [42]. They are based on codes that calculate cross-sections of dust particles using Mie theory such as [49] from the complex indices of refraction for specific materials. For example, the Heidelberg–Jena–St.Petersburg–Database of Optical Constants is a very convenient source of refractive index measurements [50, 51].

## 3.4 *Dust Condensation*

Depending on the state quantities, such as temperature and pressure, matter composed of a single component usually exists in one particular phase, e.g. gas, liquid, or solid. The Clausius–Clapeyron equation, which gives a relation between the temperature and pressure, marks a transition or boundaries between the different phases. Once the temperature drops below the condensation temperature (at a certain pressure) or the pressure exceeds the equilibrium (saturated) vapour pressure (at a particular temperature) the dust starts to condense out of the gas. Condensates may be in either the liquid or solid phase. The equilibrium vapor pressure where the transition occurs can be approximated by [52, 53]:

$$P_{\mathrm{v}}(T) = \exp(-A/T + B) \tag{16}$$

where $A$, $B$ are material-specific sublimation parameters.

Materials with low vapour pressure or high condensation temperature (refractory materials) condense first out of a hot cooling gas (or last to evaporate if the dust were heated). For a solar chemical composition these are mainly calcium and aluminum oxides such as corrundum ($Al_2O_3$), grossite ($CaAl_4O_7$) and hibonite ($CaAl_{12}O_{19}$). They are followed by titanium compounds such as perovskite ($CaTiO_3$) or $TiO_2$ at lower temperatures. The most important refractory species are usually silicates. They form two branches: pyroxenes ($Mg_xFe_{1-x}SiO_3$) and olivines ($Mg_{2y}Fe_{2-2y}SiO_4$). In each branch a fraction of magnesium atoms can be replaced by iron. Iron free pyroxene is called enstatite ($MgSiO_3$) while an iron free olivine is forsterite ($Mg_2SiO_4$). The other extreme member of the olivine family is fayalite ($Fe_2SiO_4$). Silicates are a type of glass and, as such, are quite transparent in the optical region, although they can scatter light quite efficiently. The amount of iron can affect their absorption properties significantly [54]. Other refractory dust species which might be encountered in such an environment are amorphous carbon, graphite (C), silicon carbide (SiC), Quartz ($SiO_2$), spinel ($MgAl_2O_4$), or akermanite ($Ca_2MgSi_2O_7$). At the other end of the condensation temperature scale are volatile species such as water and ammonia. In between are numerous compounds, depending on the chemical composition and pressure, for example sulfides and alkali halides, and troilite but we are not likely to observe these in such hot and close disintegrating objects.

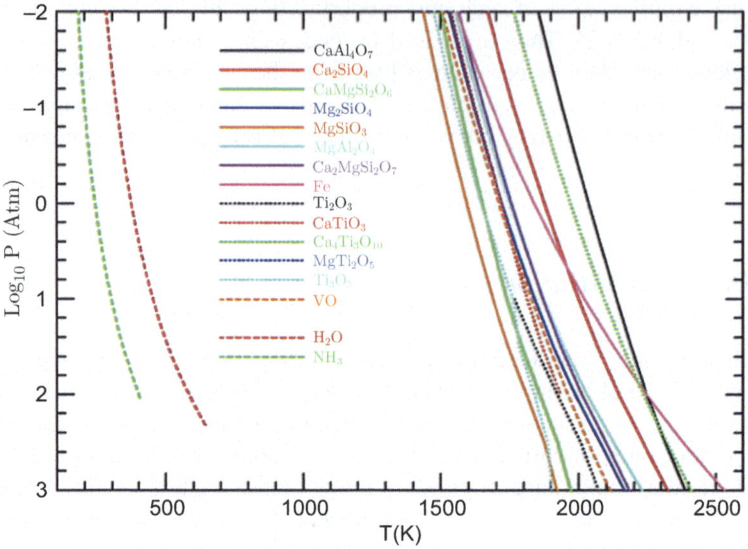


**Fig. 5** Condensation temperatures of several compounds as a function of atmospheric pressure assuming a solar composition gas. Taken from [56]. Courtesy of ApJ

Apart from the temperature, the occurrence of a particular dust component also depends critically on the abundances and the availability of the chemical elements which form the compound. The element with the lowest abundance is typically the limiting factor for the abundance of the whole compound. The solar abundances[2] of Ca, Al, and Ti are relatively small 6.34, 6.45, and 4.95, respectively [55]. That is why silicates and/or iron dust which are composed of silicon, magnesium, and iron with abundances of 7.51, 7.60, and 7.50, respectively are usually more abundant and dominate extinction processes.

The condensation properties of various compounds are nicely summarized in Fig. 5. Here the condensation curves are plotted as a function of atmospheric pressure. They were calculated mainly for the atmospheres of brown dwarfs or giant exoplanets and assume a solar chemical composition [56] but contain many dust species which are also relevant for our objects.

[2]Note that the abundances are defined, using the element number density $N$, as the number of atoms of an element per $10^{12}$ atoms of hydrogen ($\log N/H + 12$). The present-day solar photospheric abundances are generally in a good agreement with the abundances derived from the CI carbonaceous chondrite meteorites, except for a few elements such as H, He, and Li.

### 3.5 Dust Sublimation

Dust particles are also subject to sublimation [52, 53]. The mass-loss flux (rate per unit area) from a solid surfaces at temperature $T$ in vacuum is

$$J(T) = \alpha P_{\rm v}(T)\sqrt{\frac{\mu u}{2\pi k_B T}}, \tag{17}$$

where $\alpha$ is the evaporation coefficient, $P_{\rm v}(T)$ the equilibrium vapor pressure, $\mu$ the molecular weight, $u$ the atomic mass unit, and $k_B$ Boltzmann's constant. The mass-loss rate from a spherical dust grain of mass $m_{\rm g} = 4\pi r^3 \rho_{\rm g}/3$ and surface area $S = 4\pi r^2$ is then

$$\frac{dm_{\rm g}}{dt} = -SJ. \tag{18}$$

Taking into account that

$$\frac{dm_{\rm g}}{dr} = S\rho_{\rm g} \tag{19}$$

the change in the particle radius is given by:

$$\frac{dr}{dt} = \frac{dr}{dm_{\rm g}}\frac{dm_{\rm g}}{dt} = -\frac{J}{\rho_{\rm g}}. \tag{20}$$

Sublimation represents a phase transition which consumes heat and cools the particle. If that heat is not negligible one has to take it into account in computing the equilibrium temperature of the grain. In such a case the energy absorbed by the particle per unit time is balanced by the energy radiated by the particle plus the heat consumed for the phase transition. Equation 13 then reads

$$\Omega_* \int C_{\nu,{\rm a}} B_\nu(T_*) d\nu = 4\pi \int C_{\nu,{\rm a}} B_\nu(T) d\nu - \mathcal{L}\frac{dm_{\rm g}}{dt} \tag{21}$$

where $\mathcal{L}$ is the latent heat of sublimation per unit mass. The characteristic timescale for sublimation is

$$\tau = \frac{m_{\rm g}}{|dm_{\rm g}/dt|}. \tag{22}$$

## 4 Known Disintegrating Exoplanets

The great majority of exoplanets that we know of were discovered by the transit method. Nominal planet transits are symmetric and periodic without any significant variations in their shape or depth over time. This changed in 2012 when a strange object named Kepler-1520b was discovered [57].

### 4.1 *Kepler-1520b*

Kepler-1520b is an exoplanet also known as KIC 12557548b (KIC1255b). It became a prototype of a very rare new class of exoplanets called Disintegrating Exoplanets. It was found in the *Kepler* data. The host star is a $V = 16$ mag main sequence K4V type star. Its effective temperature, mass, and radius are about $T_{\rm eff} = 4440$ K, $M = 0.7$ Set visible space for consistency purpose $M_{\odot}$, and $R = 0.65$ $R_{\odot}$, respectively [57, 58]. The star is active and has spots which cause ∼1% variability with a period of about 22.9 days which enabled its rotation period to be determined [44, 59]. In its light curve, the discoverers noticed something like transits but they were highly variable, sometimes as deep as 1.2%, sometimes even missing. The strictly periodic transit signal had a very short period of about 15.7 h. Figure 6 illustrates the observed data folded with this period which yields the average light curve. One can see a significantly increased spread of fluxes in the points during the transit indicating the variability in the transit depth. Another interesting feature becomes obvious from the binned and averaged light curve (bottom panel of Fig. 6). It is highly asymmetric and features a steeper ingress and slower egress. The strict periodicity and short period of the transits indicate that they may be caused by some body orbiting the star on a very close orbit. The fact that the transits are sometimes missing implies that the body itself is very small, smaller than the Earth, otherwise it would be detected in every transit. Follow-up radial velocity measurements did not detect any reflex motion of the star which puts an upper limit on the mass of the body of 89 $M_{\oplus}$ [60, 61] which places the body deep into the planetary regime. However, what is then causing the variable asymmetric transits?

#### 4.1.1 Interpretation

The interpretation it that a body on such tight orbit around the star is heated to about 2000 K. At such temperatures even rock melts and can evaporate which may drive a thermal wind off the surface [57, 62]. Gas escapes the planet at a rate larger than 0.1 $M_{\oplus}$/Gyr dragging dust grains with it. Alternatively, the dust may condense out of the gas when it cools during or after escape from the planet. The mixture of gas and dust expands beyond the Hill sphere radius of the planet. It flows "down hill" out of the potential well of the planet through the L1 point towards the star or via the L2

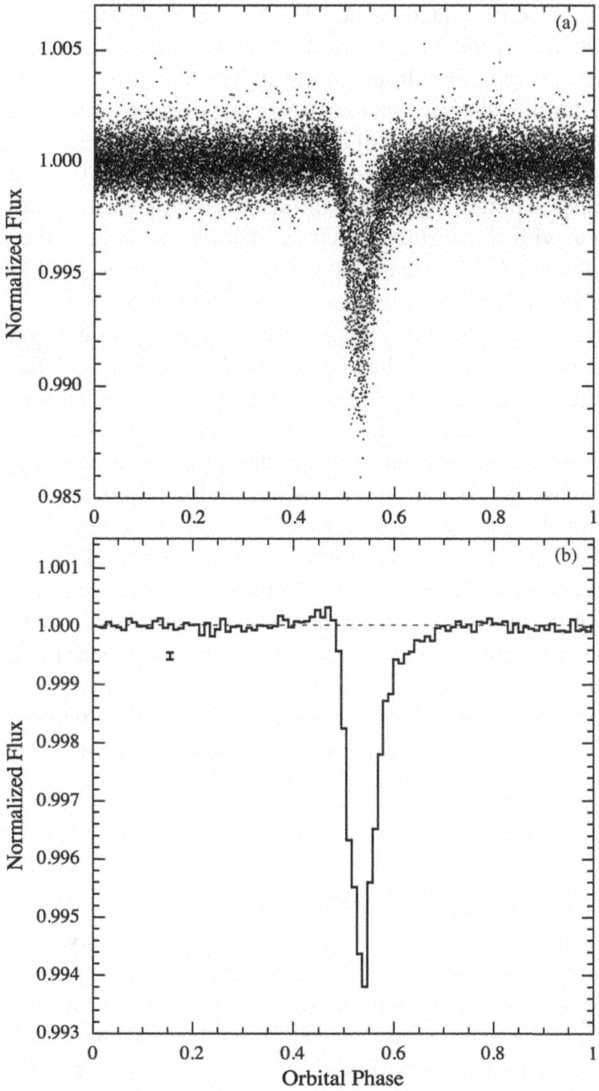


**Fig. 6** Top: First long cadence *Kepler* observations of KIC 1255 folded with the 15.7-h orbital period. Bottom: Binned and averaged light curve. Taken from [57]. Courtesy of ApJ

point away from the star. Strong radiative forces on the dust cause a weakening of the effective gravity which drives the dust into higher orbits that lag progressively behind the planet. It is this dust which is causing the transits and this is also the reason why we observe a steep ingress followed by a gradual egress.

Once such a fine dust cloud forms around and behind the planet it may not be stable and is prone to variability. For example when the dust cloud is thin the planet surface is intensively irradiated, which leads to more evaporation, outflows, and condensation, thereby producing more dust. In turn, the thick dust cloud shields the planet and the evaporation drops, limiting the production of dust, and the cloud dissipates. This limit cycle can apparently operate, even on a timescale from orbit to orbit, but there are longer intervals of order a week where the transits are reduced to a level where they are not detected (see also the following section on the variability).

Producing and maintaining a substantial outflow of gas and dust is relatively simple in bodies with the surface gravities of asteroids, where the thermal speed of the material exceeds the escape speed (see, e.g., Fig. 8 of [13]). For at least some common minerals the vapour pressure at ~2000 K is sufficiently high that for bodies below lunar size, the direct Jeans' escape mass loss rates could exceed that required to produce the inferred dust rates in KIC 1255b of ~1 $M_{\oplus}$/Gyr. For more substantial bodies, e.g., Mercury, Mars, and Earth, a Jeans' outflow of 1 $M_{\oplus}$ per Gyr of heavy molecules becomes nearly impossible [13, 57, 62]. For such massive bodies, a different escape mechanism has been proposed, namely a Parker-type hydrodynamic wind [57, 62]. Roughly speaking this requires thermal speeds that are only ~1/4 of the escape speed in order to work [57, 62]. One issue with the requirement of a planet losing 1 $M_{\oplus}$ per Gyr is that if it has of only $\sim 10^{-3} - 10^{-2}$ $M_{\oplus}$, then it will have a lifetime of only 1–10 Myr. If the lifetime of the host star is measured in Gyr, then the a prior probability of seeing one of its planets in that evaporative state are rather low. However, obviously if one surveys a large number of stars, then the odds of seeing a few such systems is non-negligible. Since such planets may have lost most of their mass their observations open a unique window into planetary interiors and their chemical composition [63, 64].

### 4.1.2 Variability

It was mentioned above that the transits are variable. They vary on a very short timescale from one orbit to another, i.e., in less than one day. This variability is strong, sometimes more than a factor of 2 from one orbit to the next one, and appears to be stochastic and associated with the deep core of the transit [57, 65]. However, a modulation of the transit depth was also found that appears to be anti-correlated with the periodic rotational variability (22.9 days) of the stellar flux[59, 66].

There is also a smooth long-term variability in the egress part of the light curve associated with the dust tail on timescales of about 1.3 year which is not seen in the core of the transit [44, 58]. There might also have been a period of decreased activity, i.e., when the transits were shallower on average, during 2013–2014 [67]. This longer-term variability in the depth and shape of the transits indicates that the dust cloud associated with the planet may not be homogeneous and has at least two components; an inner tail (or coma) and an outer tail which may behave differently (e.g., when subjected to magnetic fields or stellar winds) or have different properties (particle size, chemical composition) [44, 65]. On the contrary, [68] arrived at the conclusion that, as far as the pure shape of the average transit profile is concerned, it is well reproduced in their calculations and there is no need to invoke two such constituents. A similar long-term variability of the transit, namely a monotonic decrease of its depth over the 4-year duration of the *Kepler* mission was found in another disintegrating exoplanet, KOI-2700b [69].

The reason for the above mentioned long-term variability has not been well studied but it has been argued that it may be associated with the magnetic activity of the star and be analogous to the comet tail disconnection events observed in some of the comets in our Solar System [44, 59]. However, it has also been argued that the modulation of the transit depth with stellar rotation may be due to occultations of the stellar spots rather than the magnetic activity [66].

### 4.1.3 Pre-transit Brightening

There is a very interesting tiny feature in the transit light curve, barely visible in Fig. 6. It is a small brightening just before the transit, already noted by the discoverers [57], which is referred to as a pre-transit brightening. It is not due to the star getting brighter. It is caused by the scattering properties of the dust. As shown in Fig. 4, the dust does not scatter the light isotropically but mainly in the forward direction. For the same reason a driver gets blinded when the Sun is near, but not in, the driver's immediate field of view, but the windshield is dirty and this nonetheless scatters the sunlight into his/her eyes.

In our system, this happens mainly in the vicinity of the transit. While we cannot identify this light during or after the transit, since it is overlaid with the ongoing absorption, we can see it just before the transit (Fig. 7). This feature is sensitive to the particle size and it enables us to estimate that the size of particles in the tail is about 0.1–1 μm. At the same time this effect confirms that the transit events are caused by a dusty tail passing in front of, and close to, the star. Apart from these features in direct transits, the forward scattering effect can, in principle, be used to detect non-transiting dusty-tailed exoplanets by searching for positive bumps in the light curves [45, 46].

### 4.1.4 Particle Size and Chemical Composition

A number of authors have studied the *Kepler* light curves of KIC1255b attempting to derive the chemical composition and grain size distribution of the transiting dust material [44, 65, 70]. This problem is partially degenerate and one can fit such monochromatic[3] transits with different chemical composition and particle size. The pre-transit brightening is sensitive to the particle size and the observed brightening indicates particles 0.1–1 μm in size. On the other hand, the length of the tail is highly sensitive to the sublimation properties of the grains. Corundum and 0.2–5 μm grains are most favoured for this reason and the mass loss rate amounts to 0.6–16 Earth masses per Gyr [53, 68].

[3]In this context 'monochromatic' means transits that are observed in only a single waveband.

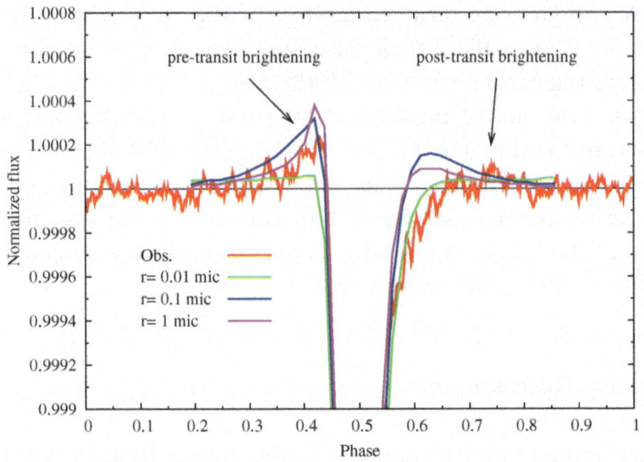


**Fig. 7** *Kepler* light curve of KIC 1255 (red) zoomed so that the pre-transit brightening is clearly visible. Models (green, blue, purple) demonstrate that this feature is sensitive to the particle size. Taken from [44] and reproduced with permission ©ESO

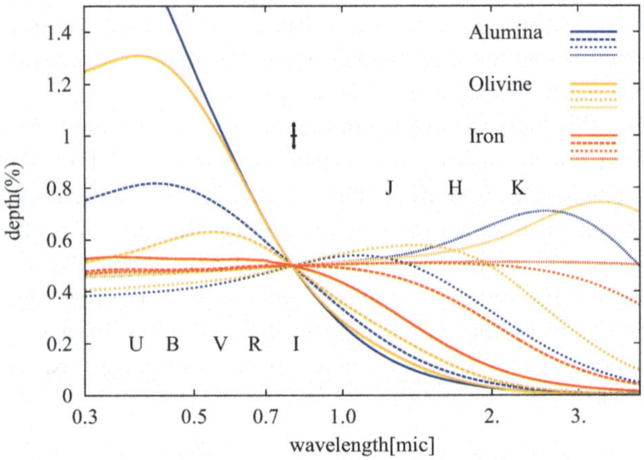


**Fig. 8** Theoretical transit depths for three species alumina, olivine, and iron; each for the particle size of 0.1 (solid), 0.16 (dashed), 0.4 (short-dashed), 1.0 (dotted) micron as a function of wavelength. Depth is normalized such that the transit in the I filter is about 0.5% deep

More information and a deeper insight can be achieved with multi-wavelength observations. This is because the opacity of dust changes with the wavelength and the behaviour is different for grains of different chemical composition and size. Consequently, under the assumption that the tail is optically thin, the transit depth would depend on the wavelength, the particle size, and the chemical composition. This is illustrated in Fig. 8 which shows theoretical transit depths for three species: corundum (alumina), olivine, and iron. One can see that the transits produced by small particles of corundum or silicates would be much deeper at the shorter wavelengths. This is because scattering dominates extinction and scattering on small particles (relative to the wavelength) is approaching the Rayleigh regime with a strong $\lambda^{-4}$ dependence. Extinction by large particles is almost grey. The problem is that the observations must be carried out at different wavelengths simultaneously because of the above mentioned strong variability of the transit depths.

Such observations in the optical and near-infrared regions have not detected a significant difference in the transit depth across these wavelengths. This implies that the dust particle size must be larger than ~0.5 μm. In a scenario where dust grains are lifted directly from the surface of the planet, this in turn implies that the planet should be less massive than Mercury otherwise its gravity would prevent such direct dust ejection [60, 67]. However, as mentioned above, dust might have also condensed later beyond the potential well of the planet. Additional multi-wavelength observations in $z'$, $g'$, $u'$ filters indicate slightly larger depths at shorter wavelengths and particle sizes of about 0.25–1 μm [71]. Recent 3D models of the dust dynamics including the sublimation and 3D radiative transfer pointed out the possibility that the tail may be optically thick. In this case the transit depth might be constant with the wavelength even for smaller particles and mass loss rates may reach 80 Earth masses per Gyr [64].

Apart from KIC 1255b, two other systems of this kind have been discovered, KOI-2700b [69] and K2-22b [72]. The first one is similar in its transit profile to KIC1255b, and the latter system is described in more detail below.

## 4.2 K2-22b

This exoplanet (also known as EPIC 201637175b) was discovered with the *Kepler* follow-on mission (*K2*) by [72]. It is in some respects similar to KIC 1255b. The host star is cooler and smaller. It is an M0V type red dwarf (r = 15.01 mag) with effective temperature, mass, and radius of about $T_{\rm eff} = 3830\,\rm K$, $M = 0.6\,\rm M_{\odot}$, and $R = 0.57\ \rm R_{\odot}$, respectively. The host star rotates with a period of 15.3 days and has a 'close' (three magnitudes fainter) companion, separated by about $2''$. The planet K2-22b is smaller than 2.5 $\rm R_{\oplus}$, is less massive than 1.4 $\rm M_J$, and has a very short orbital period of only $9.145872 \pm 0.000024\,\rm h$. It is losing mass in the form of a dusty tail at a rate $\approx 2 \times 10^{11}\,\rm g\,s^{-1}$ [72].

As in the case of KIC 1255b, the transits are asymmetric and highly variable. They are on average about 0.5% deep but the depth changes from 0 to 1.3% from transit to transit. The duration of the transits is about 50 min. The average transit shape is shown in Fig. 9. The special feature of this exoplanet is that it exhibits a post-transit brightening. Based on the lesson learned from KIC 1255b, this likely indicates that the planet also has a dusty tail but it is pointing in the opposite direction. In other words, the planet is orbiting the star with its dust tail heading forward. This is most probably due to the host star being colder and fainter than KIC 1255. Its radiation does not exert sufficient pressure on the dust grains to force them into a higher orbit, and thereby trail the star. Thus, the dust can flow from the

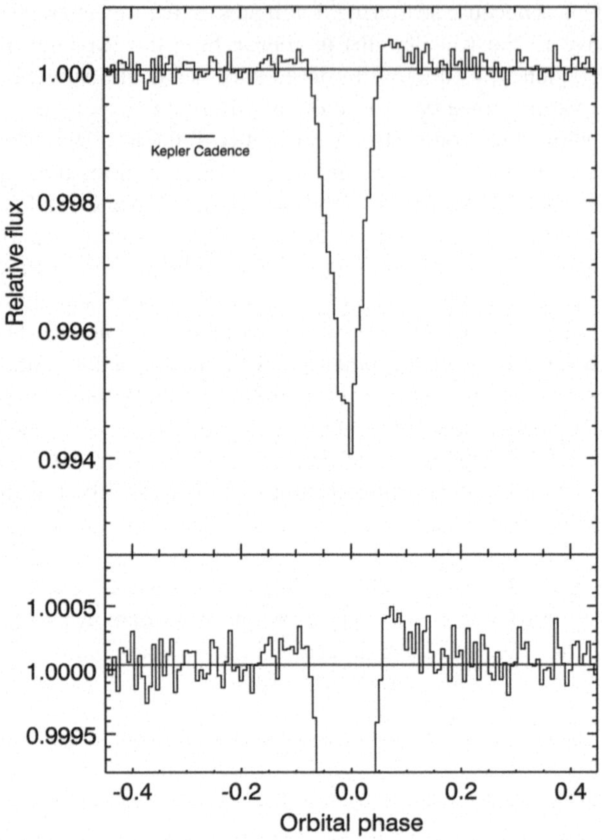


**Fig. 9** Binned and averaged *K2* light curve of K2-22 folded with the orbital period. It features a post-transit brightening likely indicative of a leading dust tail. Taken from [72]. Courtesy of ApJ

planet toward the L1 point and the host star, and then descend into the potential well of the star. Since the Keplerian velocity of these orbits is higher, these grains overtake the planet and form a leading dust tail [72].

The follow up multi-wavelength transit observations with the GTC in the visible region found no evidence for a wavelength dependence in three out of the four transits observed [72, 73]. One transit, however, did indicate that the transit depth is greater at the bluer wavelengths. This sets an upper limit on the dust grains of about 0.4–0.6 μm. The forward scattering peak indicates particle sizes of about 0.5 μm. Although the dust is the major opacity source, the gas might be detected in the cores of some strong spectral lines such as NaI in high resolution spectra. [74] searched for such gas absorption during the transits of K2-22b and Kepler-1520b but did not detect any spectral signatures.

# 5 Minor Bodies in Extrasolar Systems

As in our solar system, minor bodies are also expected to exist in extrasolar systems. While we do not yet have the capability to detect structures similar to the main asteroid belt or the Oort cloud in other planetary systems, the first extrasolar minor bodies have recently been detected.

## *5.1 Exo-Asteroids: A Debris Disk Around WD 1145+017*

In the solar system, asteroids are defined as minor bodies in the inner solar system that show significant departures from spherical shape dictated by hydrostatic equilibrium. The first extrasolar minor bodies were discovered by *K2* [75] in the form of disintegrating material orbiting the white dwarf WD 1145+017.

It has long been known that some white dwarfs have dusty debris disks around them [76, 77], and also that many of them (about 1/4–1/2) have heavy elements in their atmospheres that should have already sunk into the stellar interiors, unless they were replenished by infalling orbiting material [78–80]. Vanderburg et al. [75] observed for the first time this process in action by detecting a white dwarf being transited by 'at least one and likely multiple disintegrating planetesimals with periods ranging from 4.5 to 4.9 h. The detected transits are marked by being asymmetric, and even irregular, with respect to normal transiting planets, indicating that they do not correspond to solid spherical bodies, and can be as deep at 55%. In addition, most of the observed transits are much longer in duration than the ∼1–2 min expected transit time of an asteroid with an orbital period of 4.5 h. Note that the white dwarf radius and luminosity are quite low. Its apparent brightness, luminosity, distance, effective temperature, mass, and radius are about $g = 17.0$ mag, $L = 0.0093\,L_{\odot}$, $d = 174$ pc, $T_{\rm eff} = 15900$ K, $M = 0.6\,M_{\odot}$, and $R = 1.4\ R_{\oplus}$, respectively. The orbital period of 4.5h corresponds to a distance of about 1 $R_{\odot}$ from the star. A combination of this distance and stellar luminosity yields equilibrium temperatures of about 1400–1700 K which is similar to those of disintegrating planets.

This object has attracted the attention of exoplanet observers. Croll et al. [81] conducted ground and space follow-up observations on WD 1145+017. The observations confirmed that the white dwarf is orbited by multiple short-period objects, that egress times were longer than ingress times, and the duration of the transits was longer than expected, pointing again to cometary tail-like structures behind the debris fragments. These asteroids are nicely visualized with a 'waterfall' diagram presented in Rappaport et al. [82] showing the evolution of the phase light curve (see Fig. 10). One can easily identify several objects with slightly different periods crisscrossing the picture.

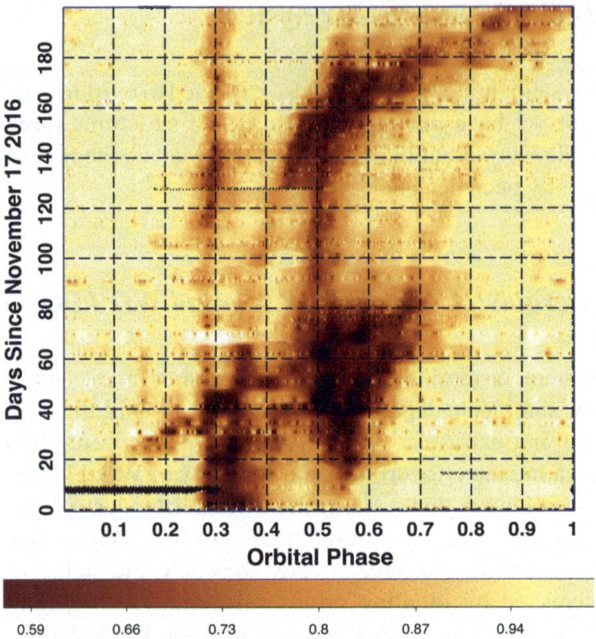


**Fig. 10** Waterfall diagram of WD 1145+017 phased with the base period of 4.49126 days. Objects with the base period follow the vertical line while objects with different periods crisscross the diagram on different tracks. Taken from [82] by permission of Oxford University Press

Croll et al. [81] also did not detect any transit chromaticity. Alonso et al [83] and Izquierdo et al. [84] used the 10-m GTC telescope to check for chromaticity but found the transits to be gray over the optical range from 480 to 920 nm (see Fig. 11), indicating that particle sizes smaller than 0.5 μm can be excluded. From their observations, Alonso et al [83] concluded that the radius of single-size particles in the tail materials must be ≈0.15 μm or larger, or ≈0.06 μm or smaller. They also report low amplitude variations in the light curves suggesting that dusty material is continuously passing in front of the stellar disk.

Zhou et al. [85] and Xu et al. [86] also observed these dips in multiple photometric bands in the visible and infrared. They find no difference in the transit depths once infrared observations are corrected for excess emission from a dusty disk. Xu et al. [86] conclude that there must be a deficit of small particles in the transiting material and that only large particles can survive without sublimating at the effective temperatures prevalent at these short orbital periods.

Xu et al. [87] found the first detection of chromaticity, showing that UV transit depths are always shallower than those in the optical. They proposed a model to explain this observations by having the transiting dust clouds block a larger fraction of the circumstellar gas than of the white dwarf and by having all of them (transiting dust, circumstellar gas, and white dwarf) aligned with respect to our line of sight.

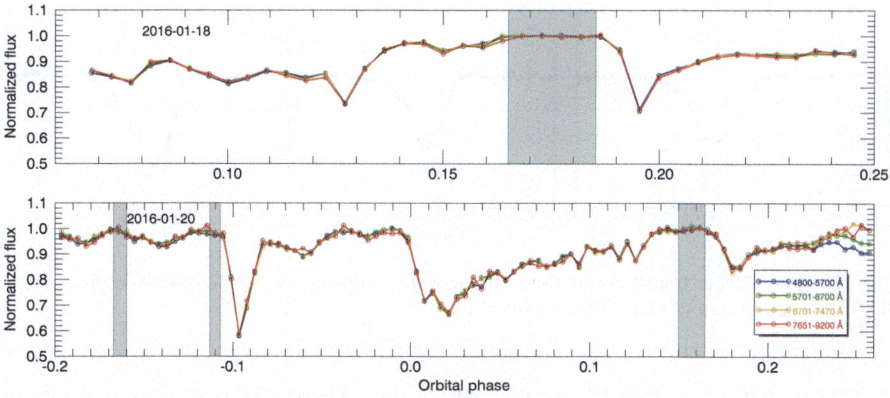


**Fig. 11** GTC light curves of WD 1145+017 taken simultaneously in four wavebands and covering several dips. The nearly identical dip profiles in the four bands can be used to constrain the dust grain sizes to larger than 0.5 μm. The divergence of the curves after phase 0.22 in the lower panel is due to atmospheric effects. Adapted from [83] and reproduced with permission ©ESO

The light curve of this object is extremely variable as shown by Rappaport et al. [82, 88] and Gänsicke et al. [89]. This is because (1) individual objects have slightly different periods, (2) the periods of some of individual objects can change slowly with time, and (3) their dust activity can change dramatically on timescales of months and years.

High resolution spectroscopic observations also revealed the presence of high-velocity gas orbiting the white dwarf [90, 91].

A more detailed review of this object can be found in [82, 92]. Very recently a second white dwarf with possibly related properties was discovered [93]. This object, ZTF J013906.17+524536.89, exhibits two deep transits separated by 110 days, but it is not yet clear if this is a periodicity.

## 5.2 *Exo-Comets*

The unprecedented precision of the *Kepler* photometry enabled the detection of even smaller objects than planets or even large asteroids. Two decades ago [94] predicted that comets orbiting other stars and emitting large dusty tails might be detected by photometry when transiting their host stars and calculated what their light curves could look like. In the *Kepler* data, [95] detected six events in the light curve of KIC 3542116 (KIC3542) and one event in the light curve of KIC 11084727 (KIC1108) which looked very much like the expected cometary transits. They were about 0.05–0.2% deep and highly asymmetric, similar in shape to the KIC1255b transits but several times more shallow (see Fig. 12). The three deeper transits of KIC3542 and that of KIC1108 lasted for about 1 day while the three shallower

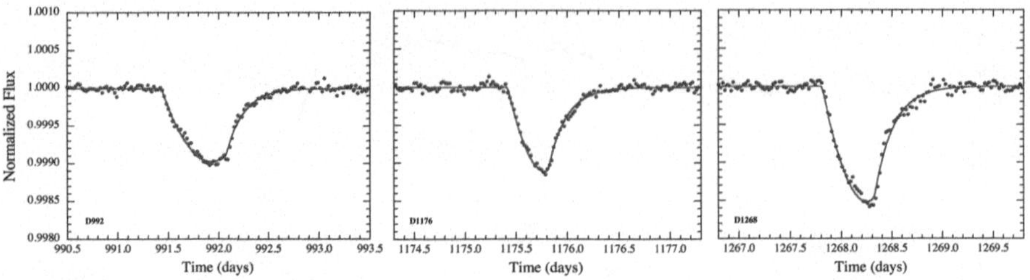


**Fig. 12** Three deeper transit events found in the *Kepler* light curve of KIC3542 by Rappaport et al. [95] by permission of Oxford University Press

transits of KIC3542 lasted for about half a day. There is no obvious periodicity to these events indicating that six transits of KIC3542 are caused by 2–6 distinct comet-like bodies. The duration of the transits corresponds to a transverse speeds of about 35–50 km s$^{-1}$ for the longer transits and about 75–90 km s$^{-1}$ for the shorter transits. This corresponds to orbital periods of ≥90 and ≥50 days, respectively. Both host stars KIC3542 and KIC1108 are relatively bright ($V = 10$ mag) and hot stars with $T_{\rm eff} = 6900$ and 6800 K, respectively. Most of the stars monitored by *Kepler* are cooler (or older) so the fact that they are hotter/younger, similar to each other, and also similar to Boyajian's star mentioned later is probably not an accident.

Recently, a single comet-like transit was found in the archival lightcurve of KIC 8027456 [96]. The *TESS* mission also detected three other dips of this kind in $\beta$ Pictoris [97]. Similar events were discovered in two stars (EPIC 205718330 and EPIC 235240266) monitored by the *K2* mission [98]. The authors call these 'little dippers' since they resemble the so-called "dipper" stars. However, contrary to dipper stars. these dips are 1–2 orders of magnitude shallower with depths of about 0.1–1%. The dips in the 'little dippers' are episodic, not periodic, lasting for about 0.5–1 days, with complicated shapes resembling more WD 1145+017 or Boyajian's star rather than the typical exocomet like profile seen in Fig. 12. Nevertheless, the authors argue that exocomets are the most likely explanation. The host stars are early-K and late-F type dwarfs, not younger than 150 and 800 Myr, respectively.

Recently, two other 'dipper' stars were discovered by K2 mission. HD 139139 is a normal early G-type star which shows a sequence of 28 transit like dips [99]. The events are about 200 ppm deep, 0.7–7 h long, are random, and do not show a significant asymmetry. EPIC 204376071 is a young M5 dwarf which shows a single 80% deep asymmetric dip [100]. There is no explanation to these phenomena yet.

## 6 Boyajian's Star

We would like to introduce another star which may be related to the above mentioned objects and which is sometimes labeled as "the most mysterious star in the Galaxy".

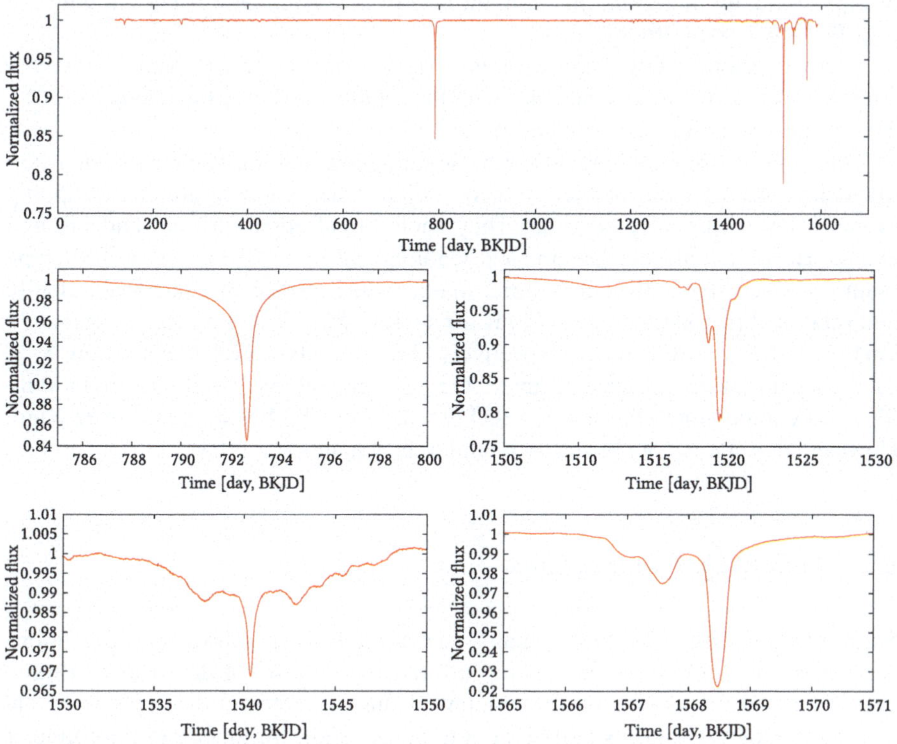


**Fig. 13** The *Kepler* light curve of Boyajian's star shows irregular dips (top). A more detailed view of the four major events is displayed in the middle and bottom panels

## *6.1 Discovery and the Kepler Light Curve*

The *Kepler* mission delivered a huge amount of high-precision photometric light curves for about 170,000 stars. A group of volunteers, the 'Planet Hunters', were reviewing the light curves by human eye and they were the first to notice that there were some very strange dips in flux from the star KIC 8462852. A more detailed analysis and follow-up observations resulted in a discovery paper led by Tabetha Boyajian [11], and since that time the star has become known as Boyajian's or Tabby's star.

So what is so special about this star? The *Kepler* light curve shows a few strong dimming events that are 10–20% deep. They are irregular with no sign of periodicity and are clustered into four main events observed near BKJD=790, 1520, 1540, 1570 days.[4] They are shown in Fig. 13. The D790 event is very smooth with a slow ingress followed by a faster egress. The D1520 and D1570 events consist of a sequence

[4]BKJD stands for the *Kepler* Barycentric Julian day which is a Julian Day minus 2454833.

of dips gradually increasing in strength. D1540 is a symmetric triple dip with the central dip being the strongest.

There is another tiny feature in the *Kepler* data at D1210 which deserves attention. It is a symmetric triple dip with the middle one being the strongest [101]. This shape resembles the D1540 event.

This kind of variability would not be anything unusual if this were a young star. Such stars are often surrounded by protostellar disks which might cause dipping events when seen nearly edge on. They show broad emission lines and infrared excess. However, this star has no such features and looks like a normal F3V type main sequence star with temperature, mass, radius, projected equatorial velocity and rotational period of $T_{\rm eff} = 6750$ K, $M = 1.43$ $M_{\odot}$, $R = 1.53$ $R_{\odot}$, $v \sin i = 78$ km s$^{-1}$, and $P_{\rm rot} = 0.88$ days, respectively [11, 102].[5] It is a relatively bright, $V = 11.7$ mag star at the distance of about 451 pc. The authors also discovered a faint M dwarf companion to the star at 1.95″ which is about 3.8 mag fainter in $H$ band. However, this star is not physically bound to Boyajian's star [104].

## *6.2 Follow Up Observations*

Soon after the discovery of Boyajian's star, a plethora of follow-up observations were performed. Observations in the infrared region did not detect any infrared excess but put constraints on the amount of dust at different distances from the star. *Spitzer*/IRAC [105], NASA/IRTF 3 m SpeX [106], Millimetre (Submillimeter Array) and submillimetre (SCUBA-2) continuum observations also did not detect any significant emission towards KIC 8462852. This places an upper limit of about $10^{-6}$ $M_{\oplus}$ of dust lying within 2–8 AU from the star, $10^{-3}$ $M_{\oplus}$ located within 26 AU, and a total overall dust budget of $<7.7$ $M_{\oplus}$ within a radius of 200 AU [107].

Since the end of the *Kepler* space mission in 2013 May the star had been relatively quiet. In 2017 May the dipping activity started again with four main events named 'Elsie', 'Celeste', 'Skara Brae', and 'Angkor' shown in Fig. 14 [108]. These dips are about 1–2.5% deep. The multiband photometry of the dips shows differential reddening favoring non-gray extinction. The data are inconsistent with dip models that invoke optically thick material, but rather they are in-line with predictions for an occulter consisting primarily of ordinary dust, where much of the material must be optically thin with particle sizes $\lesssim 1$ μm. No changes in the spectrum or polarization were detected during these events [102, 108–110]. Spectrophotometric observations of these recent dipping events with the GTC confirm that the dips are deeper in the visual than at red wavelengths. This is compatible with optically thin dust particles having sizes of $\simeq 0.0015$–$0.15$ μm. Such particles would be quickly blown away by the radiation pressure which

[5] According to [103] the 0.88 day periodicity may come from a different source, not from the target star.

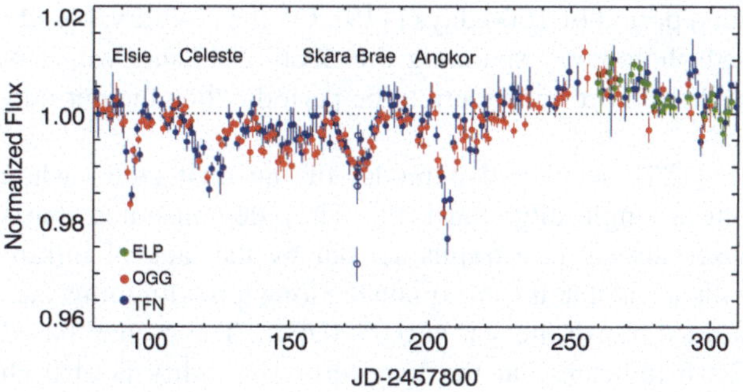


**Fig. 14** Boyajian's star became active again in May 2017. Ground based monitoring shows four dips of depth 1–2%. Taken from [108], courtesy of ApJ

indicates that the dust particles must be continuously replenished [111]. Finally, we note that the radial velocities of the host star also seem to be constant, within $2\,\sigma$ from the average value of $v_{\rm rad} = 4.21 \pm 0.02\,{\rm km\,s^{-1}}$ [102]. This sets significant constraints on any companion stars or even brown dwarfs in short orbital periods.

### 6.3 The Long Term Variability

There is evidence for a long term (secular) variability of Boyajian's star. Based on archival photographic plates from Harvard College Observatory, [112] found that the star faded at an average rate of $0.164 \pm 0.013$ magnitudes per century from 1890 to 1989. This result was questioned by Hippke et al. [113, 114]. Nevertheless, a similar study using archival photographic plates taken at the Maria Mitchell observatory during 1922–1991 found a similar trend of $0.12 \pm 0.02$ per century [115].

The star's brightness dropped significantly throughout the *Kepler* mission as well. Over the first 1000 days the star faded approximately by 0.9%. It dimmed much more rapidly in the next 200 days, with its flux dropping by more than 2% [116]. A slightly deeper 3.5% drop was found in the contemporary *GALEX* observation in the near UV [117]. These results imply $R_V \simeq 5.0$ which, in turn, indicates circumstellar rather than interstellar dust attenuation.

Follow-up observations over a wide wavelength range from the UV to the mid-infrared from 2015 October through 2016 December, using *Swift*, *Spitzer* and *AstroLAB IRIS* indicate that the star faded in a manner similar to the long-term fading seen previously in *Kepler* data. According to [118] the dimming rate for the entire period reported is "$22.1 \pm 9.7$ mmag per yr in the Swift wavebands, $21.0 \pm 4.5$ mmag in the ground-based $B$ measurements, $14.0 \pm 4.5$ mmag in $V$, $13.0 \pm 4.5$ mmag in $R$, and $5.0 \pm 1.2$ mmag per yr averaged over the two warm Spitzer bands". Continued ground based observations [119, 120] suggest that there are also brightening (not only fading) spells and that this long term variability may be periodic with a period of 1600 days [119]. On the contrary, based on the ground and space based photometry spanning the 2006–2017 interval, it was concluded [121] that if the long term trend were to be periodic then the period would have to exceed 10 years.

Wyatt et al. [122] developed a model of the dust cloud where the dust is distributed along a single elliptical orbit. They demonstrated that such a model satisfies the observational constraints set out by the lack of infrared excess and duration of the dips, and that it can explain the long term dimming. According to this model the dust must transit the star at 0.05–0.6 *au*. The ground based observations during 2015–2018 indicate that the long-term variability is also chromatic. The amplitude is largest in the *B* band, while the *VRI* flux amplitudes are progressively smaller by factors of $0.77\pm0.05$, $0.50\pm0.05$, and $0.31\pm0.05$, respectively [123]. This implies that the dust particles causing the long-term variability must be about 0.1 $\mu$m in size. Such particles will be easily blown away and must be continuously replenished. The long-term variability (dimming) has a continuum of timescales ranging from almost a century, to decades, to years, and even down to a few months. It is most probably related to the shallow dip events and caused by the same phenomena. The net result is that the star has experienced about a 12% long-term dimming over the past century. This has serious implications for the amount of the dust that must be distributed along the elliptical orbit which now amounts to at least $10^{-3}$ $M_{\oplus}$.

## 6.4 *Possible Explanation and Models*

There have been numerous models, ideas, and speculations proposed to explain the above mentioned behaviour. It is not possible to mention and discuss all the models here. An overview was presented in [124] and it concluded that intervening interstellar material (ISM) is a more plausible explanation than other natural models. The discoverers themselves discussed a number of possibilities and favoured a comet scenario. Apart from that it was proposed that KIC 8462852 might be undergoing a late heavy bombardment, but is only in its very early stages [106, 125]. It is also possible that the variability could be intrinsic to the star [126], or the dips might have been caused by matter in our Solar system [127]. According to [128] the secular dimming is the result of the inspiral of a planetary body or bodies into KIC 8462852, which took place $10 - 10^4$ yr ago. The discoverers also proposed that the dips observed with *Kepler* may be due to transits of less massive bodies placed on eccentric orbits by the Lidov–Kozai oscillations due to the outer M-dwarf companion. However, the predicted smooth decline in flux is not in agreement with the brightening episodes [119, 120], and the M-dwarf companion turned out not to be associated with Boyajian's star [104]. However, evidence is growing that the dipping phenomenon is due to circumstellar dust. In the next few sections we will mention three models that were developed to the point where they can be directly compared with the observations of the dip events.

### 6.4.1 A Swarm of Comets

This is the scenario favoured by the discoverers and developed by Bodman and Quillen [129]. In this model the deep *Kepler* dips and long-term behaviour are due to transit of a large number (70–700) of comets. Such strings of comets are known from our Solar system so it is a natural explanation. An eccentric orbit has advantages. There is a high likelihood of the transits occurring near periastron and the material spends most of the time very far from the star so it can satisfy the IR limits as well as the dynamical constraints [122, 123]. These comets must have had a common progenitor. The models can fit the *Kepler* dips very well. Unfortunately, the model has a few drawbacks. (1) It cannot reproduce the D790 event because it is very smooth and has a slow ingress and a faster recovery, while the model features just the opposite behaviour with a steeper ingress and a slower egress. (2) The symmetric triple dip, D1540, would require an accidental constellation of comets. However, there are two other events of this kind: D1210 and Skara Brae, and they would have to be the result of an accidental grouping of comets as well. (3) Comets can hardly produce and continuously replenish $\gtrsim 10^{-3}$ $M_{\oplus}$ of dust required to explain the long-term variability [123]. (4) The model requires many free parameters (related to a large number of comets) and even a perfect fit does not mean that it is correct.

### 6.4.2 Massive Asteroids Wrapped in Dust

This 'recipe' can be found in [101] but it was already considered in the discovery paper [11]. It is in many respects similar to the above mentioned scenario. According to this model there are a few massive asteroids or planetesimals surrounded by dust clouds orbiting and transiting the star on eccentric orbits. Obviously, the objects must have originated from a common progenitor as well. The orbit and the amount of the dust required to transit the star is similar to the previous model so it also satisfies the IR limits and the dynamical constraints [122, 123]. The difference is in the following. Instead of a large number of comets only four more massive objects are sufficient to explain the four major *Kepler* events. A massive object means that its gravity cannot be neglected, and it can retain a dust cloud within its Hill's sphere (contrary to a comet). It naturally explains the smooth shape of the D790 event and produces a slower ingress and faster egress. The symmetric triple dips: D1540, D1210, and Skara Brae are no longer due to an accidental constellation of objects but rather single objects surrounded by dusty disks/rings. The massive asteroids can produce and replenish $\gtrsim 10^{-3}$ $M_{\oplus}$ of dust to account for the long-term variability. It was demonstrated that if the objects were initially on exactly identical orbits, and were massive enough, then they (and their dust clouds) would mutually interact and end up on a slightly different orbits. Even though the fits are not perfect, the model requires a small number of massive objects, and hence only a handful of free parameters. One can anticipate that massive asteroids are accompanied by a large number of smaller debris which would account for the smaller dips and long-term variability.

### 6.4.3 The Lord of the Rings

We hope the reader will not mind the 'label' above. The model was proposed by Bourne et al. [130]. The authors noticed that one of the post *Kepler* dips (Skara Brae, the one that occurred around Aug 9, 2017) is very similar to the *Kepler* D1540 event, i.e., it is a symmetric triple dip with the central dip being the deepest. The similarity is indeed striking and the authors presume that it is the transit of the same body. This implies an orbital period of 1601 days. 'The Lord' is a dark and relatively massive object—a brown dwarf orbiting the star. It is accompanied by a 'fellowship' of about 9 rings which are about 0.2 AU across. With this model the authors were able to reproduce the Skara Brae and D1540 events very well. Apart from that the model explains a tentative 1600 day periodicity found in the long-term variability. The other dips observed by *Kepler* were not modelled but might be understood assuming transits of additional bodies (moons) related to the brown dwarf. The model makes a very precise and testable prediction. 'The Return of the Lord' should happen during Christmas on December 27, 2021.

A similar idea was presented earlier in [131]. The authors identified two strikingly similar events in the *Kepler* light curve which are approximately 0.1% deep and occurred at D216 and D1144. They show that these events could be explained by the occultation of the star by a giant ring system or by the transit of a string of half a dozen exocomets. These events occurred 928.25 days apart and the authors predict that the next event will occur between 3–8 October 2019.

More recent comparison and cross-correlation of *Kepler* dips and dips observed from the ground indicate a similar periodicity of 1574.4 days (4.31 yr) [132]. This period also explains a few other historical dimming events of the star in the past. It predicts the next return of the D790 event on October 17, 2019. We would like to comment that this idea presumes that the mutual gravitational interaction among the bodies orbiting the star must be negligible. It is not compatible with the brown-dwarf hypothesis.

It remains to be established whether these models are compatible with the long-term variability, infrared limits, and various other constraints including the dynamics of the system.

## 7 Ongoing and Future Space Missions

Because *Kepler* played a pioneering role in the detection of a new class of 'disintegrating' objects, we shall briefly discuss ongoing and future space missions in order to present their potential for new discoveries of this particularly interesting class of objects.

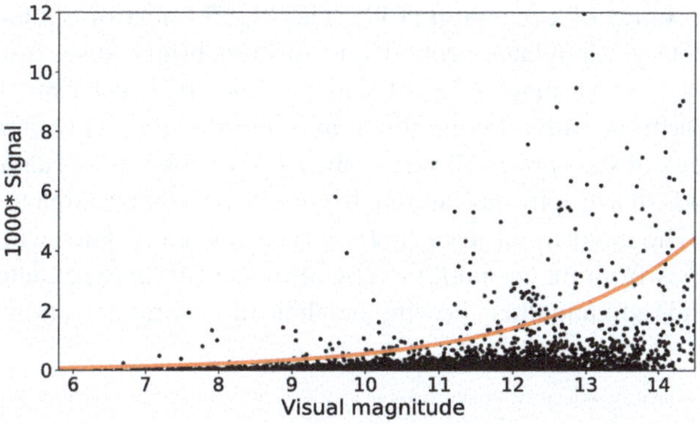


**Fig. 15** Expected distribution of *TESS* planets with the orange line representing a detection threshold for mid-sized telescopes. Figure from [138] https://iopscience.iop.org/article/10.1088/1538-3873/ab2143. ©The Astronomical Society of the Pacific Reproduced by permission of IOP Publishing. All rights reserved

## *7.1 TESS*

*TESS* is a NASA space mission successfully launched in 2018 and planned for at least 2 years of operations [133, 134]. The aim of the *TESS* mission is the detection of several thousand exoplanets, mainly Neptune- and super Earth-sized. However, several hundred Jupiter-sized planet detections are expected as well. *TESS* is delivering precise photometry down to about 200 ppm which is sufficient to detect a transiting super-Earth.[6] The first *TESS planets* were recently announced [135, 136]. There are many interesting objects discovered by *TESS*, such as a Neptune-sized planet HD 21749b with another, Earth-sized, planet HD 21749c in the same system [37] or the first *TESS* transiting brown dwarf [137]. Many of the *TESS* planets should be suitable for ground-based follow-up observations to detect exoplanetary atmospheres even with mid-sized telescopes [138] as shown in Fig. 15. Furthermore, it is expected that *TESS* will detect additional interesting systems, and among those should be the types of disintegrating and dusty objects which we described in this review.

## *7.2 The PLATO Space Mission*

The ESA M3 space mission *PLATO* (PLAnetary Transits and Oscillations of stars) will be launched in 2026. The *PLATO* space mission will consist of 26 telescopes monitoring large portion of sky (about 50%) for transits with an unprecedented

[6] https://heasarc.gsfc.nasa.gov/docs/tess/observing-technical.html.

photometric accuracy of a few ppm [139]. The *PLATO* mission should find several thousand planetary candidates around one million bright stars from naked eye brightness to $V = 11$ mag. *PLATO* will be able to detect even an Earth-like planet on an Earth-like orbit among the Solar type stars. *PLATO* will also focus on asteroseismology of stars [140]. However, the *PLATO* mission will also contribute to many other fields of astrophysics ranging from variable star research to extragalactic objects [139]. The majority of the *PLATO* targets and candidates will amenable to follow-up studies from the ground, thereby allowing for an exact determination of their masses and radii and thus allowing for their full characterization.

## *7.3 The ARIEL Space Mission*

*ARIEL*, the Atmospheric Remote-sensing Infrared Exoplanet Large-survey, is an ESA M4 mission which will be launched in 2028 and it will be dedicated to unveiling the chemical composition of a sample of about 1000 selected transiting exoplanets [141]. *ARIEL* will be equipped with an off-axis Cassegrain telescope with an elliptical primary mirror of $1.1 \times 0.7$ m. *ARIEL* will be capable of photometric monitoring in visible and infrared wavelengths between 0.50–0.55, 0.8–1.0, and 1.0–1.2 μm. A spectrograph with two medium resolving power channels of 1.95–3.9 and 3.9–7.8 μm and one low-resolution channel of 1.25–1.95 μm will be available.[7] The precision of *ARIEL* should be sufficient to detect the signature of exo-atmospheres with a precision of at least $10^{-4}$ relative to the star. The main targets will be hot (600 K and more) planets, and it is expected that species like $H_2O$, $CO_2$, $CH_4$, $NH_3$, HCN or even metallic compounds such as TiO and VO will be detected and studied.

**Acknowledgements** The authors would like to thank Prof. Saul Rappaport for carefully reviewing the manuscript and his help with several sections. We acknowledge the ERASMUS+ project 'Per aspera ad astra simul' under number 2017-1-CZ01-KA203-035562 which funded the mobilities of the authors. Apart from that Jan Budaj thanks for support the VEGA 2/0031/18 and APVV 15-0458 grants. Petr Kabáth acknowledges the support of GACR grant number 17-01752J. Enric Palle is partly financed by the Spanish Ministry of Economics and Competitiveness through projects ESP2016-80435-C2-1-R and PGC2018-098153-B-C31.

## References

1. A. Wolszczan, D.A. Frail, Nature **355**(6356), 145 (1992). https://doi.org/10.1038/355145a0
2. M. Mayor, D. Queloz, Nature **378**(6555), 355–359 (1995). https://doi.org/10.1038/378355a0
3. D. Charbonneau, T.M. Brown, D.W. Latham, M. Mayor, Astrophys. J. **529**(1), L45 (2000). https://doi.org/10.1086/312457

[7] http://sci.esa.int/ariel/59798-summary/.

4. G.Á. Bakos, *The HATNet and HATSouth Exoplanet Surveys* (Springer, Berlin, 2018), p. 111. https://doi.org/10.1007/978-3-319-55333-7_111
5. J. Pepper, K.G. Stassun, B.S. Gaudi, *KELT: The Kilodegree Extremely Little Telescope, a Survey for Exoplanets Transiting Bright, Hot Stars* (2018), p. 128. https://doi.org/10.1007/978-3-319-55333-7_128
6. A. Baglin, M. Auvergne, P. Barge, M. Deleuil, C. Catala, E. Michel, W. Weiss, COROT Team, in *The CoRoT Mission Pre-Launch Status - Stellar Seismology and Planet Finding. ESA Special Publication*, ed. by M. Fridlund, A. Baglin, J. Lochard, L. Conroy, vol. 1306 (ESA, Paris, 2006), p. 33
7. W.J. Borucki, D. Koch, G. Basri, N. Batalha, T. Brown, D. Caldwell, J. Caldwell, J. Christensen-Dalsgaard, W.D. Cochran, E. DeVore, E.W. Dunham, A.K. Dupree, T.N. Gautier, J.C. Geary, R. Gilliland, A. Gould, S.B. Howell, J.M. Jenkins, Y. Kondo, D.W. Latham, G.W. Marcy, S. Meibom, H. Kjeldsen, J.J. Lissauer, D.G. Monet, D. Morrison, D. Sasselov, J. Tarter, A. Boss, D. Brownlee, T. Owen, D. Buzasi, D. Charbonneau, L. Doyle, J. Fortney, E.B. Ford, M.J. Holman, S. Seager, J.H. Steffen, W.F. Welsh, J. Rowe, H. Anderson, L. Buchhave, D. Ciardi, L. Walkowicz, W. Sherry, E. Horch, H. Isaacson, M.E. Everett, D. Fischer, G. Torres, J.A. Johnson, M. Endl, P. MacQueen, S.T. Bryson, J. Dotson, M. Haas, J. Kolodziejczak, J. Van Cleve, H. Chandrasekaran, J.D. Twicken, E.V. Quintana, B.D. Clarke, C. Allen, J. Li, H. Wu, P. Tenenbaum, E. Verner, F. Bruhweiler, J. Barnes, A. Prsa, Science **327**, 977 (2010). https://doi.org/10.1126/science.1185402
8. W.J. Borucki, E. Agol, F. Fressin, L. Kaltenegger, J. Rowe, H. Isaacson, D. Fischer, N. Batalha, J.J. Lissauer, G.W. Marcy, D. Fabrycky, J.M. Désert, S.T. Bryson, T. Barclay, F. Bastien, A. Boss, E. Brugamyer, L.A. Buchhave, C. Burke, D.A. Caldwell, J. Carter, D. Charbonneau, J.R. Crepp, J. Christensen-Dalsgaard, J.L. Christiansen, D. Ciardi, W.D. Cochran, E. DeVore, L. Doyle, A.K. Dupree, M. Endl, M.E. Everett, E.B. Ford, J. Fortney, T.N. Gautier, J.C. Geary, A. Gould, M. Haas, C. Henze, A.W. Howard, S.B. Howell, D. Huber, J.M. Jenkins, H. Kjeldsen, R. Kolbl, J. Kolodziejczak, D.W. Latham, B.L. Lee, E. Lopez, F. Mullally, J.A. Orosz, A. Prsa, E.V. Quintana, R. Sanchis-Ojeda, D. Sasselov, S. Seader, A. Shporer, J.H. Steffen, M. Still, P. Tenenbaum, S.E. Thompson, G. Torres, J.D. Twicken, W.F. Welsh, J.N. Winn, Science **340**(6132), 587 (2013). https://doi.org/10.1126/science.1234702
9. J.N. Winn, R. Sanchis-Ojeda, S. Rappaport, New Asrton. Rev. **83**, 37 (2018). https://doi.org/10.1016/j.newar.2019.03.006
10. J.J. Lissauer, D.C. Fabrycky, E.B. Ford, W.J. Borucki, F. Fressin, G.W. Marcy, J.A. Orosz, J.F. Rowe, G. Torres, W.F. Welsh, N.M. Batalha, S.T. Bryson, L.A. Buchhave, D.A. Caldwell, J.A. Carter, D. Charbonneau, J.L. Christiansen, W.D. Cochran, J.M. Desert, E.W. Dunham, M.N. Fanelli, J.J. Fortney, I. Gautier, N. Thomas, J.C. Geary, R.L. Gilliland, M.R. Haas, J.R. Hall, M.J. Holman, D.G. Koch, D.W. Latham, E. Lopez, S. McCauliff, N. Miller, R.C. Morehead, E.V. Quintana, D. Ragozzine, D. Sasselov, D.R. Short, J.H. Steffen, Nature **470**(7332), 53 (2011). https://doi.org/10.1038/nature09760
11. T.S. Boyajian, D.M. LaCourse, S.A. Rappaport, D. Fabrycky, D.A. Fischer, D. Gandolfi, G.M. Kennedy, H. Korhonen, M.C. Liu, A. Moor, K. Olah, K. Vida, M.C. Wyatt, W.M.J. Best, J. Brewer, F. Ciesla, B. Csák, H.J. Deeg, T.J. Dupuy, G. Handler, K. Heng, S.B. Howell, S.T. Ishikawa, J. Kovács, T. Kozakis, L. Kriskovics, J. Lehtinen, C. Lintott, S. Lynn, D. Nespral, S. Nikbakhsh, K. Schawinski, J.R. Schmitt, A.M. Smith, G. Szabo, R. Szabo, J. Viuho, J. Wang, A. Weiksnar, M. Bosch, J.L. Connors, S. Goodman, G. Green, A.J. Hoekstra, T. Jebson, K.J. Jek, M.R. Omohundro, H.M. Schwengeler, A. Szewczyk, Mon. Not. R. Astron. Soc. (2016). https://doi.org/10.1093/mnras/stw218
12. K.D. Colón, G. Zhou, A. Shporer, K.A. Collins, A. Bieryla, N. Espinoza, F. Murgas, P. Pattarakijwanich, S. Awiphan, J.D. Armstrong, J. Bailey, G. Barentsen, D. Bayliss, A. Chakpor, W.D. Cochran, V.S. Dhillon, K. Horne, M. Ireland, L. Kedziora-Chudczer, J.F. Kielkopf, S. Komonjinda, D.W. Latham, T.R. Marsh, D.E. Mkrtichian, E. Pallé, D. Ruffolo, R. Sefako, C.G. Tinney, S. Wannawichian, S. Yuma, Astron. J. **156**(5), 227 (2018). https://doi.org/10.3847/1538-3881/aae31b

13. R. van Lieshout, S.A. Rappaport, *Disintegrating Rocky Exoplanets* (2018), p. 15. https://doi.org/10.1007/978-3-319-55333-7_15
14. C. Lovis, D. Fischer, *Radial Velocity Techniques for Exoplanets* (2010), pp. 27–53
15. J.T. Wright, *Radial Velocities as an Exoplanet Discovery Method* (2018), p. 4. https://doi.org/10.1007/978-3-319-55333-7_4
16. A.P. Hatzes, E.W. Guenther, M. Endl, W.D. Cochran, M.P. Döllinger, A. Bedalov, Astron. Astrophys. **437**(2), 743 (2005). https://doi.org/10.1051/0004-6361:20052850
17. J.N. Winn, (2010) e-prints arXiv:1001.2010
18. A. Léger, D. Rouan, J. Schneider, P. Barge, M. Fridlund, B. Samuel, M. Ollivier, E. Guenther, M. Deleuil, H.J. Deeg, M. Auvergne, R. Alonso, S. Aigrain, A. Alapini, J.M. Almenara, A. Baglin, M. Barbieri, H. Bruntt, P. Bordé, F. Bouchy, J. Cabrera, C. Catala, L. Carone, S. Carpano, S. Csizmadia, R. Dvorak, A. Erikson, S. Ferraz-Mello, B. Foing, F. Fressin, D. Gand olfi, M. Gillon, P. Gondoin, O. Grasset, T. Guillot, A. Hatzes, G. Hébrard, L. Jorda, H. Lammer, A. Llebaria, B. Loeillet, M. Mayor, T. Mazeh, C. Moutou, M. Pätzold, F. Pont, D. Queloz, H. Rauer, S. Renner, R. Samadi, A. Shporer, C. Sotin, B. Tingley, G. Wuchterl, M. Adda, P. Agogu, T. Appourchaux, H. Ballans, P. Baron, T. Beaufort, R. Bellenger, R. Berlin, P. Bernardi, D. Blouin, F. Baudin, P. Bodin, L. Boisnard, L. Boit, F. Bonneau, S. Borzeix, R. Briet, J.T. Buey, B. Butler, D. Cailleau, R. Cautain, P.Y. Chabaud, S. Chaintreuil, F. Chiavassa, V. Costes, V. Cuna Parrho, F. de Oliveira Fialho, M. Decaudin, J.M. Defise, S. Djalal, G. Epstein, G.E. Exil, C. Fauré, T. Fenouillet, A. Gaboriaud, A. Gallic, P. Gamet, P. Gavalda, E. Grolleau, R. Gruneisen, L. Gueguen, V. Guis, V. Guivarc'h, P. Guterman, D. Hallouard, J. Hasiba, F. Heuripeau, G. Huntzinger, H. Hustaix, C. Imad, C. Imbert, B. Johlander, M. Jouret, P. Journoud, F. Karioty, L. Kerjean, V. Lafaille, L. Lafond, T. Lam-Trong, P. Landiech, V. Lapeyrere, T. Larqué, P. Laudet, N. Lautier, H. Lecann, L. Lefevre, B. Leruyet, P. Levacher, A. Magnan, E. Mazy, F. Mertens, J.M. Mesnager, J.C. Meunier, J.P. Michel, W. Monjoin, D. Naudet, K. Nguyen-Kim, J.L. Orcesi, H. Ottacher, R. Perez, G. Peter, P. Plasson, J.Y. Plesseria, B. Pontet, A. Pradines, C. Quentin, J.L. Reynaud, G. Rolland , F. Rollenhagen, R. Romagnan, N. Russ, R. Schmidt, N. Schwartz, I. Sebbag, G. Sedes, H. Smit, M.B. Steller, W. Sunter, C. Surace, M. Tello, D. Tiphène, P. Toulouse, B. Ulmer, O. Vandermarcq, E. Vergnault, A. Vuillemin, P. Zanatta, Astron. Astrophys. **506**(1), 287 (2009). https://doi.org/10.1051/0004-6361/200911933
19. T. Barclay, J.F. Rowe, J.J. Lissauer, D. Huber, F. Fressin, S.B. Howell, S.T. Bryson, W.J. Chaplin, J.M. Désert, E.D. Lopez, G.W. Marcy, F. Mullally, D. Ragozzine, G. Torres, E.R. Adams, E. Agol, D. Barrado, S. Basu, T.R. Bedding, L.A. Buchhave, D. Charbonneau, J.L. Christiansen, J. Christensen-Dalsgaard, D. Ciardi, W.D. Cochran, A.K. Dupree, Y. Elsworth, M. Everett, D.A. Fischer, E.B. Ford, J.J. Fortney, J.C. Geary, M.R. Haas, R. Handberg, S. Hekker, C.E. Henze, E. Horch, A.W. Howard, R.C. Hunter, H. Isaacson, J.M. Jenkins, C. Karoff, S.D. Kawaler, H. Kjeldsen, T.C. Klaus, D.W. Latham, J. Li, J. Lillo-Box, M.N. Lund, M. Lundkvist, T.S. Metcalfe, A. Miglio, R.L. Morris, E.V. Quintana, D. Stello, J.C. Smith, M. Still, S.E. Thompson, Nature **494**(7438), 452 (2013). https://doi.org/10.1038/nature11914
20. D. Charbonneau, T.M. Brown, R.W. Noyes, R.L. Gilliland, Astrophys. J. **568**(1), 377 (2002). https://doi.org/10.1086/338770
21. A. Vidal-Madjar, A. Lecavelier des Etangs, J.M. Désert, G.E. Ballester, R. Ferlet, G. Hébrard, M. Mayor, Nature **422**, 143 (2003). https://doi.org/10.1038/nature01448
22. S. Redfield, M. Endl, W.D. Cochran, L. Koesterke, Astrophys. J. **673**, L87 (2008). https://doi.org/10.1086/527475
23. J.L. Bean, E. Miller-Ricci Kempton, D. Homeier, Nature **468**(7324), 669 (2010). https://doi.org/10.1038/nature09596
24. J.L. Bean, J.M. Désert, P. Kabath, B. Stalder, S. Seager, E. Miller-Ricci Kempton, Z.K. Berta, D. Homeier, S. Walsh, A. Seifahrt, Astrophys. J. **743**(1), 92 (2011). https://doi.org/10.1088/0004-637X/743/1/92
25. F. Murgas, G. Chen, E. Pallé, L. Nortmann, G. Nowak, Astron. Astrophys. **622**, A172 (2019). https://doi.org/10.1051/0004-6361/201834063

26. G. Chen, E. Pallé, L. Welbanks, J. Prieto-Arranz, N. Madhusudhan, S. Gandhi, N. Casasayas-Barris, F. Murgas, L. Nortmann, N. Crouzet, H. Parviainen, D. Gandolfi, Astron. Astrophys. **616**, A145 (2018). https://doi.org/10.1051/0004-6361/201833033
27. E. Sedaghati, H.M.J. Boffin, R.J. MacDonald, S. Gandhi, N. Madhusudhan, N.P. Gibson, M. Oshagh, A. Claret, H. Rauer, Nature **549**(7671), 238 (2017). https://doi.org/10.1038/nature23651
28. E. Pallé, M.R. Zapatero Osorio, A. García Muñoz, Astrophys. J. **728**(1), 19 (2011). https://doi.org/10.1088/0004-637X/728/1/19
29. N. Astudillo-Defru, P. Rojo, Astron. Astrophys. **557**, A56 (2013). https://doi.org/10.1051/0004-6361/201219018
30. H.M. Cegla, C. Lovis, V. Bourrier, B. Beeck, C.A. Watson, F. Pepe, Astron. Astrophys. **588**, A127 (2016). https://doi.org/10.1051/0004-6361/201527794
31. I.A.G. Snellen, S. Albrecht, E.J.W. de Mooij, R.S. Le Poole, Astron. Astrophys. **487**(1), 357 (2008). https://doi.org/10.1051/0004-6361:200809762
32. A. Wyttenbach, D. Ehrenreich, C. Lovis, S. Udry, F. Pepe, Astron. Astrophys. **577**, A62 (2015). https://doi.org/10.1051/0004-6361/201525729
33. J.V. Seidel, D. Ehrenreich, A. Wyttenbach, R. Allart, M. Lendl, L. Pino, V. Bourrier, H.M. Cegla, C. Lovis, D. Barrado, D. Bayliss, N. Astudillo-Defru, A. Deline, C. Fisher, K. Heng, R. Joseph, B. Lavie, C. Melo, F. Pepe, D. Ségransan, S. Udry, Astron. Astrophys. **623**, A166 (2019). https://doi.org/10.1051/0004-6361/201834776
34. J. Žák, P. Kabáth, H.M.J. Boffin, V.D. Ivanov, M. Skarka, Astron. J. **158**(3), 120 (2019). https://doi.org/10.3847/1538-3881/ab32ec
35. N. Casasayas-Barris, E. Palle, G. Nowak, F. Yan, L. Nortmann, F. Murgas, Astron. Astrophys. **608**, A135 (2017). https://doi.org/10.1051/0004-6361/201731956
36. S.B. Howell, C. Sobeck, M. Haas, M. Still, T. Barclay, F. Mullally, J. Troeltzsch, S. Aigrain, S.T. Bryson, D. Caldwell, W.J. Chaplin, W.D. Cochran, D. Huber, G.W. Marcy, A. Miglio, J.R. Najita, M. Smith, J.D. Twicken, J.J. Fortney, Publ. Astron. Soc. Pac. **126**(938), 398 (2014). https://doi.org/10.1086/676406
37. D. Dragomir, J. Teske, M.N. Günther, D. Ségransan, J.A. Burt, C.X. Huang, A. Vanderburg, E. Matthews, X. Dumusque, K.G. Stassun, J. Pepper, G.R. Ricker, R. Vanderspek, D.W. Latham, S. Seager, J.N. Winn, J.M. Jenkins, T. Beatty, F. Bouchy, T.M. Brown, R.P. Butler, D.R. Ciardi, J.D. Crane, J.D. Eastman, L. Fossati, J. Francis, B.J. Fulton, B.S. Gaudi, R.F. Goeke, D. James, T.C. Klaus, R.B. Kuhn, C. Lovis, M.B. Lund, S. McDermott, M. Paegert, F. Pepe, J.E. Rodriguez, L. Sha, S.A. Shectman, A. Shporer, R.J. Siverd, A. Garcia Soto, D.J. Stevens, J.D. Twicken, S. Udry, J. Villanueva, Steven, S.X. Wang, B. Wohler, X. Yao, Z. Zhan, Astrophys. J. **875**(2), L7 (2019). https://doi.org/10.3847/2041-8213/ab12ed
38. T.L. Campante, T. Barclay, J.J. Swift, D. Huber, V.Z. Adibekyan, W. Cochran, C.J. Burke, H. Isaacson, E.V. Quintana, G.R. Davies, V. Silva Aguirre, D. Ragozzine, R. Riddle, C. Baranec, S. Basu, W.J. Chaplin, J. Christensen-Dalsgaard, T.S. Metcalfe, T.R. Bedding, R. Handberg, D. Stello, J.M. Brewer, S. Hekker, C. Karoff, R. Kolbl, N.M. Law, M. Lundkvist, A. Miglio, J.F. Rowe, N.C. Santos, C. Van Laerhoven, T. Arentoft, Y.P. Elsworth, D.A. Fischer, S.D. Kawaler, H. Kjeldsen, M.N. Lund, G.W. Marcy, S.G. Sousa, A. Sozzetti, T.R. White, Astrophys. J. **799**(2), 170 (2015). https://doi.org/10.1088/0004-637X/799/2/170
39. N.M. Batalha, W.J. Borucki, S.T. Bryson, L.A. Buchhave, D.A. Caldwell, J. Christensen-Dalsgaard, D. Ciardi, E.W. Dunham, F. Fressin, I. Gautier, N. Thomas, R.L. Gilliland, M.R. Haas, S.B. Howell, J.M. Jenkins, H. Kjeldsen, D.G. Koch, D.W. Latham, J.J. Lissauer, G.W. Marcy, J.F. Rowe, D.D. Sasselov, S. Seager, J.H. Steffen, G. Torres, G.S. Basri, T.M. Brown, D. Charbonneau, J. Christiansen, B. Clarke, W.D. Cochran, A. Dupree, D.C. Fabrycky, D. Fischer, E.B. Ford, J. Fortney, F.R. Girouard, M.J. Holman, J. Johnson, H. Isaacson, T.C. Klaus, P. Machalek, A.V. Moorehead, R.C. Morehead, D. Ragozzine, P. Tenenbaum, J. Twicken, S. Quinn, J. VanCleve, L.M. Walkowicz, W.F. Welsh, E. Devore, A. Gould, Astrophys. J. **729**(1), 27 (2011). https://doi.org/10.1088/0004-637X/729/1/27

40. F. Fressin, G. Torres, J.F. Rowe, D. Charbonneau, L.A. Rogers, S. Ballard, N.M. Batalha, W.J. Borucki, S.T. Bryson, L.A. Buchhave, D.R. Ciardi, J.M. Désert, C.D. Dressing, D.C. Fabrycky, E.B. Ford, I. Gautier, N. Thomas, C.E. Henze, M.J. Holman, A. Howard, S.B. Howell, J.M. Jenkins, D.G. Koch, D.W. Latham, J.J. Lissauer, G.W. Marcy, S.N. Quinn, D. Ragozzine, D.D. Sasselov, S. Seager, T. Barclay, F. Mullally, S.E. Seader, M. Still, J.D. Twicken, S.E. Thompson, K. Uddin, Nature **482**(7384), 195 (2012). https://doi.org/10.1038/nature10780
41. P. Marigo, B. Aringer, Astron. Astrophys. **508**, 1539 (2009). https://doi.org/10.1051/0004-6361/200912598
42. J. Budaj, M. Kocifaj, R. Salmeron, I. Hubeny, Mon. Not. R. Astron. Soc. **454**, 2 (2015). https://doi.org/10.1093/mnras/stv1711
43. J.A. Cardelli, G.C. Clayton, J.S. Mathis, Astrophys. J. **345**, 245 (1989). https://doi.org/10.1086/167900
44. J. Budaj, Astron. Astrophys. **557**, A72 (2013). https://doi.org/10.1051/0004-6361/201220260
45. J. DeVore, S. Rappaport, R. Sanchis-Ojeda, K. Hoffman, J. Rowe, Mon. Not. R. Astron. Soc. **461**, 2453 (2016). https://doi.org/10.1093/mnras/stw1439
46. A. García Muñoz, J. Cabrera, Mon. Not. R. Astron. Soc. **473**, 1801 (2018). https://doi.org/10.1093/mnras/stx2428
47. M. Gillon, D.R. Anderson, A. Collier-Cameron, L. Delrez, C. Hellier, E. Jehin, M. Lendl, P.F.L. Maxted, F. Pepe, D. Pollacco, D. Queloz, D. Ségransan, A.M.S. Smith, B. Smalley, J. Southworth, A.H.M.J. Triaud, S. Udry, V. Van Grootel, R.G. West, Astron. Astrophys. **562**, L3 (2014). https://doi.org/10.1051/0004-6361/201323014
48. J. Southworth, L. Mancini, S. Ciceri, J. Budaj, M. Dominik, R. Figuera Jaimes, T. Haugbølle, U.G. Jørgensen, A. Popovas, M. Rabus, S. Rahvar, C. von Essen, R.W. Schmidt, O. Wertz, K.A. Alsubai, V. Bozza, D.M. Bramich, S. Calchi Novati, G. D'Ago, T.C. Hinse, T. Henning, M. Hundertmark, D. Juncher, H. Korhonen, J. Skottfelt, C. Snodgrass, D. Starkey, J. Surdej, Mon. Not. R. Astron. Soc. **447**, 711 (2015). https://doi.org/10.1093/mnras/stu2394
49. C.F. Bohren, D.R. Huffman, *Absorption and Scattering of Light by Small Particles* (Wiley, Weinheim, 1983)
50. T. Henning, V.B. Il'In, N.A. Krivova, B. Michel, N.V. Voshchinnikov, Astron. Astrophys. Suppl. Ser. **136**, 405 (1999). https://doi.org/10.1051/aas:1999222
51. C. Jäger, J. Dorschner, H. Mutschke, T. Posch, T. Henning, Astron. Astrophys. **408**, 193 (2003). https://doi.org/10.1051/0004-6361:20030916
52. H. Kimura, I. Mann, D.A. Biesecker, E.K. Jessberger, Icarus **159**, 529 (2002). https://doi.org/10.1006/icar.2002.6940
53. R. van Lieshout, M. Min, C. Dominik, Astron. Astrophys. **572**, A76 (2014). https://doi.org/10.1051/0004-6361/201424876
54. J. Dorschner, B. Begemann, T. Henning, C. Jaeger, H. Mutschke, Astron. Astrophys. **300**, 503 (1995)
55. M. Asplund, N. Grevesse, A.J. Sauval, P. Scott, Annu. Rev. Astron. Astrophys. **47**, 481 (2009). https://doi.org/10.1146/annurev.astro.46.060407.145222
56. A. Burrows, D. Sudarsky, I. Hubeny, Astrophys. J. **640**, 1063 (2006). https://doi.org/10.1086/500293
57. S. Rappaport, A. Levine, E. Chiang, I. El Mellah, J. Jenkins, B. Kalomeni, E.S. Kite, M. Kotson, L. Nelson, L. Rousseau-Nepton, K. Tran, Astrophys. J. **752**, 1 (2012). https://doi.org/10.1088/0004-637X/752/1/1
58. E. Schlawin, T. Hirano, H. Kawahara, J. Teske, E.M. Green, B.V. Rackham, J. Fraine, R. Bushra, Astron. J. **156**, 281 (2018). https://doi.org/10.3847/1538-3881/aaeb32
59. H. Kawahara, T. Hirano, K. Kurosaki, Y. Ito, M. Ikoma, Astrophys. J. **776**, L6 (2013). https://doi.org/10.1088/2041-8205/776/1/L6
60. B. Croll, S. Rappaport, J. DeVore, R.L. Gilliland, J.R. Crepp, A.W. Howard, K.M. Star, E. Chiang, A.M. Levine, J.M. Jenkins, L. Albert, A.S. Bonomo, J.J. Fortney, H. Isaacson, Astrophys. J. **786**, 100 (2014). https://doi.org/10.1088/0004-637X/786/2/100

61. K. Masuda, T. Hirano, H. Kawahara, B. Sato, Res. Not. Am. Astron. Soc. **2**(1), 50 (2018). https://doi.org/10.3847/2515-5172/aaba12
62. D. Perez-Becker, E. Chiang, Mon. Not. R. Astron. Soc. **433**, 2294 (2013). https://doi.org/10.1093/mnras/stt895
63. E.H.L. Bodman, J.T. Wright, S.J. Desch, C.M. Lisse, Astron. J. **156**, 173 (2018). https://doi.org/10.3847/1538-3881/aadc60
64. A.R. Ridden-Harper, C.U. Keller, M. Min, R. van Lieshout, I.A.G. Snellen, Astron. Astrophys. **618**, A97 (2018). https://doi.org/10.1051/0004-6361/201731947
65. T.I.M. van Werkhoven, M. Brogi, I.A.G. Snellen, C.U. Keller, Astron. Astrophys. **561**, A3 (2014). https://doi.org/10.1051/0004-6361/201322398
66. B. Croll, S. Rappaport, A.M. Levine, Mon. Not. R. Astron. Soc. **449**, 1408 (2015). https://doi.org/10.1093/mnras/stv297
67. E. Schlawin, T. Herter, M. Zhao, J.K. Teske, H. Chen, Astrophys. J. **826**, 156 (2016). https://doi.org/10.3847/0004-637X/826/2/156
68. R. van Lieshout, M. Min, C. Dominik, M. Brogi, T. de Graaff, S. Hekker, M. Kama, C.U. Keller, A. Ridden-Harper, T.I.M. van Werkhoven, Astron. Astrophys. **596**, A32 (2016). https://doi.org/10.1051/0004-6361/201629250
69. S. Rappaport, T. Barclay, J. DeVore, J. Rowe, R. Sanchis-Ojeda, M. Still, Astrophys. J. **784**, 40 (2014). https://doi.org/10.1088/0004-637X/784/1/40
70. M. Brogi, C.U. Keller, M. de Juan Ovelar, M.A. Kenworthy, R.J. de Kok, M. Min, I.A.G. Snellen, Astron. Astrophys. **545**, L5 (2012). https://doi.org/10.1051/0004-6361/201219762
71. J.J. Bochinski, C.A. Haswell, T.R. Marsh, V.S. Dhillon, S.P. Littlefair, Astrophys. J. **800**, L21 (2015). https://doi.org/10.1088/2041-8205/800/2/L21
72. R. Sanchis-Ojeda, S. Rappaport, E. Pallè, L. Delrez, J. DeVore, D. Gandolfi, A. Fukui, I. Ribas, K.G. Stassun, S. Albrecht, F. Dai, E. Gaidos, M. Gillon, T. Hirano, M. Holman, A.W. Howard, H. Isaacson, E. Jehin, M. Kuzuhara, A.W. Mann, G.W. Marcy, P.A. Miles-Páez, P. Montañés-Rodríguez, F. Murgas, N. Narita, G. Nowak, M. Onitsuka, M. Paegert, V. Van Eylen, J.N. Winn, L. Yu, Astrophys. J. **812**, 112 (2015). https://doi.org/10.1088/0004-637X/812/2/112
73. K.D. Colón, G. Zhou, A. Shporer, K.A. Collins, A. Bieryla, N. Espinoza, F. Murgas, P. Pattarakijwanich, S. Awiphan, J.D. Armstrong, J. Bailey, G. Barentsen, D. Bayliss, A. Chakpor, W.D. Cochran, V.S. Dhillon, K. Horne, M. Ireland, L. Kedziora-Chudczer, J.F. Kielkopf, S. Komonjinda, D.W. Latham, T.R. Marsh, D.E. Mkrtichian, E. Pallé, D. Ruffolo, R. Sefako, C.G. Tinney, S. Wannawichian, S. Yuma, Astron. J. **156**, 227 (2018). https://doi.org/10.3847/1538-3881/aae31b
74. E. Gaidos, T. Hirano, M. Ansdell, Mon. Not. R. Astron. Soc. **485**(3), 3876–3886 (2019). https://doi.org/10.1093/mnras/stz693
75. A. Vanderburg, J.A. Johnson, S. Rappaport, A. Bieryla, J. Irwin, J.A. Lewis, D. Kipping, W.R. Brown, P. Dufour, D.R. Ciardi, R. Angus, L. Schaefer, D.W. Latham, D. Charbonneau, C. Beichman, J. Eastman, N. McCrady, R.A. Wittenmyer, J.T. Wright, Nature **526**(7574), 546 (2015). https://doi.org/10.1038/nature15527
76. B. Zuckerman, C. Melis, B. Klein, D. Koester, M. Jura, Astrophys. J. **722**, 725 (2010). https://doi.org/10.1088/0004-637X/722/1/725
77. D. Koester, B.T. Gänsicke, J. Farihi, Astron. Astrophys. **566**, A34 (2014). https://doi.org/10.1051/0004-6361/201423691
78. M. Jura, Astrophys. J. **584**(2), L91 (2003). https://doi.org/10.1086/374036
79. M. Jura, J. Farihi, B. Zuckerman, Astrophys. J. **663**(2), 1285 (2007). https://doi.org/10.1086/518767
80. M. Kilic, T. von Hippel, S.K. Leggett, D.E. Winget, Astrophys. J. **646**(1), 474 (2006). https://doi.org/10.1086/504682
81. B. Croll, P.A. Dalba, A. Vanderburg, J. Eastman, S. Rappaport, J. DeVore, A. Bieryla, P.S. Muirhead, E. Han, D.W. Latham, T.G. Beatty, R.A. Wittenmyer, J.T. Wright, J.A. Johnson, N. McCrady, Astrophys. J. **836**(1), 82 (2017). https://doi.org/10.3847/1538-4357/836/1/82

82. S. Rappaport, B.L. Gary, A. Vanderburg, S. Xu, D. Pooley, K. Mukai, Mon. Not. R. Astron. Soc. **474**, 933 (2018). https://doi.org/10.1093/mnras/stx2663
83. R. Alonso, S. Rappaport, H.J. Deeg, E. Palle, Astron. Astrophys. **589**, L6 (2016). https://doi.org/10.1051/0004-6361/201628511
84. P. Izquierdo, P. Rodríguez-Gil, B.T. Gänsicke, A.J. Mustill, O. Toloza, P.E. Tremblay, M. Wyatt, P. Chote, S. Eggl, J. Farihi, D. Koester, W. Lyra, C.J. Manser, T.R. Marsh, E. Pallé, R. Raddi, D. Veras, E. Villaver, S. Portegies Zwart, Mon. Not. R. Astron. Soc. **481**(1), 703 (2018). https://doi.org/10.1093/mnras/sty2315
85. G. Zhou, L. Kedziora-Chudczer, J. Bailey, J.P. Marshall, D.D.R. Bayliss, C. Stockdale, P. Nelson, T.G. Tan, J.E. Rodriguez, C.G. Tinney, D. Dragomir, K. Colon, A. Shporer, J. Bento, R. Sefako, K. Horne, W. Cochran, Mon. Not. R. Astron. Soc. **463**, 4422 (2016). https://doi.org/10.1093/mnras/stw2286
86. S. Xu, S. Rappaport, R. van Lieshout, A. Vand erburg, B. Gary, N. Hallakoun, V.D. Ivanov, M.C. Wyatt, J. DeVore, D. Bayliss, J. Bento, A. Bieryla, A. Cameron, J.M. Cann, B. Croll, K.A. Collins, P.A. Dalba, J. Debes, D. Doyle, P. Dufour, J. Ely, N. Espinoza, M.D. Joner, M. Jura, T. Kaye, J.L. McClain, P. Muirhead, E. Palle, P.A. Panka, J. Provencal, S. Randall, J.E. Rodriguez, J. Scarborough, R. Sefako, A. Shporer, W. Strickland, G. Zhou, B. Zuckerman, Mon. Not. R. Astron. Soc. **474**(4), 4795 (2018). https://doi.org/10.1093/mnras/stx3023
87. S. Xu, N. Hallakoun, B. Gary, P.A. Dalba, J. Debes, P. Dufour, M. Fortin-Archambault, A. Fukui, M.A. Jura, B. Klein, N. Kusakabe, P.S. Muirhead, N. Narita, A. Steele, K.Y.L. Su, A. Vanderburg, N. Watanabe, Z. Zhan, B. Zuckerman, Astron. J. **157**(6), 255 (2019). https://doi.org/10.3847/1538-3881/ab1b36
88. S. Rappaport, B.L. Gary, T. Kaye, A. Vanderburg, B. Croll, P. Benni, J. Foote, Mon. Not. R. Astron. Soc. **458**, 3904 (2016). https://doi.org/10.1093/mnras/stw612
89. B.T. Gänsicke, A. Aungwerojwit, T.R. Marsh, V.S. Dhillon, D.I. Sahman, D. Veras, J. Farihi, P. Chote, R. Ashley, S. Arjyotha, S. Rattanasoon, S.P. Littlefair, D. Pollacco, M.R. Burleigh, Astrophys. J. **818**, L7 (2016). https://doi.org/10.3847/2041-8205/818/1/L7
90. S. Xu, M. Jura, P. Dufour, B. Zuckerman, Astrophys. J. **816**, L22 (2016). https://doi.org/10.3847/2041-8205/816/2/L22
91. S. Redfield, J. Farihi, P.W. Cauley, S.G. Parsons, B.T. Gänsicke, G.M. Duvvuri, Astrophys. J. **839**, 42 (2017). https://doi.org/10.3847/1538-4357/aa68a0
92. A. Vanderburg, S.A. Rappaport, *Transiting Disintegrating Planetary Debris Around WD 1145+017* (2018), p. 37. https://doi.org/10.1007/978-3-319-55333-7_37
93. Z. Vanderbosch, J.J. Hermes, E. Dennihy, B.H. Dunlap, P. Izquierdo, P.E. Tremblay, P.B. Cho, B.T. Gaensicke, K.J. Bell, M.H. Montgomery, D.E. Winget (2019), e-prints arXiv:1908.09839
94. A. Lecavelier Des Etangs, A. Vidal-Madjar, R. Ferlet, Astron. Astrophys. **343**, 916 (1999)
95. S. Rappaport, A. Vanderburg, T. Jacobs, D. LaCourse, J. Jenkins, A. Kraus, A. Rizzuto, D.W. Latham, A. Bieryla, M. Lazarevic, A. Schmitt, Mon. Not. R. Astron. Soc. **474**, 1453 (2018). https://doi.org/10.1093/mnras/stx2735
96. G.M. Kennedy, G. Hope, S.T. Hodgkin, M.C. Wyatt, Mon. Not. R. Astron. Soc. **482**, 5587 (2019). https://doi.org/10.1093/mnras/sty3049
97. S. Zieba, K. Zwintz, M.A. Kenworthy, G.M. Kennedy, Astron. Astrophys. **625**, L13 (2019). https://doi.org/10.1051/0004-6361/201935552
98. M. Ansdell, E. Gaidos, T.L. Jacobs, A. Mann, C.F. Manara, G.M. Kennedy, A. Vanderburg, M. Kenworthy, T. Hirano, D.M. LaCourse, C. Hedges, A. Frasca, Mon. Not. R. Astron. Soc. **483**, 3579 (2019). https://doi.org/10.1093/mnras/sty3289
99. S. Rappaport, A. Vanderburg, M.H. Kristiansen, M.R. Omohundro, H.M. Schwengeler, I.A. Terentev, F. Dai, K. Masuda, T.L. Jacobs, D. LaCourse, D.W. Latham, A. Bieryla, C.L. Hedges, J. Dittmann, G. Barentsen, W. Cochran, M. Endl, J.M. Jenkins, A. Mann, Mon. Not. R. Astron. Soc. **488**, 2455 (2019). https://doi.org/10.1093/mnras/stz1772

100. S. Rappaport, G. Zhou, A. Vanderburg, A. Mann, M.H. Kristiansen, K. Oláh, T.L. Jacobs, E. Newton, M.R. Omohundro, D. LaCourse, H.M. Schwengeler, I.A. Terentev, D.W. Latham, A. Bieryla, M. Soares-Furtado, L.G. Bouma, M.J. Ireland, J. Irwin, Mon. Not. R. Astron. Soc. **485**, 2681 (2019). https://doi.org/10.1093/mnras/stz537
101. L. Neslušan, J. Budaj, Astron. Astrophys. **600**, A86 (2017). https://doi.org/10.1051/0004-6361/201629344
102. M.J. Martínez González, C. González-Fernández, A. Asensio Ramos, H. Socas-Navarro, C. Westendorp Plaza, T.S. Boyajian, J.T. Wright, A. Collier Cameron, J.I. González Hernández, G. Holgado, G.M. Kennedy, T. Masseron, E. Molinari, J. Saario, S. Simón-Díaz, B. Toledo-Padrón, Mon. Not. R. Astron. Soc. **486**, 236 (2019). https://doi.org/10.1093/mnras/stz850
103. V.V. Makarov, A. Goldin, Astrophys. J. **833**, 78 (2016). https://doi.org/10.3847/1538-4357/833/1/78
104. D.P. Clemens, K. Maheshwari, R. Jagani, J. Montgomery, A.M. El Batal, T.G. Ellis, J.T. Wright, Astrophys. J. **856**, L8 (2018). https://doi.org/10.3847/2041-8213/aab492
105. M. Marengo, A. Hulsebus, S. Willis, Astrophys. J. **814**, L15 (2015). https://doi.org/10.1088/2041-8205/814/1/L15
106. C.M. Lisse, M.L. Sitko, M. Marengo, Astrophys. J. **815**, L27 (2015). https://doi.org/10.1088/2041-8205/815/2/L27
107. M.A. Thompson, P. Scicluna, F. Kemper, J.E. Geach, M.M. Dunham, O. Morata, S. Ertel, P.T.P. Ho, J. Dempsey, I. Coulson, G. Petitpas, L.E. Kristensen, Mon. Not. R. Astron. Soc. **458**, L39 (2016). https://doi.org/10.1093/mnrasl/slw008
108. T.S. Boyajian, R. Alonso, A. Ammerman, D. Armstrong, A. Asensio Ramos, K. Barkaoui, T.G. Beatty, Z. Benkhaldoun, P. Benni, R.O. Bentley, et al., Astrophys. J. **853**, L8 (2018). https://doi.org/10.3847/2041-8213/aaa405
109. I.A. Steele, C.M. Copperwheat, H.E. Jermak, G.M. Kennedy, G.P. Lamb, Mon. Not. R. Astron. Soc. **473**, L26 (2018). https://doi.org/10.1093/mnrasl/slx145
110. E. Bodman, J. Wright, T. Boyajian, T. Ellis, arXiv e-prints (2018)
111. H.J. Deeg, R. Alonso, D. Nespral, T.S. Boyajian, Astron. Astrophys. **610**, L12 (2018). https://doi.org/10.1051/0004-6361/201732453
112. B.E. Schaefer, Astrophys. J. **822**, L34 (2016). https://doi.org/10.3847/2041-8205/822/2/L34
113. M. Hippke, D. Angerhausen, M.B. Lund, J. Pepper, K.G. Stassun, Astrophys. J. **825**, 73 (2016). https://doi.org/10.3847/0004-637X/825/1/73
114. M. Hippke, P. Kroll, F. Matthai, D. Angerhausen, T. Tuvikene, K.G. Stassun, E. Roshchina, T. Vasileva, I. Izmailov, N.N. Samus, E.N. Pastukhova, I. Bryukhanov, M.B. Lund, Astrophys. J. **837**, 85 (2017). https://doi.org/10.3847/1538-4357/aa615d
115. M. Castelaz, T. Barker, J. Am. Assoc. Var. Star Observers **46**, 33 (2018)
116. B.T. Montet, J.D. Simon, Astrophys. J. **830**, L39 (2016). https://doi.org/10.3847/2041-8205/830/2/L39
117. J.R.A. Davenport, K.R. Covey, R.W. Clarke, Z. Laycock, S.W. Fleming, T.S. Boyajian, B.T. Montet, B. Shiao, C.C. Million, D.J. Wilson, M. Olmedo, E.E. Mamajek, D. Olmedo, M. Chávez, E. Bertone, Astrophys. J. **853**, 130 (2018). https://doi.org/10.3847/1538-4357/aaa413
118. H.Y.A. Meng, G. Rieke, F. Dubois, G. Kennedy, M. Marengo, M. Siegel, K. Su, N. Trueba, M. Wyatt, T. Boyajian, C.M. Lisse, L. Logie, S. Rau, S. Vanaverbeke, Astrophys. J. **847**, 131 (2017). https://doi.org/10.3847/1538-4357/aa899c
119. B. Gary, R. Bourne, Res. Not. Am. Astron. Soc. **1**(1), 22 (2017). https://doi.org/10.3847/2515-5172/aa9bdd
120. J.D. Simon, B.J. Shappee, G. Pojmański, B.T. Montet, C.S. Kochanek, J. van Saders, T.W.S. Holoien, A.A. Henden, Astrophys. J. **853**, 77 (2018). https://doi.org/10.3847/1538-4357/aaa0c1
121. M. Hippke, D. Angerhausen, Astrophys. J. **854**, L11 (2018). https://doi.org/10.3847/2041-8213/aaab44

122. M.C. Wyatt, R. van Lieshout, G.M. Kennedy, T.S. Boyajian, Mon. Not. R. Astron. Soc. **473**, 5286 (2018). https://doi.org/10.1093/mnras/stx2713
123. B.E. Schaefer, R.O. Bentley, T.S. Boyajian, P.H. Coker, S. Dvorak, F. Dubois, E. Erdelyi, T. Ellis, K. Graham, B.G. Harris, J.E. Hall, R. James, S.J. Johnston, G. Kennedy, L. Logie, K.M. Nugent, A. Oksanen, J.J. Ott, S. Rau, S. Vanaverbeke, R. van Lieshout, M. Wyatt, Mon. Not. R. Astron. Soc. **481**, 2235 (2018). https://doi.org/10.1093/mnras/sty1644
124. J.T. Wright, S. Sigurdsson, Astrophys. J. **829**, L3 (2016). https://doi.org/10.3847/2041-8205/829/1/L3
125. K.M. Punzi, J.H. Kastner, C. Melis, B. Zuckerman, C. Pilachowski, L. Gingerich, T. Knapp, Astron. J. **155**, 33 (2018). https://doi.org/10.3847/1538-3881/aa9524
126. P. Foukal, Astrophys. J. **842**, L3 (2017). https://doi.org/10.3847/2041-8213/aa740f
127. J.I. Katz, Mon. Not. R. Astron. Soc. **471**, 3680 (2017). https://doi.org/10.1093/mnras/stx1876
128. B.D. Metzger, K.J. Shen, N. Stone, Mon. Not. R. Astron. Soc. **468**, 4399 (2017). https://doi.org/10.1093/mnras/stx823
129. E.H.L. Bodman, A. Quillen, Astrophys. J. **819**, L34 (2016). https://doi.org/10.3847/2041-8205/819/2/L34
130. R. Bourne, B.L. Gary, A. Plakhov, Mon. Not. R. Astron. Soc. **475**, 5378 (2018). https://doi.org/10.1093/mnras/sty097
131. F. Kiefer, A. Lecavelier des Étangs, A. Vidal-Madjar, G. Hébrard, V. Bourrier, P.A. Wilson, Astron. Astrophys. **608**, A132 (2017). https://doi.org/10.1051/0004-6361/201731306
132. G. Sacco, L.D. Ngo, J. Modolo, J. Am. Assoc. Var. Star Observers **46**, 14 (2018)
133. G.R. Ricker, J.N. Winn, R. Vanderspek, D.W. Latham, G.Á. Bakos, J.L. Bean, Z.K. Berta-Thompson, T.M. Brown, L. Buchhave, N.R. Butler, R.P. Butler, W.J. Chaplin, D. Charbonneau, J. Christensen-Dalsgaard, M. Clampin, D. Deming, J. Doty, N. De Lee, C. Dressing, E.W. Dunham, M. Endl, F. Fressin, J. Ge, T. Henning, M.J. Holman, A.W. Howard, S. Ida, J. Jenkins, G. Jernigan, J.A. Johnson, L. Kaltenegger, N. Kawai, H. Kjeldsen, G. Laughlin, A.M. Levine, D. Lin, J.J. Lissauer, P. MacQueen, G. Marcy, P.R. McCullough, T.D. Morton, N. Narita, M. Paegert, E. Palle, F. Pepe, J. Pepper, A. Quirrenbach, S.A. Rinehart, D. Sasselov, B. Sato, S. Seager, A. Sozzetti, K.G. Stassun, P. Sullivan, A. Szentgyorgyi, G. Torres, S. Udry, J. Villasenor, in *Society of Photo-Optical Instrumentation Engineers (SPIE) Conference Series*, vol. 9143 (2014), p. 20. https://doi.org/10.1117/12.2063489
134. G.R. Ricker, J.N. Winn, R. Vanderspek, D.W. Latham, G.Á. Bakos, J.L. Bean, Z.K. Berta-Thompson, T.M. Brown, L. Buchhave, N.R. Butler, R.P. Butler, W.J. Chaplin, D. Charbonneau, J. Christensen-Dalsgaard, M. Clampin, D. Deming, J. Doty, N. De Lee, C. Dressing, E.W. Dunham, M. Endl, F. Fressin, J. Ge, T. Henning, M.J. Holman, A.W. Howard, S. Ida, J.M. Jenkins, G. Jernigan, J.A. Johnson, L. Kaltenegger, N. Kawai, H. Kjeldsen, G. Laughlin, A.M. Levine, D. Lin, J.J. Lissauer, P. MacQueen, G. Marcy, P.R. McCullough, T.D. Morton, N. Narita, M. Paegert, E. Palle, F. Pepe, J. Pepper, A. Quirrenbach, S.A. Rinehart, D. Sasselov, B. Sato, S. Seager, A. Sozzetti, K.G. Stassun, P. Sullivan, A. Szentgyorgyi, G. Torres, S. Udry, J. Villasenor, J. Astron. Telescopes Instrum. Syst. **1**(1), 014003 (2015). https://doi.org/10.1117/1.JATIS.1.1.014003
135. D. Gandolfi, O. Barragán, J.H. Livingston, M. Fridlund, A.B. Justesen, S. Redfield, L. Fossati, S. Mathur, S. Grziwa, J. Cabrera, R.A. García, C.M. Persson, V. Van Eylen, A.P. Hatzes, D. Hidalgo, S. Albrecht, L. Bugnet, W.D. Cochran, S. Csizmadia, H. Deeg, P. Eigmüller, M. Endl, A. Erikson, M. Esposito, E. Guenther, J. Korth, R. Luque, P. Montañes Rodríguez, D. Nespral, G. Nowak, M. Pätzold, J. Prieto-Arranz, Astron. Astrophys. **619**, L10 (2018). https://doi.org/10.1051/0004-6361/201834289
136. C.X. Huang, J. Burt, A. Vanderburg, M.N. Günther, A. Shporer, J.A. Dittmann, J.N. Winn, R. Wittenmyer, L. Sha, S.R. Kane, G.R. Ricker, R.K. Vand erspek, D.W. Latham, S. Seager, J.M. Jenkins, D.A. Caldwell, K.A. Collins, N. Guerrero, J.C. Smith, S.N. Quinn, S. Udry, F. Pepe, F. Bouchy, D. Ségransan, C. Lovis, D. Ehrenreich, M. Marmier, M. Mayor, B. Wohler, K. Haworth, E.H. Morgan, M. Fausnaugh, D.R. Ciardi, J. Christiansen, D. Charbonneau, D. Dragomir, D. Deming, A. Glidden, A.M. Levine, P.R. McCullough, L. Yu, N. Narita, T. Nguyen, T. Morton, J. Pepper, A. Pál, J.E. Rodriguez, K.G. Stassun,

G. Torres, A.R. Sozzetti, J.P. Doty, J. Christensen-Dalsgaard, G. Laughlin, M. Clampin, J.L. Bean, L.A. Buchhave, G.Á. Bakos, B. Sato, S. Ida, L. Kaltenegger, E. Palle, D. Sasselov, R.P. Butler, J. Lissauer, J. Ge, S.A. Rinehart, Astrophys. J. **868**(2), L39 (2018). https://doi.org/10.3847/2041-8213/aaef91

137. J. Šubjak, R. Sharma, T.W. Carmichael, M.C. Johnson, E.J. Gonzales, E. Matthews, H.M.J. Boffin, R. Brahm, P. Chaturvedi, A. Chakraborty, D.R. Ciardi, K.A. Collins, M. Esposito, M. Fridlund, T. Gan, D. Gandolfi, R.A. García, E. Guenther, A. Hatzes, D.W. Latham, C.M. Persson, H.M. Relles, J.E. Schlieder, T. Barclay, C. Dressing, I. Crossfield, A.W. Howard, F. Rodler, G. Zhou, S.N. Quinn, G.A. Esquerdo, M.L. Calkins, P. Berlind, K.G. Stassun, S. Albrecht, R. Alonso Sobrino, P. Beck, M. Blažek, J. Cabrera, I. Carleo, W.D. Cochran, S. Csizmadia, F. Dai, H.J. Deeg, J.P. de Leon, P. Eigmüller, M. Endl, A. Erikson, A. Fukui, I. Georgieva, L. González-Cuesta, S. Grziwa, D. Hidalgo, T. Hirano, M. Hjorth, E. Knudstrup, J. Korth, K.W.F. Lam, J.H. Livingston, M.N. Lund, R. Luque, S. Mathur, P. Montanes Rodríguez, F. Murgas, N. Narita, D. Nespral, P. Niraula, G. Nowak, E. Pallé, M. Pätzold, J. Prieto-Arranz, H. Rauer, S. Redfield, I. Ribas, M. Skarka, A.M.S. Smith, M. Špoková, V. Van Eylen, P. Kabáth, Astron. J. **159**(4) (2020). https://doi.org/10.3847/1538-3881/ab7245
138. P. Kabáth, J. Žák, H.M.J. Boffin, V.D. Ivanov, D. Jones, M. Skarka, Publ. Astron. Soc. Pac. **131**(1002), 085001 (2019). https://doi.org/10.1088/1538-3873/ab2143
139. H. Rauer, C. Catala, C. Aerts, T. Appourchaux, W. Benz, A. Brandeker, J. Christensen-Dalsgaard, M. Deleuil, L. Gizon, M.J. Goupil, M. Güdel, E. Janot-Pacheco, M. Mas-Hesse, I. Pagano, G. Piotto, D. Pollacco, Ċ. Santos, A. Smith, J.C. Suárez, R. Szabó, S. Udry, V. Adibekyan, Y. Alibert, J.M. Almenara, P. Amaro-Seoane, M.A.V. Eiff, M. Asplund, E. Antonello, S. Barnes, F. Baudin, K. Belkacem, M. Bergemann, G. Bihain, A.C. Birch, X. Bonfils, I. Boisse, A.S. Bonomo, F. Borsa, I.M. Brandão, E. Brocato, S. Brun, M. Burleigh, R. Burston, J. Cabrera, S. Cassisi, W. Chaplin, S. Charpinet, C. Chiappini, R.P. Church, S. Csizmadia, M. Cunha, M. Damasso, M.B. Davies, H.J. Deeg, R.F. Díaz, S. Dreizler, C. Dreyer, P. Eggenberger, D. Ehrenreich, P. Eigmüller, A. Erikson, R. Farmer, S. Feltzing, F. de Oliveira Fialho, P. Figueira, T. Forveille, M. Fridlund, R.A. García, P. Giommi, G. Giuffrida, M. Godolt, J. Gomes da Silva, T. Granzer, J.L. Grenfell, A. Grotsch-Noels, E. Günther, C.A. Haswell, A.P. Hatzes, G. Hébrard, S. Hekker, R. Helled, K. Heng, J.M. Jenkins, A. Johansen, M.L. Khodachenko, K.G. Kislyakova, W. Kley, U. Kolb, N. Krivova, F. Kupka, H. Lammer, A.F. Lanza, Y. Lebreton, D. Magrin, P. Marcos-Arenal, P.M. Marrese, J.P. Marques, J. Martins, S. Mathis, S. Mathur, S. Messina, A. Miglio, J. Montalban, M. Montalto, M.J.P.F.G. Monteiro, H. Moradi, E. Moravveji, C. Mordasini, T. Morel, A. Mortier, V. Nascimbeni, R.P. Nelson, M.B. Nielsen, L. Noack, A.J. Norton, A. Ofir, M. Oshagh, R.M. Ouazzani, P. Pápics, V.C. Parro, P. Petit, B. Plez, E. Poretti, A. Quirrenbach, R. Ragazzoni, G. Raimondo, M. Rainer, D.R. Reese, R. Redmer, S. Reffert, B. Rojas-Ayala, I.W. Roxburgh, S. Salmon, A. Santerne, J. Schneider, J. Schou, S. Schuh, H. Schunker, A. Silva-Valio, R. Silvotti, I. Skillen, I. Snellen, F. Sohl, S.G. Sousa, A. Sozzetti, D. Stello, K.G. Strassmeier, M. Švanda, G.M. Szabó, A. Tkachenko, D. Valencia, V. Van Grootel, S.D. Vauclair, P. Ventura, F.W. Wagner, N.A. Walton, J. Weingrill, S.C. Werner, P.J. Wheatley, K. Zwintz, Exp. Astron. **38**, 249 (2014). https://doi.org/10.1007/s10686-014-9383-4
140. M. Goupil, in *European Physical Journal Web of Conferences*, vol. 160 (2017), p. 01003. https://doi.org/10.1051/epjconf/201716001003
141. G. Tinetti, J.P. Beaulieu, T. Henning, M. Meyer, G. Micela, I. Ribas, D. Stam, M. Swain, O. Krause, M. Ollivier, E. Pace, B. Swinyard, A. Aylward, R. van Boekel, A. Coradini, T. Encrenaz, I. Snellen, M.R. Zapatero-Osorio, J. Bouwman, J.Y.K. Cho, V. Coudé de Foresto, T. Guillot, M. Lopez-Morales, I. Mueller-Wodarg, E. Palle, F. Selsis, A. Sozzetti, P.A.R. Ade, N. Achilleos, A. Adriani, C.B. Agnor, C. Afonso, C. Allende Prieto, G. Bakos, R.J. Barber, M. Barlow, V. Batista, P. Bernath, B. Bézard, P. Bordé, L.R. Brown, A. Cassan, C. Cavarroc, A. Ciaravella, C. Cockell, A. Coustenis, C. Danielski, L. Decin, R. De Kok, O. Demangeon, P. Deroo, P. Doel, P. Drossart, L.N. Fletcher, M. Focardi, F. Forget, S. Fossey, P. Fouqué, J. Frith, M. Galand, P. Gaulme, J.I.G. Hernández, O. Grasset, D. Grassi, J.L. Grenfell, M.J. Griffin, C.A. Griffith, U. Grözinger, M. Guedel, P. Guio, O. Hainaut, R. Hargreaves, P.H.

Hauschildt, K. Heng, D. Heyrovsky, R. Hueso, P. Irwin, L. Kaltenegger, P. Kervella, D. Kipping, T.T. Koskinen, G. Kovács, A. La Barbera, H. Lammer, E. Lellouch, G. Leto, M. Lopez Morales, M.A. Lopez Valverde, M. Lopez-Puertas, C. Lovis, A. Maggio, J.P. Maillard, J. Maldonado Prado, J.B. Marquette, F.J. Martin-Torres, P. Maxted, S. Miller, S. Molinari, D. Montes, A. Moro-Martin, J.I. Moses, O. Mousis, N. Nguyen Tuong, R. Nelson, G.S. Orton, E. Pantin, E. Pascale, S. Pezzuto, D. Pinfield, E. Poretti, R. Prinja, L. Prisinzano, J.M. Rees, A. Reiners, B. Samuel, A. Sánchez-Lavega, J.S. Forcada, D. Sasselov, G. Savini, B. Sicardy, A. Smith, L. Stixrude, G. Strazzulla, J. Tennyson, M. Tessenyi, G. Vasisht, S. Vinatier, S. Viti, I. Waldmann, G.J. White, T. Widemann, R. Wordsworth, R. Yelle, Y. Yung, S.N. Yurchenko, Exp. Astron. **34**, 311 (2012). https://doi.org/10.1007/s10686-012-9303-4

# Physical Conditions and Chemical Abundances in Photoionized Nebulae from Optical Spectra

**Jorge García-Rojas**

## Contents



**Abstract** This chapter presents a review on the latest advances in the computation of physical conditions and chemical abundances of elements present in photoionized gas (H II regions and planetary nebulae). The arrival of highly sensitive spectrographs attached to large telescopes and the development of more sophisticated and detailed atomic data calculations and ionization correction factors have helped to raise the number of ionic species studied in photoionized nebulae in the last years, as well as to reduce the uncertainties in the computed abundances. Special attention will be given to the detection of very faint lines such as heavy-element

J. García-Rojas (✉)
Instituto de Astrofísica de Canarias, La Laguna, Tenerife, Spain

Universidad de La Laguna, Depart. de Astrofísica, La Laguna, Tenerife, Spain
e-mail: jogarcia@iac.es



P. Kabáth et al. (eds.), *Reviews in Frontiers of Modern Astrophysics*,
https://doi.org/10.1007/978-3-030-38509-5_4

recombination lines of C, N and O in H II regions and planetary nebulae, and collisionally excited lines of neutron-capture elements (Z>30) in planetary nebulae.

## 1 A Very Brief Introduction on Emission Line Spectra of Photoionized Nebulae

Photoionized nebulae (i. e. H II regions and planetary nebulae) are among the most "photogenic" objects in the sky. Given their relatively high surface-brightness they are easily accessible, even for non-professional telescopes. This allowed earliest visual spectroscopic observations by William Huggins and William A. Miller [1] who obtained the first spectrum of a planetary nebula (The Cat's Eye Nebula), where they detected a bright emission line coming from a mysterious element that Margaret L. Huggins [2] called "nebulium". Several decades later, Ira S. Bowen [3] showed that this emission was produced by doubly ionized oxygen ($O^{2+}$) at extremely low densities. An historical review on the early steps of the study of the physics of gaseous nebulae was provided by Donald E. Osterbrock [4] who used a seminal paper by Bowen [5] as the starting point for a review of nebular astrophysics.

Photoionized nebulae are excited by the strong ultraviolet (UV) radiation of hot stars ($T_{\rm eff} \geq$ 25–30 kK) which produce photons with energy that could be above the ionization threshold of the gas particles and hence, ionize them releasing a free electron. The probability of occurrence of this phenomenon depends on the photoionization cross-section which, in turn, depends on the energy of the photon and the target being considered. Once ionized, the gas particles tend to recombine with the free electrons, and eventually an equilibrium stage is established in which the rate of ionization equals the rate of recombination for each species (see [6]).

The optical spectra of photoionized nebulae are dominated by emission lines, which are formed when atoms or ions make a transition from one bound electronic state to another bound state at a lower energy via spontaneous emission. These bound electrons can be excited either by free electrons colliding with the atom/ion, or by absorption of a photon. However, the background radiation field in the interstellar medium in generally not strong enough for excitation by photon absorption to be significant (see chapter 5 of [7]) and therefore, the only way of having a bound electron in an excited state is by collisional excitation from a lower state, which subsequent radiative decays to lower levels originating the collisionally excited lines (hereinafter CELs), or owing to a recombination between a free electron and an ion, which is the mechanism behind the emission of recombination lines (hereinafter RLs). Given that the abundance of H and He ions are several orders of magnitude higher than that of heavier elements, one can instinctively assume that the emission spectra will be dominated by H and He lines, which is not the case. In photoionized nebulae the peak of the energy distribution of free electrons is of the order of 1 eV. Ions of heavy atoms like N, O, Ne, S, Cl, Ar, etc. have electronic structures with low-lying electronic states in the range of fractions to few eV from the ground state

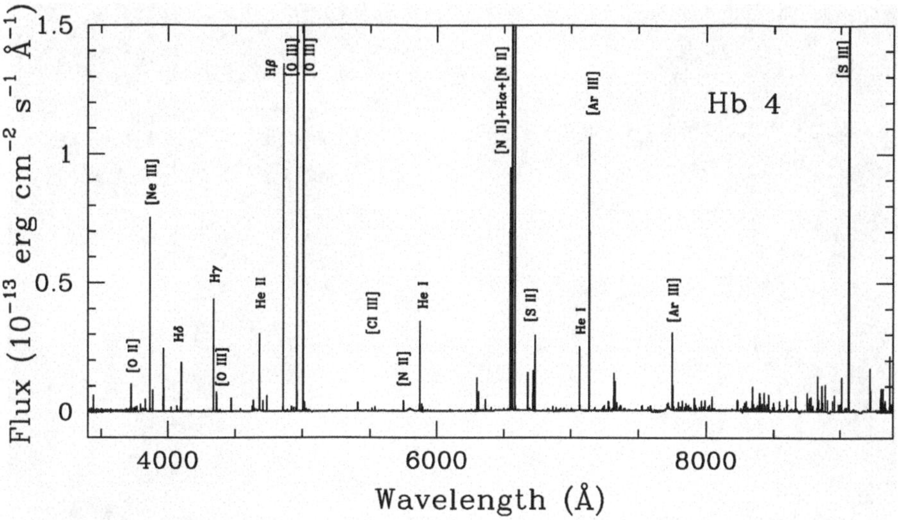


**Fig. 1** Section of the very deep spectrum of the PN Hb 4 analysed in [8, 9] showing bright RLs of H I and He I and CELs of different ions of O, N, S and Ne

and can therefore be effectively excited by collisions. On the other hand, for H and He ions, the gap between the ground state and the first excited state is very large and cannot be excited by collisions, but by recombination. Figure 1 shows a typical optical spectrum of a photoionized nebulae (in this case the planetary nebula Hb 4); remarkably bright H and He RLs and CELs of different ionic species of N, O, Ne, S, Cl and Ar are labelled.

Therefore, the spectrum of a photoionized nebula is dominated by the emission of RLs of H and He (the most abundant elements) and CELs of heavier elements. The combination of narrow-band images taken in the brightest emission lines allows to construct the beautiful coloured images of photoionized nebulae (see Fig. 2) from which we can have a first sketch of the ionization structure of the photoionized region. In Fig. 2 we show a three narrow-band filter combined image of a star-forming region in the Large Magellanic Cloud where is clear that the emission of [O III] is more internally located than the emission from [S II].

Although the emission line spectra from H II regions and planetary nebulae (hereinafter, PNe) are roughly similar, there are some remarkable differences between them. H II regions are large (tens of parsecs), massive (generally between $10^2 - 10^3$ $M_{\odot}$) regions of gas that are ionized by the ultraviolet (UV) radiation emitted by recently formed OB-type massive stars with typical effective temperatures between 25–50 kK; in general, these stars are not hot enough to ionize nebular He II, whose ionization potential is $h\nu = 54.4$ eV. However, there are exceptions to this rule, especially in the integrated spectra of blue compact dwarf galaxies (BCDs), Wolf-Rayet (WR) galaxies and a couple of nebulae in the Local Group, associated to WR stars. On the other hand, PNe are much smaller ($10^{-1}$ pc) and less massive ($\sim 10^{-1}$ $M_{\odot}$) nebulae that are excited by central stars which are generally hotter (central

**Fig. 2** "Bubbles of Brand New Stars" Composite image of a star-forming region in the Large Magellanic Cloud (LMC) captured by the Multi Unit Spectroscopic Explorer (MUSE) instrument on ESO's Very Large Telescope (VLT). The following colour code was used: [O III] λ5007 (blue). Hα (yellow), [S II] λ6731 (red). The field-of-view of the image is 7.82 × 8.00 arcminutes$^2$. North is 180.2° left of vertical. Credit: ESO, A. McLeod et al.

stars can reach temperatures as high as 250 kK); therefore, there will be ionizing photons with enough energy to ionize high-excitation species and hence, producing qualitatively different spectra than that of H II regions, showing emission lines of He II, [Ne V], [Ar V], [Fe V], and even more excited species.

## 1.1 Why Are Abundances in Photoionized Nebulae Important in Astrophysics?

The analysis of emission line spectra of photoionized nebulae allows us to determine the chemical composition of the interstellar medium (ISM) from the solar neighbourhood to the high redshift star-forming galaxies. It stands as an essential tool for our knowledge of stellar nucleosynthesis and the cosmic chemical evolution. Since the early achievements in spectrophotometry of photoionized nebulae, the quality of deep optical and near-infrared spectrophotometric data of PNe has increased significantly mainly thanks to both the development of more efficient instruments and to the advent of large aperture (8–10 m-class) ground-based telescopes. In this sense, the future installation of giant-class ones (diameters 30–50 m) opens new horizons in the field of nebular spectroscopy. The detection of very faint emission lines in ionized nebulae as auroral CELs in faint, distant or high-metallicity objects; optical recombination lines (hereinafter, ORLs) of heavy-element ions or CELs of trans-iron neutron-capture elements are becoming a routine fact and provide new information of paramount interest in many different areas of astrophysics.

H II regions can be observed at considerable distances in the Universe and hence, are crucial to determine the chemical composition of the interestellar medium (ISM) in the extragalactic domain. Since H II regions lie where star formation is occurring, chemical abundances computed in H II regions are probes to trace the present-day chemical composition of the ISM. In particular, the study of radial variations of chemical abundances along galactic discs in spiral galaxies are essential observational constraints for chemical evolution models, and precise determinations of chemical abundances in low-metallicity dwarf galaxies, can permit to determine the primordial abundance of helium owing to Big Bang nucleosynthesis (see [10] and references therein). The global picture of abundances in PNe is more complicated because for elements that are supposed to be not modified, such as O and $\alpha$-elements, the computed abundances reflect the chemical conditions in the cloud where the progenitor star was formed, while the chemical abundances of N, C, or neutron-capture elements, that could be modified during the cycle of life of low-to-intermediate mass stars allow us to constrain the nucleosynthetic processes in these stars.

## 1.2 Recent Reviews on Chemical Abundance Determinations

Recently, two tutorials focused on the determination of ionized gaseous nebulae abundances have been released [10, 11] although with different points of view. In the former, [10] give a brief review on the physics basics of abundance determinations, like local ionization and local thermal equilibrium, emission line mechanisms and on the calculation of physical conditions and ionic and elemental abundance determinations from observations; these authors also review recent results in abundance determinations in both H II regions and PNe. However the review is quite focused to the abundance discrepancy problem from the point of view of temperature fluctuations (see Sect. 4.5). In [11], the focus is on the determination of abundances in extragalactic H II regions from the direct method (when electron temperature, $T_e$, and electron density, $n_e$, diagnostic lines are available) and in the use of some strong-line methods calibrated using the direct method (see Sect. 3.2).

Further comprehensive tutorials on abundance determinations are those by Stasińska [12] and [13] where the theoretical background of photoionized nebulae is treated in more detail, and particular emphasis is given to the description of line formation mechanisms, transfer of radiation, as well as to the use of empirical diagnostics based on emission lines and determination of chemical abundances using photoionization models. It is not the scope of this chapter to repeat the basic concepts of the physics of photoionized nebulae, which have been described in different detail in the aforementioned tutorials. Moreover, for a much more detailed description of such processes, we refer the reader to the canonical book of photoionized nebulae: "Astrophysics of gaseous nebulae and active Galactic nuclei" [6].

In the following sections I will focus on recent advances in chemical abundances determinations in photoionized nebulae from the analysis of deep optical and near-infrared spectra, from an observational point of view. Due to space limitations, I refer the reader to [12, 13] for an overview on abundance determinations based on photoionization model fitting. Similarly, the strong line methods to determine abundances in the extragalactic domain (from giant H II regions to high-redshift galaxies) will be only briefly discussed in Sect. 3.2.

## 2 Observational Spectroscopic Data: The First Step to Obtain Reliable Abundances

In the last years, the number of deep high-quality spectra of photoionized nebulae has increased significantly, allowing the detection of very faint emission lines (see e.g. [8, 14–23]) and the computation of, in principle, very reliable chemical abundances. The advantages of obtaining deep and high resolution spectra of photoionized regions are clear because one can easily isolate faint lines that in lower resolution spectra would be blended and go unnoticed. As an illustration, in Fig. 3 we show an excerpt of the spectrum of the high-excitation PN H 1–50 analysed in [24] with the same spectrum downgraded to a lower resolution overplotted in red. Several permitted lines of O, N, and C would have remained hidden in the low-resolution spectra and *ad-hoc* atomic physics would have been needed to estimate their fluxes. In the last years, several groups have provided a large sample of deep, high-resolution spectra of both Galactic and extragalactic H II regions and PNe (see e.g. [24–26] and the compilation made by McNabb [27]).

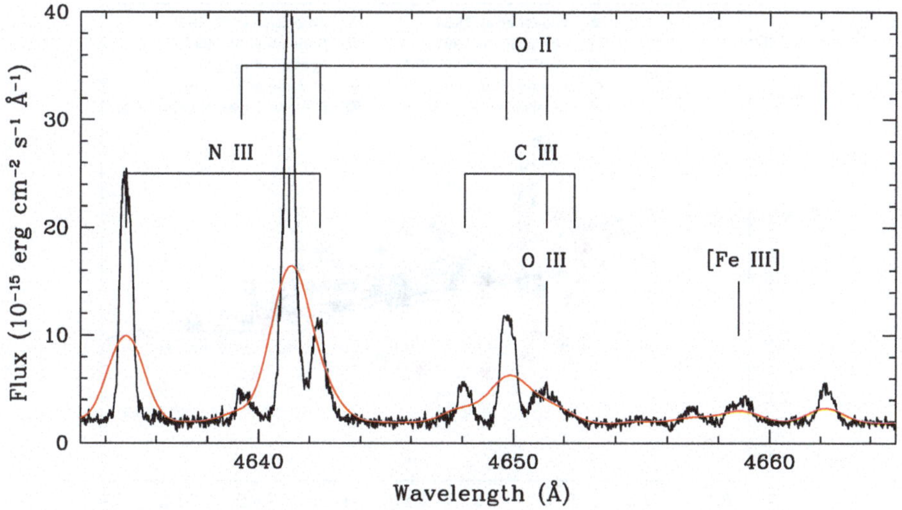


**Fig. 3** Portion of a high-resolution (R∼15000) spectra of the high-excitation PN H 1–50 showing the zone where the multiplet 1 O II lines lie. Overplotted in red is the same spectra degraded to a resolution of R∼3500. As it can be shown the high-resolution of the original spectra allows one to deblend several very close permitted emission lines of C, O, and N that would have remained hidden in the low-resolution spectra. Data originally published in [24]

However, deep, high signal-to-noise, high-resolution spectra are not the panacea. Rodríguez [28] has recently shown that the effects of observational uncertainties can be very important even making use of high quality spectra, owing to the high number of sources of uncertainty that are acting in the process, which include: assumptions in the nebular structure, atomic data (see Sect. 4.1), atmospheric differential refraction, telluric absorption and emission, flux calibration, extinction correction, blends with unknown lines, etc. Therefore, a careful data reduction procedure should be carried out to obtain reliable results. Additionally, an homogeneous analysis determining physical conditions and chemical abundances from the same set of spectra is mandatory if one want to compute precise abundances. For instance, many studies devoted to study the radial abundance gradients have made use of physical conditions derived from radio recombination lines combined with and optical or infrared lines to compute abundances; these approach has been used in several seminal papers on the Galactic abundance gradient (see Sect. 4.4), however, it can introduce systematic uncertainties owing to the different areas of the nebula covered in the different wavelength ranges. Additionally, we should use a set of appropriate lines to compute the abundances; as an example, computing $O^+/H^+$ ratios from the trans-auroral [O II] λλ7320+30 lines could introduce undesired uncertainties because these lines could be strongly affected by telluric emission, and are also very sensitive to electron density and temperature. To illustrate these effects, in Fig. 4 we show an adaptation of Fig. 5 of [29] where the radial oxygen abundance gradient making a consistent analysis of several data sets is presented.

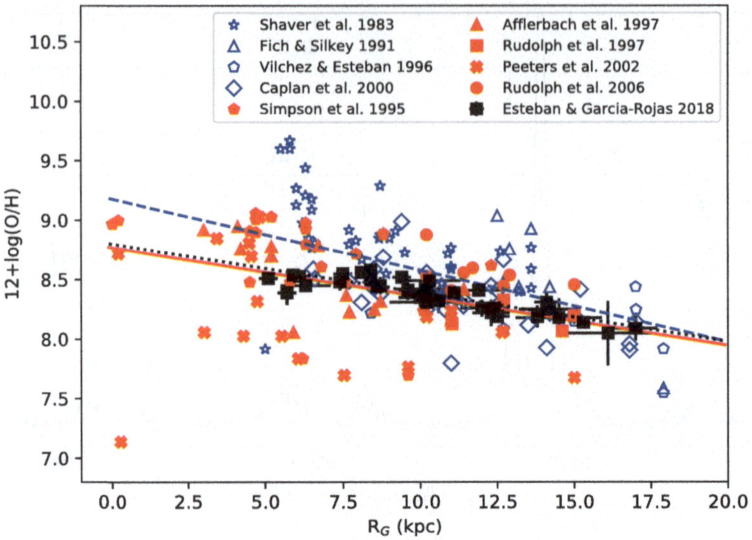


**Fig. 4** Radial oxygen abundance gradient in the MW using abundances derived from optical lines (blue points) or from far-IR lines (red points) compiled by Rudolph et al. [29]. The data from [32] obtained through deep spectra taken in 8–10 m class telescopes are overplotted (black squares). The fits for each set of data are represented by lines with the same colour than the data points. References of the original data compiled by Rudolph et al. [29] are shown in the legend

Physical conditions have been derived from both radio and optical diagnostics, and abundances have been derived using optical CELs of oxygen (blue points) or far-IR fine-structure CELs of oxygen (red points). As can be seen, both data sets show significant scatter in the oxygen abundance at a given Galactocentric distance, which can be interpreted as an "intrinsic scatter" owing to the gas not being well mixed [29–31]. However, high-quality observations seem to rule out this interpretation. In Fig. 4, the abundances computed by Esteban and García-Rojas [32] from an homogeneous analysis of optical spectrophotometric data of 35 H ii regions with direct determinations of the electron temperature have been overplotted on the [29] sample. As it is clearly shown, the scatter in the oxygen abundance is reduced significantly and is not substantially larger than the observational uncertainties, indicating that oxygen seems to be well mixed in the ISM at a given distance along the Galactic disc. Moreover, [33] showed from the analysis of high-quality spectra with high signal-to-noise auroral [O iii] line detections in H ii regions in the inner parts of M 33, a much lower scatter than that found by Rosolowsky and Simon [31]; this author also found no evidence for significant azimuthal variations in the H ii region metallicity distributions, ruling out large anomalies in the mixing of the gas.

Finally, one has to take into account some biases that the direct method can have. Stasińska [34] discussed about the limitations of the direct method to determine O abundances in giant H ii regions at metallicities larger than solar. This author used *ab-initio* photoionization models of giant H ii regions, and applied to the models the same methods as used for real objects to test the direct method. The global result of this study was that for log(O/H)+12 larger than 8.7 (i.e. larger than the solar value), the computed O/H values were below the ones implied by the photoionization models owing to strong temperature gradients present in giant H II regions. Finally, [34] propose that PNe, which are not affected by these biases, could be potential probes of the metallicity of the interstellar medium in the internal parts of spiral galaxies as well as in metal-rich elliptical galaxies. However, in Sect. 4.3 we will discuss that this idea should be taken with some caution.

# 3 Determination of Physical Conditions and Ionic Abundances

## 3.1 *The Direct Method*

The most popular way to compute the chemical abundances of the elements that are present in a photoionized gas is the so-called direct method. This method makes use of CELs intensities of different ionic species of elements like N, O, S, Ne, Cl, Ar, Ne, Fe, etc., and involves the determination of the physical conditions (temperature and electron density) in the emitting plasma. In the conditions prevailing in photoionized nebulae like H II regions and PNe, most of the observed emission lines are optically thin[1] with the exception of some resonance UV lines and some fine-structure IR lines (see [12]) making their use for abundance determinations very robust.

In the analysis of photoionized nebulae it is usually assumed that the physical conditions are homogeneous in the photoionized region. Under these assumptions one can compute electron temperature and density by using sensitive line ratios. Electron temperature ($T_e$) and density ($n_e$) in nebulae are represented by the kinetic energy and density of the free electrons in the photoionized gas. Some CEL intensity ratios of ions of common elements like O, N, S or Ar depend on the physical conditions of the gas, and are useful to calculate $T_e$ and $n_e$ (see Section 3.5 of [10] for more details). In particular, the intensity ratios of emission lines of a given ion that originate in very different energy levels, are sensitive to $T_e$ and almost independent on $n_e$, since the populations of the different atomic levels are strongly dependent on the kinetic energy of the colliding free electrons. Typical optical electron temperature diagnostics are: [N II] λ5754/λ6583, [O II] λλ7320+30/λλ3726+29, [O III] λ4363/λ5007, [Ar III] λ5191/λ7531 or [S III] λ6312/λ9531. Therefore, determination of chemical abundances making use of the direct method in optical spectra requires the detection of faint auroral lines, which correspond to transitions from the state $^1$S to $^1$D and are very $T_e$-sensitive. The

[1] An emission line is said to be optically thick if on average a photon emitted cannot pass through the ISM without absorption. Conversely, an emission line is said to be optically thin if we can see the radiation coming from behind the nebula (i.e. it is not absorbed).

detection of such lines is a relatively easy task in Galactic H II regions and PNe. However, their emissivity decreases rapidly with metallicity and with decreasing surface brightness of the objects so, detecting them is a challenging task in the extragalactic domain, especially in objects beyond the Local Group. However, the combination of high-sensitivity spectrographs with large aperture (10 m type) telescopes have allowed the detection of the auroral [O III] λ4363 Å line in at least 18 star-forming galaxies at $z > 1$ (see [35] and references therein).

On the other hand, line ratios sensitive to $n_e$ come from levels with very similar energy, so that the ratio of their populations does not depend on $T_e$. These levels show different transition probabilities or different collisionally de-excitation rates, such that the ratio between the emission lines generated is strongly dependent on the electron density of the photoionized gas. Typical optical density diagnostics are: [O II] λ3726/λ3729, [S II] λ6716/λ6731, [Cl III] λ5517/λ5537 and [Ar IV λ4711/λ4741.

A precise determination of the physical conditions is crucial to derive reliable abundances from CELs. As the abundances are computed relative to H by using the relative intensities of CELs or ORLs relative to a H I ORL (usually H$\beta$), and given the very different dependence of the emissivity of CELs and ORLs (see [10]) the abundances from CELs show a strong (exponential) dependence on $T_e$, while abundances computed from faint metallic ORLs are almost $T_e$ dependent. This has important implications for the so-called abundance discrepancy problem (see Sect. 4.5).

Once physical conditions are computed one has to decide the temperature and density structure that is going to be assumed in the nebula. The most common approach, is to assume a two-zone scheme, where the high ionization zone is characterized by $T_e$(high) (usually $T_e$([O III]), the low-ionization zone is characterized by $T_e$(low) (usually $T_e$([N II]) and the density is considered homogeneous in the whole nebula and is characterized by $n_e$([S II]). Then, each temperature is applied to compute ionic abundances of species with similar ionization potentials than the ion used in the $T_e$ diagnostic. In a typical spectra, $T_e$(low) is applied to compute abundances of $N^+$, $O^+$, $S^+$, $Cl^+$ and $Fe^+$, while $T_e$(high) is used for the remaining ionic species observed in the optical spectra.

However, recent results from [36] have shown that this scheme can be erroneous. These authors, from deep, high-resolution spectra of H II regions in the Magellanic Clouds, have proposed that for some ions, it is better to adopt other scheme in order to avoid trends with metallicity. In particular, they propose to use $T_e$([N II]) to calculate $Cl^{2+}$ and the mean of $T_e$([N II]) and $T_e$([O III]) for $S^{2+}$ and $Ar^{2+}$, finding that, in such cases, Cl/O, S/O and Ar/O are approximately constant with metallicity (see their Fig. 3) as expected for $\alpha$-elements.

In deep spectra, covering the whole optical (or even up to 1 μm) wavelength range more electron temperature and density diagnostics will be available. In such cases, $T_e$(low) and $T_e$(high) can be computed as the average of the values obtained from different diagnostics, which are generally in reasonable good agreement within the uncertainties (see e.g. [15, 19, 37]). In some cases, particularly in relatively high-density PNe ($n_e > 10^4 \mathrm{cm}^{-3}$), density stratification can be observed, with the [Ar IV] densities being larger than those computed with the other three diagnostics (see [38]). In such cases it is better to consider also a two-zone density model (see [9]). In some extreme cases of extremely young and dense PNe, with densities higher than the critical densities of the upper levels of the transitions producing the [Ar IV] lines, all the classical electron density diagnostics will be saturated and can provide inaccurate densities. An alternative density indicator is based on the analysis of [Fe III] emission lines, which are robust density diagnostics when collisional de-excitation dominates over collisional excitation. Indeed, if inappropriate density diagnostics are used, then physical conditions deduced from commonly used line ratios will be in error, leading to unreliable chemical abundances for these objects. (see [39]).

### 3.1.1 Analysis Tools

The first public code for the computation of physical conditions and ionic abundances was FIVEL [40], an interactive FORTRAN program which used a basic five-level atom approximation, which considers that only the five low-lying levels (i.e. at energies $\leq$5 eV above the ground state are physically relevant for computing the observed emission line spectrum. Later, [41] developed NEBULAR, a set of software tools (based in the FIVEL program, but extending it to an N-level atom) in the IRAF/STSDAS[2] environment that allow the user to compute diagnostic for a variety of ground-state electron configurations, and compute ionic abundances separately for up to three zones of ionization. The main advantage of NEBULAR is that it can be scripted. However, changes of atomic data sets is not trivial and computations of elemental abundances are not included.

Wesson et al. [42] developed the Nebular Empirical Analysis Tool (NEAT[3]), a very simple to use code written in FORTRAN90 which requires little or no user input to return robust results, trying to provide abundance determinations as objective as possible. One of the main advantages of this code is that it can evaluate uncertainties of the computed physical conditions and abundances by using a Monte Carlo approach. Another advantage of this code is that it also accounts for the effect of upward biasing on measurements of lines with low signal-to-noise ratios, allowing to reduce uncertainties of abundance determinations based on these lines. Finally, as atomic data for heavier elements than helium are stored externally in plain text files, the user can easily change the atomic data.

The last package to be offered in the field has been PYNEB[4] [43] which is completely written in python and is designed to be easily scripted, and is more flexible and therefore, powerful than its predecessors. This package allow the user to

[2]IRAF is distributed by National Optical Astronomy Observatories, which is operated by AURA (Association of Universities for Research in Astronomy), under cooperative agreement with NSF (National Science Foundation).

[3]https://www.nebulousresearch.org/codes/neat/.

[4]https://github.com/Morisset/PyNeb_devel.

easily change and update atomic data as well as providing tools to plot and compare atomic data from different publications.

### 3.2 Abundances in Distant Photoionized Nebulae: The Strong Line Methods

In the absence of reliable plasma diagnostics (a common fact in extragalactic objects) in giant H II regions or integrated spectra of galaxies, one needs to use alternative methods to derive accurate chemical abundances. This is especially important to estimate the metallicities of giant extragalactic H II regions as well as of local and high-redshift emission-line galaxies and hence, it has a relevant influence on the study of the chemical evolution of the Universe.

The first mention of the strong-line methods was 40 years ago, when [44] and [45] proposed a method to compute the oxygen abundance using strong lines only: the $R_{23}$ method, in which oxygen abundance is a one dimensional function of the $R_{23}$ parameter, defined as:

$$R_{23} = \frac{[\text{ O II}]\lambda3727 + [\text{ O III}]\lambda4959 + 5007}{H\beta} \tag{1}$$

This method was calibrated using the few relevant photoionization models available at that time. The problem with dealing with $R_{23}$ is that it is double valued with respect to metallicity. In fact, at low oxygen abundances $-12+\log(\text{O/H}) \lesssim 8.0$—the $R_{23}$ index increases with the abundance, while for high oxygen abundances $-12 + \log(\text{O/H}) \geq 8.25$—the efficiency of the cooling caused by metals make $R_{23}$ to drop with rising abundance. There is also a transition zone between 8.0 and 8.25 (see e.g. [46] for a detailed description of the high and low metallicity branches). This method has been refined multiple times since then and several calibrations, using data sets with abundances from the direct method (e.g. [46, 47]), using photoionization model grids (e.g. [48, 49]), or a combination of both, are now available in the literature. An overview of the most popular calibrations of strong-line methods can be found in [50]. A comprehensive critical evaluation of the different semi-empirical strong-line methods has been done by López-Sánchez et al. [51] who also develop a method for reducing systematics in the techniques to compute chemical abundances by using electron temperatures and ionization correction factors.

In the last years, mainly thanks to the increasingly easy access to supercomputing resources, new approaches have been proposed. Bayesian methods have been used by several authors to determine chemical abundances in extragalactic targets (e.g. [52]) although the priors should be selected cautiously to avoid unreliable results. On the other hand, as in most of astronomy fields, machine learning techniques are also being used to infer chemical abundances (see e.g. [53]). However, as has been pointed out by Stasinska [54], making use of an illustrative example, the use of these techniques ignoring the underlying physics can lead to unphysical inferences.

Stasinska [54] have argued in a comprehensive review that although strong-line methods are routinely used to estimate metallicities owing to their apparent simplicity, the users need to have a solid background on the physics of H II regions to understand the approximations made on the different approaches, and the limitations each calibration has, to avoid biases, misinterpretations and mistakes.

Even taking into account the drawbacks mentioned above, strong-line methods have been widely used for studying giant H II regions and emission line galaxies in large long-slit spectroscopic surveys as the Sloan Digital Sky Survey (SDSS) [55], or 2D spectroscopic surveys as MANGA (e.g. [56, 57]), CALIFA (e.g. [58]), and AMUSING (e.g. [59]).

# 4 Advances in Abundances Determinations in Photoionized Nebulae

In this section I will focus on the latest advances that have been reached in the field of photoionized nebulae. I will pay special attention to atomic data, ionization correction factors and the abundance discrepancy problem, that have been traditionally claimed as potential sources of uncertainty in chemical abundance determinations.

## *4.1 Atomic Data*

The atomic data used for computing abundances in photoionized nebulae are ususally considered as a black box by the users. Most users consider the default atomic data sets used by their favourite analysis tools or directly use the last available atomic data in the literature for each ion. In the last years large compilations of atomic data have been done in the CHIANTI[5] and NIST[6] databases, although the available atomic data in each database do not always match for a given ion. Luridiana et al. [60] and Luridiana and García-Rojas [61] discussed how to ensure that atomic data are correctly understood and used, as well as on the typical uncertainties in atomic data.

High-quality observations of photoionized nebulae are a powerful tool to check the reliability of atomic data. Copetti and Writzl [62] and Wang et al. [38] found, using a large data set of PNe spectra and comparing electron density estimates for PNe based on different density diagnostics, that the [O II] transition probabilities

[5]http://www.chiantidatabase.org.

[6]http://physics.nist.gov.

calculated by Wiese et al. [63] yielded systematically lower electron densities than those computed using the [S II] diagnostic, and that such discrepancies were caused by errors in the computed transition probabilities. Moreover, [38] found that the transition probabilities of [64] and the collision strengths of [65] were completely inconsistent with observations at the high and low density limits, respectively, and should be ruled out.

Juan de Dios and Rodríguez [66] determined chemical abundances of O, N, S, Ne, Cl and Ar for a sample of PNe and H II regions and evaluated the impact of using different sets of atomic data on the computed physical conditions and abundances. These authors used all the possible combinations of 52 different sets of transition probabilities and collision strengths to calculate physical conditions and chemical abundances, finding that different combinations of atomic data introduce differences in the derived abundances that can reach or surpass 0.6–0.8 dex at higher densities ($n_e > 10^{-4}$ cm$^{-3}$ in several abundance ratios like O/H and N/O). Removing the data sets that introduce the largest differences can reduce the total uncertainties, although they can still remain in high-density objects. Additionally, they have pointed out that special attention should be paid to the transition probabilities of the $S^+$, $O^+$, $Cl^{++}$ and $Ar^{3+}$ density diagnostic lines, and to the collision strengths of $Ar^{3+}$ which, if incorrectly selected, can lead to unreliable chemical abundances in high-density nebulae.

Finally, [54] has pointed out that the role of atomic data in strong-line method calibrations cannot be ignored. Recent changes in routinely used atomic data have revealed that they play a crucial role in direct abundance determinations and in photoionization models.

## 4.2 Ionization Correction Factors

The elemental abundance of a particular element is computed by adding up the ionic abundances of all the ions present in a nebula. However, it is usually found that not all the ions of a given element are observed, whether because they are emitted in a different spectral range than that observed or because the spectra is not deep enough to detect them. Therefore, the contribution of these unobserved ions should be estimated in someway. With this aim, the use of Ionization Correction Factors (ICFs) was proposed by Peimbert and Costero [67]. These authors proposed to use similarities between ionization potentials of different ions to construct ICFs. This approach has been used by several authors since then (see e.g. [68]). However, [12] argued that these ICFs should be treated with caution because the ionization structure in a photoionized nebula does not depend only on the ionization potential. Moreover, it has been shown that using recent photoionization models, these simple expressions are not always valid and new ICFs are needed to obtain more reliable abundances (see e.g. [69]).

The alternative is to compute ICFs using photoionization models, where the physics involved in ionized nebulae is treated with much more detail. Photoionization models allow to compute the detailed ionization structure of the various elements present in a nebula, by taking into account all the processes that govern ionization and recombination (i.e. mostly photoionization, radiative and di-electronic recombination, and charge exchange), as well as all the heating and cooling processes that determine the electron temperature [70].

Traditionally, different ICFs have been computed for H II regions and PNe, given the differences in the hardness of the radiation field and the different ionic species detected in each type of object (see Sect. 1). Several authors have derived ICFs from photoionization models for H II regions [71–77] and for PNe [70, 78, 79]. It is not the scope of this text to show the details of the different approaches used to compute ICFs from photoionization models, but I think it is worth mentioning some of the most widely used ICF schemes. Izotov et al. [75] re-evaluated empirical expressions for the abundance determination of N, O, Ne, S, Cl, Ar and Fe to compute abundances of emission-line galaxies from the Data Release 3 of the Sloan Digital Sky Survey (SDSS). They took special care in the selection of atomic data and constructed an appropriate grid of photoionization models with state-of-the art model atmospheres. In particular, these authors take care of a problem that should not be ignored in the computation of photoionization models, which is the uncertain rate of the dielectronic recombination for sulfur, chlorine and argon ions. They compared the abundances of these elements calculated with different assumed dielectronic recombination rates and could put some constraints on these rates. Additionally, following an approach that was defined by Stasińska and Izotov [80] these authors proposed different ICFs depending on the metallicity range of the nebulae. Regarding PNe, [70] constructed ICFs for He, N, O, C, Ne, S, Cl, and Ar using a large grid of photoionization models that are representative of most of the observed PNe. Besides the obvious advantage of covering a wide range of physical parameters with a large photoionization model grid, the main advantage of this work is the provision of analytical expressions to estimate the uncertainties arising from their computed ICFs.

Finally, a third scheme to compute ICFs is to derive analytical expressions obtained from observational fittings to large sets of high-quality data (see e.g. [81] for Cl, and [82] for C).

In Sect. 4.7.2 I will come back to the ICFs mentioning some works devoted to the computation of ICFs for neutron-capture elements in PNe.

## *4.3 Oxygen Enrichment in PNe*

Oxygen is the element for which more reliable abundances can be obtained and, therefore, it has been traditionally used as a proxy for metallicity. In H II regions, oxygen reflects the current abundance in the ISM, while in PNe, it is supposed to reflect the chemical composition of the environment where the star was born because its abundance remain unchanged during the life of the star [83]. However, AGB stars can modify the oxygen abundance by two mechanisms: the third dredge-up (TDU)

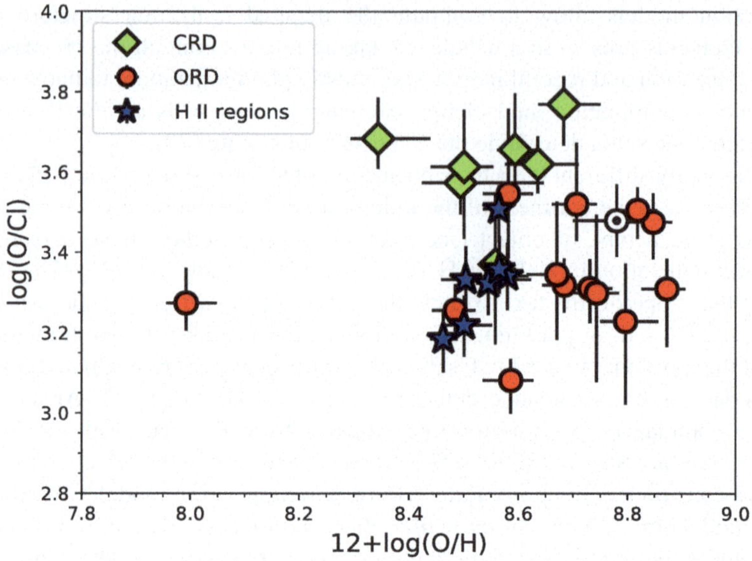


**Fig. 5** Values of O/Cl as a function of O/H for a sample of Galactic PNe and H II regions (see [83]). The red circles represent PNe with oxygen-rich dust, the green diamonds PNe with carbon-rich dust, and the blue stars the H II regions. The protosolar abundances of [88] are overplotted with the solar symbol. Plot made with data gently provided by Gloria Delgado-Inglada

and the hot bottom burning (HBB), although only nucleosynthesis models which include extra-mixing processes like diffusive convective overshooting (e.g. [84–86]) predict a significant production of oxygen.

Until recently, the only observational probes of oxygen production in AGB stars have been restricted to low-metallicity PNe (see e.g. [87]). However, using deep, high-quality optical spectra (with spectral resolution better than 1 Å) [83] recomputed accurate abundances of He, O, N, Ne, C, Ar, and Cl in 20 PNe and 7 H II regions in our Galaxy at near-solar metallicities. These authors found that all but one of the Galactic PNe with C-rich dust (the one with the highest metallicity according to Cl/H) show higher O/Cl values than the PNe with O-rich dust and the H II regions (see Fig. 5), and interpret that result as O is enriched in C-rich PNe due to an efficient third dredge-up in their progenitor stars. These results have been confirmed later by nucleosynthesis models including convective overshooting by [85, 89].

These findings confirm that oxygen is not always a good proxy of the original ISM metallicity and other chemical elements such as chlorine or argon, the abundance of which is unaltered in the evolution of low- and intermediate-mass stars, should be used instead. Additionally, as has been pointed out by García-Hernández et al. [89], the production of oxygen by low-mass stars should be thus considered in galactic-evolution models.

## 4.4 *Abundance Gradients in the Milky Way and in Nearby Spiral Galaxies from Direct Abundance Determinations*

The study of the radial distribution of the gas phase metallicity in a Galaxy (usually using oxygen as a proxy for the metallicity) is fundamental for our understanding of the evolution of Galaxies. The pioneering studies on the gradient of abundances in spiral galaxies were those of [90] and [91], which were based on the spectral differences found by Aller [92] between the H II regions in the spiral galaxy M33. Shaver et al. [93] were the first in carrying out an homogeneous study of abundance gradients in the Milky Way (hereinafter, MW) with a relatively large sample of H II regions. However, these authors rely on electron temperatures determined from radio RLs and abundances from optical lines, obtaining a relatively large scatter at a given Galactocentric distance. Since these pioneering works, several authors have computed radial abundance gradients using the direct method in our Galaxy (e.g. [29, 94–97]) and in external galaxies (see e.g. [25, 98–101] and references therein).

Regarding the MW, the relatively large scatter at a given Galactocentric distance found in several works has been claimed as a possible indication that the gas is not as well mixed as commonly thought (see e.g. [29]). Moreover, [102] found significant differences in the radial gradient of O in the MW depending on the Galactic azimuth region considered, strengthening the idea that metals are not well mixed at a given radius. However, [32] made an homogeneous analysis using a set of deep optical spectra of 35 H II regions, from which they computed accurate physical conditions and ionic and elemental abundances, finding that the scatter of the N and O abundances of H II regions is of the order of the observational uncertainties, indicating that both chemical elements seem to be well mixed in the ISM at a given Galactocentric distance (see the comparison between radial O abundance from this work and that of [29] in Fig. 5).

In the extragalactic domain, it is worth mentioning the existence of the CHemical Abundances of Spirals (CHAOS) project, which is devoted to surveying several spiral galaxies to determine precise "direct" abundances in large samples of H II regions in spiral galaxies (see [100, 101, 103]). This project has increased by more than an order of magnitude the number of H II regions with direct measurements of the chemical abundances in nearby disk galaxies (see [101]).

There are many open problems with the abundance gradients of the MW and nearby spiral Galaxies such as a possible temporal evolution [104, 105] based on the differences found in the gradients using different populations of PNe and H II regions; the existence or not of a flattening of the gradient in the outer disc of spiral galaxies, including the MW [59, 96, 106] or in the inner disc [59, 96]; distance determinations uncertainties, particularly for PNe (see [105]) or, as mentioned in Sect. 4.3, the applicability of oxygen as a reliable element to trace the metallicity in PNe [83]. Some of the limitations that, in my opinion, should be taken into account have been summarized in [107].

The determination of precise radial metallicity gradients are precious constraints for chemical evolution models of the MW in particular, and of spiral galaxies in general. The presence of a negative gradient agrees with the stellar mass growth of galaxies being inside-out (see e.g. [108]). However, additional information, such as the possible temporal evolution of the gradients (see e.g. [104, 105]) can give information about physical processes that can modify gradients, and that should be considered by chemical evolution models (see discussion by Stanghellini and Haywood [105] and references therein). As H II regions reflect the young stellar populations and, on the other hand, PNe reflect older stellar populations (with a relatively large spread in ages) a careful comparison between gradients obtained from different objects is very useful to constrain the temporal evolution of the gradient predicted by chemical evolution models [109].

## *4.5 The Abundance Discrepancy Problem*

The abundance discrepancy problem is one of the major unresolved problems in nebular astrophysics and it has been around for more than 70 years [110]. It consists in the fact that in photoionized nebulae—both H II regions and PNe—ORLs provide abundance values that are systematically larger than those obtained using CELs. Solving this problem has obvious implications for the measurement of the chemical content of nearby and distant galaxies, because this task is most often done using CELs from their ionized ISM.

For a given ion, the abundance discrepancy factor (ADF) is defined as the ratio between the abundances obtained from ORLs and CELs, i. e.,

$$\mathrm{ADF}(X^{i+}) = (X^{i+}/H^{+})_{\mathrm{ORLs}}/(X^{i+}/H^{+})_{\mathrm{CELs}}, \tag{2}$$

and is usually between 1.5 and 3, with a mean value of about 2.0 in H II regions and the bulk of PNe (see e.g. [27, 111], but in PNe it has a significant tail extending to much larger values, up to 2–3 orders of magnitude.[7] It is important to remark that the ADF is most easily determined for $O^{2+}$ owing to both CELs and RLs are straightforward to detect in the optical. ADFs can be also determined for other ions, such as $C^{2+}$, $N^{2+}$, and $Ne^{2+}$, although the obtained values are more uncertain because CELs and RLs are detected in different wavelength ranges (in the case of $C^{2+}$ and $N^{2+}$) or because RLs are intrinsically very faint (in the case of $Ne^{2+}$).

The possible origin of this discrepancy has been discussed for many years and three main scenarios have been proposed:

- Peimbert [112] was the first proposing the presence of temperature fluctuations in the gas to explain the discrepancy between $T_e$([O III]) and $T_e$(H I) derived from the Balmer jump. After that seminal work, [67] developed a scheme to correct the abundances computed from CELs for the presence of temperature

[7] An updated record of the distribution of values of the ADF in both H II regions and PNe can be found in Roger Wesson's webpage: http://nebulousresearch.org/adfs.

inhomogeneities. Later, [113] suggested that the discrepancy between ORLs and CELs abundances could be explained if spatial temperature variations over the observed volume were considered. Peimbert et al. [10] have recently summarized the mechanisms proposed to explain and maintain the presence of temperature fluctuations in photoionized nebulae.

- Torres-Peimbert et al. [114] were the first in proposing the existence of chemical inhomogeneities in the gas as a plausible mechanism to explain the abundance discrepancy. This scenario was later expanded by Liu et al. [14], who claimed that metal-rich (i.e. H-poor) inclusions could be the clue to resolve the abundance discrepancy problem; in this scenario, metal ORLs would be emitted in the metal-rich inclusions, where cooling has been enhanced, while CELs would be emitted in the "normal" metallicity (H-rich) zones. This model was tested by several authors by constructing two-phase photoionization models that, in several cases, successfully simultaneously reproduced the ORLs and CELs emissions in H II regions [115] and PNe [116]. However, at the present time, the origin of such metal-rich inclusions remains elusive, although some scenarios have been proposed for PNe [117] and H II regions [118]. In the last years, increasing evidence has been found of a link between the presence of a close central binary star at the heart of PNe and very high ($>10$) ADFs (see Sect. 4.5.1).
- A third scenario was brought into play by Nicholls et al. [119], who proposed that the departure of the free electron energy distribution from the Maxwellian distribution ($\kappa$-distribution) could explain the abundance discrepancies owing to the presence of a long tail of supra-thermal electrons that contribute to an increase in the intensity of the CELs at a given value of the kinetic temperature. However, in the last years little theoretical [120–122] or observational [123] support has been presented for this scenario in photoionized nebulae. Ferland et al. [121] have shown that the heating or cooling timescales are much longer than the timescale needed to thermalize supra-thermal electrons because they can only travel over distances that are much shorter than the distances over which heating rates change, implying that the electron velocity distribution will be close to a Maxwellian one long before the supra-thermal electrons can affect the emission of CELs and RLs. Moreover, [122] demonstrated analytically that the electron energy distribution relaxes rapidly to a steady-state distribution that is very close to a Maxwellian, having negligible effects on line ratios.

One of the most active groups in the study of the abundance discrepancy in PNe in the last two decades has been the University College London/U. Beijing group, who have developed deep medium-resolution spectrophotometry of dozens of PNe to compute the physical and chemical properties of these objects from ORLs. In one of the most detailed and comprehensive studies of this group, [124] showed that the values of the ADF deduced for the four most abundant second-row heavy elements (C, N, O and Ne) are comparable (see their Fig. 18). However, they also computed abundances from ORLs from a third-row element (Mg) and they found that no enhancement of ORL abundances relative to CEL ones is obvious for Mg: the average Mg abundances from ORLs for disk PNe remained in a range compatible to the solar photospheric value, even taking into account the small depletion expected for this element onto dust grains (less than 30%). Finally, these authors also showed that, regardless of the value of the ADF, both CEL and ORL abundances yield similar relative abundance ratios of heavy elements such as C/O, N/O and Ne/O. This has important implications, especially in the case of the C/O ratio, given the difficulties of obtaining this ratio from UV CELs (see Sect. 4.6).

Several authors have strongly argued in favour of the inhomogeneous composition of PNe and against pure temperature fluctuations (see e.g. [125] and references therein); some of the reasoning that has been presented supporting this model are: (1) far-IR [O III] CELs, which in principle, have a much lower dependence on electron temperature than optical CELs, provide abundances that are consistent with those derived from optical CELs (see e.g. [14]); (2) the analysis of the physical conditions using H, He, O and N ORLs yields electron temperatures that are much lower than those computed from classical CEL diagnostic ratios (see [20, 126–128]); additionally, ORL density diagnostics provide densities that are higher than those derived from CEL diagnostics; (3) chemically homogeneous photoionization models do not reproduce the required temperature fluctuations to match CEL and ORL abundances, while bi-abundance photoionization models including an H-poor (i.e. metal-rich) component of the gas successfully reproduce the observed intensities of both CELs and ORLs (e.g. [116]). All these arguments strongly favour the presence of a low-mass component of the gas that is much colder and denser than the "normal" gas, and that is responsible for the bulk of the ORL emission. However, we cannot rule out the possibility that different physical phenomena can contribute simultaneously to the abundance discrepancy in PNe.

Some physical phenomena have been proposed to explain the abundance discrepancy in the framework of temperature fluctuations or chemical inhomogeneities scenarios [10, 117]. However, until very recently, there was no observational proof that demonstrated a single physical process to be responsible for the abundance discrepancy. Some recent works on the Orion nebula have observationally linked the abundance discrepancy to the presence of high velocity flows [129] or to the presence of high density clumps, such as proto-planetary disks [39, 130]. On the other hand, [125] found a very extreme value of the ADF for the PN Hf 2–2 (ADF$\sim$70) and, for the first time, suggested the possibility that this large ADF could be related to the fact that the central star of the PN, which is a close-binary star, has gone through a common-envelope phase.

### 4.5.1 The Link Between Close Binary Central Stars and Large Abundance Discrepancy Factors

Several papers in recent years have confirmed the hypothesis proposed by Liu et al. [125] that the largest abundance discrepancies are reached in PNe with close-binary central stars. Corradi et al. [131] found that three PNe with known close-binary central stars showed high ADFs, with the PN Abell 46, with an ADF($O^{2+}$)$\sim$120, and as high as 300 in its inner regions, being the most extreme object. Their spectroscopic analysis supports the previous interpretation that, in addition to "standard" hot ($T_e \sim 10^4$ K) gas, a colder ($T_e \sim 10^3$ K), metal-rich, ionized component also exists in these nebulae. Both the origin of the metal-rich component and how the two gas phases are mixed in the nebulae are basically unknown. Moreover, this dual nature is not predicted by mass-loss theories. However, it seems clear that the large-ADF phenomena in PNe is linked to the presence of a close-binary central star. In fact, [23] have recently completed a survey of the ADFs in seven PNe with known close-binary central stars and they found ADFs larger than 10 for all of them, confirming the strong link between large ADFs and close-binary central stars. On the other hand, several spectroscopic studies have shown that the ORL emitting plasma is generally concentrated in the central parts of the PNe. This occurs in PNe with known close-binary central stars and large ADFs (e.g. [131, 132]), in PNe with low-to-moderate ADFs and no indication of binarity (e.g. [133, 134]) and in PNe with relatively large ADFs but no known close-binary central star (e.g. M 1–42, see [135]).

García-Rojas et al. [136] recently obtained the first direct image of the PN NGC 6778 (a PN with ADF~20) in O II recombination lines, taking advantage of the tunable filters available at the OSIRIS instrument in the 10.4m Gran Telescopio Canarias (GTC). They found that in NGC 6778, the spatial distribution of the O II λλ4649+50 ORL emission does not match that of the [O III] λ5007 CEL. García-Rojas et al. [135] found the same behaviour in Abell 46 using direct tunable filter images centred at λλ4649+51 Å.

Moreover, [135] presented preliminary results obtained from deep 2D spectroscopic observations with MUSE at the 8.2 m Very Large Telescope (VLT) of five southern large-ADF PNe, and they confirmed this behaviour in at least the PNe Hf 2-2 (ADF~84), M 1-42 and NGC 6778 (both with ADF~20). In Fig. 6 we show the MUSE emission line maps of several emission lines in the PN NGC 6778. The emission maps are ordered by increasing ionization potential of the parent ion from left to right and from top to bottom. It is clear that O II λ4649+50 ORLs emission is more centrally concentrated that [O III] λ4959 CEL emission, and seems to be emitted in a zone that correspond to a higher ionization specie. A similar result has been found by Richer et al. [137] from a kinematical analysis of several heavy metal ORLs and CELs in NGC 7009. These authors found that the kinematics of ORLs and CELs were discrepant and incompatible with the ionization structure of the nebula, unless there is an additional plasma component to the CEL emission that arises from a different volume from that giving rise to the RL emission from the parent ions within NGC 7009. Similarly, [138] found that the kinematics of the C II λ6578 line is not what expected if this line arises from the recombination of $C^{2+}$ ions or the fluorescence of $C^{+}$ ions in ionization equilibrium in a chemically homogeneous nebular plasma, but instead its kinematics are those appropriate for a volume more internal than expected.

These results clearly support the hypothesis of the existence of two separate plasmas, at least in these large-ADF PNe, with the additional indication that they are not well mixed, perhaps because they were produced in distinct ejection events related to the binary nature of the PN central star. Wesson et al. [23] propose that

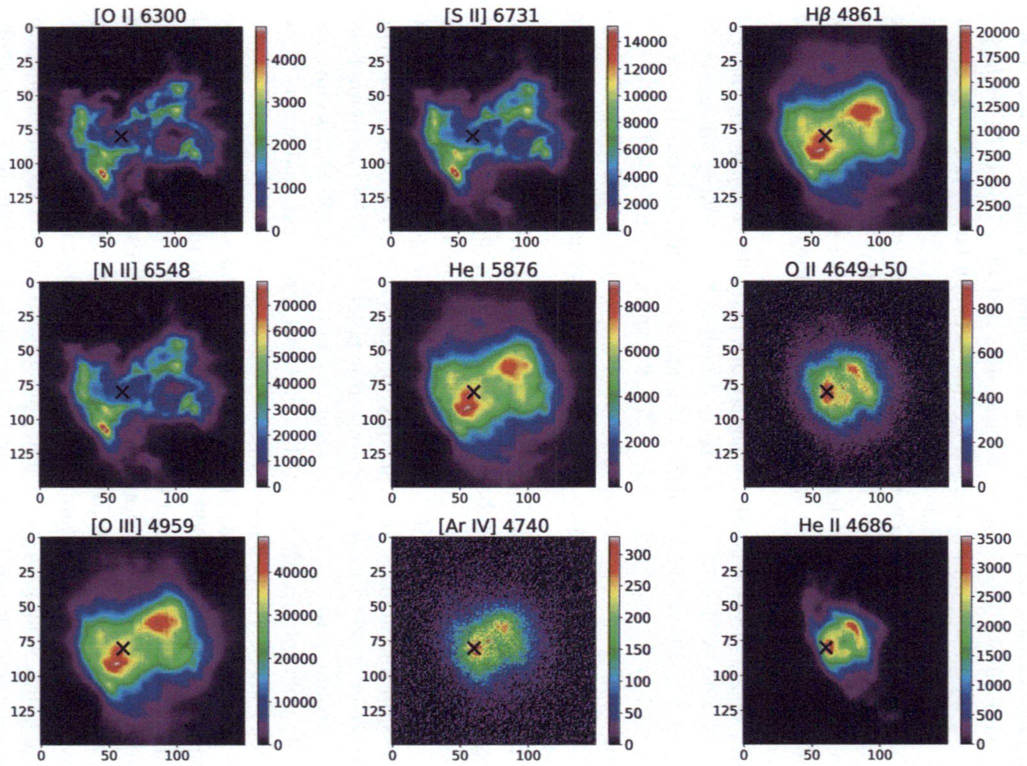


**Fig. 6** MUSE emission line maps of several lines in the PN NGC 6778, ordered by ionization potential of the parent ion: from left to right and from top to bottm: [O I] λ6300 Å, [S II] λ6731 Å, Hβλ4861 Å, [N II] λ6548 Å, He I λ5876 Å, O II λλ4649+50 Å ORLs, [O III] λ4959 Å CEL, [Ar IV] λ4740 Å and He II λ4686 Å. It is worth re-emphasising that the O II and the [O III] emission comes from the same ion: $O^{2+}$. The "x" marks a reference spaxel. x and y axis units are spaxels

a nova-like outburst from the close-binary central star could be responsible for ejecting H-deficient material into the nebulae soon after the formation of the main nebula. This material would be depleted in H, and enhanced in C,N, O, and Ne, but not in third row elements. It is worth mentioning the similarity of these plasma component with some well-known old nova shells as CP Pup and DQ Her that show low $T_e$ and strong ORLs [139, 140].

## 4.6 *The C/O Ratio from Recombination Lines*

The determination of accurate C/H and C/O ratios in H II regions is of paramount importance to constrain chemical evolution models of galaxies [141]. Furthermore, C is an important source of opacity in stars and one of the main elements found in interstellar dust and organic molecules, making it one of the biogenic elements [142]. Despite its importance, C abundances in H II regions have been poorly explored as they are traditionally derived from UV observations of the semi-forbidden C III] λ1909 and C II] λ2326 CELs from space [82, 143]. For PNe the situation is better, owing to several successful *IUE* and *HST* programs that have provided reliable measurements of these lines in dozens of objects (see e.g. [144–148] and references therein). However, the determination of reliable fluxes from these lines is difficult as they are severely affected by interstellar reddening. Moreover, the emissivities of these lines are also very dependent on the electron temperature. Finally, aperture effects owing to the different areas covered by UV and optical observations must be taken into account to guarantee the observation of the same volume of the nebula in both ranges. Alternatively, scanning techniques have been used to match UV International Ultraviolet Explorer (IUE) observations and optical spectroscopy (see e.g. [14, 133]).

The determination of C and O abundances from ORLs skips these difficulties and takes advantage of the fact that using the same type of lines (CELs or ORLs) the C/O ratios remain the same in most of the objects were both types of ratio have been computed [124, 149]. Thanks to the new CCDs with improved efficiency in the blue and the use of large telescopes, several high-quality observations of the C II λ4267 ORL in PNe have been achieved in the last years (e.g. [8, 19, 20, 24, 124, 144, 150, 151]).

### 4.6.1 C/O Ratios in H II Regions

Esteban et al. [95] computed for the first time the C/H and C/O radial gradients of the ionized gas from ORLs in the MW. Later, [152] derived these gradients in M101, and [25] in M33 and NGC 300.

The general conclusion of these works is that C abundance gradients are always steeper than those of O, producing negative C/O gradients across the galactic disks which reflect the non-primary behavior of C enrichment. Furthermore, the comparison between the C/H and C/O gradients obtained in the MW with state-of-the-art chemical evolution models revealed that the obtained C gradients can only been explained if the C produced by massive and low-intermediate mass stars depends strongly on time and on the Galactocentric distance [141].

From the C/O ratios computed from ORLs in several low-metallicity star-forming galaxies compiled by López-Sánchez [153] and Esteban et al. [142] it has been found that H II regions in star-forming dwarf galaxies have different chemical evolution histories than the inner discs of spiral galaxies and that the bulk of C in the most metal-poor extragalactic H II regions should have the same origin as in halo stars (see Figs. 8 and 9 of [142]).

### 4.6.2 C/O Ratios in PNe

As was pointed out above, there are multiple determinations of C abundance from UV lines in the literature. These determinations, when considered together with detailed analysis of optical spectra covering the same volume of the nebula, provide strong constraints to nucleosynthesis models (see e.g. [147]). However, owing to limitations in space, in this text I am going to focus on the determination of C/O ratios from ORLs.

Thanks to the new CCDs with improved efficiency in the blue and the use of large telescopes, several high-quality observations of the C II $\lambda$4267 RL in PNe have been achieved in the last years (e.g. [8, 19, 20, 24, 124, 144, 150, 151].

C/O ratios derived from ORLs combined with other abundance ratios such as N/O or He/H, can set strong constraints to the initial mass of PNe progenitors. This is because different processes occurring at the interior of AGB stars (third dredge-up episodes, hot bottom burning process) activate at different masses and can strongly modify C/O and N/O ratios (see e.g. [154] and references therein). These ratios can also provide strong constraints to the physics assumed by nucleosynthesis models, i.e. the assumption of convective overshooting into the core during the main sequence and the He-burning phases can diminish the mass limit at which He-flashes can occur at a given metallicity [89] compared to those models not considering it [154]; this has implications for the mass limit at which hot bottom burning can be activated. Additionally, C/O ratios can be used to obtain information about the efficiency of dust formation in C-rich or O-rich environments (see below) and to learn about different dust-formation mechanisms (see [85] and [24]).

## *4.7 Abundances of Heavy Elements in PNe From Faint Emission Lines*

As I have emphasized several times in this chapter, the increasing efficiency of astronomical detectors as well as the advent of large (8–10 m type) telescopes have boosted the detection of very faint emission lines in photoionized regions. In this section, I will focus on the detection of extremely faint lines in deep spectra of PNe.

### 4.7.1 Faint Emission Lines of Refractory Elements

The determination of elemental abundances of refractory elements in the spectra of Galactic PNe is not an easy task because the available lines of these elements (mainly iron and nickel) are relatively faint. Additionally, these elements are constituents of dust grains in the ISM and their abundance in photoionized nebulae do not reflect their actual abundances. Several authors have computed abundances of Fe ions in both PNe and H II regions [155–158], Delgado Inglada et al. [157] computed detailed Fe abundances for a sample of 28 PNe and found that more than 90% of Fe atoms are condensed on dust grains. These authors did not find differences between the iron abundances in C-rich and O-rich PNe, suggesting similar depletion efficiencies in both environments.

Delgado-Inglada and Rodríguez [149] combined C/O ratios derived from both UV CELs and optical ORLs (as we comment in Sect. 4.5, they seem to be equivalent) with information obtained from *Spitzer* mid-infrared spectra. They also computed Fe depletions onto dust grains, and found that the highest depletion factors are found in C-rich objects with SiC or the 30 $\mu$m feature in their infrared spectra, while the lowest depletion factors were found for some of the O-rich objects showing silicates in their infrared spectra.

Delgado-Inglada et al. [159] compiled detections of very faint [Ni II] and [Ni III] lines in deep spectra of Galactic PNe and H II regions. They determined the nickel abundance from the [Ni III] lines using an extensive grid of photoionization models to determine a reliable ionization correction factor (ICF). From the comparison of Fe/Ni ratios with the depletion factor obtained from both [Fe/H] and [Ni/H], they conclude that nickel atoms adhere to dust grains more efficiently than iron atoms in environments where dust formation or growth is more important.

### 4.7.2 Neutron-Capture Elements

Nebular spectroscopy of neutron($n$)-capture elements (atomic number $Z > 30$) is a recent field that has seen rapid development in the last 10 years, and holds promise to significantly advance our understanding of AGB $n$-capture nucleosynthesis. Nebular spectroscopy can reveal unique and complementary information to stellar spectroscopy. Observations of PNe provide the first opportunity to study the production of the lightest $n$-capture elements ($Z \leq 36$) and noble gases (Kr and Xe) in one of their sites of origin. Unlike the case of AGB stars, nucleosynthesis and convective dredge-up are complete in PNe, whose envelopes contain material from the last 2−3 thermal pulses. Accurate computations of $n$-capture elements would shed light on the different scenarios proposed for the production of these elements and would constrain the chemical yields of low- and intermediate-mass stars for these elements.

*n*-capture elements were not recognized in any astrophysical nebula until [160] identified emission lines of Br, Kr, Rb, Xe, Ba, and possibly other heavy species in the bright PN NGC 7027. Since then, a breathtaking number of *n*-capture element emission lines have been identified for the first time in near-infrared [161–165], UV [166] and optical [21, 22, 151] spectra of PNe. The new detections have led to a dedicated effort to produce atomic data needed for abundance determinations (e.g., see [163, 165, 167], and references therein). The new photoionization cross-sections and recombination coefficients have been incorporated in photoionization calculations to compute reliable ICFs (e.g., [168]). The new collisional strengths have been used for abundance determinations of newly detected ions (see [163], for [Rb IV], [164], for [Se III] and [Kr VI], and [165] for [Te III] and [Br V]). Thanks to the fast advances in observations, atomic data determinations and numerical modelling, this field has grown from just 3 PNe with *n*-capture element abundances in 2001 to more than 100 Galactic PNe in 2019 [8, 21, 22, 151, 162–165, 169].

The importance of deep, high-resolution optical spectrophotometry of PNe to detect faint *n*-capture elements can be understood when comparing the works by [151] and [21]. In the first case, several *n*-capture emission lines were discovered in the spectra of 5 PNe, but even at a resolution of ∼22,000, many features were not unambiguously detected. [21] took advantage of the very high-resolution (R∼40,000) spectrum of NGC 3918 to clearly identify several ions of Kr, Xe, Rb and Se, testing for the first time the complete set of ICFs for Kr created by Sterling et al. [168].

The combination of deep optical and near-infrared spectra of PNe has been proved to be a powerful tool to test the predictions of modern AGB nucleosynthesis models as well as to test the accuracy of new atomic data computations and ICF prescriptions. Madonna et al. [22] have combined a deep optical spectrum and a near-infrared spectrum of NGC 5315, testing for the first time the complete set of ICFs for Se created by Sterling et al. [168].

Finally, the determination of precise elemental abundances ratios of *n*-capture elements like Kr/Xe or Te/Se from optical and near-infrared spectra are potential indicators of the time integrated neutron flux in the intershell region between the H- and He-burning shells, giving strong constraints to nucleosynthesis models in the thermally pulsing AGB phase [170].

**Acknowledgements** First of all, I want to acknowledge support of the Erasmus+ programme of the European Union under grant number 2017-1-CZ01-KA203-035562. I also acknowledge support from an Advanced Fellowship from the Severo Ochoa excellence program (SEV-2015-0548) and support from the State Research Agency (AEI) of the Spanish Ministry of Science, Innovation and Universities (MCIU) and the European Regional Development Fund (FEDER) under grant AYA2017-83383-P, and support under grant P/308614 financed by funds transferred from the Spanish Ministry of Science, Innovation and Universities, charged to the General State Budgets and with funds transferred from the General Budgets of the Autonomous Community of the Canary Islands by the Ministry of Economy, Industry, Trade and Knowledge. I am grateful to

Dr. Gloria Delgado-Inglada, who kindly provided me the data and the script to produce Fig. 5 for this chapter.

## References

1. W. Huggins, W.A. Miller, Philos. Trans. R. Soc. Lond. I **154**, 437 (1864)
2. M.L. Huggins, Astrophys. J. **8**, 54 (1898). https://doi.org/10.1086/140540
3. I.S. Bowen, Nature **120**(3022), 473 (1927). https://doi.org/10.1038/120473a0
4. D.E. Osterbrock, Publ. Astron. Soc. Pac. **100**, 412 (1988). https://doi.org/10.1086/132188
5. I.S. Bowen, Publ. Astron. Soc. Pac. **39**(231), 295 (1927). https://doi.org/10.1086/123745
6. D.E. Osterbrock, G.J. Ferland, *Astrophysics of Gaseous Nebulae and Active Galactic Nuclei* (University Science Books, Sausalito, 2006)
7. S. Kwok, *Physics and Chemistry of the Interstellar Medium* (Cambridge University Press, Cambridge, 2007)
8. J. García-Rojas, M. Peña, C. Morisset, A. Mesa-Delgado, M.T. Ruiz, Astron. Astrophys. **538**, A54 (2012). https://doi.org/10.1051/0004-6361/201118217
9. J. García-Rojas, M. Peña, C. Morisset, G. Delgado-Inglada, A. Mesa-Delgado, M.T. Ruiz, Astron. Astrophys. **558**, A122 (2013). https://doi.org/10.1051/0004-6361/201322354
10. M. Peimbert, A. Peimbert, G. Delgado-Inglada, Publ. Astron. Soc. Pac. **129**(978), 082001 (2017). https://doi.org/10.1088/1538-3873/aa72c3
11. E. Pérez-Montero, Publ. Astron. Soc. Pac. **129**(974), 043001 (2017). https://doi.org/10.1088/1538-3873/aa5abb
12. G. Stasińska, in *Proceedings of the XIII Canary Islands Winter School of Astrophysics*. "Cosmochemistry. The melting pot of the elements" astro-ph/0207500 (2002)
13. G. Stasińska, (2007). e-prints arXiv:0704.0348
14. X.W. Liu, P.J. Storey, M.J. Barlow, I.J. Danziger, M. Cohen, M. Bryce, Mon. Not. R. Astron. Soc. **312**(3), 585 (2000). https://doi.org/10.1046/j.1365-8711.2000.03167.x
15. Y.G. Tsamis, M.J. Barlow, X.W. Liu, I.J. Danziger, P.J. Storey, Mon. Not. R. Astron. Soc. **338**(3), 687 (2003). https://doi.org/10.1046/j.1365-8711.2003.06081.x
16. Y.G. Tsamis, M.J. Barlow, X.W. Liu, I.J. Danziger, P.J. Storey, Mon. Not. R. Astron. Soc. **345**(1), 186 (2003). https://doi.org/10.1046/j.1365-8711.2003.06972.x
17. C. Esteban, M. Peimbert, J. García-Rojas, M.T. Ruiz, A. Peimbert, M. Rodríguez, Mon. Not. R. Astron. Soc. **355**(1), 229 (2004). https://doi.org/10.1111/j.1365-2966.2004.08313.x
18. J. García-Rojas, C. Esteban, M. Peimbert, M. Rodríguez, M.T. Ruiz, A. Peimbert, Astrophys. J. Suppl. Ser. **153**(2), 501 (2004). https://doi.org/10.1086/421909
19. R. Wesson, X.W. Liu, M.J. Barlow, Mon. Not. R. Astron. Soc. **362**(2), 424 (2005). https://doi.org/10.1111/j.1365-2966.2005.09325.x
20. X. Fang, X.W. Liu, Mon. Not. R. Astron. Soc. **415**(1), 181 (2011). https://doi.org/10.1111/j.1365-2966.2011.18681.x
21. J. García-Rojas, S. Madonna, V. Luridiana, N.C. Sterling, C. Morisset, G. Delgado-Inglada, L. Toribio San Cipriano, Mon. Not. R. Astron. Soc. **452**(3), 2606 (2015). https://doi.org/10.1093/mnras/stv1415
22. S. Madonna, J. García-Rojas, N.C. Sterling, G. Delgado-Inglada, A. Mesa-Delgado, V. Luridiana, I.U. Roederer, A.L. Mashburn, Mon. Not. R. Astron. Soc. **471**(2), 1341 (2017). https://doi.org/10.1093/mnras/stx1585
23. R. Wesson, D. Jones, J. García-Rojas, H.M.J. Boffin, R.L.M. Corradi, Mon. Not. R. Astron. Soc. **480**(4), 4589 (2018). https://doi.org/10.1093/mnras/sty1871
24. J. García-Rojas, G. Delgado-Inglada, D.A. García-Hernández, F. Dell'Agli, M. Lugaro, A.I. Karakas, M. Rodríguez, Mon. Not. R. Astron. Soc. **473**(4), 4476 (2018). https://doi.org/10.1093/mnras/stx2519

25. L. Toribio San Cipriano, J. García-Rojas, C. Esteban, F. Bresolin, M. Peimbert, Mon. Not. R. Astron. Soc. **458**(2), 1866 (2016). https://doi.org/10.1093/mnras/stw397
26. L. Toribio San Cipriano, G. Domínguez-Guzmán, C. Esteban, J. García-Rojas, A. Mesa-Delgado, F. Bresolin, M. Rodríguez, S. Simón-Díaz, Mon. Not. R. Astron. Soc. **467**(3), 3759 (2017). https://doi.org/10.1093/mnras/stx328
27. I.A. McNabb, X. Fang, X.W. Liu, R.J. Bastin, P.J. Storey, Mon. Not. R. Astron. Soc. **428**(4), 3443 (2013). https://doi.org/10.1093/mnras/sts283
28. M. Rodríguez, in *Chemical Abundances in Gaseous Nebulae: Open Problems in Nebular Astrophysics* (2019)
29. A.L. Rudolph, M. Fich, G.R. Bell, T. Norsen, J.P. Simpson, M.R. Haas, E.F. Erickson, Astrophys. J. Suppl. Ser. **162**(2), 346 (2006). https://doi.org/10.1086/498869
30. A. Afflerbach, E. Churchwell, M.W. Werner, Astrophys. J. **478**(1), 190 (1997). https://doi.org/10.1086/303771
31. E. Rosolowsky, J.D. Simon, Astrophys. J. **675**(2), 1213 (2008). https://doi.org/10.1086/527407
32. C. Esteban, J. García-Rojas, Mon. Not. R. Astron. Soc. **478**(2), 2315 (2018). https://doi.org/10.1093/mnras/sty1168
33. F. Bresolin, Astrophys. J. **730**(2), 129 (2011). https://doi.org/10.1088/0004-637X/730/2/129
34. G. Stasińska, Astron. Astrophys. **434**(2), 507 (2005). https://doi.org/10.1051/0004-6361:20042216
35. R.L. Sanders, A.E. Shapley, N.A. Reddy, M. Kriek, B. Siana, A.L. Coil, B. Mobasher, I. Shivaei, W.R. Freeman, M. Azadi, S.H. Price, G. Leung, T. Fetherolf, L. de Groot, T. Zick, F.M. Fornasini, G. Barro, (2019). e-prints arXiv:1907.00013
36. G. Domínguez-Guzmán, M. Rodríguez, C. Esteban, J. García-Rojas, (2019). e-prints arXiv:1906.02102
37. J. García-Rojas, C. Esteban, M. Peimbert, M.T. Costado, M. Rodríguez, A. Peimbert, M.T. Ruiz, Mon. Not. R. Astron. Soc. **368**(1), 253 (2006). https://doi.org/10.1111/j.1365-2966.2006.10105.x
38. W. Wang, X.W. Liu, Y. Zhang, M.J. Barlow, Astron. Astrophys. **427**, 873 (2004). https://doi.org/10.1051/0004-6361:20041470
39. A. Mesa-Delgado, M. Núñez-Díaz, C. Esteban, J. García-Rojas, N. Flores-Fajardo, L. López-Martín, Y.G. Tsamis, W.J. Henney, Mon. Not. R. Astron. Soc. **426**(1), 614 (2012). https://doi.org/10.1111/j.1365-2966.2012.21230.x
40. M.M. De Robertis, R.J. Dufour, R.W. Hunt, J. R. Astron. Soc. Can. **81**, 195 (1987)
41. R.A. Shaw, R.J. Dufour, Publ. Astron. Soc. Pac. **107**, 896 (1995). https://doi.org/10.1086/133637
42. R. Wesson, D.J. Stock, P. Scicluna, Mon. Not. R. Astron. Soc. **422**(4), 3516 (2012). https://doi.org/10.1111/j.1365-2966.2012.20863.x
43. V. Luridiana, C. Morisset, R.A. Shaw, Astron. Astrophys. **573**, A42 (2015). https://doi.org/10.1051/0004-6361/201323152
44. B.E.J. Pagel, M.G. Edmunds, D.E. Blackwell, M.S. Chun, G. Smith, Mon. Not. R. Astron. Soc. **189**, 95 (1979). https://doi.org/10.1093/mnras/189.1.95
45. D. Alloin, S. Collin-Souffrin, M. Joly, L. Vigroux, Astron. Astrophys. **78**, 200 (1979)
46. L.S. Pilyugin, T.X. Thuan, Astrophys. J. **631**(1), 231 (2005). https://doi.org/10.1086/432408
47. L.S. Pilyugin, J.M. Vílchez, T.X. Thuan, Astrophys. J. **720**(2), 1738 (2010). https://doi.org/10.1088/0004-637X/720/2/1738
48. S.S. McGaugh, Astrophys. J. **380**, 140 (1991). https://doi.org/10.1086/170569
49. L.J. Kewley, M.A. Dopita, Astrophys. J. Suppl. Ser. **142**(1), 35 (2002). https://doi.org/10.1086/341326
50. Á.R. López-Sánchez, C. Esteban, Astron. Astrophys. **517**, A85 (2010). https://doi.org/10.1051/0004-6361/201014156
51. Á.R. López-Sánchez, M.A. Dopita, L.J. Kewley, H.J. Zahid, D.C. Nicholls, J. Scharwächter, Mon. Not. R. Astron. Soc. **426**(4), 2630 (2012). https://doi.org/10.1111/j.1365-2966.2012.21145.x

52. N. Vale Asari, G. Stasińska, C. Morisset, R. Cid Fernandes, Mon. Not. R. Astron. Soc. **460**(2), 1739 (2016). https://doi.org/10.1093/mnras/stw971
53. I.T. Ho, Mon. Not. R. Astron. Soc. **485**(3), 3569 (2019). https://doi.org/10.1093/mnras/stz649
54. G. Stasinska, (2019). e-prints arXiv:1906.04520
55. G. Stasińska, Astron. Astrophys. **454**(3), L127 (2006). https://doi.org/10.1051/0004-6361:20065516
56. T. Parikh, D. Thomas, C. Maraston, K.B. Westfall, J. Lian, A. Fraser-McKelvie, B.H. Andrews, N. Drory, S. Meneses-Goytia, Mon. Not. R. Astron. Soc. **483**(3), 3420 (2019). https://doi.org/10.1093/mnras/sty3339
57. L. Sánchez-Menguiano, J. Sánchez Almeida, C. Muñoz-Tuñón, S.F. Sánchez, M. Filho, H.C. Hwang, N. Drory, (2019). e-prints arXiv:1904.03930
58. S.F. Sánchez, F.F. Rosales-Ortega, J. Iglesias-Páramo, M. Mollá, J. Barrera-Ballesteros, R.A. Marino, E. Pérez, P. Sánchez-Blazquez, R. González Delgado, R. Cid Fernand es, Astron. Astrophys. **563**, A49 (2014). https://doi.org/10.1051/0004-6361/201322343
59. L. Sánchez-Menguiano, S.F. Sánchez, I. Pérez, T. Ruiz-Lara, L. Galbany, J.P. Anderson, T. Krühler, H. Kuncarayakti, J.D. Lyman, Astron. Astrophys. **609**, A119 (2018). https://doi.org/10.1051/0004-6361/201731486
60. V. Luridiana, J. García-Rojas, K. Aggarwal, M. Bautista, M. Bergemann, F. Delahaye, G. del Zanna, G. Ferland, K. Lind, A. Manchado, (2011). e-prints arXiv:1110.1873
61. V. Luridiana, J. García-Rojas, in *IAU Symposium*, vol. 283 (2012), pp. 139–143. https://doi.org/10.1017/S1743921312010848
62. M.V.F. Copetti, B.C. Writzl, Astron. Astrophys. **382**, 282 (2002). https://doi.org/10.1051/0004-6361:20011621
63. W.L. Wiese, J.R. Fuhr, T.M. Deters, *Atomic Transition Probabilities of Carbon, Nitrogen, and Oxygen: A Critical Data Compilation* (AIP Press, Melville, 1996)
64. C.J. Zeippen, Astron. Astrophys. **173**, 410 (1987)
65. B.M. McLaughlin, K.L. Bell, J. Phys. B: At. Mol. Phys. **31**(19), 4317 (1998). https://doi.org/10.1088/0953-4075/31/19/017
66. L. Juan de Dios, M. Rodríguez, Mon. Not. R. Astron. Soc. **469**(1), 1036 (2017). https://doi.org/10.1093/mnras/stx916
67. M. Peimbert, R. Costero, Boletin de los Observatorios Tonantzintla y Tacubaya **5**, 3 (1969)
68. M. Peimbert, S. Torres-Peimbert, Mon. Not. R. Astron. Soc. **179**, 217 (1977). https://doi.org/10.1093/mnras/179.2.217
69. G. Delgado-Inglada, A. Medina-Amayo, G. Stasińska, (2019). e-prints arXiv:1905.01191
70. G. Delgado-Inglada, C. Morisset, G. Stasińska, Mon. Not. R. Astron. Soc. **440**(1), 536 (2014). https://doi.org/10.1093/mnras/stu341
71. G. Stasińska, Astron. Astrophys. **66**, 257 (1978)
72. J.S. Mathis, M.R. Rosa, Astron. Astrophys. **245**, 625 (1991)
73. Y.I. Izotov, T.X. Thuan, V.A. Lipovetsky, Astrophys. J. **435**, 647 (1994). https://doi.org/10.1086/174843
74. D.R. Garnett, G.A. Shields, M. Peimbert, S. Torres-Peimbert, E.D. Skillman, R.J. Dufour, E. Terlevich, R.J. Terlevich, Astrophys. J. **513**(1), 168 (1999). https://doi.org/10.1086/306860
75. Y.I. Izotov, G. Stasińska, G. Meynet, N.G. Guseva, T.X. Thuan, Astron. Astrophys. **448**(3), 955 (2006). https://doi.org/10.1051/0004-6361:20053763
76. E. Pérez-Montero, G.F. Hägele, T. Contini, Á.I. Díaz, Mon. Not. R. Astron. Soc. **381**(1), 125 (2007). https://doi.org/10.1111/j.1365-2966.2007.12213.x
77. O.L. Dors, E. Pérez-Montero, G.F. Hägele, M.V. Cardaci, A.C. Krabbe, Mon. Not. R. Astron. Soc. **456**(4), 4407 (2016). https://doi.org/10.1093/mnras/stv2995
78. R.L. Kingsburgh, M.J. Barlow, Mon. Not. R. Astron. Soc. **271**, 257 (1994). https://doi.org/10.1093/mnras/271.2.257
79. M. Rodríguez, R.H. Rubin, Astrophys. J. **626**(2), 900 (2005). https://doi.org/10.1086/429958
80. G. Stasińska, Y. Izotov, Astron. Astrophys. **397**, 71 (2003). https://doi.org/10.1051/0004-6361:20021510

81. C. Esteban, J. García-Rojas, V. Pérez-Mesa, Mon. Not. R. Astron. Soc. **452**(2), 1553 (2015). https://doi.org/10.1093/mnras/stv1367
82. D.A. Berg, D.K. Erb, R.B.C. Henry, E.D. Skillman, K.B.W. McQuinn, Astrophys. J. **874**(1), 93 (2019). https://doi.org/10.3847/1538-4357/ab020a
83. G. Delgado-Inglada, M. Rodríguez, M. Peimbert, G. Stasińska, C. Morisset, Mon. Not. R. Astron. Soc. **449**(2), 1797 (2015). https://doi.org/10.1093/mnras/stv388
84. P. Marigo, Astron. Astrophys. **370**, 194 (2001). https://doi.org/10.1051/0004-6361:20000247
85. D.A. García-Hernández, P. Ventura, G. Delgado-Inglada, F. Dell'Agli, M. Di Criscienzo, A. Yagüe, Mon. Not. R. Astron. Soc. **461**(1), 542 (2016). https://doi.org/10.1093/mnras/stw1349
86. M. Pignatari, F. Herwig, R. Hirschi, M. Bennett, G. Rockefeller, C. Fryer, F.X. Timmes, C. Ritter, A. Heger, S. Jones, U. Battino, A. Dotter, R. Trappitsch, S. Diehl, U. Frischknecht, A. Hungerford, G. Magkotsios, C. Travaglio, P. Young, Astrophys. J. Suppl. Ser. **225**(2), 24 (2016). https://doi.org/10.3847/0067-0049/225/2/24
87. P. Leisy, M. Dennefeld, Astron. Astrophys. **456**(2), 451 (2006). https://doi.org/10.1051/0004-6361:20053063
88. K. Lodders, Astrophys. Space Sci. Proc. **16**, 379 (2010). https://doi.org/10.1007/978-3-642-10352-0_8
89. D.A. García-Hernández, P. Ventura, G. Delgado-Inglada, F. Dell'Agli, M. Di Criscienzo, A. Yagüe, Mon. Not. R. Astron. Soc. **458**(1), L118 (2016). https://doi.org/10.1093/mnrasl/slw029
90. L. Searle, Astrophys. J. **168**, 327 (1971). https://doi.org/10.1086/151090
91. B.E.J. Pagel, M.G. Edmunds, Annu. Rev. Astron. Astrophys. **19**, 77 (1981). https://doi.org/10.1146/annurev.aa.19.090181.000453
92. L.H. Aller, Astrophys. J. **95**, 52 (1942). https://doi.org/10.1086/144372
93. P.A. Shaver, R.X. McGee, L.M. Newton, A.C. Danks, S.R. Pottasch, Mon. Not. R. Astron. Soc. **204**, 53 (1983). https://doi.org/10.1093/mnras/204.1.53
94. L. Deharveng, M. Peña, J. Caplan, R. Costero, Mon. Not. R. Astron. Soc. **311**(2), 329 (2000). https://doi.org/10.1046/j.1365-8711.2000.03030.x
95. C. Esteban, J. García-Rojas, M. Peimbert, A. Peimbert, M.T. Ruiz, M. Rodríguez, L. Carigi, Astrophys. J. Lett. **618**(2), L95 (2005). https://doi.org/10.1086/426889
96. C. Esteban, X. Fang, J. García-Rojas, L. Toribio San Cipriano, Mon. Not. R. Astron. Soc. **471**(1), 987 (2017). https://doi.org/10.1093/mnras/stx1624
97. A. Fernández-Martín, E. Pérez-Montero, J.M. Vílchez, A. Mampaso, Astron. Astrophys. **597**, A84 (2017). https://doi.org/10.1051/0004-6361/201628423
98. D.R. Garnett, G.A. Shields, E.D. Skillman, S.P. Sagan, R.J. Dufour, Astrophys. J. **489**(1), 63 (1997). https://doi.org/10.1086/304775
99. A. Zurita, F. Bresolin, Mon. Not. R. Astron. Soc. **427**(2), 1463 (2012). https://doi.org/10.1111/j.1365-2966.2012.22075.x
100. K.V. Croxall, R.W. Pogge, D.A. Berg, E.D. Skillman, J. Moustakas, Astrophys. J. **808**(1), 42 (2015). https://doi.org/10.1088/0004-637X/808/1/42
101. D.A. Berg, E.D. Skillman, K.V. Croxall, R.W. Pogge, J. Moustakas, M. Johnson-Groh, Astrophys. J. **806**(1), 16 (2015). https://doi.org/10.1088/0004-637X/806/1/16
102. D.S. Balser, R.T. Rood, T.M. Bania, L.D. Anderson, Astrophys. J. **738**(1), 27 (2011). https://doi.org/10.1088/0004-637X/738/1/27
103. K.V. Croxall, R.W. Pogge, D.A. Berg, E.D. Skillman, J. Moustakas, Astrophys. J. **830**(1), 4 (2016). https://doi.org/10.3847/0004-637X/830/1/4
104. W.J. Maciel, R.D.D. Costa, RevMexAA **49**, 333 (2013)
105. L. Stanghellini, M. Haywood, Astrophys. J. **862**(1), 45 (2018). https://doi.org/10.3847/1538-4357/aacaf8
106. R.B.C. Henry, K.B. Kwitter, A.E. Jaskot, B. Balick, M.A. Morrison, J.B. Milingo, Astrophys. J. **724**(1), 748 (2010). https://doi.org/10.1088/0004-637X/724/1/748
107. J. García-Rojas, (2018). e-prints arXiv:1810.09755

108. E. Pérez, R. Cid Fernandes, R.M. González Delgado, R. García-Benito, S.F. Sánchez, B. Husemann, D. Mast, J.R. Rodón, D. Kupko, N. Backsmann, A.L. de Amorim, G. van de Ven, J. Walcher, L. Wisotzki, C. Cortijo-Ferrero, CALIFA Collaboration, Astrophys. J. Lett. **764**(1), L1 (2013). https://doi.org/10.1088/2041-8205/764/1/L1
109. M. Mollá, Á.I. Díaz, O. Cavichia, B.K. Gibson, W.J. Maciel, R.D.D. Costa, Y. Ascasibar, C.G. Few, Mon. Not. R. Astron. Soc. **482**(3), 3071 (2019). https://doi.org/10.1093/mnras/sty2877
110. A.B. Wyse, Astrophys. J. **95**, 356 (1942). https://doi.org/10.1086/144409
111. J. García-Rojas, C. Esteban, Astrophys. J. **670**(1), 457 (2007). https://doi.org/10.1086/521871
112. M. Peimbert, Astrophys. J. **150**, 825 (1967). https://doi.org/10.1086/149385
113. S. Torres-Peimbert, M. Peimbert, E. Daltabuit, Astrophys. J. **238**, 133 (1980). https://doi.org/10.1086/157966
114. S. Torres-Peimbert, M. Peimbert, M. Pena, Astron. Astrophys. **233**, 540 (1990)
115. Y.G. Tsamis, D. Péquignot, Mon. Not. R. Astron. Soc. **364**(2), 687 (2005). https://doi.org/10.1111/j.1365-2966.2005.09595.x
116. H.B. Yuan, X.W. Liu, D. Péquignot, R.H. Rubin, B. Ercolano, Y. Zhang, Mon. Not. R. Astron. Soc. **411**(2), 1035 (2011). https://doi.org/10.1111/j.1365-2966.2010.17732.x
117. W.J. Henney, G. Stasińska, Astrophys. J. **711**(2), 881 (2010). https://doi.org/10.1088/0004-637X/711/2/881
118. G. Stasińska, G. Tenorio-Tagle, M. Rodríguez, W.J. Henney, Astron. Astrophys. **471**(1), 193 (2007). https://doi.org/10.1051/0004-6361:20065675
119. D.C. Nicholls, M.A. Dopita, R.S. Sutherland, Astrophys. J. **752**(2), 148 (2012). https://doi.org/10.1088/0004-637X/752/2/148
120. C. Mendoza, M.A. Bautista, Astrophys. J. **785**(2), 91 (2014). https://doi.org/10.1088/0004-637X/785/2/91
121. G.J. Ferland, W.J. Henney, C.R. O'Dell, M. Peimbert, RevMexAA **52**, 261 (2016)
122. B.T. Draine, C.D. Kreisch, Astrophys. J. **862**(1), 30 (2018). https://doi.org/10.3847/1538-4357/aac891
123. Y. Zhang, B. Zhang, X.W. Liu, Astrophys. J. **817**(1), 68 (2016). https://doi.org/10.3847/0004-637X/817/1/68
124. W. Wang, X.W. Liu, Mon. Not. R. Astron. Soc. **381**(2), 669 (2007). https://doi.org/10.1111/j.1365-2966.2007.12198.x
125. X.W. Liu, M.J. Barlow, Y. Zhang, R.J. Bastin, P.J. Storey, Mon. Not. R. Astron. Soc. **368**(4), 1959 (2006). https://doi.org/10.1111/j.1365-2966.2006.10283.x
126. Y.G. Tsamis, M.J. Barlow, X.W. Liu, P.J. Storey, I.J. Danziger, Mon. Not. R. Astron. Soc. **353**(3), 953 (2004). https://doi.org/10.1111/j.1365-2966.2004.08140.x
127. Y. Zhang, X.W. Liu, R. Wesson, P.J. Storey, Y. Liu, I.J. Danziger, Mon. Not. R. Astron. Soc. **351**(3), 935 (2004). https://doi.org/10.1111/j.1365-2966.2004.07838.x
128. Y. Zhang, H.B. Yuan, C.T. Hua, X.W. Liu, J. Nakashima, S. Kwok, Astrophys. J. **695**(1), 488 (2009). https://doi.org/10.1088/0004-637X/695/1/488
129. A. Mesa-Delgado, C. Esteban, J. García-Rojas, V. Luridiana, M. Bautista, M. Rodríguez, L. López-Martín, M. Peimbert, Mon. Not. R. Astron. Soc. **395**(2), 855 (2009). https://doi.org/10.1111/j.1365-2966.2009.14554.x
130. Y.G. Tsamis, J.R. Walsh, J.M. Vílchez, D. Péquignot, Mon. Not. R. Astron. Soc. **412**(2), 1367 (2011). https://doi.org/10.1111/j.1365-2966.2010.17991.x
131. R.L.M. Corradi, J. García-Rojas, D. Jones, P. Rodríguez-Gil, Astrophys. J. **803**(2), 99 (2015). https://doi.org/10.1088/0004-637X/803/2/99
132. D. Jones, R. Wesson, J. García-Rojas, R.L.M. Corradi, H.M.J. Boffin, Mon. Not. R. Astron. Soc. **455**(3), 3263 (2016). https://doi.org/10.1093/mnras/stv2519
133. X.W. Liu, S.G. Luo, M.J. Barlow, I.J. Danziger, P.J. Storey, Mon. Not. R. Astron. Soc. **327**(1), 141 (2001). https://doi.org/10.1046/j.1365-8711.2001.04676.x
134. D.R. Garnett, H.L. Dinerstein, Astrophys. J. **558**(1), 145 (2001). https://doi.org/10.1086/322452
135. J. García-Rojas, R.L.M. Corradi, H.M.J. Boffin, H. Monteiro, D. Jones, R. Wesson, A. Cabrera-Lavers, P. Rodríguez-Gil, in *Planetary Nebulae: Multi-Wavelength Probes of*

*Stellar and Galactic Evolution, IAU Symposium*, ed. by X. Liu, L. Stanghellini, A. Karakas vol. 323 (2017), pp. 65–69. https://doi.org/10.1017/S1743921317000849
136. J. García-Rojas, R.L.M. Corradi, H. Monteiro, D. Jones, P. Rodríguez-Gil, A. Cabrera-Lavers, Astrophys. J. Lett. **824**(2), L27 (2016). https://doi.org/10.3847/2041-8205/824/2/L27
137. M.G. Richer, L. Georgiev, A. Arrieta, S. Torres-Peimbert, Astrophys. J. **773**(2), 133 (2013). https://doi.org/10.1088/0004-637X/773/2/133
138. M.G. Richer, G. Suárez, J.A. López, M.T. García Díaz, Asrton. J. **153**(3), 140 (2017). https://doi.org/10.3847/1538-3881/aa5f53
139. R.E. Williams, N.J. Woolf, E.K. Hege, R.L. Moore, D.A. Kopriva, Astrophys. J. **224**, 171 (1978). https://doi.org/10.1086/156362
140. R.E. Williams, Astrophys. J. **261**, 170 (1982). https://doi.org/10.1086/160328
141. L. Carigi, M. Peimbert, C. Esteban, J. García-Rojas, Astrophys. J. **623**(1), 213 (2005). https://doi.org/10.1086/428491
142. C. Esteban, J. García-Rojas, L. Carigi, M. Peimbert, F. Bresolin, A.R. López-Sánchez, A. Mesa-Delgado, Mon. Not. R. Astron. Soc. **443**(1), 624 (2014). https://doi.org/10.1093/mnras/stu1177
143. D.A. Berg, E.D. Skillman, R.B.C. Henry, D.K. Erb, L. Carigi, Astrophys. J. **827**(2), 126 (2016). https://doi.org/10.3847/0004-637X/827/2/126
144. Y. Liu, X.W. Liu, S.G. Luo, M.J. Barlow, Mon. Not. R. Astron. Soc. **353**(4), 1231 (2004). https://doi.org/10.1111/j.1365-2966.2004.08155.x
145. L. Stanghellini, T.H. Lee, R.A. Shaw, B. Balick, E. Villaver, Astrophys. J. **702**(1), 733 (2009). https://doi.org/10.1088/0004-637X/702/1/733
146. R.J. Dufour, K.B. Kwitter, R.A. Shaw, R.B.C. Henry, B. Balick, R.L.M. Corradi, Astrophys. J. **803**(1), 23 (2015). https://doi.org/10.1088/0004-637X/803/1/23
147. R.B.C. Henry, B.G. Stephenson, M.M. Miller Bertolami, K.B. Kwitter, B. Balick, Mon. Not. R. Astron. Soc. **473**(1), 241 (2018). https://doi.org/10.1093/mnras/stx2286
148. T.R. Miller, R.B.C. Henry, B. Balick, K.B. Kwitter, R.J. Dufour, R.A. Shaw, R.L.M. Corradi, Mon. Not. R. Astron. Soc. **482**(1), 278 (2019). https://doi.org/10.1093/mnras/sty2517
149. G. Delgado-Inglada, M. Rodríguez, Astrophys. J. **784**(2), 173 (2014). https://doi.org/10.1088/0004-637X/784/2/173
150. M. Peimbert, A. Peimbert, M.T. Ruiz, C. Esteban, Astrophys. J. Suppl. Ser. **150**(2), 431 (2004). https://doi.org/10.1086/381090
151. B. Sharpee, Y. Zhang, R. Williams, E. Pellegrini, K. Cavagnolo, J.A. Baldwin, M. Phillips, X.W. Liu, Astrophys. J. **659**(2), 1265 (2007). https://doi.org/10.1086/511665
152. C. Esteban, F. Bresolin, M. Peimbert, J. García-Rojas, A. Peimbert, A. Mesa-Delgado, Astrophys. J. **700**(1), 654 (2009). https://doi.org/10.1088/0004-637X/700/1/654
153. Á.R. López-Sánchez, C. Esteban, J. García-Rojas, M. Peimbert, M. Rodríguez, Astrophys. J. **656**(1), 168 (2007). https://doi.org/10.1086/510112
154. A.I. Karakas, M. Lugaro, Astrophys. J. **825**(1), 26 (2016). https://doi.org/10.3847/0004-637X/825/1/26
155. M. Rodriguez, Astron. Astrophys. **313**, L5 (1996)
156. M. Rodríguez, Astrophys. J. **590**(1), 296 (2003). https://doi.org/10.1086/374925
157. G. Delgado Inglada, M. Rodríguez, A. Mampaso, K. Viironen, Astrophys. J. **694**(2), 1335 (2009). https://doi.org/10.1088/0004-637X/694/2/1335
158. Y. Zhang, X. Fang, W. Chau, C.H. Hsia, X.W. Liu, S. Kwok, N. Koning, Astrophys. J. **754**(1), 28 (2012). https://doi.org/10.1088/0004-637X/754/1/28
159. G. Delgado-Inglada, A. Mesa-Delgado, J. García-Rojas, M. Rodríguez, C. Esteban, Mon. Not. R. Astron. Soc. **456**(4), 3855 (2016). https://doi.org/10.1093/mnras/stv2961
160. D. Pequignot, J.P. Baluteau, Astron. Astrophys. **283**, 593 (1994)
161. H.L. Dinerstein, Astrophys. J. Lett. **550**(2), L223 (2001). https://doi.org/10.1086/319645
162. N.C. Sterling, H.L. Dinerstein, Astrophys. J. Suppl. Ser. **174**(1), 158 (2008). https://doi.org/10.1086/520845
163. N.C. Sterling, H.L. Dinerstein, K.F. Kaplan, M.A. Bautista, Astrophys. J. Lett. **819**(1), L9 (2016). https://doi.org/10.3847/2041-8205/819/1/L9

164. N.C. Sterling, S. Madonna, K. Butler, J. García-Rojas, A.L. Mashburn, C. Morisset, V. Luridiana, I.U. Roederer, Astrophys. J. **840**(2), 80 (2017). https://doi.org/10.3847/1538-4357/aa6c28
165. S. Madonna, M. Bautista, H.L. Dinerstein, N.C. Sterling, J. García-Rojas, K.F. Kaplan, M. del Mar Rubio-Díez, N. Castro-Rodríguez, F. Garzón, Astrophys. J. Lett. **861**(1), L8 (2018). https://doi.org/10.3847/2041-8213/aaccef
166. N.C. Sterling, H.L. Dinerstein, C.W. Bowers, Astrophys. J. Lett. **578**(1), L55 (2002). https://doi.org/10.1086/344473
167. N.C. Sterling, M.C. Witthoeft, Astron. Astrophys. **529**, A147 (2011). https://doi.org/10.1051/0004-6361/201116718
168. N.C. Sterling, R.L. Porter, H.L. Dinerstein, Astrophys. J. Suppl. Ser. **218**(2), 25 (2015). https://doi.org/10.1088/0067-0049/218/2/25
169. A.L. Mashburn, N.C. Sterling, S. Madonna, H.L. Dinerstein, I.U. Roederer, T.R. Geballe, Astrophys. J. Lett. **831**(1), L3 (2016). https://doi.org/10.3847/2041-8205/831/1/L3
170. S. Madonna, PhD Thesis. *Neutron-capture Elements in Planetary Nebulae as Tracers of Nucleosynthesis in AGB Stars* (2019)

# Observational Constraints on the Common Envelope Phase

**David Jones**

## Contents



**Abstract** The common envelope phase was first proposed more than 40 years ago to explain the origins of evolved, close binaries like cataclysmic variables. It is now believed that the phase plays a critical role in the formation of a wide variety of other phenomena ranging from type Ia supernovae through to binary black holes, while common envelope mergers are likely responsible for a range of enigmatic transients and supernova imposters. Yet, despite its clear importance, the common envelope phase is still rather poorly understood. Here, we outline some of the basic principles involved, the remaining questions as well as some of the recent observational hints from common envelope phenomena—namely planetary nebulae and luminous red novae—which may lead to answering these open questions.

D. Jones (✉)
Instituto de Astrofísica de Canarias, La Laguna, Tenerife, Spain

Departamento de Astrofísica, Universidad de La Laguna, La Laguna, Tenerife, Spain
e-mail: djones@iac.es



P. Kabáth et al. (eds.), *Reviews in Frontiers of Modern Astrophysics*,
https://doi.org/10.1007/978-3-030-38509-5_5

## 1 Preface

It is important to highlight from the outset that the common envelope (CE) and its progeny have been the subject of constant and vibrant study, both observational and theoretical, for many years. As a result, several excellent reviews have already been written [44, 50, 118, 129]. The review presented here is intended to be complementary to these, with an emphasis on the (sometimes puzzling) observations which may hold the key to understanding the CE.

## 2 Introduction

The Roche model (named for the nineteenth century French astronomer Édouard Roche) describes the gravitational potential of a close-binary system assuming that the two stars are well represented by synchronously-rotating point masses in a circular orbit [65]. While rather simplistic, this model is still a relatively good representation of many close binaries with detailed numerical simulations generally required to complement and verify the model [22, 23].

The Roche model predicts the existence of five local minima of gravitational potential surrounding the binary, known as the Lagrangian points (named for Joseph-Louis Lagrange who discovered the fourth and fifth points shortly after Leonhard Euler discovered the first three). Furthermore, the Roche model also calculates the existence of an equipotential surface enclosing each star which represents the largest extent at which a point mass could be gravitationally bound to that star (rather than to the binary system as a whole or unbound completely). These so-called Roche lobes meet at the first Lagrangian point. The configuration of the Roche lobes and five Lagrangian points is highlighted in Fig. 1. The extent of the Roche lobe is a function of the orbital separation and mass ratio of the binary components (being larger for the more massive component). A commonly used approximation for the Roche lobe radius, $R_{1,\mathrm{RL}}$, is

$$R_{1,\mathrm{RL}} = a\, r_{1,\mathrm{RL}} \tag{1}$$

$$r_{1,\mathrm{RL}} \approx \frac{0.49q^{2/3}}{0.6q^{2/3} + \ln(1 + q^{1/3})} \tag{2}$$

where

$$q = \frac{M_1}{M_2} \tag{3}$$

and $a$ is the separation of the stellar centres of mass, $M_1$ is the mass of the star for which we are calculating the Roche lobe radius and $M_2$ is the mass of its companion [26]. This Roche lobe radius is conceptually important in understanding the CE as it

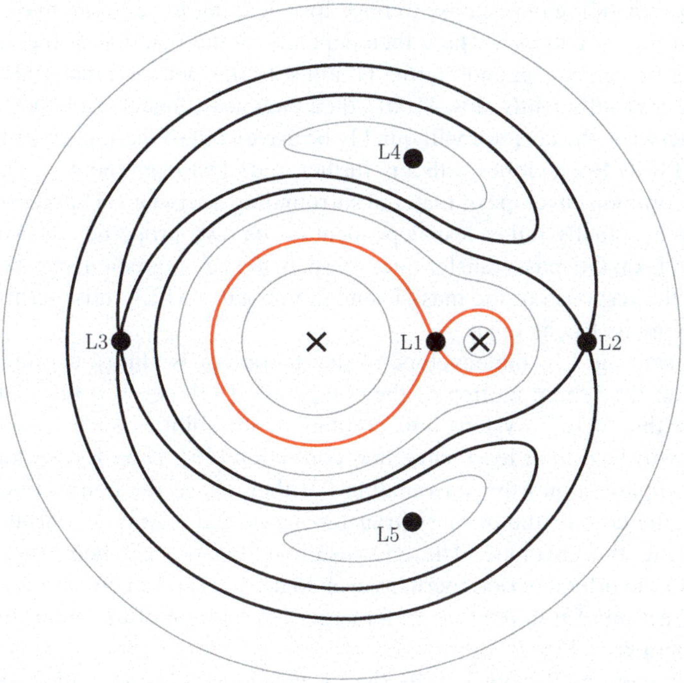


**Fig. 1** The lines of equipotential, including the Roche lobe (marked in red), around a binary system with a mass ratio of 5 (i.e. the left-most star of the binary is five times as massive as its companion). The Langrangian points are also marked (L1 through L5) as are the centres of mass of the two stars (black crosses)

represents the critical stellar radius, beyond which the star will begin to transfer material on to its companion via $L_1$. A process known as Roche-lobe overflow (RLOF). This can occur at various stages during the evolution of the star depending on the component masses as well as the binary orbital period (more massive stars at shorter orbital periods are more likely to fill their Roche lobes earlier during their evolution).

If RLOF occurs while the donor is on the main sequence, this is generally referred to as Case A RLOF. If the star is on its first ascent of the giant branch (i.e. a red giant), then it would be Case B, and while on the second ascent (asymptotic giant branch) it is Case C. Given the change in stellar radius between these evolutionary stages, Case A occurs at much shorter orbital periods than Case B and Case C (see Sect. 3 for further discussion). These are important definitions as the evolutionary phase of the donor can have an important impact on its reaction to the mass loss. For example, a star with a large, radiative envelope is likely to shrink in response to mass loss (and thus, perhaps, recede away from filling its Roche lobe, ending the mass transfer), while those with deep convective envelopes are more likely to expand [27]. In this case, a positive feedback loop is initiated with the star continually expanding in response to mass loss, leading to yet more mass loss. The response of the system as a whole then depends on the reaction of the accretor. If the companion can accept and thermally-adjust to this accreted material (i.e. if the mass transfer is sufficiently slow [100]), then the mass transfer could potentially be stable. Otherwise, the accretor will quickly be driven out of thermal equilibrium and expand to fill its Roche lobe, with any further mass lost from the donor's envelope forming a common envelope of material surrounding the binary. The response of the companion is actually rather less dependent on its own properties or evolutionary state than it is on the mass transfer rate. As such, the key ingredient in forming a CE (or not) is the response of the mass-losing donor star—which must act to continue to overflowing its Roche lobe.

Once inside the CE, the envelope of the donor star is almost certainly not co-rotating with the orbital motion of the companion [89]—even if tidal interactions had driven the pre-CE system into corotation, instabilities such as the Darwin instability will lead to at least some non-corotation [15]. Drag forces between the orbiting companion and the surrounding CE then cause the companion to spiral in towards the core of the primary, transferring orbital energy and orbital angular momentum to the envelope. The end result of this process being the dramatic reduction of the orbital period (perhaps even to merger) and ejection of the envelope (or some fraction of it in the case of a merger). A cartoon of the main steps in this process is shown in Fig. 2.

As pointed out by Paczynski [89], the CE provides a clear evolutionary pathway towards the formation of close binaries with an evolved component which would

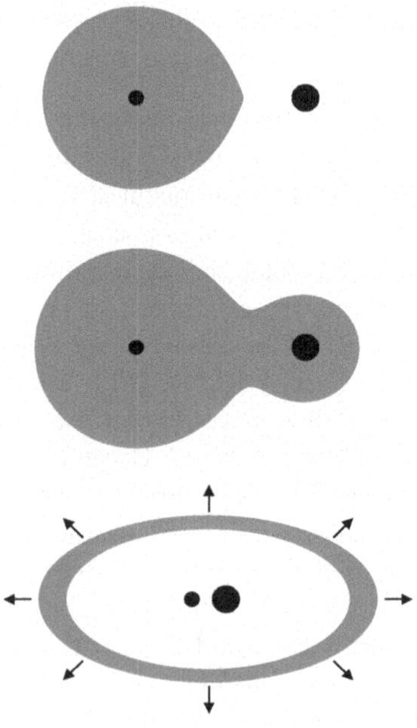

**Fig. 2** A toy model of the common envelope. The phase begins when the primary (left) begins to overflow its Roche lobe transferring material to its companion (right), as shown by the configuration in the top panel. If neither the primary nor the secondary can adjust to the mass transfer, the secondary will also fill its Roche lobe resulting in the formation of a common envelope of material surrounding the secondary and the primary's core (middle panel). Drag forces then transfer orbital energy and angular momentum from the binary to the envelope leading to its ejection while reducing the binary orbital separation (bottom panel)

have been too large while on the giant branch to exist in the current orbital configuration—highlighting the example of V471 Tau which comprises a K-type main sequence star in a 12.5-h orbit with a $0.8\,M_{\odot}$ white dwarf (WD). A system which would have necessitated a $\sim$10 year orbit in order to accommodate the full asymptotic giant branch radius of the WD progenitor.

In discussing the CE hypothesis, [89] concluded that following its ejection:

> We are left with two small stars accreting whatever hydrogen rich matter is left within their Roche lobes. At this time the degenerate core with the remaining envelope has a structure which is identical to that of a nucleus of a planetary nebula. This hot star will ionize the expanding envelope. As a result we should see a planetary nebula with a close binary as its nucleus.

Indeed, the discovery of the first close-binary planetary nebula (PN) nucleus, which they considered "important support for the evolutionary scenario" of the CE, was found later that year [5]. We will return to the importance of PNe in understanding the CE later, but first we must cover more of the theory and mathematical prescriptions used to study the CE.

## 3 Conditions for a Common Envelope

For a binary to experience a CE, clearly, one component must initiate the process by filling its Roche lobe—this can occur for a number of reasons. Dynamical interactions, for example with a third body leading to Kozai–Lidov interactions [111, 117], can shrink the orbital separation sufficiently that one star becomes Roche lobe filling. In the majority of cases, however, the CE will be initiated due to the radius evolution of the star. This gives us important clues as to when a given star could fill its Roche lobe and initiate the CE, for example, as the radius evolution on the main sequence (MS) is minimal one would expect the vast majority of CE's to occur when the primary has evolved off the MS. Similarly, the much smaller MS radii would also imply far smaller orbital separations, again restricting the likelihood of a MS CE event.

Looking at the evolution of stellar radii (see e.g. Fig. 3), one can see that more massive stars reach larger maximum radii at the tip of the AGB than their lower mass counterparts—thus increasing the range of orbital periods for which such a star will fill its Roche lobe and thus the likelihood of experiencing a CE event.[1] Perhaps more interestingly, lower mass stars tend to reach larger radii while on the RGB, with their maximum RGB radii being rather comparable to their maximum radii while on the AGB. This means that for lower mass stars, the vast majority of CE events will occur while on the RGB while the envelope itself is more massive and more bound—likely impacting on the energetics of the CE and the likelihood of ejection/binary survival

[1]This is roughly assuming that the orbital period distribution is not strongly dependent on the primary mass, which may not be the case [79].

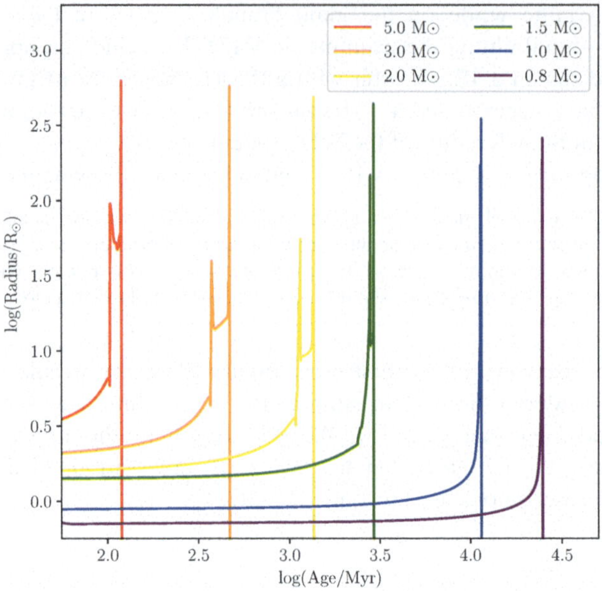


**Fig. 3** The evolution of stellar radius as a function of time for a variety of initial stellar masses, (based on MIST tracks [11, 94]). Note that while the maximum radius (always at the tip of the AGB) increases as a function of initial mass, lower mass stars tend to achieve larger radii while on the RGB greatly increasing the likelihood of experiencing a CE while on the RGB

(see Sect. 4). More massive stars, however reach significantly larger radii while on the AGB compared to the RGB, thus these stars are more likely to experience AGB CE events particularly given the observed orbital period distribution [79].

Somewhat obviously, the likelihood of entering a CE is not only a function of the primary's radius, but also on the parameters of the binary—namely the orbital separation and mass ratio (as shown in Eq. 1). To highlight these dependencies, in Fig. 4, the Roche lobe radius is plotted as a function of orbital period for a range of binary configurations. It is clear that all masses of primary will be Roche lobe filling at longer orbital periods for smaller mass ratios. In terms of primary mass, more massive primaries have larger Roche lobe radii but also reach larger maximum radii on the AGB counteracting the effect—in the systems considered this leads to a similar spread of maximum orbital periods for which the systems will experience Roche lobe overflow, however this is not a general rule. It is important to also consider the possible influence of orbital eccentricity, which acts to reduce the orbital separation at periastron passage (by a factor $1 - e$) and increase the likelihood of Roche lobe overflow. Note, however, that as the system evolves tidal dissipation should act to reduce orbital eccentricity unless some eccentricity pumping mechanism is at work in the system [107, 127].

Thus far, we have considered only the conditions required for a star to fill its Roche lobe but this is not the only necessity for a CE. The resulting mass transfer

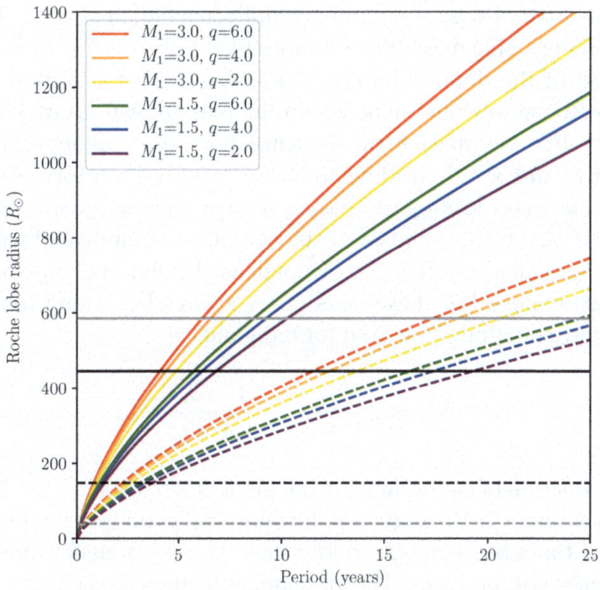


**Fig. 4** The Roche lobe radius as a function of orbital period for various combinations of primary mass ($M_1$) and mass ratio ($q$). The solid lines show configurations with zero eccentricity while the dashed lines are for eccentricity, $e = 0.5$. The horizontal lines represent the maximum RGB (dashed) and AGB (solid) radii for the primary masses considered, with the gray lines being $M_1 = 3.0\,\mathrm{M}_\odot$ while the black lines show $M_1 = 1.5\,\mathrm{M}_\odot$

must also be dynamically unstable—a runaway process where the reaction of the primary to mass loss is to continue to overflow its Roche lobe. If the mass transfer is conservative, the primary's Roche-lobe radius evolves only with the mass ratio and thus the stability of mass transfer only depends on the mass ratio and how the primary's radius reacts to the mass loss. Often, the primary's radius is considered to be proportional to some exponent of its mass, $R_\mathrm{d} \propto M_1^{\zeta_*}$, with the Roche lobe radius behaving similarly but with a different exponent, $R_\mathrm{L} \propto M_1^{\zeta_\mathrm{RL}}$. The mass transfer would thus be unstable if $\zeta_\mathrm{RL} > \zeta_*$.

Given that the Roche lobe radius is principally dependent on the mass ratio but also on the total system mass and orbital period, the value of $\zeta_\mathrm{RL}$ depends heavily on whether the mass transfer is conservative. In the fully conservative case, $\zeta_\mathrm{RL}$ is proportional to the mass ratio and often approximated by [27]:

$$\zeta_\mathrm{RL} \approx 2.13q - 1.67. \tag{4}$$

As such, a condition for unstable mass transfer becomes:

$$q > \frac{\zeta_*}{2.13} + 0.79. \tag{5}$$

The non-conservative case is rather more complicated but, in general, leads to lower values of $\zeta_*$ making stable mass transfer more likely.

The response of the donor's radius, characterised by $\zeta_*$, is more complicated than that of its Roche lobe and depends on the mass transfer time scale. For fully convective stars, like low-mass main sequence stars, one can approximate the star as a polytrope of index 1.5, leading to $\zeta_* = -1/3$. Therefore, such stars will always respond to mass loss by increasing in size, thereby guaranteeing unstable mass transfer. However, RGB and AGB stars are more complicated with convective envelopes which necessitate the use of condensed polytropes or, more properly, complete stellar models. For these stars, $\zeta_*$ is always larger and quite frequently positive. An approximation often used for red giants is:

$$3\zeta_* = 2\frac{q_c}{1-q_c} - \frac{1-q_c}{1+2q_c}, \tag{6}$$

where $q_c$ is the fraction of stellar mass in the giant donor's core [113]. However, this approximation assumes hydrostatic equilibrium—an assumption which only holds if the dynamical timescale is much shorter than the mass transfer timescale (which is almost certainly not the case). This dramatically alters the stability condition for the mass transfer. Binary stellar evolution modelling by Woods et al. [134] showed that for a 5 $M_\odot$ giant donor with a 0.86 $M_\odot$ core, the mass transfer was found to be stable for mass ratios up to $q = 1.47$, while using Eq. 6 one would predict unstable mass transfer for $q > 0.75$. As such, strong conclusions with regards mass transfer stability cannot be drawn except for specific cases which have been modelled using detailed stellar evolutionary codes. In any case, it seems likely that only rather large mass ratios, $q \gtrsim 2$, could lead to a CE [91, 93].

## 4 Common Envelope Energetics

The outcome of the CE—be that merger or a surviving short-period binary—depends on whether the energy transferred to the envelope was sufficient to unbind it. This permits us to define a parameter, known as the common envelope efficiency, which relates the change in orbital energy due as a result of the CE to the binding energy of the envelope [124, 128]. One can write the change in orbital energy (i.e. the sum of both gravitational potential and kinetic energies of both bodies) as

$$\Delta E_{\mathrm{orb}} = G\left(\frac{M_{1,f}M_{2,f}}{2a_f} - \frac{M_{1,i}M_{2,i}}{2a_i}\right) \tag{7}$$

where the subscripts 1 and 2 refer to the Roche-lobe-filling star (primary/donor) and its companion (secondary/accretor), respectively, while $i$ and $f$ denote initial (pre-CE) and final (post-CE) values. The prescription used here defines the initial orbital energy as being that between the entirety of the overflowing star and its companion at the initial separation [128]. Alternative prescriptions define it as that of the primary's core and the secondary (i.e. the $M_{1,i}$ term in Eq. 7 becomes the primary's core mass, $M_c$, only), essentially assuming that the envelope has already engulfed the system and is now bound to the secondary and the primary's core rather than to the primary only [44, 137]. This definition is sometimes referred to as the Iben–Livio–Yungelson formulation [139] and, as we will shortly see, also implies an alternative prescription for the envelopes binding energy.

Assuming that the companion does not grow in mass during the CE (i.e. $M_{2,f} = M_{2,i} = M_2$), and that the all of the overflowing star's envelope ($M_e$) is ejected, leaving behind only its core ($M_c$), the change in orbital energy can be rewritten

$$\Delta E_{orb} = G\left(\frac{M_c M_2}{2a_f} - \frac{(M_c + M_e)M_2}{2a_i}\right). \tag{8}$$

Some fraction, $\alpha$, of this liberated orbital energy is transferred to the envelope. Thus, assuming that there are no other potential sources of energy (an assumption which likely does not hold and which we will discuss in Sect. 4.1), this fraction ($\alpha\Delta E_{orb}$) should at least be equal to the binding energy of the envelope ($E_b$) in order for the binary to exit the CE without merging,

$$E_b = \alpha\Delta E_{orb}. \tag{9}$$

As mentioned before, the definition of the binding energy is dependent on the formalism employed. For the aforementioned Iben–Livio–Yungelson formulation, the binding energy is taken to be the gravitational energy between the primary's envelope and the combined mass of the primary's core and the companion. The more-commonly-used alternative is the Podsialowski-Rappaport-Han formulation, which is more consistent with the definition of the change of orbital energy in Eqs. 7 and 8. Under this formulation, the envelope is considered to be bound only to the primary, and is often approximated as the gravitational energy between the envelope and primary mass,

$$E_b = G\frac{M_1 M_e}{\lambda r_{1,RL}} \tag{10}$$

where $\lambda$ is of order unity and describes the radial mass distribution of the primary's envelope. Here, the radius is assumed to be the Roche lobe radius as the star must be overflowing at the start of the CE. More formally, rather than estimating the binding energy of the envelope, one can integrate envelope mass from stellar structure. Using such models, the following approximation can be derived [20],

$$E_b = G\frac{(\frac{M_e}{2} + M_c)M_e}{\lambda r_{1,RL}}, \tag{11}$$

similar but subtly different to the more simplistic version shown in Eq. 10.

Combining Eqs. 8, 9 and 11 gives:

$$\frac{(\frac{M_e}{2} + M_c)M_e}{\lambda r_{1,\mathrm{RL}}} = \alpha\left(\frac{M_c M_2}{2a_f} - \frac{(M_c + M_e)M_2}{2a_i}\right) \tag{12}$$

in which $M_c$ (the core mass of the primary, assumed to also be the post-CE remnant mass), $M_2$ (the secondary mass, assumed to be unchanged by the CE) and $a_f$ (the post-CE orbital separation[2]) are all observables (or can at least be derived from observations by, for example, combined light and radial velocity curve modelling). Furthermore, using stellar evolutionary models, one can find possible progenitors for the primary which would present with consistent core masses and pre-CE radii, constraining $M_e$ ($\equiv M_1 - M_c$), $r_{1,\mathrm{RL}}$ and $a_f$ [17, 20, 139]. The only remaining variable needed in order to derive the CE efficiency for a given system is the stellar structure parameter, $\lambda$, which is frequently assumed to be a constant (e.g. 0.5 [18]) or simply left incorporated in the efficiency as the product $\alpha\lambda$ [85, 139]. Alternatively, the parameter can be estimated based on stellar evolutionary models [20, 24]. Ultimately, this permits a "reconstruction" of the CE phase and the estimation of the efficiency, $\alpha$, for individual systems (as opposed to using a population synthesis approach [121]).

## 4.1 Other Factors That May Impact α

In the previous section, we have assumed that only orbital energy is responsible for the unbinding of the CE (as has traditionally been considered). However, there are likely many other factors which may play a role in helping to remove the envelope. The most obvious of these is the thermal energy of the envelope itself. According to the Virial theorem (applied to the envelope alone rather than the whole star), the thermal energy, that reduces the value of the potential energy and makes the envelope closer to being unbound, is approximately one half the envelope's binding energy [24], thus adding a factor $\frac{1}{2}$ to the left hand side of Eq. 12 and halving the derived values of $\alpha$ [20]. However, the Virial theorem is based on the assumption of hydrostatic equilibrium, which may not apply if the CE happens on timescales much shorter than the stellar dynamical timescale. As such, while it is important to account for this thermal energy when considering the binding energy of the envelope, one must exercise caution in including in it the calculation of the total energy and thus whether the envelope is ultimately unbound or not [20].

[2] The assumption that the current separation is equal to the immediately post-CE separation only holds for "young" systems, which have experienced negligible angular momentum loss since the CE. For older systems, one must account for the influence of disrupted magnetic braking on the observed orbital period [17].

As well as the envelope's thermal energy, its rotational energy is similarly neglected in the treatment outlined earlier. This seems to be a reasonable assumption given that it should be negligible compared to its gravitational binding energy [129]. A perhaps more questionable assumption is the treatment of the primary's core and the companion as inert masses, which do not gain or lose energy nor mass during the process. It seems plausible to consider that the companion might accrete (appreciably) during the CE, however it has generally been thought that, due to the highly supersonic in-spiral and the large entropy barrier that forms between the secondary and the far less dense envelope, the total accretion is rather limited [43, 108]. More recent models seem to indicate that this might not be the case providing there is some way of releasing pressure and thus maintaining steady flows during in-spiral [9, 105], the most obvious being the formation of jets which might also help to "clear out" the envelope increasing the ejection efficiency [68, 112]. Furthermore, if the companion is itself a compact object then nuclear burning (perhaps of accreted material) on its surface could potentially provide a further energy source within the CE that might help to unbind the envelope [44].

As the CE is ejected preferentially in the orbital plane (due to the conservation of angular momentum), a pressure gradient develops which leads to material above and below the plane flowing inwards to replace that which has been ejected. This circulation could plausibly lead to mixing that increases the primary's nuclear burning rate, providing additional energy which might contribute to the envelope's ejection [44]. However, it is seemingly unlikely that, in the face of strong entropy forces, the dynamical penetration of this mixing could reach deep enough to impact on the burning region [129].

The deposition of energy into the envelope (even just via gravitational drag) can have profound effects on its stability beyond simply lifting and ejecting it. Modelling efforts have demonstrated that the envelope itself can rapidly become unstable and develop large-amplitude pulsations on relatively short timescales [12, 52]. In some cases, the shocks associated with these pulsations may be sufficient to dynamically eject up to 10% of the primary's envelope [12].

Perhaps the most contentious additional energy source in the CE is the inclusion of the envelope's recombination energy [35, 52, 83, 116]. This energy source is significant, being proportional to the mass of the envelope and of order a few times $10^{46}$ ergs c.f $\sim 10^{47}$ for the binding energy of the envelope [49, 83]. However, it has been argued that only a small fraction of this energy ($\sim$10%) might actually be able to contribute to the removal of the envelope, with the majority simply radiated away [35]. The true importance of recombination energy is still a matter of intense debate [48, 116]. Recent studies of the impact of convection on the CE indicate that the associated energy transfer timescale is shorter than dynamical timescales, such that recombination energy may be convectively carried to the outer parts of the envelope where it is unable to aid with ejection [133]. However, they also find that the inclusion of convection in their models could reduce the need for additional energy sources, as it leads the binary orbit to shrink significantly before orbital energy can be tapped for ejection [133].

Hydrodynamic models of the CE, that do not include recombination energy, invariably fail to unbind the entirety of the envelope, in the majority of cases leaving a significant fraction "lifted" away from the central binary but not completely unbound [9, 45, 88, 92, 102, 108]. This led [34] to propose an intriguing solution, just as in single AGB stars [67], radiation pressure on grains which form in the now cooling (but still bound) envelope could lead to dust-driven winds capable of efficiently unbinding the envelope. The authors also highlight that such dust formation could also aid in trapping any recombination energy from the envelope, adding weight to its possible importance in the CE ejection process. While dust forming in the expanding common envelope likely has the ability to increase the mass-loss rate from the envelope and help with unbinding it, a question remains surrounding the timescales. Dust can form rapidly in the relatively high density common envelope gas. However, fall-back of bound material can happen equally rapidly [66]. The post-in-spiral environment would therefore see a complex competition of mechanisms including: the release of recombination energy, dust formation, fall-back of material. Somehow the competition of all these processes must result in the full envelope ejection, at least in some cases.

## 4.2 The Gamma Prescription

An alternative to the $\alpha$ prescription has also been proposed based on the conservation of angular momentum [85, 86]. Here, the same energy conservation as in the $\alpha$ formalism is implicit, although not restricted simply to orbital and binding energy. This prescription can be written:

$$\frac{\Delta J}{J} = \gamma \frac{\Delta M}{M} = \gamma \frac{M_{\rm e}}{M_{\rm e} + M_{\rm c} + M_2} \tag{13}$$

where $J$ and $\Delta J$ are the total angular momentum of the binary and the change in total angular momentum, respectively.

This use of this formulation was driven by the apparent difficulty in explaining double WD binaries, which were thought to have formed from two consecutive CE episodes (though this has since been shown to not be necessary; see Sect. 5.2 [135]). Initial results were particularly encouraging, with reconstruction methods finding that a single value of $\gamma$ could well reproduce a number of systems [85, 86]. The question then became, is this formulation intrinsically "better" than its $\alpha$ counterpart? And, moreover, could the apparently universal value of $\gamma$ be used to constrain the physics of the CE? Unfortunately, it has been shown beyond doubt that the energy conserving prescription places far more constraints on the CE outcome [129]. Ultimately, the limited range of $\gamma$ which can be used to successfully reconstruct all post-CE systems [139], is rather more a short-coming of the formalism than a sign of insight into CE physics. This is highlighted by the small range of $\gamma$ required to account for essentially all possible post-CE configurations

[129]. This was shown mathematically by Woods et al. [134] who demonstrated that the ratio of initial to final orbital separations is extremely sensitive to small changes in $\gamma$—with the range of $1.5 \lesssim \gamma \lesssim 1.75$ found by Nelemans et al. [86] encompassing values that would lead to merger during the first CE and values that could lead to a double WD system. For further, in-depth discussion of failings of the $\gamma$ formalism as a replacement for $\alpha$, the reader is referred to Sect. 5.2 of [50].

### *4.3 Grazing Envelope Evolution*

We have already mentioned the possibility that jets may help remove the envelope, as well as acting as a pressure-release valve and allowing for appreciable accretion during the CE [9]. However jets could fundamentally alter the general picture of the CE as presented in Fig. 2. As outlined by Soker [114], one might envisage a situation where jets launched at the onset of the CE remove enough of the envelope to prevent engulfment and a full-blown CE event, rather maintaining the system in a constant state of "just entering the CE phase". This hypothesis is known as the Grazing Envelope (GE).

The GE evolution could prevent a CE entirely or postpone it, removing a significant amount of the envelope prior to engulfment. Clearly, this has a dramatic effect on the energetics of the CE as described above [114], principally providing a significant additional energy source proportional to the amount of mass accreted onto the secondary. As such, the GE offers a, perhaps, more reasonable explanation for longer period post-CE binaries, such as the 16-d period binary central star of the PN NGC 2346 [7] which [114] argue must have experienced a GE for a significant fraction of its evolution. Indeed, hydrodynamical simulations of the CE including jets do lead to greater final separations than those without jets and also lead to a greater fraction of the envelope being unbound, however they still fail to completely unbind the entirety of the envelope [112].

Beyond the additional energy source, the most significant difference between the CE and GE are the evolutionary timescales, with the GE expected to last tens to hundreds of years c.f. the days to months long duration of the standard CE.

## 5 Planetary Nebulae Nuclei

As previously highlighted, the immediate product of the CE ejection is expected to be a PN with a close-binary nucleus.[3] As such, we may look to these objects for clues towards understanding the CE process. It is only now, following the recent

[3]This may only be the case if the CE occurs on the AGB. If the CE occurs while on the RGB, there are doubts as to whether the post-CE evolution of the exposed core would be fast enough to ionise the expanding envelope in time for it to be visible as a PN. However recent theoretical and observational efforts seem to indicate that these doubts are unfounded and that post-RGB PNe do indeed exist [36, 42].

leaps in sample size (due to, for example, the OGLE survey [74] and recent targeted surveys [55, 56, 59, 109]), that one can begin to contemplate this idea [54]. In this section, I will try to discuss some of the more interesting findings to have come from the study of PNe in terms of the CE phase.

## *5.1 Morphologies*

The PNe surrounding post-CE central stars are thought to principally comprise the ejected envelope itself, thus they offer a unique window into the ejection process. Studying the morphologies of post-CE PNe currently represents the *only* way to observe the form of the CE when ejected in its entirety.[4] There is an added complexity in inferring the structure of CE ejecta from post-CE PNe in that the observed morphologies are the result of a complex interplay between the initial ejecta and subsequent fast wind and ionisation front originating from the emerging pre-WD core [33]. As such it is perhaps no surprise that post-CE PNe display a wide range of morphologies, but with a few over-arching trends that can be used to understand the CE. Most importantly, the vast majority of post-CE PNe present with bipolar morphologies [75]. Spatio-kinematical modelling of these bipolar structures has revealed that their symmetry axes are always found to lie perpendicular to the orbital plane of the surviving post-CE central binary [41]. This is a clear confirmation that the CE is, indeed, preferentially ejected in the orbital plane, with the subsequent equatorial over-density going on the form the waist of the resulting bipolar nebulae—in many cases leading to the formation of a ring or torus (Figs. 5 and 6 [4, 13]).

Post-CE PNe have also been found to show a prevalence of jet-like structures as well as low-ionisation filaments or knots. Jets are rather clearly a consequence of mass transfer which, given the relatively short nebular visibility times ($\tau \approx$ 30,000 years), must have occurred around the time of the CE (either just before, during or just after—a question we will return to in Sect. 5.3). The low-ionisation filaments and knots are slightly more challenging to understand, but might perhaps be related to instabilities in the envelope at the time of ejection (such as those described in [12, 52]), or to a later fast tenuous wind originating from the central star ploughing into the ejected material.

[4]The ejecta from luminous red novae, as discussed in Sect. 6, could also be used to study the CE ejection. However, these objects represent "failed" CEs, where the binary merged inside the CE (likely ejecting only a fraction of the envelope).

**Fig. 5** Image of the Necklace Nebula (Credit: Romano L. M. Corradi, IPHAS), shown to host a post-CE binary central star in which the companion is a carbon dwarf [13, 76]

**Fig. 6** FORS2 image of the post-CE PN Fg 1 (Credit: ESO/H. Boffin). The observed jets have been shown to pre-date the central nebula by a few thousand years, while the central star is a double-degenerate binary with an orbital period of 1.195 d [4]

## 5.2 *Double Degenerates*

Unfortunately, while the number of PNe known to host post-CE central stars has grown dramatically, in the majority of cases little more is known beyond the orbital period. Most systems were discovered via photometric monitoring, with the periodicity revealing the orbital period[5] and the type of variability offering some hints towards the evolutionary phase of the secondary. The majority of photometrically-discovered post-CE central stars display variability due to irradiation or ellipsoidal modulation, sometimes with eclipses superimposed if the orbital inclination is high enough. All ellipsoidally-modulated central stars subjected to further detailed study have been found to be double-degenerate (DD) systems [39, 109], where the companion is also an evolved star (WD or post-AGB star). If one assumes that all the known post-CE central stars displaying ellipsoidal modulation in their light curves are indeed DDs, then they should comprise at least 20% of the total population—likely far greater as such DD systems will only display photometric variability at very short orbital periods (indeed a number of further systems have been discovered via radial-velocity monitoring which do not display any photometric variability [4, 78]).

Such a high fraction of DD central stars is particularly intriguing for a number of reasons. As noted in Sect. 4.2, there is some debate over how such systems form—either via consecutive CE episodes or through stable mass transfer followed by a single CE, or perhaps even via a GE evolution. The first possibility—consecutive CEs—clearly presents a challenge, with the first CE already spiralling-in the binary and vastly reducing the orbital energy available to unbind a second CE (particularly problematic given that due to the reduced separation the second CE is more likely to occur with the donor on the RGB rather than AGB, where the envelope is more bound [53]). It has been shown that under certain circumstances the first CE could be avoided, with the initially more massive component losing its envelope through a phase of stable, non-conservative mass transfer allowing the binary to continue towards being DD via a single CE episode [135]. However, it is unclear what range of initial conditions could lead to this evolution, and whether this range is sufficiently broad to account for the large DD fraction observed. GE evolution can relatively easily explain such DD systems, however it is unclear what initial configurations (if any) could lead to some form of GE evolution.

A high DD fraction is particularly interesting in the context of understanding type Ia supernovae (SNe) which—in spite of being successfully employed as standard candles in probing the increased expansion rate of the Universe [98, 106] which ultimately led to the award of the Nobel prize in Physics 2011—still have rather uncertain origins. DD mergers may represent the main, or even sole, pathway by which SNe Ia occur [73], however to-date no bona-fide progenitor system has been discovered. As such, a high DD fraction, as well as the observation that many SN Ia are found to explode in circumstellar environments consistent with a remnant PN shell [123], could be construed as support for this DD merger hypothesis. Furthermore, two of the strongest candidate SN Ia progenitors have been found to reside inside PNe. However, neither has been unambiguously

[5] With lower fidelity data it can be difficult to distinguish between variability due to irradiation and variability due to ellipsoidal modulation, plausibly leading to a derived orbital period which is discrepant by a factor of two [3, 72].

shown to satisfy the criteria of being both super-Chandrasekhar mass and in a close enough orbit to merge within the age of the Universe. The central star of TS 01 was found to be a short-period DD but the total mass of the system is rather uncertain—encompassing both sub- and super-Chandrasekhar solutions [122]. Somewhat similarly, simultaneous light- and radial-velocity curve modelling of the central star of Hen 2-428 led [109] to conclude that the system was a DD with total mass 1.76±0.26 $M_{\odot}$ that would merge in roughly 700 million years. While recent analyses have confirmed the DD classification [30], they have brought into question its super-Chandrasekhar nature with different spectral lines seemingly presenting with different radial velocity amplitudes [103], and thus differing mass solutions (some of which are sub-Chandrasekhar). In any case, both Hen 2-428 and TS 01 represent plausible SN Ia progenitor candidates, and emphasise the possible importance that the high DD fraction among post-CE PNe may hold in evaluating the DD merger scenario for SNe Ia in general.

### 5.2.1 Mass and Period Distributions

Only a handful of post-CE central stars have been subjected to the detailed modelling required to derive their stellar parameters [3, 54, 59]. However, those that have present with some intriguing properties—when one considers main sequence companions (see Table 1 for a list of post-CE central stars with MS secondaries and well-constrained masses, temperatures and radii), for example, all but one are found to present with rather low masses. Indeed, other than in the case of Sp 1, the primaries are still more massive than the companions even after the ejection of the CE. This is consistent with the studies highlighted in Sect. 3 which indicate that, in order to experience dynamical Roche lobe overflow, the initial mass ratio ($q = M_1/M_2$) must be rather high. A similar dearth of more massive companions is also found in the general post-CE population [17], with some suggesting that this may be due to the intrinsic difficulty in identifying white-dwarf-main-sequence (WDMS) binaries with a massive (and optically bright) MS component as a result of the large brightness difference between the two stars [90]. This is not such an issue for PN central stars, where the pre-WD would be more luminous than a typical field WD which has already reached the cooling curve. Furthermore, the presence of the nebula itself makes the identification of such systems more likely, acting as a signpost for the existence of a central white dwarf which would present with colours very different to that of earlier-type (AFG or even K type, for example) MS stars—thus such MS stars stand-out when found at the centre of a PN. Several wide-binary central stars have been identified through this methodology, initially being labelled "peculiar" due to the discovery of the optically-bright companion before the identification of the nebular progenitor [2, 58, 70]. Only one such system has since been shown to be a post-CE binary—the central star of NGC 2346—which is made even more unusual by playing host to a ~3.5 $M_{\odot}$ sub-giant companion [7], making it the most massive post-CE secondary known (not just among post-CE PNe but in general).

**Table 1** Close-binary WDMS central stars with well-constrained masses, radii and temperatures derived from simultaneous light and radial velocity curve modelling

| PN | Period (days) | $M_{CS}$ ($M_\odot$) | $R_{CS}$ ($R_\odot$) | $T_{CS}$ (kK) | $M_S$ ($M_\odot$) | $R_S$ ($R_\odot$) | $T_S$ (kK) | $i$ (°) | Ref |
|---|---|---|---|---|---|---|---|---|---|
| Abell 46 | 0.47 | 0.51±0.05 | 0.15±0.02 | 49.5±4.5 | 0.15±0.02 | 0.46±0.02 | 3.9±0.4 | 80.3±0.1 | [1] |
| Abell 63 | 0.47 | 0.63±0.05 | 0.35±0.01 | 78±3 | 0.29±0.03 | 0.56±0.02 | 6.1±0.2 | 87.1±0.2 | [1] |
| Abell 65 | 1.00 | 0.56±0.04 | 0.056±0.008 | 110±10 | 0.22±0.04 | 0.41±0.05 | 5.0±1.0 | 61±5 | [40] |
| DS 1 | 0.36 | 0.63±0.03 | 0.16±0.01 | 77±3 | 0.23±0.01 | 0.40±0.01 | 3.4±1 | 62.5±1.5 | [38] |
| ESO 330-9 | 0.30 | 0.38–0.45 | 0.03–0.07 | 55–65 | 0.3–0.5 | 0.35–0.50 | ≤4.5 | 7–13 | [42] |
| HaTr 7 | 0.32 | 0.50–0.56 | 0.13–0.18 | 90–100 | 0.14–0.20 | 0.3–0.4 | ≤5 | 45–50 | [42] |
| Hen 2-155 | 0.15 | 0.62±0.05 | 0.31±0.02 | 90±5 | 0.13±0.02 | 0.30±0.03 | 3.5±0.5 | 68.8 ±0.8 | [56] |
| LTNF 1 | 2.29 | 0.70±0.07 | 0.08±0.01 | 105±5 | 0.36±0.07 | 0.72±0.05 | 5.8±0.3 | 84±1 | [29] |
| M 3-1 | 0.13 | 0.65[a] | 0.41±0.02 | $48^{+17}_{-10}$ | 0.17±0.02 | 0.23±0.02 | 5–12 | 75.5±2 | [59] |
| NGC 6337 | 0.17 | 0.56[a] | 0.045–0.085 | 115±5 | 0.14–0.35 | 0.30–0.42 | 4.5±0.5 | 17–23 | [41] |
| Sp 1 | 2.91 | 0.52–0.60 | 0.20–0.35 | 80±10 | 0.52–0.90 | 1.05–1.60 | 3.5–4.6 | 7–11 | [41] |

[a]Fixed in the modelling

The distribution of primary masses is also interesting, with a number presenting masses consistent with being post-RGB objects [42]. This is perhaps in-keeping with the radius evolution properties of low-mass stars as discussed in Sect. 3. However, such stars were thought likely to evolve too slowly following the ejection of their envelopes and thus never produce a visible PN (only reaching the temperature required to ionise the envelope long after it has dissipated into the surrounding interstellar medium). Recent theoretical studies of the end-of-CE structures of such stars have challenged this interpretation, indicating that core masses as low as $0.3\,M_{\odot}$ may well be capable of producing an observable PN [36]. As such, these post-RGB stars offer a promising avenue to understand the properties of such stars upon leaving the CE and thus the physics of RGB CEs.

In terms of the post-CE PN period distribution, it is generally very similar to that of the general WDMS population [74], though perhaps with some minor differences which result from small number statistics and detection biases [3]. In both cases, the period distribution shows a strong peak at around 8 h, with a paucity of longer period systems (greater than a few days). Only IK Peg, FF Aqr, V1379 Aql and the central stars of NGC 2346 and MyCn 18 present with confirmed post-CE periods longer than 5 days[6] [7, 16, 77]. IK Peg, FF Aqr, V1379 Aql and NGC 2346 also have the most massive secondaries known, perhaps hinting at a connection between secondary and/or primary mass and CE ejection efficiency (see [45]). Some attempts to reconstruct the CE phase for the general population of WDMS binaries with known masses and periods (roughly as described in Sect. 4) do indeed show statistically significant correlations between primary mass and ejection efficiency [17], and between the mass ratio and the efficiency [20]. However, these are not universal findings, with other studies claiming that there are no dependencies [99, 139]—see [133] for a critical overview of the possible correlations determined, and some reconciliation of the apparently contradictory results from previous studies. In any case, population synthesis models seem to indicate that the efficiency must be low in order to explain the absence of longer period systems [8, 121], which is very unlikely to be solely due to observational bias [19, 84].

## *5.3 Pre-common Envelope Mass Transfer*

As highlighted in Sect. 5.2.1, only a handful of post-CE central stars with MS secondaries have been subjected to detailed modelling, but in spite of these they hold even more surprises beyond their mass distribution. In all but one case, the MS companions were found to be greatly inflated (sometimes by a factor of two or more)

[6]The double-degenerate central star of NGC 1360 has an orbital period of ~142 days [78], but may not be the result of a CE. Instead, such systems may evolve through stable, non-conservative mass transfer [121, 135] as described in Sect. 5.2.

with respect to isolated stars of similar masses [56], and even though the exception to this rule shows a fairly typical radius for its mass it was found to be Roche-lobe filling and thus could not be inflated without transferring material back onto the WD primary [59]. While some of this inflation could be due to the high levels of irradiation from the hot, nebular progenitors [19], it is now generally accepted that it is principally a consequence of rapid mass transfer on to the MS star either during or just prior to the CE phase [56]. This mass transfer knocks the star out of thermal equilibrium causing them to "puff up"—a state in which they remain given that the thermal timescale of these stars is orders of magnitude longer than the timescale of the CE ejection as well as the PN visibility time. In this way, the post-CE central stars of PNe are often referred to as "fresh-out-of-the-oven", as the relatively short PN lifetime (a few tens of thousands of years) guarantees that the central binary has had little time to adjust following the CE ejection.

Further support for significant accretion onto main sequence companions in post-CE central stars comes from the spectacular Necklace nebula (Fig. 5). Discovered as part of the IPHAS survey [25], the central star was later shown to be a photometrically-variable post-CE binary with a period of 1.16 d [13]. Later spectroscopic observations taken around the photometric minimum revealed that the MS companion in the system was greatly enriched in carbon [76]. Such carbon dwarfs are either primordially-enriched in carbon or are the product of chemical contamination via accretion from a more evolved companion—the latter hypothesis being supported by the large number of dwarf carbon stars which are also found to be X-ray bright (considered a strong sign of post-accretion activity). In any case, the carbon dwarf in the necklace is highly likely to have been contaminated via accretion given that nebula also presents with a remarkable pair of polar outflows or jets, which are almost certainly also a consequence of mass transfer between the binary components.

A significant fraction of post-CE PNe are found to display polar outflows similar to the Necklace [75], the properties of which can be used to probe the mass transfer chronology and, for example, the magnetic fields associated with the accretion disc (assuming that magnetic fields are responsible for angular momentum transport and jet launching). Kinematical studies of the jets reveal that in almost all cases the jets pre-date the central nebular regions by a few thousand years.[7] Given that the central regions are thought to represent the remnant of the ejected CE, this is strong evidence that the jets originate from a phase of pre-CE mass transfer. This hypothesis is supported by apparent precession rate of the jets of Fg 1 (see Fig. 6). The central star was shown to be a post-CE DD binary with an orbital period of 1.195 d, while hydrodynamic models indicate that the precession rate of the bipolar rotating episodic jets is inconsistent with such a short period binary instead being associated with the pre-CE orbital period [4]. The magnetic fields strengths (a few

[7]The ages referred to here are kinematical ages and, as such, represent the minimum ages for each component (i.e. the age assuming that the material was ejected ballistically and has not been slowed by interaction with the surrounding interstellar medium).

Gauss) and accretion rates ($10^{-5}$–$10^{-6}\,M_{\odot}\,yr^{-1}$) associated with the formation of these pre-CE jets are consistent with wind accretion shortly before the onset of Roche lobe overflow [119], providing important constraints on the pre-CE evolution.

It thus seems clear that most, if not all, post-CE central stars must have experienced some form of pre-CE mass transfer episode. With this in mind, it is perhaps interesting to return some of the previously highlighted results—particularly that the "long-period" post-CE systems (IK Peg, FF Aqr, V1379 Aql, NGC 2346) highlighted in Sect. 5.2.1 would have had initial mass ratios much closer to unity than, for example, those short-period systems in Table 1. We have already discussed in Sect. 3 that an extreme mass ratio is likely required for the RLOF to be unstable and thus for a CE to occur, however the indication that pre-CE mass transfer occurs in a majority of systems seems to hint that the RLOF may initially be at least somewhat stable (although non-conservative), perhaps via wind RLOF. It is thus perhaps not unreasonable to surmise that the closer to unity the initial mass ratio, the longer this pseudo-stable, non-conservative pre-CE RLOF could be. The longer this phase, the more mass could be lost in the form of jets or via the outer Lagrange points [71, 102] or, at the very least, more mass that will be redistributed within the system. It has already been suggested that such extended phases of pre-CE mass transfer could greatly impact on the in-spiral, leading to wider binary systems just as observed [45].

## *5.4 Chemistry*

The chemical properties of post-CE PNe can also potentially be used to probe the CE phase. Their chemical abundances trace the abundances in the envelope and, if measured with sufficient precision, could feasibly be used to probe the evolutionary phase of the donor upon entry into the CE. Indeed, in a handful of cases, the abundance patterns of post-CE PNe have been shown to be consistent with the CE cutting short the AGB evolution of the nebular progenitor [19, 55], helping to constrain the pre-CE configuration.

It has recently been shown that some post-CE PNe display highly anomalous abundances depending on the emission lines used to derive them. For more than 70 years, it has been clear that the abundances of ionised nebulae differ depending on whether they are measured using recombination lines or the much brighter collisionally-excited lines [136]—becoming known as the "abundance discrepancy problem". In the general PN population, abundances from recombination lines are found to be a factor of 2–3 greater than those from collisionally-excited lines [130]. Post-CE PNe, however, tend to show even larger abundance discrepancy factors [132], with some even reaching up to several hundred [14]. Multiple explanations have been considered for the smaller discrepancies found in the general population of PNe and HII regions, ranging from temperature variations to non-thermal electron energy distributions [87, 95], however in the most extreme (post-CE) cases chemical inhomogeneities play a dominant role (for a more-detailed explanation of the derivation of chemical abundances in astrophysical nebulae, as well as the possible explanations for the abundance discrepancy, the reader is referred to the excellent review chapter by Jorge García-Rojas in this volume).

The chemical inhomogeneities in short-period post-CE PNe manifest themselves as a second, lower-temperature, higher-metallicity gas phase in addition to a more standard temperature and metallicity phase consistent with that of the general PN population. In a majority of cases, this higher-metallicity gas is found to be centrally concentrated and closer to the central star [32, 132]. This has led some authors to consider that it may represent a form of post-CE eruptive event which leads to the ejection of reprocessed material [57]—a particularly intriguing prospect given that the overall abundance pattern of the higher metallicity ejecta is reminiscent of that of Neon novae [131]. If this were the case then, presumably, the two gas phases would show differing kinematical properties. Unfortunately, studying the kinematics of the second gas phase is challenging due to the intrinsic faintness of the recombination lines in which it is brightest, however a preliminary study comparing lines across different chemical species did find evidence for discrepant kinematics between the two gas phases [104]. Returning to the possibility that the high-metallicity gas originates from some form of reprocessing event, it is unclear what could lead to this eruptive event. However, it could speculatively occur as a result of fall-back of CE material [66]. What does seem to be clear is that whatever process leads to the ejection of this higher-metallicity material, it only occurs in the very shortest-period post-CE binaries [132], with longer-period post-CE PNe ($P_{\rm orb} \gtrsim 1.2$ days) tending to present with less extreme abundance discrepancies.

## 6 Mergers

Thus far, we have focused on systems which survived the CE phase as binaries, but a significant number will instead lead to mergers [64]. To date, only one PN central star can be considered a strong post-merger candidate—that of NGC 6826, found to display a rotation rate too high to have originated from a single star [21, 37]. However, another class of post-merger phenomena exist—luminous red novae[8] (LRNe)—slowly-evolving red transients, the peak brightness of which is brighter than classical novae but fainter than supernovae ($10^{39}$–$10^{41}$ ergs s$^{-1}$ [63]). At the turn of the century, only a handful of such transients had been identified and to relatively little fanfare—one (M 31-RV) being classified simply as "a nova of unusual type" which did not "comfortably fit into the standard scenarios for eruptive events on white dwarfs" [80]. However, our understanding of these transients

[8]Continuing with the unfortunate misnomers surrounding CE-related phenomena, just as planetary nebulae have no relation to planets, lumninous red novae are completely unrelated to classical novae or supernovae.

**Fig. 7** Hubble Space Telescope image of the light echo around V838 Mon taken in October 2002 some 9 months after eruption [6]. At this stage in its evolution, the central merger remnant presented with a temperature and radius consistent with an L-type supergiant [28]. Image credit: NASA, ESA and H.E. Bond (STScI)

was greatly advanced following the 2002 eruption of V838 Monocerotis (Fig. 7) which, being detected early and residing in our own Galaxy, could be studied in exquisite detail [6, 10, 28, 81, 82]. The observed evolution of V838 Mon (as well as the other members of the class M 31-RV and V4332 Sgr)—a brightening of several magnitudes followed by a slow decline, all the while developing redder and redder colours (V838 Mon resembling an L-type supergiant less than a year after its discovery [28])—was found to only be consistent with a merger scenario [115, 125]. Neither a nova-like event (comprising some form of thermonuclear runaway on the surface of a WD) nor a helium shell flash or very late thermal pulse associated with a born-again event could be reconciled with the observed post-eruption colour/temperature evolution [125].

Following the LRN eruption of V1309 Sco, the merger scenario for these objects was confirmed with the pre-eruption light curve from OGLE showing clear evidence that the progenitor was a contact binary (red inset of Fig. 8), the orbital period of which was exponentially decaying [126]. These pre-merger observations offered a unique window into the processes that led up to the dynamical CE event in this system. In the 2 years prior to eruption, the light curve evolved from showing the typical ellipsoidal modulations of a contact binary through to showing only a single peak. This is likely due to obscuration of the binary by mass lost from the second Lagrange point [97]. Furthermore, roughly 200 days prior to the outburst, variability

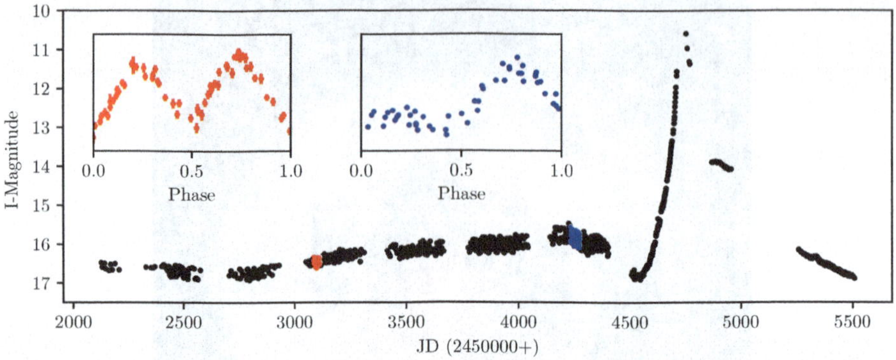


**Fig. 8** The OGLE light curve of V1309 Sco showing the slow rise before eruption (data originally presented in [126]. Note that the final outburst was even brighter than shown here, reaching lower than 8th magnitude. The insets show phase folded light curves corresponding to the regions of the same colour in the main plot, highlighting the evolution from double-peaked (the typical ellipsoidal modulation of a Roche-lobe filling binary) through to single-peaked (where the previous secondary peak has now been obscured by the mass lost through the $L_2$ point [97])

associated with the orbital period was no longer detected, instead being replaced by a systematic slow brightening, which again could be associated with mass loss through the outer Lagrange point during the final stages of orbital decay [96]. The subsequent rapid brightening (∼4 mag in ∼5 days) was then likely due to the final, dynamical merger of the two components and the liberation of recombination energy associated with expulsion of a shell of CE material [51]. Intriguingly, the total mass loss associated with the gradual in-spiral phase (several hundredths of a solar mass over the course of a few thousand orbits) is comparable to the amount thought to be ejected during the final merger phase [97]. This is a clear demonstration that pre-CE interactions may play an important role in the outcome of the CE itself—helping to remove the envelope or at least dramatically altering the initial conditions (mass ratio, envelope structure, etc.) prior to the dynamical interaction as compared to those typically employed in hydrodynamical simulations (see e.g. [102]). Furthermore, magnetic fields could be generated or amplified via shearing motions as a result of loss of corotation during the pre-CE phase of period decay [97, 101]. These magnetic fields could feasibly launch jets removing further mass from the system prior to the dynamic event [119].

Beyond pre-merger observations such as those serendipitously obtained for V1309 Sco, their post-CE light curves can also provide important constraints on the CE process. Even a relatively simple model, whereby the emission from a spherically-symmetric ejection is controlled by a recombination front as the material cools, was shown to match the observed colour and luminosity evolution of the handful of LRNe known at the time [51]. More realistic models of this process—combining both three-dimensional magneto-hydrodynamics and radiation transport—have the potential to reveal much more about the CE process [31, 47, 138]. However, such models are computationally particularly challenging, not only due to the difficulty in incorporating radiation transport into the chosen hydrodynamic modelling code but also due to the rather long wall clock times that are required [31]. In spite of this, important progress has been made, strongly indicating that continued effort may prove key in understanding the long-term post-ejection behaviour of LRNe [47].

Finally, late-time observations of LRNe can be used to directly measure the amount of mass ejected during the merger, as well as its morphology and kinematics [62]. Similarly, such late-time observations can also be used to probe the nucleosynthesis which occurs during merger via the astrochemistry of the ejected envelope [61], placing constraints on how mergers could impact the enrichment of the interstellar medium.

## 7 Discussion

In this chapter, we have introduced some of the fundamental concepts important for our understanding of the CE—one of the most critical, yet poorly understood phases of close-binary evolution. Subsequently, we have discussed some of the recent observational findings related to two CE phenomena—post-CE PNe and LRNe. Both show strong evidence for (appreciable) mass transfer/loss before the dynamical CE event, which may prove to be a critical ingredient in deriving self-consistent and complete models of the CE process [71], particularly given that most hydrodynamical modelling efforts focus on the dynamical in-spiral and begin with the companion at the surface of the giant [88, 92].

Post-CE PNe, which represent the direct progeny of CE events where the envelope was successfully ejected without the cores merging, also have more to tell us about the CE phase. The spatio-chemical properties of the nebulae offer some indication that the final stages of CE ejection, particularly in systems with small final orbital separations, may result in some form of reprocessing and the ejection of chemically-enriched material into the expanding envelope (giving rise to the extreme abundance discrepancies observed in these systems). This similarly indicates that modelling efforts need, not only to begin before the dynamical in-spiral phase but also, to extend out towards the nebular phase [102]. Whatever the process behind the ejection of this enriched material, it could feasibly play a role in unbinding the remaining envelope and successfully terminating the CE—a problem faced by a majority of hydrodynamic models which generally fail to unbind the entirety of the envelope, instead leaving a fraction "lifted" but still bound to the central binary [46].

The mass distribution of the companions inside post-CE PNe (as well as the general post-CE WDMS population) seems to indicate that only systems with much more massive donors will experience, or at least survive, a CE—helping constrain the conditions for dynamically unstable Roche lobe overflow. Understanding the initial parameter space which could lead to a CE is of critical importance, along with constraining the efficiency (or refining whichever prescription is chosen to derive the end result of a CE), for the population synthesis efforts which will prove essential in interpreting the awaiting deluge of close-binary phenomena that will be revealed by The Rubin Observatory Legacy Survey of Space and Time (LSST [69]) and the Laser Interferometer Space Antenna (*LISA* [60, 110, 120]).

While we are still far from understanding the CE, the (thus far) limited observational studies of CE phenomena like post-CE PNe and LRNe have already provided valuable insight. Continued, deeper study of these phenomena will, without doubt, further refine our understanding of the CE and, when combined with continued theoretical and computational modelling efforts, perhaps lead to a unified picture of the processes at work in the phase.

**Acknowledgements** DJ would like to thank the referee, Orsola De Marco, for her comprehensive report which improved both the clarity and completeness of this review.

DJ acknowledges support from the State Research Agency (AEI) of the Spanish Ministry of Science, Innovation and Universities (MCIU) and the European Regional Development Fund (FEDER) under grant AYA2017-83383-P. DJ also acknowledges support under grant P/308614 financed by funds transferred from the Spanish Ministry of Science, Innovation and Universities, charged to the General State Budgets and with funds transferred from the General Budgets of the Autonomous Community of the Canary Islands by the Ministry of Economy, Industry, Trade and Knowledge.

## References

1. M. Afşar, C. Ibanoğlu, Mon. Not. R. Astron. Soc. **391**(2), 802 (2008). https://doi.org/10.1111/j.1365-2966.2008.13927.x
2. A. Aller, B. Montesinos, L.F. Miranda, E. Solano, A. Ulla, Mon. Not. R. Astron. Soc. **448**(3), 2822 (2015). https://doi.org/10.1093/mnras/stv196
3. H.M.J. Boffin, D. Jones, *The Importance of Binaries in the Formation and Evolution of Planetary Nebulae*. SpringerBriefs in Astronomy (Springer, Berlin, 2019). https://doi.org/10.1007/978-3-030-25059-1
4. H.M.J. Boffin, B. Miszalski, T. Rauch, D. Jones, R.L.M. Corradi, R. Napiwotzki, A.C. Day-Jones, J. Köppen, Science **338**(6108), 773 (2012). https://doi.org/10.1126/science.1225386
5. H.E. Bond, Publ. Astron. Soc. Pac. **88**, 192 (1976). https://doi.org/10.1086/129923
6. H.E. Bond, A. Henden, Z.G. Levay, N. Panagia, W.B. Sparks, S. Starrfield, R.M. Wagner, R.L.M. Corradi, U. Munari, Nature **422**(6930), 405 (2003). https://doi.org/10.1038/nature01508
7. A.J. Brown, D. Jones, H.M.J. Boffin, H. Van Winckel, Mon. Not. R. Astron. Soc. **482**(4), 4951 (2019). https://doi.org/10.1093/mnras/sty2986
8. J. Camacho, S. Torres, E. García-Berro, M. Zorotovic, M.R. Schreiber, A. Rebassa-Mansergas, A. Nebot Gómez-Morán, B.T. Gänsicke, Astron. Astrophys. **566**, A86 (2014). https://doi.org/10.1051/0004-6361/201323052
9. L. Chamandy, A. Frank, E.G. Blackman, J. Carroll-Nellenback, B. Liu, Y. Tu, J. Nordhaus, Z. Chen, B. Peng, Mon. Not. R. Astron. Soc. **480**(2), 1898 (2018). https://doi.org/10.1093/mnras/sty1950
10. O. Chesneau, F. Millour, O. De Marco, S.N. Bright, A. Spang, D.P.K. Banerjee, N.M. Ashok, T. Kamiński, J.P. Wisniewski, A. Meilland, Astron. Astrophys. **569**, L3 (2014). https://doi.org/10.1051/0004-6361/201424458

11. J. Choi, A. Dotter, C. Conroy, M. Cantiello, B. Paxton, B.D. Johnson, Astrophys. J. **823**, 102 (2016). https://doi.org/10.3847/0004-637X/823/2/102
12. M. Clayton, P. Podsiadlowski, N. Ivanova, S. Justham, Mon. Not. R. Astron. Soc. **470**(2), 1788 (2017). https://doi.org/10.1093/mnras/stx1290
13. R.L.M. Corradi, L. Sabin, B. Miszalski, P. Rodríguez-Gil, M. Santander-García, D. Jones, J.E. Drew, A. Mampaso, M.J. Barlow, M.M. Rubio-Díez, Mon. Not. R. Astron. Soc. **410**(2), 1349 (2011). https://doi.org/10.1111/j.1365-2966.2010.17523.x
14. R.L.M. Corradi, J. García-Rojas, D. Jones, P. Rodríguez-Gil, Astrophys. J. **803**(2), 99 (2015). https://doi.org/10.1088/0004-637X/803/2/99
15. G.H. Darwin, Proc. R. Soc. Lond. Ser. I **29**, 168 (1879)
16. P.J. Davis, U. Kolb, B. Willems, Mon. Not. R. Astron. Soc. **403**(1), 179 (2010). https://doi.org/10.1111/j.1365-2966.2009.16138.x
17. P.J. Davis, U. Kolb, C. Knigge, Mon. Not. R. Astron. Soc. **419**(1), 287 (2012). https://doi.org/10.1111/j.1365-2966.2011.19690.x
18. M. de Kool, E.P.J. van den Heuvel, E. Pylyser, Astron. Astrophys. **183**, 47 (1987)
19. O. De Marco, T.C. Hillwig, A.J. Smith, Astron. J. **136**(1), 323 (2008). https://doi.org/10.1088/0004-6256/136/1/323
20. O. De Marco, J.C. Passy, M. Moe, F. Herwig, M.M. Mac Low, B. Paxton, Mon. Not. R. Astron. Soc. **411**(4), 2277 (2011). https://doi.org/10.1111/j.1365-2966.2010.17891.x
21. O. De Marco, J. Long, G.H. Jacoby, T. Hillwig, M. Kronberger, S.B. Howell, N. Reindl, S. Margheim, Mon. Not. R. Astron. Soc. **448**, 3587 (2015). https://doi.org/10.1093/mnras/stv249
22. T. Dermine, A. Jorissen, L. Siess, A. Frankowski, Astron. Astrophys. **507**(2), 891 (2009). https://doi.org/10.1051/0004-6361/200912313
23. R. Deschamps, L. Siess, P.J. Davis, A. Jorissen, Astron. Astrophys. **557**, A40 (2013). https://doi.org/10.1051/0004-6361/201321509
24. J.D.M. Dewi, T.M. Tauris, Astron. Astrophys. **360**, 1043 (2000)
25. J.E. Drew, R. Greimel, M.J. Irwin, A. Aungwerojwit, M.J. Barlow, R.L.M. Corradi, J.J. Drake, B.T. Gänsicke, P. Groot, A. Hales, Mon. Not. R. Astron. Soc. **362**(3), 753 (2005). https://doi.org/10.1111/j.1365-2966.2005.09330.x
26. P.P. Eggleton, Astrophys. J. **268**, 368 (1983). https://doi.org/10.1086/160960
27. P. Eggleton, *Close Binary Stars* (Institute of Physics Publishing, Bristol, 2000). https://doi.org/10.1888/0333750888/1624
28. A. Evans, T.R. Geballe, M.T. Rushton, B. Smalley, J.T. van Loon, S.P.S. Eyres, V.H. Tyne, Mon. Not. R. Astron. Soc. **343**(3), 1054 (2003). https://doi.org/10.1046/j.1365-8711.2003.06755.x
29. D.H. Ferguson, J. Liebert, S. Haas, R. Napiwotzki, T.A. James, Astrophys. J. **518**(2), 866 (1999). https://doi.org/10.1086/307289
30. N.L. Finch, N. Reindl, M.A. Barstow, S.L. Casewell, S. Geier, M.M.M. Bertolami, S. Taubenberger, Open Astron. **27**(1), 57 (2018). https://doi.org/10.1515/astro-2018-0017
31. P. Galaviz, O. De Marco, J.C. Passy, J.E. Staff, R. Iaconi, Astrophys. J. Suppl. Ser. **229**(2), 36 (2017). https://doi.org/10.3847/1538-4365/aa64e1
32. J. García-Rojas, R.L.M. Corradi, H. Monteiro, D. Jones, P. Rodríguez-Gil, A. Cabrera-Lavers, Astrophys. J. Lett. **824**(2), L27 (2016). https://doi.org/10.3847/2041-8205/824/2/L27
33. G. García-Segura, P.M. Ricker, R.E. Taam, Astrophys. J. **860**(1), 19 (2018). https://doi.org/10.3847/1538-4357/aac08c
34. H. Glanz, H.B. Perets, Mon. Not. R. Astron. Soc. **478**(1), L12 (2018). https://doi.org/10.1093/mnrasl/sly065
35. A. Grichener, E. Sabach, N. Soker, Mon. Not. R. Astron. Soc. **478**(2), 1818 (2018). https://doi.org/10.1093/mnras/sty1178
36. P.D. Hall, C.A. Tout, R.G. Izzard, D. Keller, Mon. Not. R. Astron. Soc. **435**(3), 2048 (2013). https://doi.org/10.1093/mnras/stt1422
37. G. Handler, R.K. Prinja, M.A. Urbaneja, V. Antoci, J.D. Twicken, T. Barclay, Mon. Not. R. Astron. Soc. **430**, 2923 (2013). https://doi.org/10.1093/mnras/stt092

38. R.W. Hilditch, T.J. Harries, G. Hill, Mon. Not. R. Astron. Soc. **279**(4), 1380 (1996). https://doi.org/10.1093/mnras/279.4.1380
39. T.C. Hillwig, H.E. Bond, M. Afşar, O. De Marco, Astron. J. **140**(2), 319 (2010). https://doi.org/10.1088/0004-6256/140/2/319
40. T.C. Hillwig, D.J. Frew, M. Louie, O. De Marco, H.E. Bond, D. Jones, S.C. Schaub, Astron. J. **150**(1), 30 (2015). https://doi.org/10.1088/0004-6256/150/1/30
41. T.C. Hillwig, D. Jones, O. De Marco, H.E. Bond, S. Margheim, D. Frew, Astrophys. J. **832**(2), 125 (2016). https://doi.org/10.3847/0004-637X/832/2/125
42. T.C. Hillwig, D.J. Frew, N. Reindl, H. Rotter, A. Webb, S. Margheim, Astron. J. **153**(1), 24 (2017). https://doi.org/10.3847/1538-3881/153/1/24
43. M.S. Hjellming, R.E. Taam, Astrophys. J. **370**, 709 (1991). https://doi.org/10.1086/169854
44. I. Iben, M. Livio, Publ. Astron. Soc. Pac. **105**, 1373 (1993). https://doi.org/10.1086/133321
45. R. Iaconi, O. De Marco, Mon. Not. R. Astron. Soc. **490**(2), 2550 (2019). https://doi.org/10.1093/mnras/stz2756
46. R. Iaconi, T. Reichardt, J. Staff, O. De Marco, J.C. Passy, D. Price, J. Wurster, F. Herwig, Mon. Not. R. Astron. Soc. **464**(4), 4028 (2017). https://doi.org/10.1093/mnras/stw2377
47. R. Iaconi, K. Maeda, O. De Marco, T. Nozawa, T. Reichardt, Mon. Not. R. Astron. Soc. **489**(3), 3334 (2019). https://doi.org/10.1093/mnras/stz2312
48. N. Ivanova, Astrophys. J. **858**(2), L24 (2018). https://doi.org/10.3847/2041-8213/aac101
49. N. Ivanova, J.L.A. Nandez, Mon. Not. R. Astron. Soc. **462**(1), 362 (2016). https://doi.org/10.1093/mnras/stw1676
50. N. Ivanova, S. Justham, X. Chen, O. De Marco, C.L. Fryer, E. Gaburov, H. Ge, E. Glebbeek, Z. Han, X.D. Li, G. Lu, T. Marsh, P. Podsiadlowski, A. Potter, N. Soker, R. Taam, T.M. Tauris, E.P.J. van den Heuvel, R.F. Webbink, Astron. Astrophys. Rev. **21**, 59 (2013). https://doi.org/10.1007/s00159-013-0059-2
51. N. Ivanova, S. Justham, J.L. Avendano Nandez, J.C. Lombardi, Science **339**(6118), 433 (2013). https://doi.org/10.1126/science.1225540
52. N. Ivanova, S. Justham, P. Podsiadlowski, Mon. Not. R. Astron. Soc. **447**(3), 2181 (2015). https://doi.org/10.1093/mnras/stu2582
53. D. Jones, *The Importance of Binarity in the Formation and Evolution of Planetary Nebulae*. Cambridge Astrophysics (Cambridge University Press, Cambridge, 2019), pp. 106–127. https://doi.org/10.1017/9781108553070.010
54. D. Jones, H.M.J. Boffin, Nat. Astron. **1**, 0117 (2017). https://doi.org/10.1038/s41550-017-0117
55. D. Jones, H.M.J. Boffin, B. Miszalski, R. Wesson, R.L.M. Corradi, A.A. Tyndall, Astron. Astrophys. **562**, A89 (2014). https://doi.org/10.1051/0004-6361/201322797
56. D. Jones, H.M.J. Boffin, P. Rodríguez-Gil, R. Wesson, R.L.M. Corradi, B. Miszalski, S. Mohamed, Astron. Astrophys. **580**, A19 (2015). https://doi.org/10.1051/0004-6361/201425454
57. D. Jones, R. Wesson, J. García-Rojas, R.L.M. Corradi, H.M.J. Boffin, Mon. Not. R. Astron. Soc. **455**(3), 3263 (2016). https://doi.org/10.1093/mnras/stv2519
58. D. Jones, H. Van Winckel, A. Aller, K. Exter, O. De Marco, Astron. Astrophys. **600**, L9 (2017). https://doi.org/10.1051/0004-6361/201730700
59. D. Jones, H.M.J. Boffin, P. Sowicka, B. Miszalski, P. Rodríguez-Gil, M. Santand er-García, R.L.M. Corradi, Mon. Not. R. Astron. Soc. Lett. **482**(1), L75 (2019). https://doi.org/10.1093/mnrasl/sly142
60. B. Kalomeni, L. Nelson, S. Rappaport, M. Molnar, J. Quintin, K. Yakut, Astrophys. J. **833**(1), 83 (2016). https://doi.org/10.3847/1538-4357/833/1/83
61. T. Kamiński, K.M. Menten, R. Tylenda, A. Karakas, A. Belloche, N.A. Patel, Astron. Astrophys. **607**, A78 (2017). https://doi.org/10.1051/0004-6361/201731287
62. T. Kamiński, W. Steffen, R. Tylenda, K.H. Young, N.A. Patel, K.M. Menten, Astron. Astrophys. **617**, A129 (2018). https://doi.org/10.1051/0004-6361/201833165
63. M.M. Kasliwal, Publ. Astron. Soc. Aust. **29**(4), 482 (2012). https://doi.org/10.1071/AS11061

64. C.S. Kochanek, S.M. Adams, K. Belczynski, Mon. Not. R. Astron. Soc. **443**(2), 1319 (2014). https://doi.org/10.1093/mnras/stu1226
65. Z. Kopal, *Close Binary Systems* (Wiley, Hoboken, 1959)
66. R.L. Kuruwita, J. Staff, O. De Marco, Mon. Not. R. Astron. Soc. **461**(1), 486 (2016). https://doi.org/10.1093/mnras/stw1414
67. E. Lagadec, A.A. Zijlstra, Mon. Not. R. Astron. Soc. **390**(1), L59 (2008). https://doi.org/10.1111/j.1745-3933.2008.00535.x
68. D. López-Cámara, F. De Colle, E. Moreno Méndez, Mon. Not. R. Astron. Soc. **482**(3), 3646 (2019). https://doi.org/10.1093/mnras/sty2959
69. LSST Science Collaboration, P.A. Abell, J. Allison, S.F. Anderson, J.R. Andrew, J.R.P. Angel, L. Armus, D. Arnett, S.J. Asztalos, T.S. Axelrod, et al. (2009). e-prints arXiv:0912.0201
70. J.H. Lutz, Astron. Astrophys. **60**, 93 (1977)
71. M. MacLeod, E.C. Ostriker, J.M. Stone, Astrophys. J. **868**(2), 136 (2018). https://doi.org/10.3847/1538-4357/aae9eb
72. R. Manick, B. Miszalski, V. McBride, Mon. Not. R. Astron. Soc. **448**(2), 1789 (2015). https://doi.org/10.1093/mnras/stv074
73. D. Maoz, F. Mannucci, G. Nelemans, Annu. Rev. Astron. Astrophys. **52**, 107 (2014). https://doi.org/10.1146/annurev-astro-082812-141031
74. B. Miszalski, A. Acker, A.F.J. Moffat, Q.A. Parker, A. Udalski, Astron. Astrophys. **496**(3), 813 (2009). https://doi.org/10.1051/0004-6361/200811380
75. B. Miszalski, A. Acker, Q.A. Parker, A.F.J. Moffat, Astron. Astrophys. **505**(1), 249 (2009). https://doi.org/10.1051/0004-6361/200912176
76. B. Miszalski, H.M.J. Boffin, R.L.M. Corradi, Mon. Not. R. Astron. Soc. **428**(1), L39 (2013). https://doi.org/10.1093/mnrasl/sls011
77. B. Miszalski, R. Manick, J. Mikołajewska, H. Van Winckel, K. Iłkiewicz, Publ. Astron. Soc. Aust. **35**, e027 (2018). https://doi.org/10.1017/pasa.2018.23
78. B. Miszalski, R. Manick, J. Mikołajewska, K. Iłkiewicz, D. Kamath, H. Van Winckel, Mon. Not. R. Astron. Soc. **473**, 2275 (2018). https://doi.org/10.1093/mnras/stx2501
79. M. Moe, R. Di Stefano, Astrophys. J. Suppl. Ser. **230**(2), 15 (2017). https://doi.org/10.3847/1538-4365/aa6fb6
80. J. Mould, J. Cohen, J.R. Graham, D. Hamilton, K. Matthews, A. Picard, N. Reid, M. Schmidt, T. Soifer, C. Wilson, Astrophys. J. Lett. **353**, L35 (1990). https://doi.org/10.1086/185702
81. U. Munari, A. Henden, A. Vallenari, H.E. Bond, R.L.M. Corradi, L. Crause, S. Desidera, E. Giro, P.M. Marrese, S. Ragaini, Astron. Astrophys. **434**(3), 1107 (2005). https://doi.org/10.1051/0004-6361:20041751
82. U. Munari, R.L.M. Corradi, A. Henden, H. Navasardyan, M. Valentini, R. Greimel, P. Leisy, T. Augusteijn, A.A. Djupvik, L. Glowienka, Astron. Astrophys. **474**(2), 585 (2007). https://doi.org/10.1051/0004-6361:20077837
83. J.L.A. Nandez, N. Ivanova, Mon. Not. R. Astron. Soc. **460**(4), 3992 (2016). https://doi.org/10.1093/mnras/stw1266
84. A. Nebot Gómez-Morán, B.T. Gänsicke, M.R. Schreiber, A. Rebassa-Mansergas, A.D. Schwope, J. Southworth, A. Aungwerojwit, M. Bothe, P.J. Davis, U. Kolb, M. Müller, C. Papadaki, S. Pyrzas, A. Rabitz, P. Rodríguez-Gil, L. Schmidtobreick, R. Schwarz, C. Tappert, O. Toloza, J. Vogel, M. Zorotovic, Astron. Astrophys. **536**, A43 (2011). https://doi.org/10.1051/0004-6361/201117514
85. G. Nelemans, C.A. Tout, Mon. Not. R. Astron. Soc. **356**(2), 753 (2005). https://doi.org/10.1111/j.1365-2966.2004.08496.x
86. G. Nelemans, F. Verbunt, L.R. Yungelson, S.F. Portegies Zwart, Astron. Astrophys. **360**, 1011 (2000)
87. D.C. Nicholls, M.A. Dopita, R.S. Sutherland, Astrophys. J. **752**(2), 148 (2012). https://doi.org/10.1088/0004-637X/752/2/148
88. S.T. Ohlmann, F.K. Röpke, R. Pakmor, V. Springel, Astrophys. J. **816**(1), L9 (2016). https://doi.org/10.3847/2041-8205/816/1/L9

89. B. Paczynski, in *Structure and Evolution of Close Binary Systems, IAU Symposium*, vol. 73, ed. by P. Eggleton, S. Mitton, J. Whelan (1976), p. 75
90. S.G. Parsons, A. Rebassa-Mansergas, M.R. Schreiber, B.T. Gänsicke, M. Zorotovic, J.J. Ren, Mon. Not. R. Astron. Soc. **463**(2), 2125 (2016). https://doi.org/10.1093/mnras/stw2143
91. J.C. Passy, F. Herwig, B. Paxton, Astrophys. J. **760**(1), 90 (2012). https://doi.org/10.1088/0004-637X/760/1/90
92. J.C. Passy, O. De Marco, C.L. Fryer, F. Herwig, S. Diehl, J.S. Oishi, M.M. Mac Low, G.L. Bryan, G. Rockefeller, Astrophys. J. **744**(1), 52 (2012). https://doi.org/10.1088/0004-637X/744/1/52
93. K. Pavlovskii, N. Ivanova, Mon. Not. R. Astron. Soc. **449**(4), 4415 (2015). https://doi.org/10.1093/mnras/stv619
94. B. Paxton, L. Bildsten, A. Dotter, F. Herwig, P. Lesaffre, F. Timmes, Astrophys. J. Suppl. Ser. **192**(1), 3 (2011). https://doi.org/10.1088/0067-0049/192/1/3
95. M. Peimbert, S. Torres-Peimbert, Astrophys. J. **168**, 413 (1971). https://doi.org/10.1086/151097
96. O. Pejcha, Astrophys. J. **788**(1), 22 (2014). https://doi.org/10.1088/0004-637X/788/1/22
97. O. Pejcha, B.D. Metzger, J.G. Tyles, K. Tomida, Astrophys. J. **850**(1), 59 (2017). https://doi.org/10.3847/1538-4357/aa95b9
98. S. Perlmutter, G. Aldering, G. Goldhaber, R.A. Knop, P. Nugent, P.G. Castro, S. Deustua, S. Fabbro, A. Goobar, D.E. Groom, Astrophys. J. **517**(2), 565 (1999). https://doi.org/10.1086/307221
99. M. Politano, K.P. Weiler, Astrophys. J. **665**(1), 663 (2007). https://doi.org/10.1086/518997
100. D. Prialnik, M. Livio, Mon. Not. R. Astron. Soc. **216**, 37 (1985). https://doi.org/10.1093/mnras/216.1.37
101. E. Regős, C.A. Tout, Mon. Not. R. Astron. Soc. **273**(1), 146 (1995). https://doi.org/10.1093/mnras/273.1.146
102. T.A. Reichardt, O. De Marco, R. Iaconi, C.A. Tout, D.J. Price, Mon. Not. R. Astron. Soc. **484**(1), 631 (2019). https://doi.org/10.1093/mnras/sty3485
103. N. Reindl, N. Finch, V. Schaffenroth, M. Barstow, S. Casewell, S. Geier, M. Miller Bertolami, S. Taubenberger, Galaxies **6**(3), 88 (2018). https://doi.org/10.3390/galaxies6030088
104. M.G. Richer, G. Suárez, J.A. López, M.T. García Díaz, Astron. J. **153**(3), 140 (2017). https://doi.org/10.3847/1538-3881/aa5f53
105. P.M. Ricker, R.E. Taam, Astrophys. J. Lett. **672**(1), L41 (2008). https://doi.org/10.1086/526343
106. A.G. Riess, A.V. Filippenko, P. Challis, A. Clocchiatti, A. Diercks, P.M. Garnavich, R.L. Gilliland, C.J. Hogan, S. Jha, R.P. Kirshner, Astron. J. **116**(3), 1009 (1998). https://doi.org/10.1086/300499
107. M.I. Saladino, O.R. Pols, Astron. Astrophys. **629**, 103 (2019). https://doi.org/10.1051/0004-6361/201935625
108. E.L. Sandquist, R.E. Taam, X. Chen, P. Bodenheimer, A. Burkert, Astrophys. J. **500**(2), 909 (1998). https://doi.org/10.1086/305778
109. M. Santander-García, P. Rodríguez-Gil, R.L.M. Corradi, D. Jones, B. Miszalski, H.M.J. Boffin, M.M. Rubio-Díez, M.M. Kotze, Nature **519**(7541), 63 (2015). https://doi.org/10.1038/nature14124
110. S. Shah, G. Nelemans, Astrophys. J. **790**(2), 161 (2014). https://doi.org/10.1088/0004-637X/790/2/161
111. B.J. Shappee, T.A. Thompson, Astrophys. J. **766**(1), 64 (2013). https://doi.org/10.1088/0004-637X/766/1/64
112. S. Shiber, R. Iaconi, O. De Marco, N. Soker, Mon. Not. R. Astron. Soc. **488**(4), 5615 (2019). https://doi.org/10.1093/mnras/stz2013
113. G.E. Soberman, E.S. Phinney, E.P.J. van den Heuvel, Astron. Astrophys. **327**, 620 (1997)
114. N. Soker, Astrophys. J. **800**(2), 114 (2015). https://doi.org/10.1088/0004-637X/800/2/114
115. N. Soker, R. Tylenda, Mon. Not. R. Astron. Soc. **373**(2), 733 (2006). https://doi.org/10.1111/j.1365-2966.2006.11056.x

116. N. Soker, A. Grichener, E. Sabach, Astrophys. J. **863**(1), L14 (2018). https://doi.org/10.3847/2041-8213/aad736
117. A.P. Stephan, S. Naoz, A.M. Ghez, M.R. Morris, A. Ciurlo, T. Do, K. Breivik, S. Coughlin, C.L. Rodriguez, Astrophys. J. **878**(1), 58 (2019). https://doi.org/10.3847/1538-4357/ab1e4d
118. R.E. Taam, E.L. Sandquist, Annu. Rev. Astron. Astrophys. **38**, 113 (2000). https://doi.org/10.1146/annurev.astro.38.1.113
119. J. Tocknell, O. De Marco, M. Wardle, Mon. Not. R. Astron. Soc. **439**(2), 2014 (2014). https://doi.org/10.1093/mnras/stu079
120. O. Toloza, E. Breedt, D. De Martino, J. Drake, B. Gansicke, M. Green, A. Ederoclite, J. Jennifer, K. Juna, C. Knigge, T. Kupfer, K. Long, T. Marsh, A. Pala, S. Parsons, T. Prince, R. Raddi, A. Rebassa-Manserga, P. Rodriguez-Gil, S. Scaringi, L. Schmidtobreick, M. Schreiber, A. Schwope, K. Shen, D. Steeghs, P. Szkody, C. Tappert, S. Toonen, D. Townsley, M. Zorotovic, in *Bulletin of the American Astronomical Society (BAAS)*, vol. 51 (2019), p. 168
121. S. Toonen, G. Nelemans, Astron. Astrophys. **557**, A87 (2013). https://doi.org/10.1051/0004-6361/201321753
122. G. Tovmassian, L. Yungelson, T. Rauch, V. Suleimanov, R. Napiwotzki, G. Stasińska, J. Tomsick, J. Wilms, C. Morisset, M. Peña, Astrophys. J. **714**(1), 178 (2010). https://doi.org/10.1088/0004-637X/714/1/178
123. D. Tsebrenko, N. Soker, Mon. Not. R. Astron. Soc. **447**(3), 2568 (2015). https://doi.org/10.1093/mnras/stu2567
124. A. Tutukov, L. Yungelson, in *Mass Loss and Evolution of O-Type Stars, IAU Symposium*, vol. 83, ed. by P.S. Conti, C.W.H. De Loore (1979), pp. 401–406
125. R. Tylenda, N. Soker, Astron. Astrophys. **451**(1), 223 (2006). https://doi.org/10.1051/0004-6361:20054201
126. R. Tylenda, M. Hajduk, T. Kamiński, A. Udalski, I. Soszyński, M.K. Szymański, M. Kubiak, G. Pietrzyński, R. Poleski, Ł. Wyrzykowski, Astron. Astrophys. **528**, A114 (2011). https://doi.org/10.1051/0004-6361/201016221
127. J. Vos, R.H. Østensen, P. Marchant, H. Van Winckel, Astron. Astrophys. **579**, A49 (2015). https://doi.org/10.1051/0004-6361/201526019
128. R.F. Webbink, Astrophys. J. **277**, 355 (1984). https://doi.org/10.1086/161701
129. R.F. Webbink, in *Astrophysics and Space Science Library*, vol. 352, ed. by E.F. Milone, D.A. Leahy, D.W. Hobill (2008), p. 233. https://doi.org/10.1007/978-1-4020-6544-6_13
130. R. Wesson, X.W. Liu, M.J. Barlow, Mon. Not. R. Astron. Soc. **362**(2), 424 (2005). https://doi.org/10.1111/j.1365-2966.2005.09325.x
131. R. Wesson, M.J. Barlow, X.W. Liu, P.J. Storey, B. Ercolano, O. De Marco, Mon. Not. R. Astron. Soc. **383**(4), 1639 (2008). https://doi.org/10.1111/j.1365-2966.2007.12683.x
132. R. Wesson, D. Jones, J. García-Rojas, H.M.J. Boffin, R.L.M. Corradi, Mon. Not. R. Astron. Soc. **480**(4), 4589 (2018). https://doi.org/10.1093/mnras/sty1871
133. E.C. Wilson, J. Nordhaus, Mon. Not. R. Astron. Soc. **485**(4), 4492 (2019). https://doi.org/10.1093/mnras/stz601
134. T.E. Woods, N. Ivanova, M. van der Sluys, S. Chaichenets, in *Evolution of Compact Binaries*, ed. by L. Schmidtobreick, M.R. Schreiber, C. Tappert. Astronomical Society of the Pacific Conference Series, vol. 447 (2011), p. 127
135. T.E. Woods, N. Ivanova, M.V. van der Sluys, S. Chaichenets, Astrophys. J. **744**(1), 12 (2012). https://doi.org/10.1088/0004-637X/744/1/12
136. A.B. Wyse, Astrophys. J. **95**, 356 (1942). https://doi.org/10.1086/144409
137. L.R. Yungelson, M. Livio, A.V. Tutukov, R.A. Saffer, Astrophys. J. **420**, 336 (1994). https://doi.org/10.1086/173563
138. J. Zhu, P.M. Ricker, F. Timmes, R. Taam, R.F. Webbink, in *American Astronomical Society Meeting Abstracts*, vol. 51 (2019), p. 207.04
139. M. Zorotovic, M.R. Schreiber, B.T. Gänsicke, A. Nebot Gómez-Morán, Astron. Astrophys. **520**, A86 (2010). https://doi.org/10.1051/0004-6361/200913658

# A Modern Guide to Quantitative Spectroscopy of Massive OB Stars

**Sergio Simón-Díaz**

## Contents



**Abstract** Quantitative spectroscopy is a powerful technique from which we can extract information about the physical properties and surface chemical composition of stars. In this chapter, I guide the reader through the main ideas required to get initiated in the learning process to become an expert in the application of state-of-the-art quantitative spectroscopic techniques to the study of massive OB stars.

---

S. Simón-Díaz (✉)
Instituto de Astrofísica de Canarias, La Laguna, Tenerife, Spain

Departamento de Astrofísica, Universidad de La Laguna, La Laguna, Tenerife, Spain
e-mail: ssimon@iac.es



P. Kabáth et al. (eds.), *Reviews in Frontiers of Modern Astrophysics*,
https://doi.org/10.1007/978-3-030-38509-5_6

# 1 Introduction

Quantitative spectroscopy is one of the most rewarding fields of stellar astrophysics. It allows the courageous researcher who has decided to devote time to this fascinating enterprise, to have first hand access to an important set of empirical information about the investigated stars which, in most cases, cannot be obtained by any other means. This mainly comprises several key stellar parameters (including, e.g., spin rates, effective temperatures, and gravities), as well as surface abundances of those elements which have left their imprint in the observed piece of the stellar spectrum that will be analysed. In addition, the analysis process can help to highlight (and better characterize) the presence of stellar winds, circumstellar material, spots and surface magnetic fields, faint companions in binary systems, and/or some sources of stellar variability/activity, specially when multi-epoch spectroscopy is considered.

I write *courageous researcher* because becoming a reliable expert on quantitative stellar spectroscopy requires mastering several skills comprising observational, theoretical, modeling and programming aspects, as well as having medium to high knowledge of radiative transfer and atomic physics. Also, from a practical point of view, my own expertise gives me confidence to remark that having a detail oriented profile certainly helps to avoid providing erroneous and/or spurious information from the analysed spectra.

But, what does quantitative spectroscopy means? Paraphrasing my good friend M. A. Urbaneja from the University of Innsbruck, quantitative spectroscopy can be defined in a simple way as the *inference of the physical parameters that (uniquely and completely?) characterize an astronomical object based on three tools: an observed spectrum, a set of theoretical spectra, and a given comparison metric*.

Obviously, as the reader can imagine, this topic is so broad that it could lead to several books. In this chapter, I take the opportunity that the kind invitation[1] to participate in this book offers me to provide a broad overview of the main quantitative spectroscopic techniques which are presently applied to the study of the so-called massive OB stars.

The chapter is intended to serve to young students as a first approach to a field which has attracted my attention during the last 20 years. I should note that, despite its importance, at present, the number of real experts in the field around the world is limited to less than 50 people, and about one third of them are close to retirement. Hence, I consider that this is a good moment to write a summary text on the subject to serve as guideline for the next generations of students interested in joining the massive star crew.

If you are one of them, please, use this chapter as a first working notebook. Do not stop here. Dig also, for further details, into the literature I quote along the text. And, once there, dig even deeper to find all the original sources explaining in more

[1]By D. Jones, J. García-Rojas and Petr Kabáth (co-PI's of the *ERASMUS+ project "Per aspera ad astra simul"*).

detail the physical and technical concepts that are presently incorporated into our modern (almost) automatized tools.

Someone posed me the following question long time ago: *why a student needs to learn how to compute the square root of a 10-digit number if a calculator can easily do it?* If you know the answer to this question, I'm sure you can become one of the next experts in quantitative spectroscopy of massive OB stars. Go ahead and do a good job! You are really lucky to start in a fascinating time in which you will have easy access to thousands of observed and theoretical spectra, as well as to powerful computers allowing to run—in a fast and efficient way—analysis tools incorporating various types of comparison metrics. But, please, never forget the most important rule to enjoy what you are doing and make real progress: *don't use any of these nice tools as black boxes*.

This chapter is structured in two main sections as follows. First, I put massive OB stars in context. Then, I describe the main tools and techniques presently used for quantitative spectroscopy of this important, but complex, group of stars.

# 2 Setting the Scene: Massive OB Stars, from Observations to Empirical Quantities

Quantitative stellar spectroscopy is an intricate tool which allow us to jump from observations to a set of empirical quantities defining a given star. Despite this is a general statement that can be applied to any type of star, the first thing one must realize is that the details of the intermediate steps defining a specific quantitative spectroscopic analysis—as well as the outcome of such analysis—depends on the domain of stellar parameters characterizing the star under study and the available piece of stellar spectrum. In this section, I describe the main ingredients and ideas that must be taken into account to understand the strengths and limitations of the main state-of-the-art tools and techniques used for the quantitative analysis of optical spectra of massive OB stars.

## *2.1 Massive OB Stars in Context*

The term OB stars is commonly used in the literature with several different (but related) meanings. Generally speaking, all of them refer to any given sample of stars with O and B spectral types which define a specific group of interest to investigate a particular astrophysical question involving stars of this type.

However, the considered range in spectral type, as well as the inclusion of luminosity classes other than dwarfs, varies from one study to another. For example, this term is used in some studies of stellar abundances in late-O and early-B dwarfs stars (e.g., [1, 2]), but also in other works investigating the O and B star population of the Milky Way (e.g. [3–5]) or other galaxies in the Local Group (e.g., [6–8]), or performing any quantitative empirical study of the physical properties of different subsamples including stars of this type (e.g., [9–15]). As a consequence, the range of mass and evolutionary status covered by the investigated sample of stars may differ. This fact can create some confusion between different communities and, hence, it is important to be highlighted from the very beginning.

Along this chapter, I will follow the original definition by Morgan [16], which identifies OB stars as a spectroscopic "natural group" which, at intermediate and high spectral resolution may be defined by the detection of helium lines in absorption. As indicated by N. R. Walborn in Chapter 3 of the book *Stellar spectral classification* by Gray and Corbally [17], the low temperature boundary of this group is a diagonal in the Hertzsprung-Russell diagram (HRD) running from B2 V, through somewhat later types at intermediate luminosity classes, to the latest B Ia supergiants.

To put this group of stars in a broader context, I will use the schematic representation of the realm of massive stars in the so-called spectroscopic Hertzsprung-Rusell diagram[2] (sHRD) created by my former PhD student, G. Holgado, for his thesis [19]. Figure 1 is an adaptation of a figure included in [20], where they presented for the first time the observational distribution of Galactic massive stars in the sHRD. In its original form, [20] presented a density map of stars in the uppermost part of the sHRD using a compilation of spectroscopically derived effective temperatures and gravities for almost 600 stars. Figure 1 also includes, for reference purposes, the evolutionary tracks resulting from the non-rotating stellar evolution models for stars with masses in the range 9–120 $M_\odot$ computed by Ekström et al. [21]; and, overplotted as colored regions, the approximate location of various types of stellar objects associated with different evolutionary stages of massive stars.[3]

This figure illustrates how the original definition of OB stars by Morgan [16] does not only define a "natural group" from a spectral classification point of view, but also nicely covers the first part of the evolution of massive stars, fully including the main sequence (MS) as well as some early phases of the post-MS evolution.[4] The first thing to note is the broad range in mass ($M \approx 9$–$90\,M_\odot$), effective temperature ($T_{\rm eff} \approx 10{,}000$–$55{,}000$ K), and gravity ($\log g = 4.4 - 1.2$ dex) covered by OB stars. Although not represented in this diagram, these stars also cover a broad range in luminosities[5] ($L \approx 10^{3.5} - 10^6\,L_\odot$) and projected rotation velocities (values of $v \sin i$

[2]This diagram can be considered as an equivalent to the HRD, but only using stellar parameters derived spectroscopically (see also [18]).

[3]Poelarends et al. [22] propose a fiducial value of 9 $M_\odot$ for the minimum initial mass of massive stars at solar metallicity, where massive star is defined as a star that is massive enough to form a collapsing core at the end of its life and, thus, avoid the while dwarf fate [23].

[4]Actually, it is not yet completely clear whether the group of stars marked as B supergiants (B Sgs) are post-MS stars, MS stars or, even, some of them are post red supergiant stars (see Section 6.1 in [23] and references therein).

[5]Due to their high temperatures and luminosities, OB stars are sometimes also quoted as blue massive stars, and the O and B supergiants, as blue supergiants.

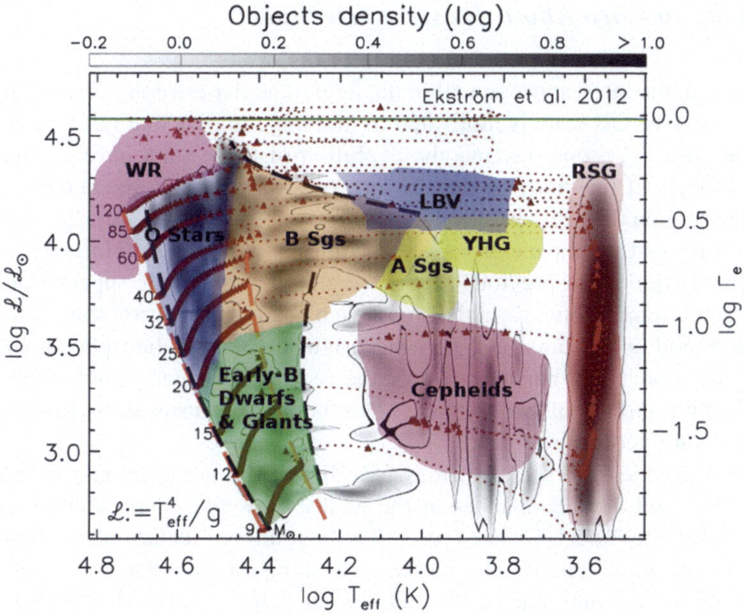


**Fig. 1** Schematic representation of the realm of massive stars in the so-called spectroscopic Hertzsprung-Russell diagram (sHRD, [18]). Colored regions depicts the approximate location of the various types of stellar objects (resulting from a phenomenological classification of their optical spectra) found to be associated with different evolutionary stages of stars born with masses above $\sim 9\,M_{\odot}$. Figure is based on the plot from [20] adapted by G. Holgado [19]. See [20] for a detail description of the various lines and symbols depicted in the figure

can reach up to 450 km s$^{-1}$). They also develop radiatively driven winds [24], which become directly observable in their spectral energy distributions and spectral lines above $\approx 10^4\,L_{\odot}$ (or, equivalently, the 15 $M_{\odot}$ track), and are mainly characterized by two global parameters: the terminal velocity ($v_{\infty}$) and the rate of mass loss ($\dot{M}$). All these extreme conditions must be taken into account when modelling the atmospheres of OB stars, a necessary step to perform any quantitative analysis of their spectra (see Sect. 3.2).

From an evolutionary point of view, as mentioned above (see also Fig. 1), OB stars represent the early evolutionary stages of massive star evolution, where early-B dwarfs/giants and O stars (including all luminosity classes) cover a different range in mass in the Main Sequence, while B Sgs are the evolved descendants of the O-type stars. Other stellar objects associated with later phases of massive star evolution (depending on the initial mass) are the A Sgs, the yellow hypergiants (YHG) and the luminous blue variables (LBV), and, last, the Wolf-Rayet stars (WR), the Red Supergiants (RSG) and the Cepheids.

Although any deeper mention to massive star evolution is out of the scope of this chapter, I refer the interested reader to a recent review by N. Langer [23] about pre-supernova evolution of massive single and binary stars as starting point. Most of the important references to learn further about this subject can be also found there.

## 2.2 Why to Care About Massive OB Stars?

In addition to the interest per se within the field of stellar astrophysics, any in-depth study of massive OB stars is motivated by the huge impact that our knowledge of the basic physical properties and the evolution of these stars have on many and diverse aspects of the study of the Cosmos (e.g. star formation, chemodynamical evolution of galaxies, re-ionization of the Universe; see [25–30]). They are also the progenitors of the most extreme stellar objects known in the Universe, some of them already quoted in previous section (e.g., hyper-energetic supernovae, Wolf-Rayet stars, luminous blue variables, massive black holes, neutron stars, magnetars, massive X and $\gamma$-ray binaries), and the origin of new studied phenomena such as long duration $\gamma$-ray bursters [31] or the recently detected gravitational waves produced by a merger of two massive black holes or neutron stars ([32, 33]; incl. LIGO and Virgo collaborations).

From a practical perspective, massive OB stars have become valuable indicators of present-day abundances in the Milky Way and other external galaxies, even beyond the Local Group (e.g., [34–36]). In particular, they cannot only be considered as a reliable alternative to H II regions as abundance indicators, but also these stars are superior to nebulae in that they are not affected neither by depletion into dust grains, nor the long-standing problem of the discrepancy resulting from the computation of nebular abundances using collisional emission lines or recombination lines (see, e.g. [37–39]).

In addition, in recent years, blue supergiants have been promoted to the hall of fame of the "standard candles", traditionally including cepheid and RR Lyrae variables, novae, Type Ia and Type II supernovae, as well as globular clusters and planetary nebulae [40, 41]. Indeed, as highlighted by Kudritzki [42], these stars are ideal stellar objects for the determination of extragalactic distances, in particular, because they are the brightest stars in the Universe and the perennial uncertainties troubling most of the other stellar distance indicators—namely, interstellar extinction and metallicity—do not affect them.

Last, the interpretation of the light emitted by close-by and distant H II regions and starburts galaxies relies on our knowledge of the effect that the strong ionizing radiation emitted by the O-type stars produce in the surrounding interstellar medium [43–45]. Also, any empirical information extracted from the spectra of OB stars about spin rates, mass loss rates and wind terminal velocities, photospheric abundances, binarity, and/or stellar variability associated with any type of pulsational-type phenomena is of ultimate importance to step forward in our understanding of the evolution and final fate of massive stars (e.g. [14, 20, 46–54]), as well as of the chemodynamical impact that these extreme stellar objects have on the surrounding interstellar medium at different scales.

## 2.3 Spectroscopy of Massive OB Stars

There are three main spectral windows which are commonly considered to extract spectroscopic information about massive OB stars:

- (far-)UV: as provided by spectrographs on board the *IUE*, *FUSE* and *HST* space missions, covering some pieces of the 900–2200 Å spectral range;
- optical: covering either the full range between 3800 and 9000 Å, or several key windows including the main set of diagnostic lines; and
- (near-)IR: mainly covering the *H*-, *K*- and *L*- bands at $\approx$1.62–1.77, 2.07–2.2, and 3.7–4.1 μm, respectively.

Figure 2 depicts three pieces of a typical optical spectrum of a mid-O dwarf (HD 199579, black line) and an early-B supergiant (HD 2905, grey line). These stars have been selected to illustrate how the different characteristics of stars in the OB star domain affect their spectra. For example, while in both cases the hydrogen Balmer lines (including $H_\gamma$, $H_\beta$, and $H_\alpha$ at λλ4341, 4860, and 6561 Å, respectively) are among the most prominent spectroscopic features, some particular characteristics of these diagnostic lines depend on the specific combination of surface gravity and effective temperature of the stars, as well as the existence of a more or less prominent stellar wind. For example, the larger the surface gravity for a given effective temperature, the more extended the wings of the Balmer lines; or, the stronger the wind density, the more remarkable the filling (in emission) of the $H_\alpha$ line. Also, while the presence of the He lines in absorption (including He Iλλ4387,

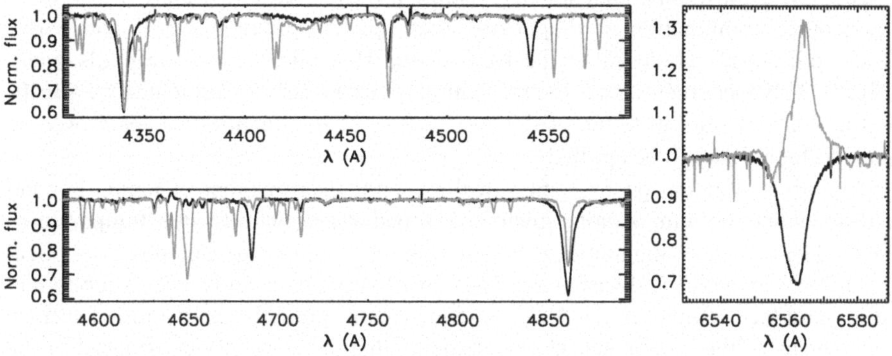


**Fig. 2** High quality spectra of two Galactic OB stars obtained with the HERMES spectrograph attached to the Mercator 1.2-m telescope in the Roque de los Muchachos observatory (La Palma, Spain). Black and grey lines correspond to the mid-O dwarf HD 199579 and the early-B supergiant HD 2905, respectively. The differences between both spectra rely on the different physical properties and chemical composition of the outermost layers of each of the two stars. Quantitative spectroscopy is a powerful tool which allows to extract this information by comparing an observed spectrum with a grid of synthetic spectra obtained by means of a stellar atmosphere code

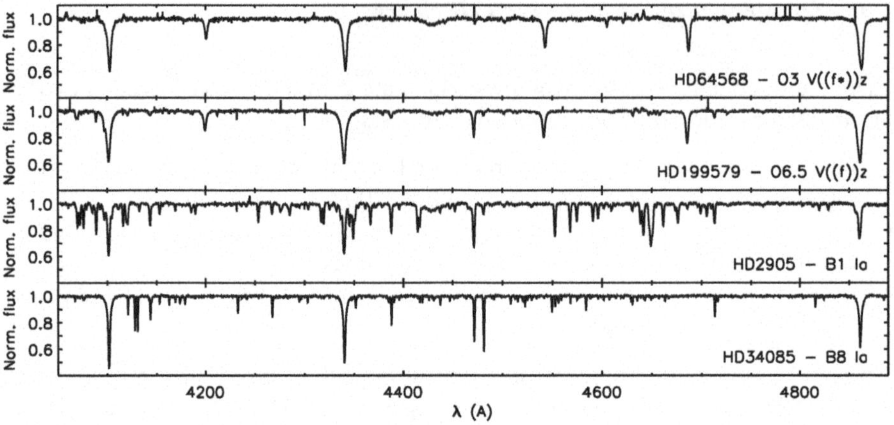


**Fig. 3** From top to bottom, illustrative examples of optical spectra of an early-O, a mid-O, an early-B and a late-B star. Some representative lines of H I, He I–II, O II, Si II–III–IV, Mg II, N V and C II are indicated in the corresponding spectrum where they are stronger

4471, 4713 Å, and He IIλλ4541, 4686 Å) is the main identifier of an OB star, most of the He II lines disappear in the B-type stars.

The later has important consequences for the spectroscopic determination of effective temperatures, for which having access to lines from two consecutive ions is compulsory. In this sense, as will be further described in Sect. 3.4, while the ratio of line strengths (or equivalent widths) of He II and He I lines has traditionally been considered as the main diagnostic to constrain the effective temperature of O4–O9.7 stars, other combination of diagnostic lines must be used in the B and early-O star domains,[6] where either the He II or the He I lines, respectively, disappear (Fig. 3). This fact creates a natural separation between early-O, mid/late-O, early-B, and mid/late-B stars in terms of the specificities of the quantitative spectroscopic analysis techniques to be applied.

A similar situation occurs when dealing with lines of other elements beyond hydrogen and helium. Again, given the broad range in effective temperatures covered by OB stars, the number and strength of metal lines populating the optical spectra of, e.g. a mid-O dwarf, an early-B dwarf/supergiant, and a late-B supergiant varies a lot (see, again, Fig. 3). As a consequence, the potential estimation of surface abundances of the typical set of key elements that are normally considered in the study of OB stars (mainly He, C, N, O, Si, and Mg, but also Ne, S and Fe) must be based on lines from different ions depending on the effective temperature of the star under analysis.[7] Indeed, in O-type stars and mid/late B Supergiants, the

[6]For example, N III–IV–V and Si IV–III–II lines in the early-O and the early-B stars, respectively (see Sect. 3.4).

[7]Also, different implementations of the associated model atoms—including a more or less detailed description of the energy levels and transitions of specific ions—are required.

number of metals with available diagnostic lines in the optical spectrum is much more limited than in the early-B star domain, hence hampering the determination of the corresponding abundances.

What about the other two spectral ranges? Despite the number of studies found in the literature performing quantitative spectroscopic analyses of OB stars in the optical range is much larger than those based on UV and/or IR spectra, the later contain important empirical information about the winds of these stars which is not directly accessible from the analysis of the optical spectrum (see below). In addition, they provide complementary information about effective temperatures, surface gravities, and abundances. Last, the quantitative spectroscopy in the (near)-IR range has been proposed as an important alternative to investigate the physical properties and chemical abundances in massive stars in highly obscured star forming regions (e.g. in the galactic center of the Milky Way), where the stars are much fainter in optical and UV wavelengths; hence, making more difficult to have access to high quality (mostly in terms of signal-to-noise ratio) spectra.

Next sections will be mainly devoted to the description of some of the tools and techniques presently used to perform quantitative spectroscopic analysis of OB stars based on their optical spectra. However, before entering into details, a few notes on these two other spectral ranges are worthwhile. The reader is also referred to the works performed by F. Najarro, M. Garcia, and T. Repolust (including a battery of papers and their PhD thesis), as well as [55–58], and [59], among others, for further details on quantitative spectroscopic analysis performed in the UV and IR.

Figure 4 provides a panchromatic view of the O9 Ia star HD 30614 ($\alpha$ Cam). The spectral windows depicted in Figs. 2 and 3 are now complemented with other portions of the spectrum of this star, including the UV range covered by the SWP spectrograph on board on the *IUE* satellite (top panel), the H, K and L bands in the IR (bottom panels), as well as two intermediate regions of the spectrum which have been elusively utilized for quantitative spectroscopy until recently.

The *Gaia* RVS range (optical IV) has been included for completeness, and to illustrate how "boring" (and mostly useless) is this spectral range for the case of O- and early B-type stars.[8] This is not the case for the spectral window between 5400 and 5900 Å (optical II), in which it can be find very useful diagnostic lines for the study of O-type stars such as, e.g., He II$\lambda$5411 Å, O III$\lambda$5591 Å, C III$\lambda$5696 Å, C IV$\lambda\lambda$5801, 5811 Å, as well as one line that can be used as a powerful diagnostic to detect double line spectroscopic binaries: He I$\lambda$5875 Å line.

Regarding the UV and IR parts of the spectrum, I want to specially highlight the three P-Cygni profiles[9] found in the range ~1200–1600 Å, as well as some of the hydrogen and helium lines located in the IR (e.g., $Br_\gamma$, $Pf_\gamma$, and $Br_\alpha$ at 2.17, 3.74, and 4.05 μm, respectively; He I$\lambda\lambda$1.70, 2.11, 3.70 μm; He II$\lambda\lambda$1,69, 2.19 μm). Most of these lines, as well as other few spectroscopic features present in other regions of the UV serve as key diagnostics to obtain information about

[8]For this type of stars, the Gaia RVs range is basically populated by a few Paschen lines.

[9]Corresponding to the transitions N V$\lambda\lambda$1239/43 Å, Si IV$\lambda\lambda$1394/403 Å, and C IV$\lambda\lambda$1548/51 Å.

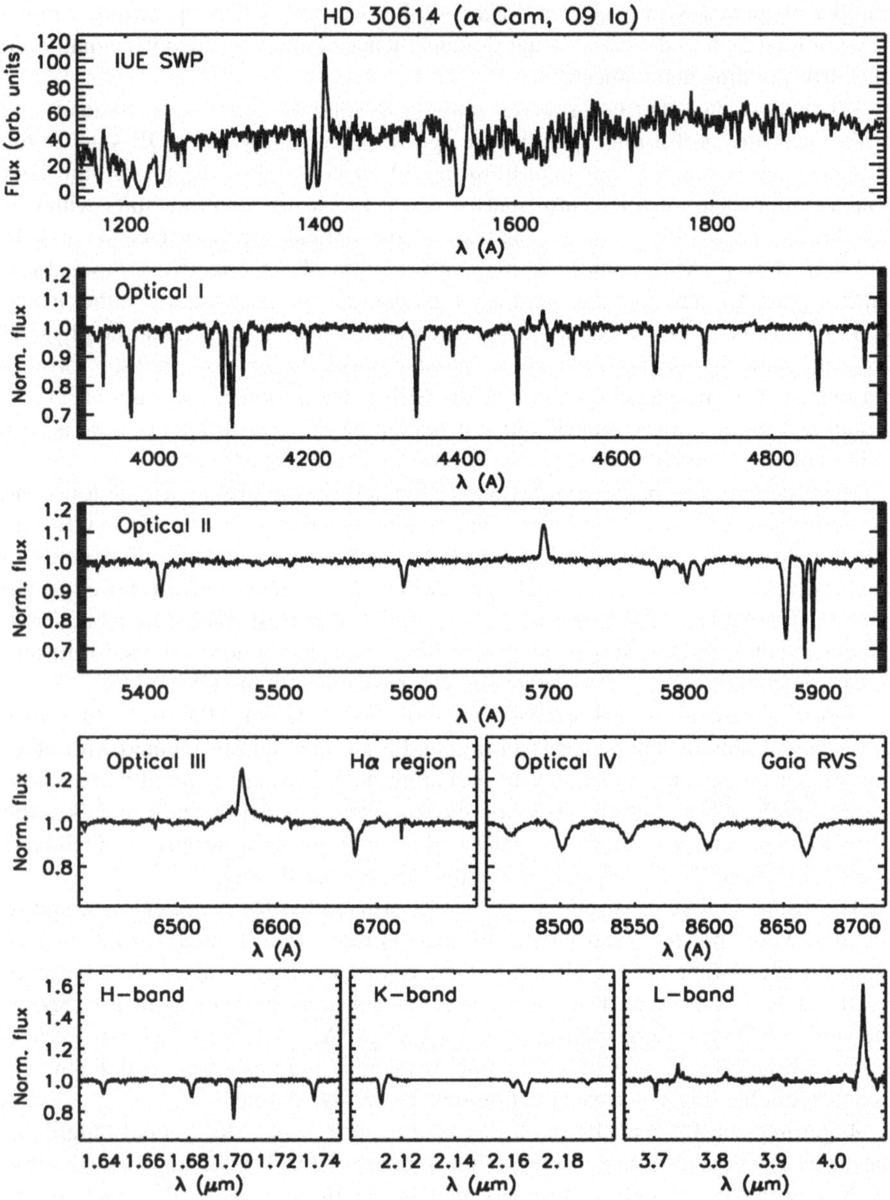


**Fig. 4** Panchromatic view of the O9 Ia star HD 30614 (α Cam). The figure has been created combining observations obtained with different spectrographs attached to several ground based telescope facilities (optical and IR) and the *IUE* space mission (UV). Spectra kindly provided by F. Najarro and M. Garcia (Centro de Astrobiolgía, Madrid)

the main physical properties of the stellar wind developed by some OB stars, such as the terminal velocity, the mass loss rate, the clumping factor and the presence of shocks in the wind (see, e.g., references quoted above). Indeed, these lines are much more sensitive to all these factors than the three main wind diagnostic lines found in the optical (i.e., $H_\alpha$, He I$\lambda$5875 Å, and He II$\lambda$4686 Å). As a consequence, any quantitative spectroscopic analysis of an OB star with an important stellar wind contribution should ideally consider the full UV+optical-IR range.

## 3 Tools and Techniques Used for Quantitative Spectroscopy of Massive OB Stars

This section is aimed at providing the reader a basic guide to the various steps that are commonly followed to perform the quantitative spectroscopic analysis of different types of OB stars based on their optical spectra,[10] as well as to the main presently available tools and techniques. Some further reading on the subject can be found below, separated by different type of analysis:

- Line-broadening parameters in O and B stars: [60–63];
- Spectroscopic parameters in O stars: [64–73];
- Abundances in O stars: [71, 74–78];
- Spectroscopic parameters and abundances in late-O and early-B stars: [2, 79–88];
- Spectroscopic parameters and abundances in B-Sgs: [34, 36, 89–94].

I also recommend the reader to have a look to:

- Chapter 1 in the book *Oxygen in the Universe* by Stasiǹska et al. [95],

as well as to the interesting reviews on:

- *Winds from hot stars* by R.-P. Kudritzki and J. Puls [24],
- *Modeling the atmospheres of massive stars* by J. Puls [96],
- *Parameters and winds of hot massive stars* by R.-P. Kudritzki and M. A. Urbaneja [97]
- *Non-LTE Model Atom Construction* by N. Przybilla [98],
- *UV, optical and near-IR diagnostics on massive stars* by F. Martins [55], and
- *Highly accurate quantitative spectroscopy of massive stars in the Galaxy* by M. F. Nieva and N. Przybilla [84].

[10]Most of the ideas presented along this section can be easily extrapolated to any quantitative spectroscopic analysis of the UV and IR spectral windows, with the only difference that other diagnostic lines, model atoms, and physical assumptions in the modeling of the stellar wind must be considered. Also some parameters and abundances may be more difficult (or even impossible in some cases) to be constrained just using the UV and/or IR part of the spectrum.

### *3.1 Quantitative Spectroscopy of Massive OB Stars in a Nutshell*

I have always considered that an efficient strategy to acquire new knowledge and skills about a given topic starts by having access to a quick, rough overview of the all the main points of the subject one wants to learn about. Then, once you have a more or less clear idea of where you want to go, you can come back—in sequential order and in more detail—to all those steps needed to fulfil your final objectives.

Let's then apply this strategy to learn about quantitative spectroscopy of massive OB stars! I enumerate below the complete list of intermediate milestones one has to pursue to perform a complete quantitative spectroscopic analysis of an OB star:

1. Acquisition of the observed spectrum.
2. Pre-processing of the spectrum, including a first qualitative visual assessment, the continuum normalization and the radial velocity correction.
3. Determination of the line-broadening parameters. This is the basic step to have access to *projected rotational velocities*.
4. Identification of the stellar atmosphere code and atomic models best suited for the analysis of the star under study.
5. Creation of a grid of stellar atmosphere models, also including the corresponding synthetic spectra and equivalent widths for the main set of diagnostic lines needed for the specific analysis one wants to perform.
6. Identification of the analysis strategy best suited to extract information from the observed spectrum of the star under study (e.g., spectral synthesis, use of equivalent widths).
7. Determination of the main set of spectroscopic parameters accessible through the analysis of the observed piece of spectrum (e.g., basically the effective temperature and surface gravity, but also the microturbulence, the abundance of the diagnostic lines used to estimate the effective temperature, and the wind strength parameter). In this case we refer to *stellar parameters determination*.
8. Determination of surface abundances of interest (among those elements with available diagnostic lines in the observed spectrum). This task is also called *chemical abundance analysis*.

In addition to these eight points, a complete characterization of the main physical properties of the star requires another two steps which must incorporate some extra empirical information not directly accessible from the analysis of the optical spectrum, namely:

1. the absolute magnitude of the star, in order to obtain estimates for the stellar luminosity, radius and mass.
2. the terminal velocity of the stellar wind, in order to obtain the mass loss rate.

And now, let's go back to the beginning and enter into more details.

## 3.2 Getting Ready!

We are entering in an era in which many of the new students (and a large fraction of the stellar community) will not have the necessity of preparing and undertaking any observing campaign to have access to all the spectroscopic observations required for their PhD studies. However, this does not mean that they should forget about learning (at least) some basic concepts of observational stellar spectroscopy.

Figures 2, 3, and 4 show illustrative examples of superb quality spectra (in terms of resolving power, S/N and wavelength coverage) of several types of OB stars. These are ideal spectra for a comprehensive and highly accurate quantitative spectroscopic analysis. However, in many situations it will not be possible to gather spectra with such a high quality and, hence, one will have to find a compromise between quality and number of stars with spectra good enough for the purposes of the study to be developed. For example, in some cases it can be more important to have access to high S/N spectra even sacrificing spectral resolution (e.g., when performing a chemical abundance analysis of extragalactic B Sgs; [89]), but in other situations is more critical to gather high-resolution data even if the S/N is somewhat poorer (e.g., when dealing with measurements of projected rotational velocities in O stars and B-Sgs, or when obtaining stellar parameters in O stars with spectra contaminated by nebular emission from the associated H II region; [61, 62, 70]).

It is also important for the beginner to realize that, in many cases, the optical spectrum of an OB star may be contaminated with some other spectroscopic features which are not directly associated with the star itself (from narrow interstellar lines and diffuse interstellar bands to telluric lines from the Earth atmosphere and/or nebular emission lines). In addition, since in most cases the starting point of a quantitative analysis is a normalized spectrum, the normalization process may have introduced spurious effects on some of the diagnostic lines (e.g., in the global shape of the wings of the Balmer lines which, as indicated in Sect. 3.6, are the main diagnostics to constraint the surface gravity in OB stars). Last, specially if one wants to extract information about the radial velocity of the star—either from a single snap shot spectrum or a time series—it is important to check whether the spectrum has been corrected from heliocentric/barycentric velocity; and, if not, learn how to do it.

All these questions will definitely affect the scope, accuracy, and reliability of any type of quantitative spectroscopic analysis, as well as its outcome. So, my first two advises before going ahead are (1) do not forget to *incorporate to your list of learnt skills* the main technical concepts about observational spectroscopy, and (2) do not start the quantitative spectroscopic analysis before performing a *qualitative (visual) assessment* of the observed spectrum to understand what you have in your hands. These two initial steps certainly help to establish the best strategy to follow, as well as to avoid over-interpretations of the outcome of the analysis. For example, a double line spectroscopic binary cannot be analysed in the same way as an isolated star, or the results for a chemical abundance analysis of an early-B star may be erroneous if one does not realize that is dealing with a Be star with a circumstellar emitting disc.

### 3.3 Radial Velocity Correction

Even if the radial velocity of the star is not a piece of information required by the study one wants to develop, the observed spectrum must be corrected by Doppler shift to ensure that all diagnostic lines are located in the laboratory position. While this correction is not very critical in those parts of the analysis based on equivalent widths, it may have non-negligible consequences in the determination of spectroscopic parameters by means of any type of line-profile fitting technique. The later is specially critical when establishing the surface gravity in the case of O- and B-type stars, since this is based in the fitting of the wings of the hydrogen Balmer lines.

There are several standard techniques that can be applied to perform the radial velocity correction, including, e.g., identification of the core of one or several diagnostic lines, either visually or using a gaussian fit to the line profile, and/or cross-correlation with a template. The most important warning to take into account when dealing with OB stars, however, is that one must remember that some lines may be affected by stellar winds (e.g. He I$\lambda$5875 Å and He II$\lambda$4686 Å; or even some metal lines in cases of stars with very strong winds), and hence the use of these lines may led to erroneous results. In addition, the usual techniques based in the identification of the core of the line may fail in those stars with a high projected rotational velocity or important asymmetries due to stellar oscillations.

### 3.4 Line-Broadening Parameters

One of the most straightforward and cheapest ways (from an observational point of view) to obtain information about stellar spins is based on the effect that rotation produces on the spectral lines: stellar rotation broadens the spectral lines. However, this is not the only line-broadening mechanism acting in O- and B-type stars. As reviewed by A. Herrero (2019) in Sect. 2.13 of the book *Radiative transfer in stellar and planetary atmospheres*, there are, at least, another five mechanisms to be taken into account in these hot, massive stars: the natural, thermal, collisional (mainly Stark, both linear[11] and quadratic[12]), microturbulent, and macroturbulent (pulsational?) broadenings, respectively. Therefore, the first step of the quantitative spectroscopic analysis consists in inferring the projected[13] component of the equatorial rotational velocity ($v \sin i$) by disentangling the effect that rotation produces on the line-profile from any other comparable effect produced

[11]Linear Stark broadening mainly affect the wings of the H and, to a less extent, the He II lines; indeed, this effect is mainly used to constrain the surface gravity (see Sect. 3.6).

[12]Quadratic Stark broadening, which is much less pronounced than the linear one, mainly affects the shape of the He I lines.

[13]Into the line-of-sight.

by the remaining broadening mechanism. And, as the reader can imagine, an important part of the process will be the selection of the best suited lines for the line broadening analysis, taking into account that the less affected the diagnostic line by the other (non-rotational) broadenings or by blends with other lines, the better. For example, it is always better to use a well isolated photospheric metal line than a hydrogen or helium line.

From a technical point of view, as in those other steps of the quantitative analysis process in which we are extracting information from the shape of the line-profile (vs. use of equivalent widths), the resolving power of the observed spectrum (as well as the number of points defining the line) is one important limiting factor of the accuracy we can reach in the determination of $v \sin i$. For example, as a general rule of thumb, a spectral resolution $R$ implies a rough minimum limit in a reliable determination of $v \sin i$ of $c/R$, where $c$ is the speed of light in the same units as $v \sin i$. On the other hand, when the Fourier transform method is used to estimate the projected rotational velocity, due to the Nyquist theorem, the spectral dispersion ($\Delta\lambda$, in Å/pix) of the stellar spectrum imposes a limit in the lowest $v \sin i$ that can be derived, roughly given by $1.320\, c\Delta\lambda/\lambda$ (see [61]).

I refer the reader to [61, 62] and references therein for a thorough description of the various methods that have been routinely applied in the last 60 years for the determination of projected rotational velocities in OB stars, also including a discussion of the pros, cons and limitations of each method.

At present, the combined use of the Fourier transform (FT) and a goodness-of-fit (GOF) methods (Fig. 5) has become a standard strategy to disentangle the effect of rotation from the other main sources of broadening shaping the line-profiles of OB stars. In brief, and following Gray ([99]; see also the latest edition of the book,

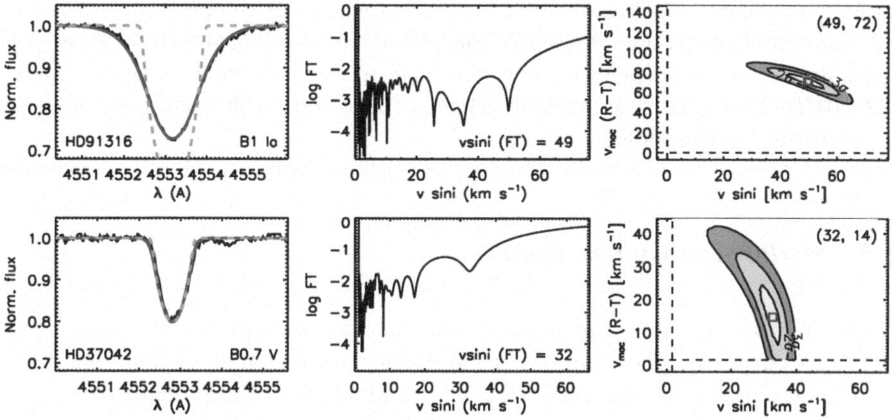


**Fig. 5** Combined FT+GOF line-broadening analysis of the Si III$\lambda$4552 Å line in the early-B dwarf HD 37042 (bottom) and the early-B supergiant HD 91316 (top). [Left panels] The best fitting synthetic profile (solid gray) and the profile corresponding to $v \sin i$(FT) and $v_{\mathrm{mac}} = 0$ (dashed gray) are over plotted to the observed profile (solid black). [Middle panels] Fourier transform of the observed profile. [Right panels] $\chi^2$-2D-map resulting from the GOF analysis

published in 2005), the Fourier transform method for the determination of $v \sin i$ is based on the intrinsic property of the rotational broadening function which develops zeroes in its Fourier transform. As firstly described by Carroll [100], the position of these zeroes in frequency space depends on the $v \sin i$ of the star, so that the frequency of the first zero ($\sigma_1$) is related to the rotational velocity through:

$$\frac{\lambda_0}{c} \, v \sin i \; \sigma_1 \;=\; A \tag{1}$$

where $\lambda_0$ is position in wavelength of the core of the line-profile, and $A$ is a constant that depends on the limb-darkening coefficient.[14]

Since the FT method only provides an estimate of $v \sin i$, but its clear that this is not the only line-broadening agent even in the case of photospheric metal lines (see, e.g. the case of the early B-Sg HD 91316 in Fig. 5), it needs to be complemented with a goodness-of-fit method, in which a $\chi^2$ fitting strategy if followed. In the later, an intrinsic profile[15]—which can be a $\delta$-function or a synthetic line resulting from a stellar atmosphere code—is convolved with a rotational and a macroturbulent profile, and both line-broadening parameters ($v \sin i$ and $v_{\rm mac}$) are obtained from the minimum value of the $\chi^2$-2D-map resulting from this GOF analysis.

The combined FT+GOF method provides a powerful and straightforward strategy to have access to the projected rotational velocity of a given star. In an ideal case, both determinations of $v \sin i$, as resulting from the FT and the GOF methods should be in agreement. However, this is not always the case, implying that the situation for this specific star is more complex than initially expected due to the presence on the line profile of effects originated by, e.g., some types of stellar oscillations or spots and/or chemical inhomogeneities in the stellar surface [101].

Overall, the FT method has been recently proven [61] to be a better suited strategy to obtain actual estimates of projected rotational velocities in the whole OB star domain than other previously considered methods [60, 102–105]. However, some limitations and caveats to be further investigated still remain (e.g., [62, 101, 106, 107]). This open new interesting lines of research for the new generation of massive star spectroscopists.

## 3.5 *Stellar Atmosphere Codes*

Stellar atmosphere codes are unavoidable tools when dealing with quantitative stellar spectroscopy. They are one of the major outcomes from an important, but complex research field which has led to a large number of texts in the literature.

[14] The most common value used for $A$ is 0.660, which corresponds to a limb-darkening coefficient of $\epsilon = 0.6$ (see, however, Fig. 3 in [101]).

[15] With the same equivalent width as the observed profile.

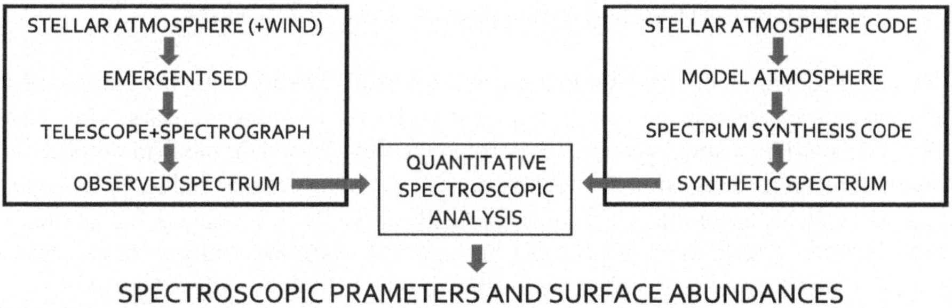


**Fig. 6** Schematic representation of the main actors of quantitative stellar spectroscopy

Although I assume that the reader of this chapter has acquired basic knowledge on radiative transfer in stellar atmospheres as part of her/his university studies, I encourage any young student willing to devote his/her career to quantitative spectroscopy of massive OB stars to deepen further into the specificities of stellar atmosphere modeling of hot stars already in the early stages of her/his career. The most recent version of the book *Theory of Stellar Atmospheres* by D. Mihalas is a perfect starting point. I also recommend Chapter 4 of the book *Radiative transfer in stellar and planetary atmopsheres*, written by J. Puls, for a more in-depth description of the methods developed for the modeling of expanding atmospheres of early-type stars.

Before providing a brief overview of the main available stellar atmosphere codes for quantitative spectroscopy of OB stars, I quote below a few basic ideas allowing the non-expert in the field to easily understand the importance of model atmospheres for quantitative spectroscopy (see the sketch presented Fig. 6):

- A stellar atmosphere is a thin layer in the surface of a star which does not have its own energy sources. Only redistribution of radiative energy takes place.
- A (one-dimensional, 1D) model atmosphere is a large table which describes the temperatures, pressures and many other properties of the gas as they vary with depth below the stellar surface.
- A stellar atmosphere is the part of the star where the emergent spectral energy distribution (including the continuum and the line spectrum) is formed.
- A spectrum synthesis code is a computational tool that allows the calculation of the stellar emergent (synthetic) spectrum from a given model atmosphere.
- Quantitative stellar spectroscopy allows to extract information about the physical properties and chemical composition of a stellar atmosphere from the comparison of an observed spectrum and a grid of spectra computed with a spectrum synthesis code coupled to a stellar atmosphere code.[16]

[16]In many cases, a stellar atmosphere code includes the computation of the emergent spectrum.

### 3.5.1 State-of-the-Art Stellar Atmosphere Codes for OB Stars

The last decades of the twentieth century witnessed an enormous progress in the development of adequate stellar atmosphere codes for hot massive stars. The great efforts devoted by a small group of experts in the field, based on the firm theoretical foundations on radiative transfer laid by e.g. V. V. Sobolev, D. Mihalas, L. H. Auer, L. B. Lucy, P. M. Solomon, J. I. Castor, D. G. Hummer, H. J. Lamers, J. P. Cassinelli (among others), have made it possible the massive star community to have access to a modern generation of stellar atmosphere codes which are allowing to perform reliable quantitative spectroscopic analyses of medium to large samples of O- and B-type stars in a reasonable amount of time.

This ambitious enterprise implied the inclusion of a realistic description of physical processes occurring in the outer layers of these extreme stellar objects such as, e.g., departure from the local thermodynamic equilibrium (non-LTE), line-blanketing and, in some cases, line-driven stellar winds. In particular, the later required the consideration of geometries departing from the simple plane-parallel approach, as well as the development of intricate computational techniques to deal with radiative transfer in (rapidly) expanding atmospheres (i.e. stellar winds).

At present, the stellar atmosphere codes most commonly used for the quantitative spectroscopic analysis of OB stars are:

- ATLAS [108] coupled with DETAIL/SURFACE [109]
- TLUSTY [110], coupled with SYNSPEC [111]
- CMFGEN [112]
- FASTWIND [113–115]
- PoWR [116–118]
- WM-*basic* [119]

While all of them are 1-D, non-LTE, line-blanketed codes, they differ in how they treat geometry (plane parallel/spherical), hydrostatic equilibrium/mass outflows, line blanketing/blocking, micro- and macro-clumping (vs. unclumped winds), as well as the considered strategy to solve the complex, interwined set of equations of radiative transfer, and how they deal with information about atomic data.[17] As a consequence, not all these codes are equally optimized to analyse different types of OB stars or specific windows of the stellar spectrum. For example, both TLUSTY and DETAIL/SURFACE calculate occupation numbers/spectra on top of hydrostatic, plane parallel atmospheres; hence, they are "only" suited for the analysis of stars with negligible winds. Also, despite CMFGEN, FASTWIND, PoWR and WM-*basic* can, all of them, deal with spherically extended atmosphered with

[17] Model atoms—including information about energy levels and the main collisional and radiative transitions between levels and/or the continuum—are a very important ingredient of stellar atmosphere code. They will be only occasionally mentioned along this chapter; however, basic knowledge of how models atoms are implemented and used in stellar atmosphere and diagnostic codes is the forth pillar a quantitative stellar spectroscopist should dominate, along with basic concepts of observational stellar spectroscopy, radiative transfer and stellar atmosphere modeling.

winds, the later (WM-*basic*) is mainly applicable to the analysis of the UV range. Last, due to the different approximation considered by these codes for the treatment o line blanketing/blocking, the amount of computational time required varies from one code to other, ranging from less than 1 h in the case of FASTWIND and DETAIL/SURFACE to several hours for CMFGEN, POWR, TLUSTY and WM- *basic* models.

Further notes on state-of-the-art approaches to model the atmospheres of hot, massive stars, as well as improvements occurred in this field in the last years can be found in [96] and [120], respectively. In particular, Table 1 in [96] provides a nice overview of the main characteristics and range of applicability of all the stellar atmosphere codes quoted above.

### 3.5.2 Grids of Models for Quantitative Spectroscopy of OB Stars

At this point, we are almost ready to proceed with the determination of those stellar parameters that can be directly obtained through the spectroscopic analysis (see Sect. 3.6). But, before, we need to spend a few time on the design and computation of a grid of stellar atmosphere models. Or, in same cases, we will be able to use directly any of the pre-computed grids which the developers (or their direct collaborators) have made publicly available. Some examples of the later can be found in the webpages of TLUSTY,[18] CMFGEN,[19] or POWR.[20] However, these are not the only available grid of models; many others have not been done public, but could be available with permission of the owners. This is, e.g. the case of the vast grid of FASTWIND models covering the O star domain (for solar and half solar metallicity) computed at the Instituto de Astrofísica de Canarias and which is presently incorporated to the IACOB grid based automatized tool (IACOB-GBAT, see [69, 70, 73]).

Regardless of using a pre-computed grid, or creating a new one, there are a few key points that must be carefully checked before performing the quantitative spectroscopic analysis:

- The stellar atmosphere code used to compute the grid must consider all important physical processes occurring in the star under study (see Sect. 3.5.1). The same consideration must be taken into account for the other geometrical and dynamical aspects of the modeling.
- All the key diagnostic lines must be properly included and treated in the computation of the associated synthetic spectra. Obviously, the wavelength coverage of the grid of synthetic spectra include the observed spectrum.

[18] nova.astro.umd.edu.

[19] http://kookaburra.phyast.pitt.edu/hillier/web/CMFGEN.htm.

[20] http://www.astro.physik.uni-potsdam.de/~wrh/PoWR/powrgrid1.php.

- Those spectroscopic parameters that we want to determine must be considered as free parameters in the computed grid of models. If some of them are kept fixed in the modeling process or the creation of the grid, one must evaluate in detail the consequences it has for the specific quantitative spectroscopic analysis to be performed.
- Always check carefully the various model atoms considered as input for the computations (specially in the case of those elements that will be included in the chemical abundance analysis, but also when for those elements/lines which are used to constrain the effective temperature).
- The step size for the various free parameters considered in the grid of models must be appropriately suited for the accuracy we want to reach in the analysis process.

The best way to acquire the necessary skills and confidence to go through all these points is to learn from someone with previous expertise or from those papers explaining the adapted strategy depending on the stars under study. Some examples of the later are provided along the next sections.

## 3.6 Spectroscopic Parameters

From here onwards, things apparently become a bit more straightforward from a practical point of view. However, only expertise and a detailed and careful management of the techniques described below will allow to extract reliable information from the quantitative spectroscopic analysis to be performed.

Once the observed spectrum is ready to be analysed (Sects. 3.2 and 3.3), and the line-broadening parameters have been determined (Sect. 3.4), the next step is the determination of the so-called spectroscopic parameters using a suitable grid of stellar atmosphere models (Sect. 3.5.2). In the case of the analysis of optical spectra of OB stars, these basically include the effective temperature ($T_{\rm eff}$), the surface gravity ($\log g$), and the wind-strength $Q$-parameter.[21] In addition, there are other secondary parameters—such as the helium abundance ($Y_{\rm He}$), the microturbulence ($\xi_{\rm t}$), the exponent of the wind velocity law ($\beta$), and the abundance of the element whose ionizing equilibrium is used to determine the effective temperature (e.g., silicon in the early-B type stars)—which need to be also determined at the same time during the analysis process.

While in the whole OB star domain the wings of the Balmer lines are the main diagnostic to estimate the surface gravity, the set of diagnostic lines that is used to constrain the effective temperature of the star depends on its spectral type. As indicated in Sect. 2.3 the He I and He II lines have traditionally been considered

[21] $\log Q = \log \dot{M} - 1.5 \log R - 1.5 \log v_\infty$ [121]. This parameter is used as a proxy of the wind properties in the optical analyses because this spectral window does not include any diagnostic line reacting exclusively (or mainly) to the mass-loss rate ($\dot{M}$) or the wind terminal velocity ($v_\infty$).

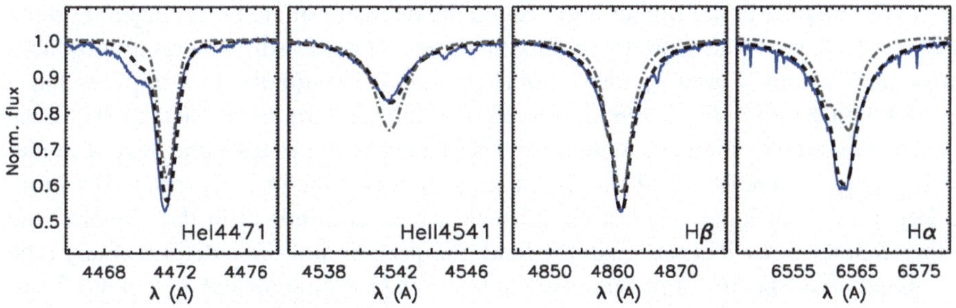


**Fig. 7** Solid blue line: Four representative H and He I–II lines of the observed spectrum of the O9 V star HD 214680. Dashed and dashed-dotted lines: Two synthetic spectra computed with the FASTWIND stellar atmosphere code, corresponding to the best fitting model resulting from the IACOB-GBAT analysis ($T_{\rm eff} = 35{,}000$ K, $\log g = 3.9$ dex, $\log Q = -14.0$) of the observed spectrum, and a model with $T_{\rm eff} = 37{,}000$ K, $\log g = 3.7$, and $\log Q = -12.5$, respectively

for the analysis of mid and late O-type stars (e.g. [64, 66, 73]). The basics of this type of analysis is summarized in Fig. 7, where an illustrative set of H and He I–II lines for the O9 V star HD 214680 (10 Lac) is depicted. In addition to the observed spectrum, two synthetic spectra computed with FASTWIND are overplotted. One of them is the best fitting model resulting from the IACOB-GBAT analysis[22] (see [73]); the second one represent a model in which the associated values for $T_{\rm eff}$, $\log g$, and $\log Q$ have been shifted from the best fitting values to illustrate the effect on the various diagnostic lines. On the one hand, this figure serves to realize the quality of the fits our state-of-the-art models are reaching; on the other hand, it shows how the second model does not fit any of the lines since it has a too low gravity (wings of $H_\beta$ less extended than in the observed spectrum), a too high effective temperature (the He II 4541 line in the model is too strong), and a too high value of the wind-strength $Q$-parameter ($H_\alpha$ is in emission in the model while it is not in the observed spectrum).

The basics of the strategy followed for the determination of the spectroscopic parameters in O-type stars can be easily understood with the simple example above. However, it is important to note that the situation is a bit more complex since, actually, there is not an unique, separated dependence of the various diagnostic lines with the different parameters. For example, an increase in $\log g$ for a given $T_{\rm eff}$ produces—in addition to a more pronounce Stark broadening of the Balmer lines due to the larger electron density in the photosphere—weaker He II lines and stronger He I lines. The larger electron density favours the recombination of higher ions into lower ionization stages, hence increasing the relative population of lower

[22] IACOB-GBAT [69] is a grid-based automatic tool for the quantitative spectroscopic analysis of O-stars. The tool consists of an extensive grid of FASTWIND models, and a variety of programs implemented in IDL to handle the observations, perform the automatic analysis, and visualize the results. The tool provides a fast and objective way to determine the stellar parameters and the associated uncertainties of large samples of O-type stars within a reasonable computational time.

ions with respect to the higher ones. Since the intensity of He I lines depend on the number $He^+$ ions (these are recombination lines), a larger surface gravity produces stronger He I lines (and opposite for He II lines). Eventually, this implies that a model with a larger $T_{\rm eff}$ is required to recover the same ratios of He II to He I lines when compared to a model with a lower surface gravity. In a similar way, a model with a larger value of log $Q$ (needed, e.g., to fit a $H_\alpha$ line in emission) will require a larger value of log $g$ (w.r.t. a model with a weaker wind) to fit the wings of the other Balmer lines. Therefore, in the final interpretation of the outcome of any type of quantitative spectroscopic analysis (not only in O-type stars, but also in the B star domain), it is important to remember that there exist important covariances between some of the spectroscopic parameters.

In early O-type stars (O2 and O3), the He I lines become too weak and the determination of the effective temperature is hence based on N IV–V lines (e.g. [74, 75]). Similarly, the He II lines are absent in B-type stars, where Si IV–III and/or Si II–III are utilized instead (e.g. [35, 46, 82, 122]). Figure 8 shows the behavior of the equivalent widths of several diagnostic lines of N and Si (in addition to He) which are commonly used to determine temperatures in OB stars. From inspection of this figure one can easily understand why different line ratios are needed depending on the specific range in $T_{\rm eff}$. Note, however, that I only represent the dependence of the equivalent width of the various lines with $T_{\rm eff}$, while some of these line may also present some dependences with other parameters. Note that, whenever possible (as is always the case in the spectral type range O2–B3), the ratio

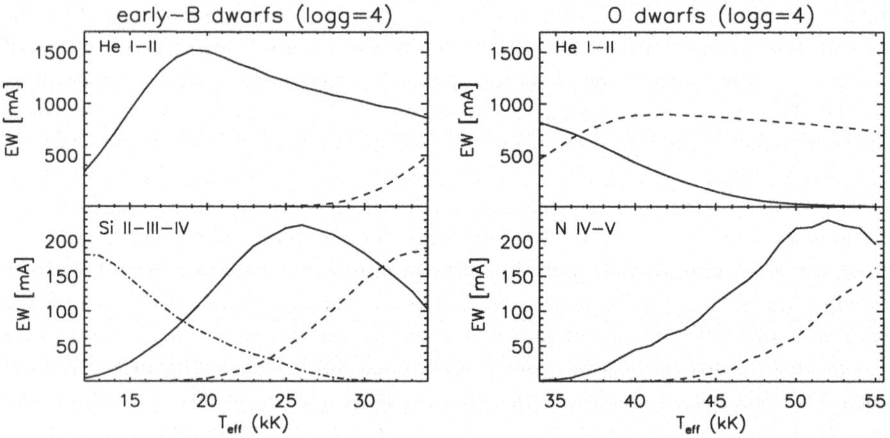


**Fig. 8** Top panels: Behaviour of the equivalent width of the He I$\lambda$4471 Å and He II$\lambda$4541 Å lines with $T_{\rm eff}$ in the early-B and O dwarf domain. As illustrated by the figures, the He I–II ionization balance cannot be used below $T_{\rm eff} \sim 30$ kK and above 47 kK. Bottom panels: Alternative diagnostic lines used to constraint $T_{\rm eff}$ in early-B (Si II–IV) and O stars (N IV–V), respectively. In this specific case, the figures depicts the behaviour with $T_{\rm eff}$ of the following lines: Si II$\lambda$4128 Å Si III$\lambda$4552 Å Si IV$\lambda$4116 Å N IV$\lambda$4058 Å and N V$\lambda$4603 Å. Note: the equivalent width of all considered lines have been obtained from a grid of FASTWIND models at solar metallicity computed by the author

of equivalent widths of two ions from the same element should be the preferred diagnostic. In this way, we eliminate the dependence of the line ratio with the abundance of the considered element.

I stop here due to space limitations of the chapter, but some further reading to deepen in this part of the quantitative spectroscopic analysis can be found in any of the references quoted along this subsection (see also Sect. 3). I also refer to [123] for an interesting discussion about why the spectroscopic approach to determine effective temperatures and surface gravities in early-B stars should be always preferred to the use of photometric indices, a common practice in the past, when we did not have the adequate tools to perform a proper quantitative spectroscopic analysis available.

## 3.7 Photospheric Abundances

There are two different approaches for the abundance analysis in stellar objects: the *curve of growth* method and the *spectral synthesis* method. The curve of growth method is based on the behaviour of the line strength with an increase in the chemical abundance, also incorporating the effect of microturbulence.[23] This method uses line equivalent widths, and hence does not require any knowledge of the exact rotational and macroturbulent broadening mechanisms affecting the line profiles.

Figure 9 provides a quick overview of the various steps followed in the oxygen abundance analysis of a narrow line early-B type star by means of the curve of growth method. I always recommend to any novice in the business to start by inspecting in detail what is summarized in the figure or, even better, perform the analysis of a similar star from scratch. The main reason is that the optical spectra of narrow line, early-B type stars include a lot of isolated O II and Si IV–III (or Si III–II lines). Hence, performing such an exercise allows the new spectroscopist to understand all the critical points which can affect the outcome of any abundance analysis using a well behave case, before jumping to less optimal cases in which, e.g., there is one or two available lines, or those critical cases cannot be easily identified given the adopted strategy.[24]

In the curve of growth method, once the stellar parameters have been established, a grid of stellar atmosphere models whereby the abundance for the studied element and the microturbulence are varied (the remaining parameters are kept fixed) is computed. In this way, the curves of growth for each line can be constructed

[23]Microturbulence ($\xi_t$) is a free parameter that was included in the stellar abundance analyses to solve the discrepancy found in the line abundances from weak and strong lines. Its physical meaning is supposed to be related to the small scale turbulent motions of the stellar plasma which could mainly affect the strong lines close to saturation.

[24]This is the case for the spectral synthesis method, where the effect of microturbulence or the existence of wrongly modeled lines (see [82]) is not so easily identified.

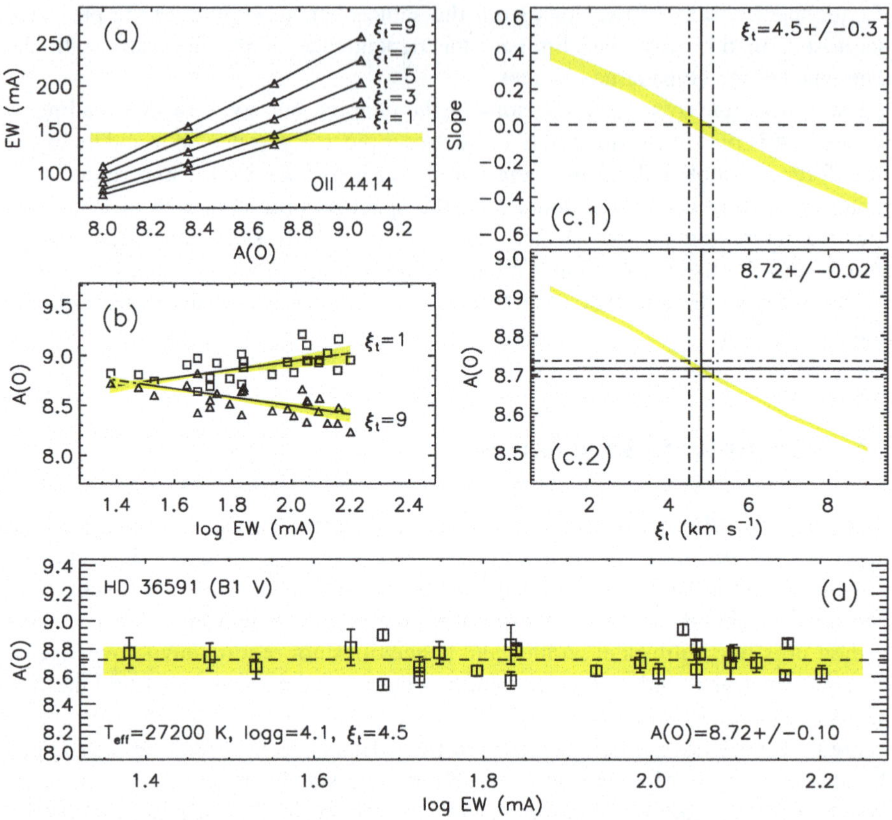


**Fig. 9** Visual summary of the various steps comprising the oxygen abundance determination—via the curve of growth method—of the B1 V star HD 36591. Up to 40 O II lines are available in the optical spectrum (mainly between 4000 and 5100 Å), but only 27 of them have been finally considered as reliable. See text for explanation, and more details about the observed spectrum considered and the analysis process in [82]

by plotting the theoretical equivalent width for each value of $\xi_t$ as a function of abundance (see Fig. 9a). From the observed equivalent width and its error, an abundance (and its uncertainty) can be derived for each line and each value of $\xi_t$. The individual line abundances are dependent on the microturbulence which affects more the strong lines than weak lines. Figure 9b shows the $A(\mathrm{O}) - \log EW$ diagrams for two different values of $\xi_t$. The value of $\xi_t$ that minimises the dependence of the line abundances on he line strength in the $A(\mathrm{O}) - \log EW$ diagrams (i.e. produces a zero slope will be the adopted microturbulence). Figure 9c.1 and c.2 show the dependence of the slope of the $A(\mathrm{O})$ vs. $\log EW$ relation and of $A(\mathrm{O})$ on $\xi_t$. In the last step, abundance values for each line as well as their uncertainties are calculated for the adopted microturbulence (Fig. 9d). The final abundance value is estimated through a weighted mean of the linear individual line abundance.

The curve of growth method, as described above, also allows a straightforward computation of the final uncertainty taking into account three different sources of errors: those associated with the line-to-line abundance dispersion, those derived from the error in the determined microturbulence and, finally, those referred to the uncertainties in the stellar parameters. In addition, diagrams as the one depicted in Fig. 9d can be used as a powerful diagnostic tool to check the reliability of the various lines available for the abundance determination.

The applicability of the curve of growth method is limited to those cases when the equivalent widths of individual lines can be measured. When the projected rotational velocity of the star is high (i.e. fast rotators), or when the spectral resolution is not good enough for resolving individual lines (e.g. in extragalactic studies beyond the Local Group), a different approach must be considered: the spectral synthesis method, one of the few techniques that can be applied when blending is severe. This method is based on the computation of a grid of synthetic spectra including all the observed lines, which is then directly compared to the observed spectrum to find the best fitting model. Basically, this method follows a very similar strategy as the one illustrated in Fig. 7, but including many other diagnostic lines for those elements under study. In contrast to the curve of growth method, this method requires a correct broadening of the line profiles and making sure that all the elements whose lines are present in the blending are included in the line formation calculation. In addition, the final results are more sensitive to other subtleties such as, e.g., a correct radial velocity correction of the observed spectrum.

As I said, my recommendation is to always start the learning process with a benchmark case as the one presented in Fig. 9. That way, the new spectroscopists will consolidate a strong critical sense to avoid misinterpretations of results in those cases in which the number of diagnostic lines is more limited or the quality of the observed spectrum is worse. For example, when there is only 1 or 2 lines available—as is, e.g., the case of nitrogen in O-type stars, or magnesium in B-type stars—the determination of the microturbulence is more critical and one will have to make a decision on the value to use.

Some illustrative examples of studies following different types of strategies for the chemical abundance analysis of OB stars can be found in the references quoted in Sect. 3 or in, e.g., [1, 35, 91, 124–134].

### 3.8 The Comparison Metric: From Visual Fitting to PCA and MCMC

We have seen in Sects. 3.6 and 3.7 that the process of determination of spectroscopic parameters and abundances via quantitative stellar spectroscopy basically consist of finding the synthetic spectrum computed with a stellar atmosphere code which result in the best possible fit to an observed spectrum.[25] The basic idea is simple, but two important questions quickly pop up as soon as one wants to provide results from the analysis:

1. What defines the best possible fit to the observed spectrum?
2. Is the solution unique, or we can reach a similarly acceptable solution with different combinations of stellar parameters and/or abundances?

I am sure that the reader is presently in a good position to assert with confidence that, given the multidimensionality of the parameter space considered during the modeling process, the existence of significant covariances between parameters[26] and taking into account some technical limitations related to the quality of the observed spectrum (in terms, e.g., of signal-to-noise ratio), the answer to the second question is "no, the solution is not unique". Indeed, a proper identification of the range of acceptable parameters/abundances (i.e. definition of the associated uncertainties) is as important as the determination of the best fitting or central values.

Regarding the first question, the considered strategy to define the central values and the associated uncertainties has gained in complexity and robustness in the last decades. Not so long ago, given the computational limitations, most of the spectroscopic analyses were made based on small grids of stellar atmosphere models, and the determination of the final solution was a subjective by eye decision, sometimes supported by some more quantitative (but still simple) arguments (e.g., [64, 135]). However, the continuously increasing amount of high-quality spectroscopic observations of massive OB stars provided by different surveys during the first decade of the twenty-first century (e.g., [136–141]) made it clear the necessity to develop more objective, semi-automatized techniques which allow for the extraction of information about stellar parameters and abundances (and the associated uncertainties) from large spectroscopic datasets in a reasonable computational time.

Some notes on various of the techniques proposed to date can be found in [36, 46, 68, 69, 89], and [85]. Most of them are based on specific grids of pre-computed models and a $\chi^2$ algorithm which allow to find the best fitting solution (or central values for each of the considered free parameters) and the associated uncertainties. However, the use of other strategies based on, e.g., projection and/or pattern recognition methods (in contrast to the minimum distance methods, as the $\chi^2$ algorithms) are slowly but surely started to be explored. In addition, some works are already exploiting strategies based on the application of Genetic Algorithms (GA), principal component analysis (PCA), Gaussian process regression and Monte Carlo Markov Chains (MCMC) techniques. In particular, the later three are certainly envisaged as a promising way to minimize the computational time needed to

[25]Either directly or by using equivalent widths of a selected sample of diagnostic lines.

[26]For example, effective temperature and surface gravity, abundance and microturbulence, mass loss rate and the $\beta$ parameter.

create optimal grids of models, and to speed up the process of exploration of the multidimensional parameter space.

**Acknowledgements** I want to warmly thank all those friends and colleagues from which I've been able to learn and discuss about quantitative spectroscopy since my first years as PhD student at the Instituto de Astrofísica de Canarias. Special thanks to A. Herrero, M. A. Urbaneja, C. Villamariz, F. Najarro, C. Trundle, D. J. Lennon, J. Puls, N. Castro, M. Garcia, F. Nieva, C. Sabín-Sanjulian, K. Rübke, G. Holgado, S. Berlanas, and A. de Burgos. Let this text serves to spread part of the knowledge I've acquired from all of you during these years.

## References

1. S. Daflon, K. Cunha, Astrophys. J. **617**(2), 1115 (2004). https://doi.org/10.1086/425607
2. G.A. Bragança, S. Daflon, T. Lanz, K. Cunha, T. Bensby, P.J. McMillan, C.D. Garmany, J.W. Glaspey, M. Borges Fernandes, M.S. Oey, I. Hubeny, Astron. Astrophys. **625**, A120 (2019). https://doi.org/10.1051/0004-6361/201834554
3. H. Bouy, J. Alves, Astron. Astrophys. **584**, A26 (2015). https://doi.org/10.1051/0004-6361/201527058
4. M. Mohr-Smith, J.E. Drew, R. Napiwotzki, S. Simón-Díaz, N.J. Wright, G. Barentsen, J. Eislöffel, H.J. Farnhill, R. Greimel, M. Monguió, V. Kalari, Q.A. Parker, J.S. Vink, Mon. Not. R. Astron. Soc. **465**(2), 1807 (2017). https://doi.org/10.1093/mnras/stw2751
5. D. Russeil, C. Adami, J.C. Bouret, A. Hervé, Q.A. Parker, A. Zavagno, F. Motte, Astron. Astrophys. **607**, A86 (2017). https://doi.org/10.1051/0004-6361/201629870
6. M. Garcia, A. Herrero, F. Najarro, D.J. Lennon, M. Alejandro Urbaneja, Astrophys. J. **788**(1), 64 (2014). https://doi.org/10.1088/0004-637X/788/1/64
7. I. Camacho, M. Garcia, A. Herrero, S. Simón-Díaz, Astron. Astrophys. **585**, A82 (2016). https://doi.org/10.1051/0004-6361/201425533
8. M.S. Oey, J. Dorigo Jones, N. Castro, P. Zivick, G. Besla, H.C. Januszewski, M. Moe, N. Kallivayalil, D.J. Lennon, Astrophys. J. Lett. **867**(1), L8 (2018). https://doi.org/10.3847/2041-8213/aae892
9. P. Degroote, Asteroseismology of OB stars with the Corot Space mission. Ph.D. Thesis, Institute of Astronomy, Katholieke Universiteit Leuven, Belgium, 2010
10. F. Najarro, M.M. Hanson, J. Puls, Astron. Astrophys. **535**, A32 (2011). https://doi.org/10.1051/0004-6361/201016003
11. N. Przybilla, M.F. Nieva, M. Firnstein, K. Butler, in *EAS Publications Series*, vol. 64, ed. by K. Pavlovski, A. Tkachenko, G. Torres (2013), pp. 37–45. https://doi.org/10.1051/eas/1364005
12. G.A. Wade, J. Grunhut, E. Alecian, C. Neiner, M. Aurière, D.A. Bohlender, A. David-Uraz, C. Folsom, H.F. Henrichs, O. Kochukhov, S. Mathis, S. Owocki, V. Petit, Petit, in *Magnetic Fields throughout Stellar Evolution, IAU Symposium*, vol. 302, ed. by P. Petit, M. Jardine, H.C. Spruit (2014), pp. 265–269. https://doi.org/10.1017/S1743921314002233
13. L. Fossati, N. Castro, M. Schöller, S. Hubrig, N. Langer, T. Morel, M. Briquet, A. Herrero, N. Przybilla, H. Sana, F.R.N. Schneider, A. de Koter, BOB Collaboration, Astron. Astrophys. **582**, A45 (2015). https://doi.org/10.1051/0004-6361/201526725
14. S. Simón-Díaz, C. Aerts, M.A. Urbaneja, I. Camacho, V. Antoci, M. Fredslund Andersen, F. Grundahl, P.L. Pallé, Astron. Astrophys. **612**, A40 (2018). https://doi.org/10.1051/0004-6361/201732160
15. L. Hennicker, J. Puls, N.D. Kee, J.O. Sundqvist, Astron. Astrophys. **616**, A140 (2018). https://doi.org/10.1051/0004-6361/201731858
16. W.W. Morgan, Publ. Mich. Obs. **10**, 33 (1951)

17. R.O. Gray, J.C. Corbally, *Stellar Spectral Classification* (Princeton University Press, Princeton, 2009)
18. N. Langer, R.P. Kudritzki, Astron. Astrophys. **564**, A52 (2014). https://doi.org/10.1051/0004-6361/201423374
19. G. Holgado, Spectroscopic and physical characterization of the Galactic O-type stars targeted by the IACOB and OWN surveys. Ph.D. Thesis, 2019
20. N. Castro, L. Fossati, N. Langer, S. Simón-Díaz, F.R.N. Schneider, R.G. Izzard, Astron. Astrophys. **570**, L13 (2014). https://doi.org/10.1051/0004-6361/201425028
21. S. Ekström, C. Georgy, P. Eggenberger, G. Meynet, N. Mowlavi, A. Wyttenbach, A. Granada, T. Decressin, R. Hirschi, U. Frischknecht, C. Charbonnel, A. Maeder, Astron. Astrophys. **537**, A146 (2012). https://doi.org/10.1051/0004-6361/201117751
22. A.J.T. Poelarends, F. Herwig, N. Langer, A. Heger, Astrophys. J. **675**(1), 614 (2008). https://doi.org/10.1086/520872
23. N. Langer, Annu. Rev. Astron. Astrophys. **50**, 107 (2012). https://doi.org/10.1146/annurev-astro-081811-125534
24. R.P. Kudritzki, J. Puls, Annu. Rev. Astron. Astrophys. **38**, 613 (2000). https://doi.org/10.1146/annurev.astro.38.1.613
25. B.G. Elmegreen, C.J. Lada, Astrophys. J. **214**, 725 (1977). https://doi.org/10.1086/155302
26. T. Preibisch, H. Zinnecker, in *Triggered Star Formation in a Turbulent ISM, IAU Symposium*, vol. 237, ed. by B.G. Elmegreen, J. Palous (2007), pp. 270–277. https://doi.org/10.1017/S1743921307001597
27. N. Prantzos, in *EAS Publications Series*, vol. 32, ed. by C. Charbonnel, J.P. Zahn (2008), pp. 311–356. https://doi.org/10.1051/eas:0832009
28. G. Tenorio-Tagle, C. Muñoz-Tuñón, E. Pérez, S. Silich, E. Telles, Astrophys. J. **643**(1), 186 (2006). https://doi.org/10.1086/502644
29. V. Bromm, N. Yoshida, L. Hernquist, C.F. McKee, Nature **459**(7243), 49 (2009). https://doi.org/10.1038/nature07990
30. B.E. Robertson, R.S. Ellis, J.S. Dunlop, R.J. McLure, D.P. Stark, Nature **468**(7320), 49 (2010). https://doi.org/10.1038/nature09527
31. S.E. Woosley, J.S. Bloom, Annu. Rev. Astron. Astrophys. **44**(1), 507 (2006). https://doi.org/10.1146/annurev.astro.43.072103.150558
32. B.P. Abbott, R. Abbott, T.D. Abbott, M.R. Abernathy, F. Acernese, K. Ackley, C. Adams, T. Adams, P. Addesso, R.X. Adhikari, Phys. Rev. Lett. **116**(6), 061102 (2016). https://doi.org/10.1103/PhysRevLett.116.061102
33. B.P. Abbott, R. Abbott, T.D. Abbott, F. Acernese, K. Ackley, C. Adams, T. Adams, P. Addesso, R.X. Adhikari, V.B. Adya, C. Affeldt, M. Afrough, B. Agarwal, M. Agathos, K. Agatsuma, N. Aggarwal, O.D. Aguiar, L. Aiello, A. Ain, P. Ajith, B. Allen, G. Allen, A. Allocca, P.A. Altin, A. Amato, A. Ananyeva, S.B. Anderson, W.G. Anderson, S.V. Angelova, S. Antier, S. Appert, K. Arai, M.C. Araya, Phys. Rev. Lett. **119**(16), 161101 (2017). https://doi.org/10.1103/PhysRevLett.119.161101
34. M.A. Urbaneja, A. Herrero, F. Bresolin, R.P. Kudritzki, W. Gieren, J. Puls, N. Przybilla, F. Najarro, G. Pietrzyński, Astrophys. J. **622**(2), 862 (2005). https://doi.org/10.1086/427468
35. M.A. Urbaneja, A. Herrero, R.P. Kudritzki, F. Najarro, S.J. Smartt, J. Puls, D.J. Lennon, L.J. Corral, Astrophys. J. **635**(1), 311 (2005). https://doi.org/10.1086/497528
36. N. Castro, M.A. Urbaneja, A. Herrero, M. Garcia, S. Simón-Díaz, F. Bresolin, G. Pietrzyński, R.P. Kudritzki, W. Gieren, Astron. Astrophys. **542**, A79 (2012). https://doi.org/10.1051/0004-6361/201118253
37. M. Peimbert, Astrophys. J. **150**, 825 (1967). https://doi.org/10.1086/149385
38. G. Stasińska (2002). e-prints. arXiv:astro-ph/0207500
39. J. García-Rojas, C. Esteban, Astrophys. J. **670**(1), 457 (2007). https://doi.org/10.1086/521871
40. D. Alloin, W. Gieren, *Stellar Candles for the Extragalactic Distance Scale*, vol. 635 (Springer, Berlin, 2003). https://doi.org/10.1007/b13985
41. R.P. Kudritzki, F. Bresolin, N. Przybilla, Astrophys. J. Lett. **582**(2), L83 (2003). https://doi.org/10.1086/367690

42. R.P. Kudritzki, M.A. Urbaneja, Astrophys. Space Sci. **341**(1), 131 (2012). https://doi.org/10.1007/s10509-012-1016-7
43. G. Stasińska, C. Leitherer, Astrophys. J. Suppl. Ser. **107**, 661 (1996). https://doi.org/10.1086/192377
44. G. Stasińska, D. Schaerer, Astron. Astrophys. **322**, 615 (1997)
45. C. Leitherer, D. Schaerer, J.D. Goldader, R.M.G. Delgado, C. Robert, D.F. Kune, D.F. de Mello, D. Devost, T.M. Heckman, Astrophys. J. Suppl. Ser. **123**(1), 3 (1999). https://doi.org/10.1086/313233
46. K. Lefever, J. Puls, C. Aerts, Astron. Astrophys. **463**(3), 1093 (2007). https://doi.org/10.1051/0004-6361:20066038
47. J.S. Vink, I. Brott, G. Gräfener, N. Langer, A. de Koter, D.J. Lennon, Astron. Astrophys. **512**, L7 (2010). https://doi.org/10.1051/0004-6361/201014205
48. I. Brott, C.J. Evans, I. Hunter, A. de Koter, N. Langer, P.L. Dufton, M. Cantiello, C. Trundle, D.J. Lennon, S.E. de Mink, S.C. Yoon, P. Anders, Astron. Astrophys. **530**, A116 (2011). https://doi.org/10.1051/0004-6361/201016114
49. I. Brott, S.E. de Mink, M. Cantiello, N. Langer, A. de Koter, C.J. Evans, I. Hunter, C. Trundle, J.S. Vink, Astron. Astrophys. **530**, A115 (2011). https://doi.org/10.1051/0004-6361/201016113
50. F. Martins, E. Depagne, D. Russeil, L. Mahy, Astron. Astrophys. **554**, A23 (2013). https://doi.org/10.1051/0004-6361/201321282
51. F. Martins, A. Palacios, Astron. Astrophys. **560**, A16 (2013). https://doi.org/10.1051/0004-6361/201322480
52. N. Markova, J. Puls, N. Langer, Astron. Astrophys. **613**, A12 (2018). https://doi.org/10.1051/0004-6361/201731361
53. M.G. Pedersen, S. Chowdhury, C. Johnston, D.M. Bowman, C. Aerts, G. Handler, P. De Cat, C. Neiner, A. David-Uraz, D. Buzasi, A. Tkachenko, S. Simón-Díaz, E. Moravveji, J. Sikora, G.M. Mirouh, C.C. Lovekin, M. Cantiello, J. Daszyńska-Daszkiewicz, A. Pigulski, R.K. Vanderspek, G.R. Ricker, Astrophys. J. Lett. **872**(1), L9 (2019). https://doi.org/10.3847/2041-8213/ab01e1
54. D.M. Bowman, S. Burssens, M.G. Pedersen, C. Johnston, C. Aerts, B. Buysschaert, M. Michielsen, A. Tkachenko, T.M. Rogers, P.V.F. Edelmann, R.P. Ratnasingam, S. Simón-Díaz, N. Castro, E. Moravveji, B.J.S. Pope, T.R. White, P. De Cat, Nat. Astron. **3**, 720 (2019). https://doi.org/10.1038/s41550-019-0768-1
55. F. Martins, Bull. Soc. Roy. Sci. Liege **80**, 29 (2011)
56. J.C. Bouret, D.J. Hillier, T. Lanz, A.W. Fullerton, Astron. Astrophys. **544**, A67 (2012). https://doi.org/10.1051/0004-6361/201118594
57. A. Lenorzer, M.R. Mokiem, A. de Koter, J. Puls, Astron. Astrophys. **422**, 275 (2004). https://doi.org/10.1051/0004-6361:20047174
58. M.F. Nieva, N. Przybilla, A. Seifahrt, K. Butler, H.U. Käufl, A. Kaufer, Bull. Soc. Roy. Sci. Liège **80**, 175 (2011)
59. W.L.F. Marcolino, J.C. Bouret, T. Lanz, D.S. Maia, M. Audard, Mon. Not. R. Astron. Soc. **470**(3), 2710 (2017). https://doi.org/10.1093/mnras/stx1191
60. R.S.I. Ryans, P.L. Dufton, W.R.J. Rolleston, D.J. Lennon, F.P. Keenan, J.V. Smoker, D.L. Lambert, Mon. Not. R. Astron. Soc. **336**(2), 577 (2002). https://doi.org/10.1046/j.1365-8711.2002.05780.x
61. S. Simón-Díaz, A. Herrero, Astron. Astrophys. **468**(3), 1063 (2007). https://doi.org/10.1051/0004-6361:20066060
62. S. Simón-Díaz, A. Herrero, Astron. Astrophys. **562**, A135 (2014). https://doi.org/10.1051/0004-6361/201322758
63. G.A. Bragança, S. Daflon, K. Cunha, T. Bensby, M.S. Oey, G. Walth, Astron. J. **144**(5), 130 (2012). https://doi.org/10.1088/0004-6256/144/5/130
64. A. Herrero, R.P. Kudritzki, J.M. Vilchez, D. Kunze, K. Butler, S. Haser, Astron. Astrophys. **261**, 209 (1992)

65. A. Herrero, M. Garcia, J. Puls, K. Uytterhoeven, F. Najarro, D.J. Lennon, J.G. Rivero-González, Astron. Astrophys. **543**, A85 (2012). https://doi.org/10.1051/0004-6361/201118383
66. T. Repolust, J. Puls, A. Herrero, Astron. Astrophys. **415**, 349 (2004). https://doi.org/10.1051/0004-6361:20034594
67. N. Markova, J. Puls, T. Repolust, H. Markov, Astron. Astrophys. **413**, 693 (2004). https://doi.org/10.1051/0004-6361:20031463
68. M.R. Mokiem, A. de Koter, J. Puls, A. Herrero, F. Najarro, M.R. Villamariz, Astron. Astrophys. **441**(2), 711 (2005). https://doi.org/10.1051/0004-6361:20053522
69. S. Simón-Díaz, N. Castro, A. Herrero, J. Puls, M. Garcia, C. Sabín-Sanjulián, J. Phys.: Conf. Ser. **328**, 012021 (2011). https://doi.org/10.1088/1742-6596/328/1/012021
70. C. Sabín-Sanjulián, S. Simón-Díaz, A. Herrero, N.R. Walborn, J. Puls, J. Maíz Apellániz, C.J. Evans, I. Brott, A. de Koter, M. Garcia, N. Markova, F. Najarro, O.H. Ramírez-Agudelo, H. Sana, W.D. Taylor, J.S. Vink, Astron. Astrophys. **564**, A39 (2014). https://doi.org/10.1051/0004-6361/201322798
71. F. Martins, A. Hervé, J.C. Bouret, W. Marcolino, G.A. Wade, C. Neiner, E. Alecian, J. Grunhut, V. Petit, Astron. Astrophys. **575**, A34 (2015). https://doi.org/10.1051/0004-6361/201425173
72. O.H. Ramírez-Agudelo, H. Sana, A. de Koter, F. Tramper, N.J. Grin, F.R.N. Schneider, N. Langer, J. Puls, N. Markova, J.M. Bestenlehner, N. Castro, P.A. Crowther, C.J. Evans, M. García, G. Gräfener, A. Herrero, B. van Kempen, D.J. Lennon, J. Maíz Apellániz, F. Najarro, C. Sabín-Sanjulián, S. Simón-Díaz, W.D. Taylor, J.S. Vink, Astron. Astrophys. **600**, A81 (2017). https://doi.org/10.1051/0004-6361/201628914
73. G. Holgado, S. Simón-Díaz, R.H. Barbá, J. Puls, A. Herrero, N. Castro, M. Garcia, J. Maíz Apellániz, I. Negueruela, C. Sabín-Sanjulián, Astron. Astrophys. **613**, A65 (2018). https://doi.org/10.1051/0004-6361/201731543
74. J.G. Rivero González, J. Puls, F. Najarro, I. Brott, Astron. Astrophys. **537**, A79 (2012). https://doi.org/10.1051/0004-6361/201117790
75. J.G. Rivero González, J. Puls, P. Massey, F. Najarro, Astron. Astrophys. **543**, A95 (2012). https://doi.org/10.1051/0004-6361/201218955
76. N.J. Grin, O.H. Ramírez-Agudelo, A. de Koter, H. Sana, J. Puls, I. Brott, P.A. Crowther, P.L. Dufton, C.J. Evans, G. Gräfener, A. Herrero, N. Langer, D.J. Lennon, J.T. van Loon, N. Markova, S.E. de Mink, F. Najarro, F.R.N. Schneider, W.D. Taylor, F. Tramper, J.S. Vink, N.R. Walborn, Astron. Astrophys. **600**, A82 (2017). https://doi.org/10.1051/0004-6361/201629225
77. L.P. Carneiro, J. Puls, T.L. Hoffmann, Astron. Astrophys. **615**, A4 (2018). https://doi.org/10.1051/0004-6361/201731839
78. L.P. Carneiro, J. Puls, T.L. Hoffmann, G. Holgado, S. Simón-Díaz, Astron. Astrophys. **623**, A3 (2019). https://doi.org/10.1051/0004-6361/201833738
79. J. Kilian, Astron. Astrophys. **262**, 171 (1992)
80. I. Hunter, P.L. Dufton, S.J. Smartt, R.S.I. Ryans, C.J. Evans, D.J. Lennon, C. Trundle, I. Hubeny, T. Lanz, Astron. Astrophys. **466**(1), 277 (2007). https://doi.org/10.1051/0004-6361:20066148
81. C. Trundle, P.L. Dufton, I. Hunter, C.J. Evans, D.J. Lennon, S.J. Smartt, R.S.I. Ryans, Astron. Astrophys. **471**(2), 625 (2007). https://doi.org/10.1051/0004-6361:20077838
82. S. Simón-Díaz, Astron. Astrophys. **510**, A22 (2010). https://doi.org/10.1051/0004-6361/200913120
83. M.F. Nieva, N. Przybilla, Astron. Astrophys. **539**, A143 (2012). https://doi.org/10.1051/0004-6361/201118158
84. M.F. Nieva, N. Przybilla, in *The Lives and Death-Throes of Massive Stars, IAU Symposium*, vol. 329, ed. by J.J. Eldridge, J.C. Bray, L.A.S. McClelland, L. Xiao (2017), pp. 81–88. https://doi.org/10.1017/S1743921317003167
85. A. Irrgang, N. Przybilla, U. Heber, M. Böck, M. Hanke, M.F. Nieva, K. Butler, Astron. Astrophys. **565**, A63 (2014). https://doi.org/10.1051/0004-6361/201323167

86. C. Cazorla, T. Morel, Y. Nazé, G. Rauw, T. Semaan, S. Daflon, M.S. Oey, Astron. Astrophys. **603**, A56 (2017). https://doi.org/10.1051/0004-6361/201629841
87. C. Cazorla, Y. Nazé, T. Morel, C. Georgy, M. Godart, N. Langer, Astron. Astrophys. **604**, A123 (2017). https://doi.org/10.1051/0004-6361/201730680
88. P.L. Dufton, A. Thompson, P.A. Crowther, C.J. Evans, F.R.N. Schneider, A. de Koter, S.E. de Mink, R. Garland, N. Langer, D.J. Lennon, C.M. McEvoy, O.H. Ramírez-Agudelo, H. Sana, S. Símon Díaz, W.D. Taylor, J.S. Vink, Astron. Astrophys. **615**, A101 (2018). https://doi.org/10.1051/0004-6361/201732440
89. M.A. Urbaneja, R.P. Kudritzki, F. Bresolin, N. Przybilla, W. Gieren, G. Pietrzyński, Astrophys. J. **684**(1), 118 (2008). https://doi.org/10.1086/590334
90. N. Przybilla, K. Butler, S.R. Becker, R.P. Kudritzki, Astron. Astrophys. **445**(3), 1099 (2006). https://doi.org/10.1051/0004-6361:20053832
91. P.L. Dufton, R.S.I. Ryans, C. Trundle, D.J. Lennon, I. Hubeny, T. Lanz, C. Allende Prieto, Astron. Astrophys. **434**(3), 1125 (2005). https://doi.org/10.1051/0004-6361:20042530
92. N. Markova, J. Puls, Astron. Astrophys. **478**(3), 823 (2008). https://doi.org/10.1051/0004-6361:20077919
93. K. Lefever, J. Puls, T. Morel, C. Aerts, L. Decin, M. Briquet, Astron. Astrophys. **515**, A74 (2010). https://doi.org/10.1051/0004-6361/200911956
94. C.M. McEvoy, P.L. Dufton, C.J. Evans, V.M. Kalari, N. Markova, S. Simón-Díaz, J.S. Vink, N.R. Walborn, P.A. Crowther, A. de Koter, S.E. de Mink, P.R. Dunstall, V. Hénault-Brunet, A. Herrero, N. Langer, D.J. Lennon, J. Maíz Apellániz, F. Najarro, J. Puls, H. Sana, F.R.N. Schneider, W.D. Taylor, Astron. Astrophys. **575**, A70 (2015). https://doi.org/10.1051/0004-6361/201425202
95. G. Stasińska, N. Prantzos, G. Meynet, S. Simón-Díaz, C. Chiappini, M. Dessauges-Zavadsky, C. Charbonnel, H.-G. Ludwig, C. Mendoza, N. Grevesse, M. Arnould, B. Barbuy, Y. Lebreton, A. Decourchelle, V. Hill, P. Ferrando, G. Hébrard, F. Durret, M. Katsuma, C.J. Zeippen, in *EAS Publications Series*, vol. 54 (2012), pp. 3–63. https://doi.org/10.1051/eas/1254001
96. J. Puls, Commun. Asteroseismol. **158**, 113 (2009)
97. R.P. Kudritzki, M.A. Urbaneja, *Parameters and Winds of Hot Massive Stars* (Cambridge University Press, Cambridge, 2009), pp. 126–151. https://doi.org/10.1017/CBO9780511770593.010
98. N. Przybilla, in *EAS Publications Series*, vol. 43, ed. by R. Monier, B. Smalley, G. Wahlgren, P. Stee (2010), pp. 115–133. https://doi.org/10.1051/eas/1043008
99. D.F. Gray, *The Observation and Analysis of Stellar Photospheres* (Cambridge University Press, Cambridge, 1976)
100. J.A. Carroll, Mon. Not. R. Astron. Soc. **93**, 478 (1933). https://doi.org/10.1093/mnras/93.7.478
101. C. Aerts, S. Simón-Díaz, P.J. Groot, P. Degroote, Astron. Astrophys. **569**, A118 (2014). https://doi.org/10.1051/0004-6361/201424012
102. A. Slettebak, I. Collins, G. W., P.B. Boyce, N.M. White, T.D. Parkinson, Astrophys. J. Suppl. Ser. **29**, 137 (1975). https://doi.org/10.1086/190338
103. P.S. Conti, D. Ebbets, Astrophys. J. **213**, 438 (1977). https://doi.org/10.1086/155173
104. L.R. Penny, Astrophys. J. **463**, 737 (1996). https://doi.org/10.1086/177286
105. I.D. Howarth, K.W. Siebert, G.A.J. Hussain, R.K. Prinja, Mon. Not. R. Astron. Soc. **284**(2), 265 (1997). https://doi.org/10.1093/mnras/284.2.265
106. S. Simón-Díaz, M. Godart, N. Castro, A. Herrero, C. Aerts, J. Puls, J. Telting, L. Grassitelli, Astron. Astrophys. **597**, A22 (2017). https://doi.org/10.1051/0004-6361/201628541
107. J.O. Sundqvist, S. Simón-Díaz, J. Puls, N. Markova, Astron. Astrophys. **559**, L10 (2013). https://doi.org/10.1051/0004-6361/201322761
108. R.L. Kurucz, SAO Special Report, vol. 309 (1970)
109. J.R. Giddings, Ph.D. Thesis, 1981
110. I. Hubeny, Comput. Phy. Commun. **52**(1), 103 (1988). https://doi.org/10.1016/0010-4655(88)90177-4

111. I. Hubeny, T. Lanz, Synspec: General Spectrum Synthesis Program (2011)
112. D.J. Hillier, D.L. Miller, Astrophys. J. **496**(1), 407 (1998). https://doi.org/10.1086/305350
113. A.E. Santolaya-Rey, J. Puls, A. Herrero, Astron. Astrophys. **323**, 488 (1997)
114. J. Puls, M.A. Urbaneja, R. Venero, T. Repolust, U. Springmann, A. Jokuthy, M.R. Mokiem, Astron. Astrophys. **435**(2), 669 (2005). https://doi.org/10.1051/0004-6361:20042365
115. J. Puls, in *The Lives and Death-Throes of Massive Stars, IAU Symposium*, vol. 329, ed. by J.J. Eldridge, J.C. Bray, L.A.S. McClelland, L. Xiao (2017), pp. 435–435. https://doi.org/10.1017/S1743921317000229
116. G. Gräfener, L. Koesterke, W.R. Hamann, Astron. Astrophys. **387**, 244 (2002). https://doi.org/10.1051/0004-6361:20020269
117. W.R. Hamann, G. Gräfener, Astron. Astrophys. **427**, 697 (2004). https://doi.org/10.1051/0004-6361:20040506
118. A. Sander, T. Shenar, R. Hainich, A. Gímenez-García, H. Todt, W.R. Hamann, Astron. Astrophys. **577**, A13 (2015). https://doi.org/10.1051/0004-6361/201425356
119. A.W.A. Pauldrach, T.L. Hoffmann, M. Lennon, Astron. Astrophys. **375**, 161 (2001). https://doi.org/10.1051/0004-6361:20010805
120. J. Puls, M. Asplund (2015). e-prints. arXiv:1512.06972
121. J. Puls, R.P. Kudritzki, A. Herrero, A.W.A. Pauldrach, S.M. Haser, D.J. Lennon, R. Gabler, S.A. Voels, J.M. Vilchez, S. Wachter, A. Feldmeier, Astron. Astrophys. **305**, 171 (1996)
122. N.D. McErlean, D.J. Lennon, P.L. Dufton, Astron. Astrophys. **349**, 553 (1999)
123. M.F. Nieva, Astron. Astrophys. **550**, A26 (2013). https://doi.org/10.1051/0004-6361/201219677
124. S. Daflon, K. Cunha, S.R. Becker, V.V. Smith, Astrophys. J. **552**(1), 309 (2001). https://doi.org/10.1086/320460
125. S. Daflon, K. Cunha, K. Butler, V.V. Smith, Astrophys. J. **563**(1), 325 (2001). https://doi.org/10.1086/323795
126. M.A. Urbaneja, A. Herrero, F. Bresolin, R.P. Kudritzki, W. Gieren, J. Puls, Astrophys. J. Lett. **584**(2), L73 (2003). https://doi.org/10.1086/368406
127. T. Morel, K. Butler, C. Aerts, C. Neiner, M. Briquet, Astron. Astrophys. **457**(2), 651 (2006). https://doi.org/10.1051/0004-6361:20065171
128. T. Morel, S. Hubrig, M. Briquet, Astron. Astrophys. **481**(2), 453 (2008). https://doi.org/10.1051/0004-6361:20078999
129. M.F. Nieva, N. Przybilla, Astron. Astrophys. **481**(1), 199 (2008). https://doi.org/10.1051/0004-6361:20078203
130. M.F. Nieva, S. Simón-Díaz, Astron. Astrophys. **532**, A2 (2011). https://doi.org/10.1051/0004-6361/201116478
131. F. Martins, S. Simón-Díaz, A. Palacios, I. Howarth, C. Georgy, N.R. Walborn, J.C. Bouret, R. Barbá, Astron. Astrophys. **578**, A109 (2015). https://doi.org/10.1051/0004-6361/201526130
132. F. Martins, S. Foschino, J.C. Bouret, R. Barbá, I. Howarth, Astron. Astrophys. **588**, A64 (2016). https://doi.org/10.1051/0004-6361/201628288
133. F. Martins, S. Simón-Díaz, R.H. Barbá, R.C. Gamen, S. Ekström, Astron. Astrophys. **599**, A30 (2017). https://doi.org/10.1051/0004-6361/201629548
134. F. Martins, L. Mahy, A. Hervé, Astron. Astrophys. **607**, A82 (2017). https://doi.org/10.1051/0004-6361/201731593
135. M.R. Villamariz, A. Herrero, Astron. Astrophys. **442**(1), 263 (2005). https://doi.org/10.1051/0004-6361:20052848
136. C.J. Evans, S.J. Smartt, J.K. Lee, D.J. Lennon, A. Kaufer, P.L. Dufton, C. Trundle, A. Herrero, S. Simón-Díaz, A. de Koter, W.R. Hamann, M.A. Hendry, I. Hunter, M.J. Irwin, A.J. Korn, R.P. Kudritzki, N. Langer, M.R. Mokiem, F. Najarro, A.W.A. Pauldrach, N. Przybilla, J. Puls, R.S.I. Ryans, M.A. Urbaneja, K.A. Venn, M.R. Villamariz, Astron. Astrophys. **437**(2), 467 (2005). https://doi.org/10.1051/0004-6361:20042446

137. C.J. Evans, W.D. Taylor, V. Hénault-Brunet, H. Sana, A. de Koter, S. Simón-Díaz, G. Carraro, T. Bagnoli, N. Bastian, J.M. Bestenlehner, A.Z. Bonanos, E. Bressert, I. Brott, M.A. Campbell, M. Cantiello, J.S. Clark, E. Costa, P.A. Crowther, S.E. de Mink, E. Doran, P.L. Dufton, P.R. Dunstall, K. Friedrich, M. Garcia, M. Gieles, G. Gräfener, A. Herrero, I.D. Howarth, R.G. Izzard, N. Langer, D.J. Lennon, J. Maíz Apellániz, N. Markova, F. Najarro, J. Puls, O.H. Ramirez, C. Sabín-Sanjulián, S.J. Smartt, V.E. Stroud, J.T. van Loon, J.S. Vink, N.R. Walborn, Astron. Astrophys. **530**, A108 (2011). https://doi.org/10.1051/0004-6361/201116782
138. R.H. Barbá, R. Gamen, J.I. Arias, N. Morrell, J. Maíz Apellániz, E. Alfaro, N. Walborn, A. Sota, in *Revista Mexicana de Astronomia y Astrofisica Conference Series*, vol. 38 (2010), pp. 30–32
139. R.H. Barbá, R. Gamen, J.I. Arias, N.I. Morrell, in *The Lives and Death-Throes of Massive Stars, IAU Symposium*, vol. 329, ed. by J.J. Eldridge, J.C. Bray, L.A.S. McClelland, L. Xiao (2017), pp. 89–96. https://doi.org/10.1017/S1743921317003258
140. S. Simón-Díaz, I. Negueruela, J. Maíz Apellániz, N. Castro, A. Herrero, M. Garcia, J.A. Pérez-Prieto, N. Caon, J.M. Alacid, I. Camacho, R. Dorda, M. Godart, C. González-Fernández, G. Holgado, K. Rübke, in *Highlights of Spanish Astrophysics VIII* (Springer, Berlin, 2015), pp. 576–581
141. G.A. Wade, C. Neiner, E. Alecian, J.H. Grunhut, V. Petit, B.d. Batz, D.A. Bohlender, D.H. Cohen, H.F. Henrichs, O. Kochukhov, J.D. Land street, N. Manset, F. Martins, S. Mathis, M.E. Oksala, S.P. Owocki, T. Rivinius, M.E. Shultz, J.O. Sundqvist, R.H.D. Townsend, A. ud-Doula, J.C. Bouret, J. Braithwaite, M. Briquet, A.C. Carciofi, A. David-Uraz, C.P. Folsom, A.W. Fullerton, B. Leroy, W.L.F. Marcolino, A.F.J. Moffat, Y. Nazé, N.S. Louis, M. Aurière, S. Bagnulo, J.D. Bailey, R.H. Barbá, A. Blazère, T. Böhm, C. Catala, J.F. Donati, L. Ferrario, D. Harrington, I.D. Howarth, R. Ignace, L. Kaper, T. Lüftinger, R. Prinja, J.S. Vink, W.W. Weiss, I. Yakunin, Mon. Not. R. Astron. Soc. **456**(1), 2 (2016). https://doi.org/10.1093/mnras/stv2568

# The Explosion Mechanism of Core-Collapse Supernovae and Its Observational Signatures

**Ondřej Pejcha**

**Contents**



**Abstract** The death of massive stars is shrouded in many mysteries. One of them is the mechanism that overturns the collapse of the degenerate iron core into an explosion, a process that determines the supernova explosion energy, properties of the surviving compact remnant, and the nucleosynthetic yields. The number of core-collapse supernova observations has been growing with an accelerating pace thanks to modern time-domain astronomical surveys and new tests of the explosion mechanism are becoming possible. We review predictions of parameterized supernova explosion models and compare them with explosion properties inferred from observed light curves, spectra, and neutron star masses.

## 1 Introduction

Massive stars develop iron or oxygen-neon-magnesium cores, which eventually experience instability and collapse to reach nuclear densities. The initial stellar mass separating white dwarf or neutron star formation is commonly placed at around $8\,M_\odot$ e.g. [95], but depends on metallicity and other parameters e.g. [41]. Stars with helium core masses $\gtrsim 30\,M_\odot$ are destabilized by creation of electron–positron pairs,

O. Pejcha (✉)
Institute of Theoretical Physics, Faculty of Mathematics and Physics, Charles University, Praha 8, Czech Republic
e-mail: pejcha@utf.mff.cuni.cz

 
P. Kabáth et al. (eds.), *Reviews in Frontiers of Modern Astrophysics*,
https://doi.org/10.1007/978-3-030-38509-5_7

which reduces their mass in one or more mass ejection episodes and which can even completely disrupt the star e.g. [116]. The collapse of the inner regions of the core stabilizes when the repulsive part of the strong nuclear force becomes important and the equation of state stiffens. The core overshoots this new equilibrium, bounces, and forms a shock wave propagating outward from a nascent proto-neutron star. The outgoing shock stalls into an accretion shock due to energy losses from neutrino emission and photodissociation of infalling nuclei. A prolonged period of accretion ensues and lasts many dynamical timescales of the system, $\gtrsim$100 ms. During this phase, the proto-neutron star grows in mass and the region below the accretion shock becomes unstable to turbulence and standing-accretion shock instability e.g. [13, 15, 37]. The instabilities are partially driven by neutrinos emanating from the proto-neutron star and the accretion region. This complex system of instabilities in a region semi-transparent to neutrinos presents a great challenge to theoretical understanding.

The evolution of stalled accretion shock likely bifurcates into two possible outcomes. The accretion can continue until the central object collapses into a black hole with much of the rest of the star following its fate. This evolutionary path can be accompanied by a transient brightening fainter than a typical supernova [54]. In a majority of stars, however, it is believed that the combined action of neutrinos and instabilities overturns the accretion into explosion e.g. [18, 44]. It is worth noting that the concept of neutrino mechanism assisted by instabilities is far from proven. There are other less-explored paths to explosion involving magneto-rotational processes e.g. [17, 50] and energy transfer by waves e.g. [16, 35].

In that case, the accretion shock starts traveling out in radius and eventually reaches the stellar surface. Since interactions of neutrinos with matter are important, it is illuminating to frame the bifurcation between a failed and a successful explosion within the context of critical neutrino luminosity required to overturn the accretion into explosion e.g. [14, 68, 121]. This framework enables assessment of the impact of various physical processes on the explosion and provides a physically-motivated "antesonic" explosion condition [80, 89]. There are other, more or less related conditions based on outward acceleration [10], timescales of heating and advection e.g. [65, 68, 105, 106], and the dynamics of the net neutrino heating region [42].

After the accretion shuts off, there can be a period of simultaneous accretion and outflow from the proto-neutron star. Eventually, the proto-neutron star develops a wind driven by absorption of neutrinos. These processes set the baryonic mass of the remnant neutron star. The shock propagating through the star heats up the stellar interior. Gas heated above $\approx 5 \times 10^9$ K undergoes nuclear burning to iron-group elements. After the shock leaves the surface of the star, we observe the hot and expanding ejecta as a core-collapse supernova. Part of the light output of the supernova comes from the radioactive decay of the newly synthesized elements, especially $^{56}$Ni. The asymptotic energy of the supernova ejecta, about $10^{51}$ ergs, is $\sim$1% of the neutron star binding energy, and the energy radiated in the optical and nearby wavelengths is $\sim$1% of the asymptotic ejecta energy. Ultimately, all of the supernova explosion energy is radiated away as the ejecta decelerates and mixes with the interstellar medium.

### 1.1 Scope of This Review

When and how does the collapse of the stellar core turn to explosion has been a major unsolved problem in theoretical astrophysics. There have been wide-ranging efforts in both theory and multi-messenger observations to make progress on this issue. An eloquent summary of much of this work is available in a recently published Handbook of Supernovae [3], especially the chapters of [6, 30, 32, 43, 46, 52, 71, 72, 84, 94, 124]. We will not repeat the covered subjects here. We suggest the reader to consult this resource for more details and more complete lists of references.

In this review, we focus on a small niche of the core-collapse supernova problem: how do we compare predictions of supernova theory with observations in the era of massive time-domain surveys? For example, the *Large Synoptic Survey Telescope* (LSST) will begin discovering $\sim 10^5$ core-collapse supernovae every year starting in 2023. It is tempting to utilize this wealth of data to learn about the explosion mechanism. However, the benefit of this rapidly expanding dataset is not immediately obvious, because the majority of the expected supernova discoveries will provide only limited information for each supernova: a sparsely sampled light curve and an occasional low-resolution classification spectrum. In fact, even nowadays the majority of transient discoveries remain spectroscopically unclassified. This type of data naturally leads to a shift of attention from individual objects to population studies, which presents new challenges. Supernova theory has long been driven by very computationally intensive simulations, which could be run only for a limited sample of initial models. Such works have been revealing fascinating complications of the explosion mechanism, but their predictive power for populations of stars has been limited. In the past few years, however, theoretical and observational efforts have been converging to a point, where mutual comparison is possible. This interaction of observations with theory is what we aim to capture in this review.

## 2 Theoretical Predictions

In this Section, we describe the theoretical efforts to quantify the progenitor–supernova connection for a large number of progenitor models. The results, when available, are summarized also graphically. Figure 1 shows predictions of whether each progenitor explodes or not and whether the explosion exhibits substantial fallback. Figure 2 shows the predictions of explosion energy $E_{\rm exp}$ as a function of ejecta mass $M_{\rm ej}$. Figure 3 displays the correlation between $E_{\rm exp}$ and the mass of radioactive $^{56}$Ni synthesized in the supernova explosion, $M_{\rm Ni}$. Finally, Fig. 4 provides predictions for mass distributions of neutron star remnants.

Can we predict the outcome of core collapse without running prohibitively large number of expensive multi-dimensional simulations? Unfortunately, supernova simulations in spherical symmetry consistently fail to explode except for the lowest-

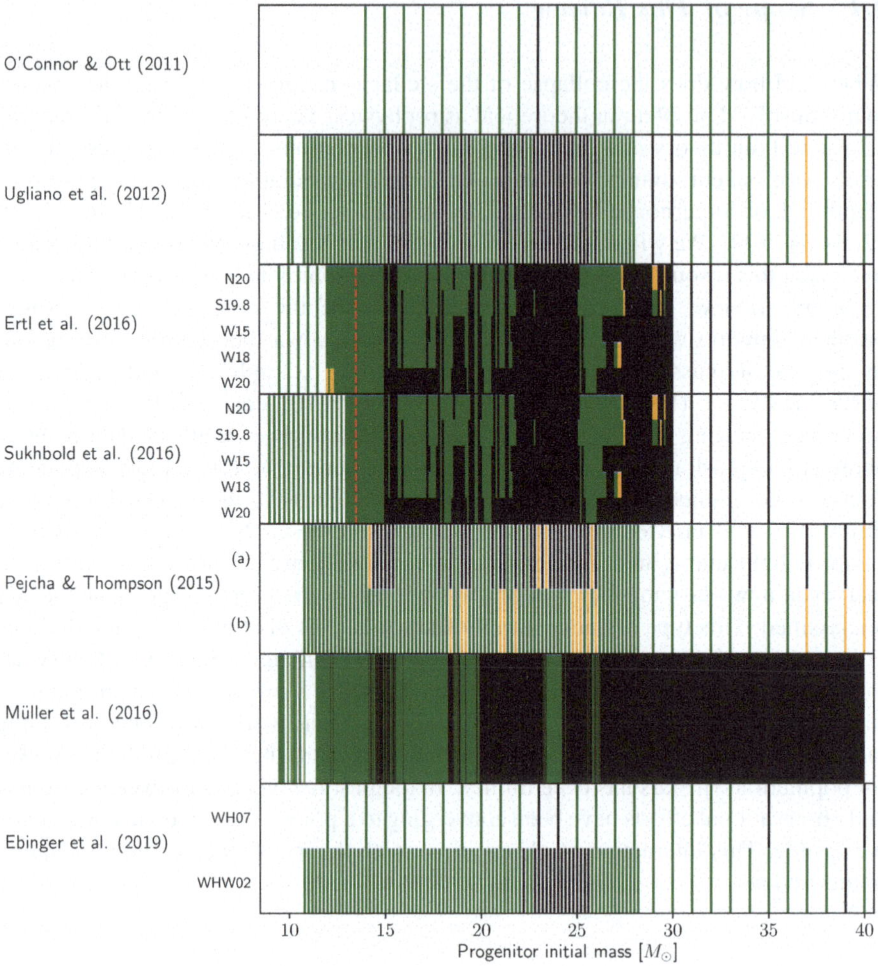


**Fig. 1** Outcome of the collapse of the core of a solar-metallicity massive star as a function of its initial mass. We show successful explosions leaving behind a neutron star (green), explosions with significant fallback (orange), and direct collapse to a black hole (black). Each bar corresponds to one progenitor model. To illustrate the differences between progenitor grids, the bar width is fixed at 0.1 $M_\odot$ except for [66], where it is set to 0.01 $M_\odot$. The results of [74] are for [117], WH07 progenitor set and LS220 equation of state. Ugliano et al. [108] used Woosley et al. [120], WHW02 progenitors. Ertl et al. [28] and Sukhbold et al. [102] used a mixture of progenitors, which were exploded with five different calibrations of the supernovae engine (N20, S19.8, W15, W18, W20), as indicated in the figure. The results are identical for progenitors above 13.5 $M_\odot$, which is marked with a vertical red dashed line. Pejcha and Thompson [81] used two different parameterizations for WHW02 progenitors, where (**a**) has a fraction of non-exploding progenitors and (**b**) has explosions for all progenitors. The results of [66] are for a custom grid of progenitors and their method cannot diagnose fallback explosions. The results of [24] are for WHW02 and WH07 progenitor sets

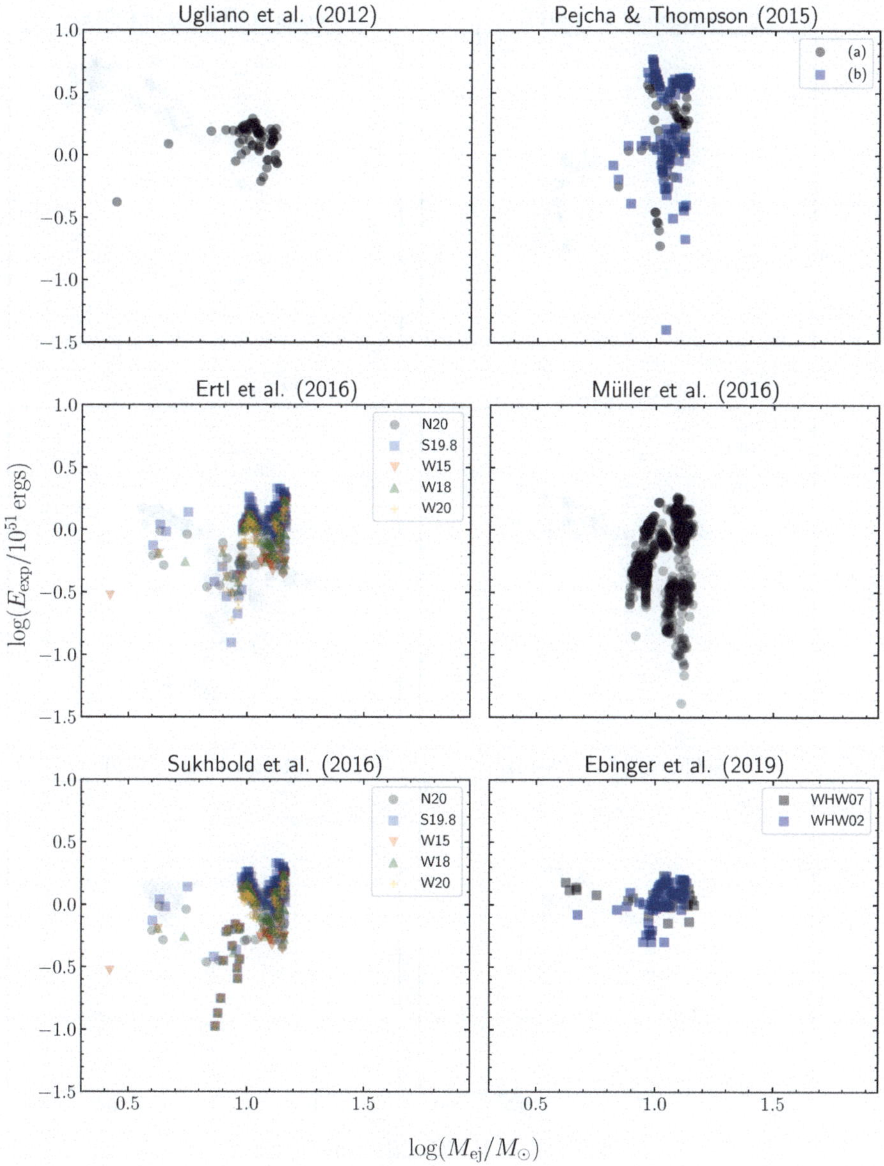


**Fig. 2** Explosion energy $E_{\mathrm{exp}}$ as a function of ejecta mass $M_{\mathrm{ej}}$. The ejecta mass was determined as the final mass of the progenitor before the explosion with the remnant mass subtracted. Note that the final progenitor mass strongly depends on very uncertain wind mass loss rates. This is different from quantities like $E_{\mathrm{exp}}$ and $M_{\mathrm{Ni}}$, which are set by the final core structure, which is not strongly affected by processes in the envelope. For WHW07 progenitors in [24] we took the final mass from the progenitor set of [28]. Labels (**a**) and (**b**) refer to the two explosion parameterizations of [81]. More detailed explanation of individual models is given in Fig. 1

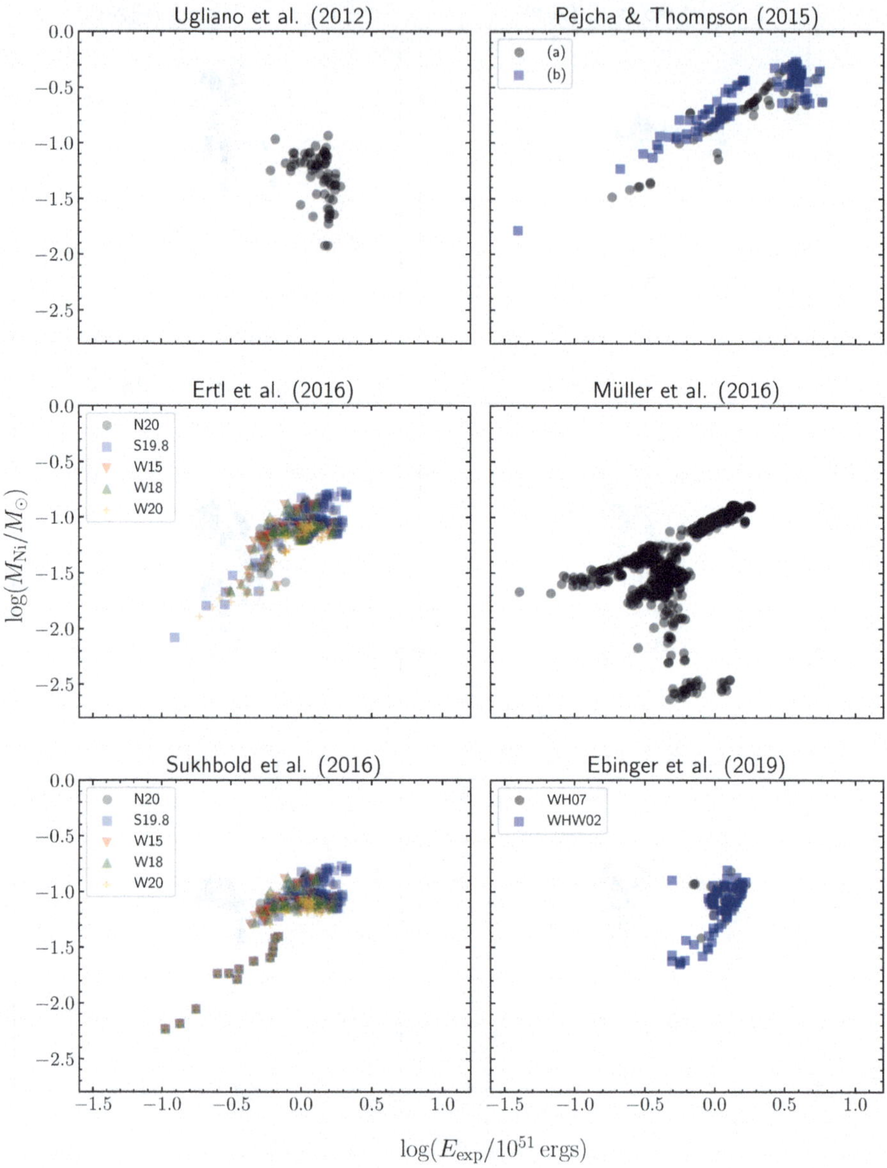


**Fig. 3** Mass of synthesized nickel as a function of explosion energy. For [28] and [102] we used the recommended value of the nickel plus half of the mass of tracer particles. Labels (**a**) and (**b**) refer to the two explosion parameterizations of [81]

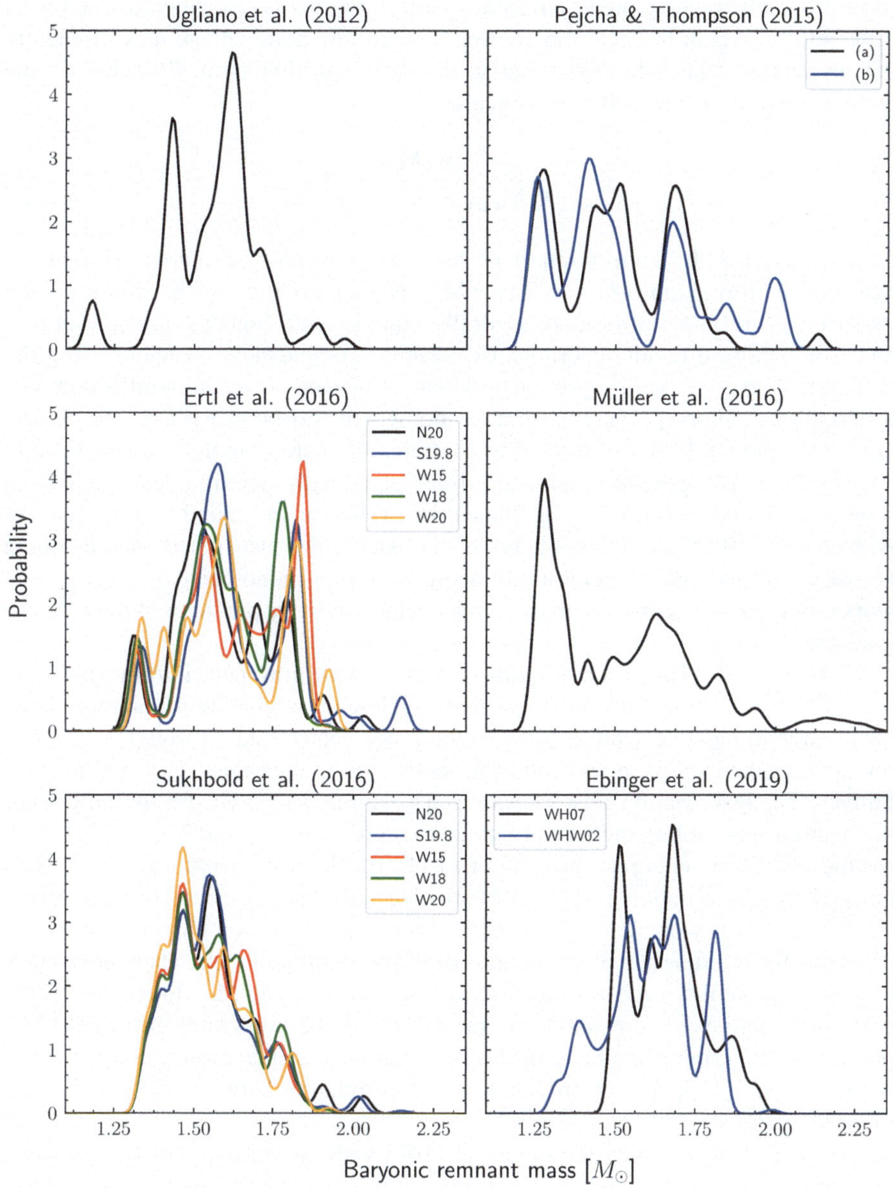


**Fig. 4** Probability distribution of baryonic masses of remnant neutron stars. The theoretical results are obtained by integrating over the provided progenitor grid, $M_i$ with weight given to $i$-th progenitor as $M_i^{-2.35}(M_{i+1} - M_{i-1})/2$, where the first factor is the [92] initial mass function. The progenitor grid is assumed to be uniformly-spaced at its edges. The remnant mass distribution from each progenitor is described by a Gaussian with a width of 0.03 $M_\odot$. Labels (**a**) and (**b**) refer to the two explosion parameterizations of [81]

mass progenitors with tenuous envelopes. Still, spherical 1D calculations can reveal progenitor trends in the neutrino cooling of accreting material and how efficiently do the neutrinos heat the layers below the shock. In this spirit, [74] defined the compactness of the pre-collapse progenitor

$$\xi_M = \frac{M/\mathrm{M}_\odot}{R(\mathrm{M}_{\mathrm{bary}} = M)/1000\,\mathrm{km}}, \tag{1}$$

which measures the concentration of mass at progenitor positions relevant for neutron star formation, $M \approx 1.2$ to $3\,\mathrm{M}_\odot$. Higher $\xi_M$ implies the iron-core and its surrounding material is more centrally condensated, higher binding energies, and hence more difficult supernova explosions. Compactness changes during the collapse of the star so a common moment is needed to meaningfully compare different progenitors: [74] argued that the moment of core bounce is a natural choice. O'Connor and Ott [74] also studied how artificially increasing the neutrino heating rate in their 1D general-relativistic simulations with neutrino leakage/heating scheme with code GR1D [73] facilitates the explosion and found that progenitors with $\xi_{2.5} \lesssim 0.45$ are more likely to explode. They also noted that mapping between $\xi_M$ and the progenitor initial mass is non-monotonic. In most of the works summarized below, compactness is a relatively good predictor of the collapse outcome.

Ugliano et al. [108] used 1D Eulerian code with gray neutrino transport and replaced the neutron star core with an inner boundary condition parameterizing the contraction and neutrino cooling of the proto-neutron star. Through a series of remappings the explosion was followed to $10^{15}$ cm to determine $E_{\mathrm{exp}}$ and to track fallback. Nuclear reaction network was used to estimate $M_{\mathrm{Ni}}$. The free parameters of the neutron star cooling model were tuned to reproduce $E_{\mathrm{exp}}$ and $M_{\mathrm{Ni}}$ of SN1987A for the s19.8 red supergiant progenitor of [120]. The same parameters were then utilized for the remaining about 100 solar-metallicity progenitor models of the same series. This work showed for the first time that explosion properties can vary dramatically as a function of initial mass of the progenitor models. There are islands of non-explodability even at progenitor masses as low as $15\,\mathrm{M}_\odot$. Interestingly, the scale of variations is comparable to the density of the progenitor grid of $0.1\,\mathrm{M}_\odot$ pointing to deterministic chaos in the pre-supernova stellar evolution [100, 101]. Ugliano et al. [108] do not predict a strong correlation between $E_{\mathrm{exp}}$ and ejecta mass $M_{\mathrm{ej}}$ or $M_{\mathrm{Ni}}$.

Ertl et al. [28] revisited the model of [108] with an updated equation of state, nuclear reaction network, larger set of progenitor models, and five different models of SN1987A progenitors chosen as calibrators: S19.8 from [120], W15 from [118], W18 with rotation from unpublished results of Woosley, W20 from [119], and N20 from [93]. Furthermore, the model for the excised core was modified for progenitors with initial masses $\lesssim 13.5\,\mathrm{M}_\odot$ so that these models exhibit weak explosions. The empirical support for this modifications is SN1054 with estimated progenitor mass of $10\,\mathrm{M}_\odot$ and explosion energy of only $10^{50}$ ergs [97, 122]. Ertl et al. [28] found that the combination of mass coordinate and its derivative at a location of entropy per baryon equal to 4 outperforms the compactness in predicting the outcome of the collapse.

Sukhbold et al. [102] used essentially the same basic setup as [28], but with updated progenitor models below $13.5\,M_{\odot}$ and different explosion calibration in this mass range (red dashed line in Fig. 1). They also post-processed the results with a hydrodynamics code KEPLER to provide nucleosynthetic yields for up to 2000 nuclei and light curves of the supernovae. Recently, [29] further improved their method and applied it to helium progenitors evolved with mass loss, which should approximate effects of binary star evolution. They find increased fraction of fallback explosions leading to small population of remnants in the "mass gap" between 2 and $5\,M_{\odot}$.

Pejcha and Thompson [81] evolved the accretion phase with GR1D code [73] for several seconds to obtain consistent runs of proto-neutron star mass, radius, and neutrino luminosity and energy. Based on these trajectories, they estimated the time-evolution of the critical neutrino luminosity under the assumption of quasi steady-state evolution. The actual neutrino luminosities never crossed the critical value so [81] parametrically modified the critical neutrino luminosity and consistently applied the changes to the progenitor suite of [120]. They estimated $E_{\rm exp}$ as a time-integrated power of neutrino-driven wind and $M_{\rm Ni}$ as a mass in volume exposed to sufficiently high temperatures. Their results are qualitatively similar to other works from parameterized explosions, with [81] showing that the pattern does not dramatically change with different choices of the parameterization.

Müller et al. [66] constructed a model of accretion phase evolution using only ordinary differential equations, which heuristically accounts for simultaneous accretion and outflows. With only little computing time required, [66] applied the method to over 2000 progenitor models and, similarly to earlier results, they estimated remnant mass, $E_{\rm exp}$, and $M_{\rm Ni}$. For the correlations visualized here, the results are in qualitative agreement with other works except for relatively small population of explosions with very small $M_{\rm Ni}$ and normal $E_{\rm exp}$.

Ebinger et al. [24] presented results of artificial explosions based on the PUSH method. This setup was initially implemented by Perego et al. [83] and utilizes a general relativistic hydrodynamics code in spherical symmetry [51] with an isotropic diffusion source approximation for the transport of electron neutrinos and antineutrinos. Parameterized explosions are triggered by introducing a heating term proportional to the luminosity of $\mu$ and $\tau$ neutrinos, which normally affect the evolution less than electron neutrinos and antineutrinos. The magnitude of the heating is scaled by the progenitor compactness. Ebinger et al. [24] provide the basic explosion predictions for solar-metallicity progenitors, while followup work gives predictions of detailed nucleosynthesis [21] and for low-metallicity progenitors [25]. The results of this approach are in broad agreement with other results except that the "bar-code" pattern in the explodability is not as prominent. However, other parameters such as $E_{\rm exp}$ are also not monotonic with the progenitor initial mass.

The predictions of parameterized explosions have been continuously evolving by including additional physics. Couch et al. [20] added modified mixing length theory of convection and turbulence to spherically-symmetric hydrodynamical equations and studied explodability of 138 solar-metallicity progenitors as a function of the mixing length parameterization. Similarly to most other works, they find complex landscape of explosions and failures, and provide predictions of explosion energies and remnant masses. A similar approach was implemented by Mabanta et al. [56] using the turbulence model of [55]. The appropriate form of the turbulence model is still debated [64]. Nakamura et al. [70] performed 2D simulations of neutrino-driven explosions of [120] progenitors and found that most of them are exploding. The explosion properties like $E_{\rm exp}$, $M_{\rm Ni}$, and neutron star mass correlate with the compactness.

It is worth noting that many of the parameterized explosion schemes and explicitly tied to the observed properties of SN1987A and SN1054. SN1987A was a peculiar explosion from a blue supergiant progenitor, which might have experienced a merger shortly before the supernova explosion e.g. [19, 84]. Kinetic energy in the Crab remnant is low, which contrasts with the apparently normal light curve e.g. [38, 97]. It is unclear how are the theoretical predictions affected by peculiarities of the calibration source. The observed properties of calibrators are never exactly matched by the theoretical models and the observations always have a margin of uncertainty. In a sense, the calibration models serve as a zero-point for the entire population. If the calibration model properties were vastly different from a typical progenitor with roughly the same mass, the supernova *population* would be discrepant in quantities like total amount of synthesized $^{56}$Ni, the results supernovae would be of too high or too low luminosity when compared to observations, etc. As we will illustrate below, this does not seem to be the case with the currently available observations and theoretical predictions. Still, some of these uncertainties might be absorbed in internal tunable model parameters.

Finally, we emphasize that parameterized models predominantly explore physical effects that were explicitly included in their construction and that the possibility of discovering new effects or their combinations is narrower than what is possible in more-complete multi-dimensional simulations with less prior assumptions. Nonetheless, one way to declare that features of the most complicated simulation are fully understood is that it is possible to replicate these results within some kind of parameterized model.

## 3 Observational Efforts

Theory predicts discrete islands of successful and failed explosions within a population of massive stars. A straightforward way to observationally verify these predictions is to witness a massive star collapsing to a black hole without an accompanying supernova. This avenue has been pursued with both ground-based [1, 49] and space-based data [90] and have finally yielded a single candidate [2, 33] with a mass of $\sim 25\,{\rm M}_\odot$. In a similar spirit, archival pre-explosion images of supernova positions can be used to infer distribution of progenitor masses of known supernovae. These observations have suggested an unexpected lack of red supergiant progenitors with initial masses $\gtrsim 17\,{\rm M}_\odot$ [49, 95, 96]. Both the disappearing star and progenitor non-detections are naturally explained by a lack of explosions in this mass range seen in parameterized models (Fig. 1). Masses of exploding stars can be also estimated from the age of the surrounding population, although the precision is lower than for directly imaged progenitors [5, 8, 45, 115], likely too rough to resolve the complex landscape of supernova explosions.

Observational studies relying on direct imaging are only feasible in a relatively small volume with correspondingly low rate of supernova explosions. But supernovae themselves are visible from great distances and their light could provide complementary observational evidence. The luminosity, duration, and shape of the supernova light curve depend on the progenitor envelope mass, size, structure, and composition, the explosion energy, and the amount of radioactive nuclei and how they are mixed within the envelope. Hereafter, we narrow the focus on the relatively common Type II-P supernovae, which are explosions of red supergiant progenitors with hydrogen envelopes. Type II-P supernovae are useful as theoretical benchmarks, because the progenitor structure should closely match the result of evolution of single stars. This does not mean that binary interactions did not occur in the previous evolution, in fact, a fraction of Type II-P supernovae are possibly the result of stellar mergers or other mass exchanges in binary stars [125].

Light curves of Type II-P supernova in the first 100 days or so are dominated by the diffusion of light out of a recombining hydrogen envelope, which leads to a "plateau" of relatively constant bolometric luminosity. The dramatic change of opacity at hydrogen ionization implies that the supernova has a relatively well-defined photosphere, which moves inward in the Lagrangian mass coordinate of the ejecta. Simple analytical estimates show that the three principle observables—plateau luminosity, duration, and spectroscopic expansion velocity—are determined by three intrinsic characteristics—progenitor radius, ejecta mass, and supernova explosion energy. Therefore, it is tempting to turn the observables into quantities interesting for the theory of the explosion mechanism e.g. [47, 85, 102].

After the hydrogen envelope fully recombines, the luminosity drops and starts to closely track energy input from the decay of radioactive elements synthesized in the explosion. Early on, the most important radioactive chain starts with $^{56}$Ni, although the actual decaying element powering the radioactive tail is its decay product $^{56}$Co. The initial mass of $^{56}$Ni is thus the relevant quantity to study. Assuming full trapping of the radioactive decay products, the normalization of the exponentially decaying light curve yields an estimate of $M_{\rm Ni}$. The temperatures required for synthesis of $^{56}$Ni imply that its formation occurred within few thousand km of the center of the supernova explosion, which makes $M_{\rm Ni}$ a useful probe of the explosion development and internal structure of the progenitors.

Here, we summarize a subset of efforts to estimate physical parameters out of Type II-P supernova light curves and spectra. There is a great range of techniques ranging from simple scaling relations to full radiation hydrodynamical models and spectrum fitting. Figure 5 shows $E_{\rm exp}$ as a function of ejecta or progenitor mass. The two masses should differ by 1–2 $M_\odot$ for explosions leaving behind a neutron star. Figure 6 explores the correlation between $E_{\rm exp}$ and $M_{\rm Ni}$. These figures should be directly compared with theoretical results in Figs. 2 and 3.

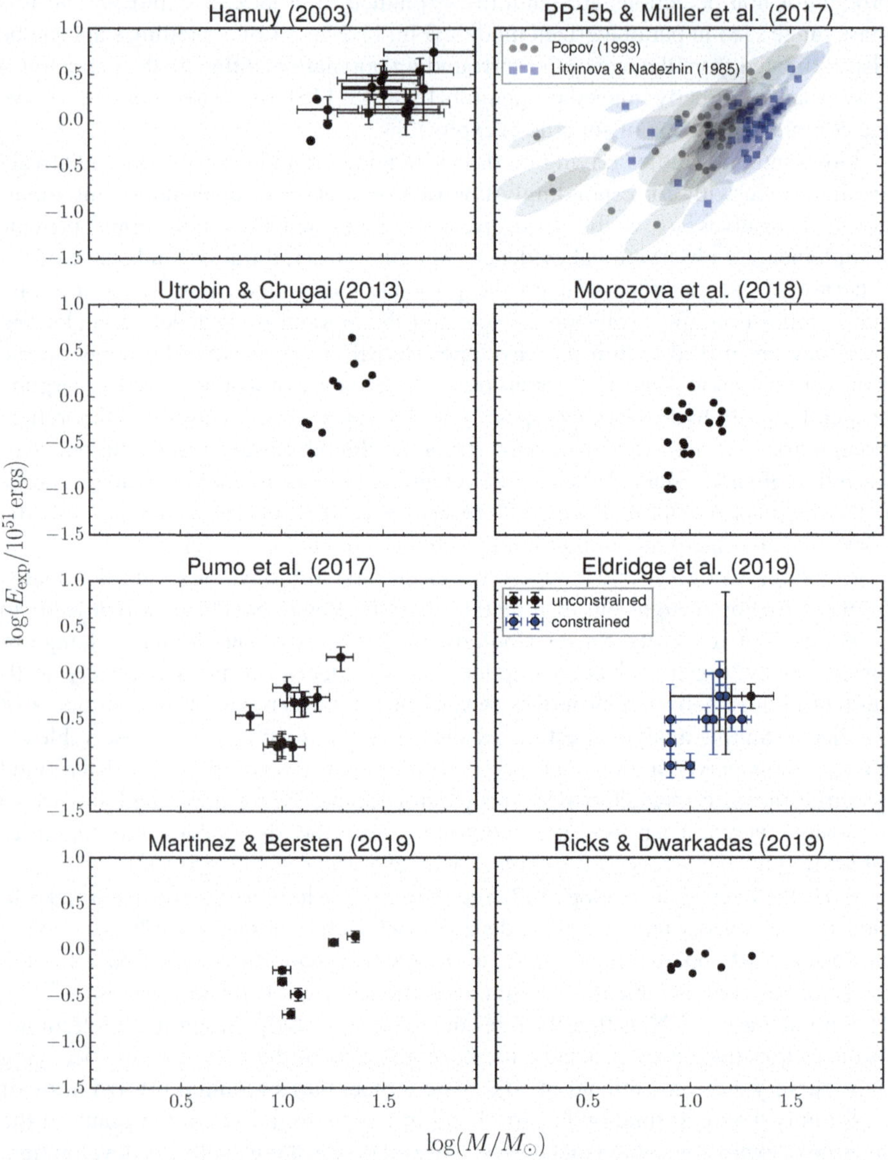


**Fig. 5** Explosion energy $E_{\rm exp}$ as a function of ejecta mass $M_{\rm ej}$ or progenitor initial mass $M$, as inferred from modeling of non-interacting hydrogen-rich (mostly Type II-P) supernova light curves and expansion velocities. Works based on analytic scaling relations [36, 67, 79] report values for $M$, which are related to ejecta mass. Utrobin and Chugai [114], Pumo et al. [87], and Morozova et al. [63] report ejected envelope mass. Martinez and Bersten [58] report stellar mass just prior to the explosion. Eldridge et al. [27] report progenitor initial mass. The caveat of Fig. 2 concerning the uncertain wind mass loss rates affecting the inferences of initial progenitor mass applies here as well

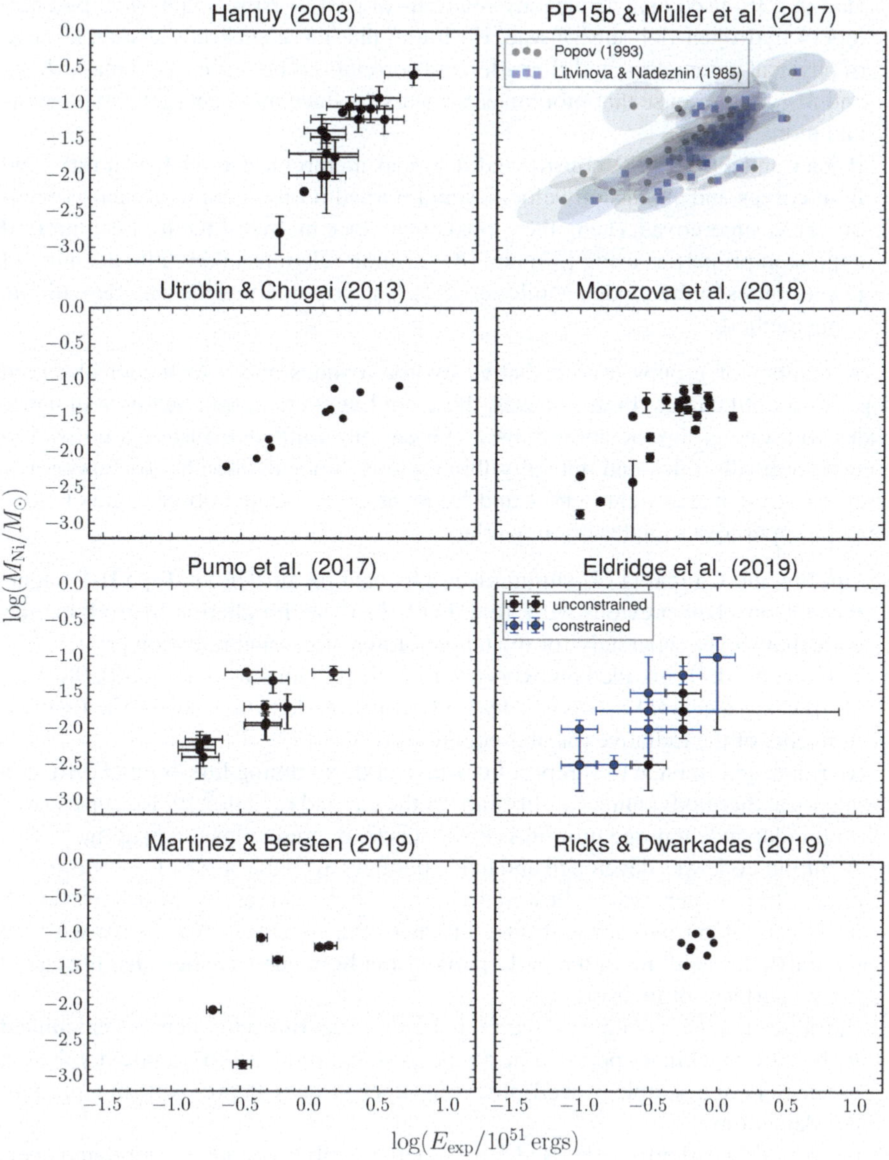


**Fig. 6** Nickel mass $M_{\rm Ni}$ and explosion energy $E_{\rm exp}$ from modeling of non-interacting hydrogen-rich (mostly Type II-P) supernovae light curves

The simplest approach to estimating explosion parameters relies on combining consistently inferred basic properties of the light curves and velocities with analytic scaling relations.

- Hamuy [36] used analytic scaling relations of [53] to estimate physical parameters of hydrogen-rich supernovae. He found that these supernovae span a range of physical parameters, and characterized correlations between ejecta mass, $E_{\rm exp}$, and $M_{\rm Ni}$ in the sense that more massive stars produce more energetic explosions and more $^{56}$Ni.
- Pejcha and Prieto [78] constructed a global hierarchical model of multi-band light curves and expansion velocities and trained it on a sample of nearby well-observed supernovae. Using the global covariance matrix of the fit, [79] inferred explosion properties using [53] and [85] scaling relations taking into account all uncertainties in the model. Müller et al. [67] nearly doubled the sample with the same method.

A number of groups used radiative hydrodynamics codes to model observed supernova light curves. In most cases, these are Lagrangian, spherically-symmetric codes with one group radiation transfer. Often, flux-limited diffusion is utilized to connect optically-thick and optically-thin regions. Since the number of supernovae with densely-covered light curves and frequent spectroscopic observations is limited, the supernova samples often overlap.

- Utrobin and Chugai [114] summarized physical parameters of Type II-P supernovae from their previous work (e.g. [111–113]) with radiation hydrodynamics code that solves separately for the temperatures of gas and radiation [109].
- Pumo et al. [87] provided an overview of their previous work (e.g. [40, 98, 103, 104]) using a general-relativistic hydrodynamics code, which solves the first two moments of the radiative transfer equations [86].
- Martinez and Bersten [58] report fits with a code assuming flux-limited diffusion and local thermodynamic equilibrium for the gas and radiation [9] for supernovae with independent progenitor mass estimates from pre-explosion imaging.
- A similar code was developed and made publicly available as SNEC by Morozova et al. [61], which was followed by fits to light curves by Morozova et al. [63], also taking into account interactions of the supernova blast wave with the circumstellar medium in the early parts of the light curves. They also presented the $\chi^2$ surfaces of their fits.
- Eldridge et al. [27] compared their database of supernova light curves calculated with SNEC and binary population synthesis stellar models [26] to infer explosion parameters of a sample of well-observed supernovae with pre-explosion progenitor detections.
- Ricks and Dwarkadas [91] used MESA stellar evolution code to simulate supernova progenitors, explode them, and calculate the resulting optical light curves with multi-group radiative hydro code STELLA [11, 12, 76, 77]. The dynamical range of the inferred parameters is more limited than in other works, which is probably due to relatively small sample size.

Finally, Fig. 7 shows the observed distribution of a subset of neutron star masses, which should be close to the birth mass. This includes double neutron star binaries and non-recycled pulsars [75]. The observations should be compared

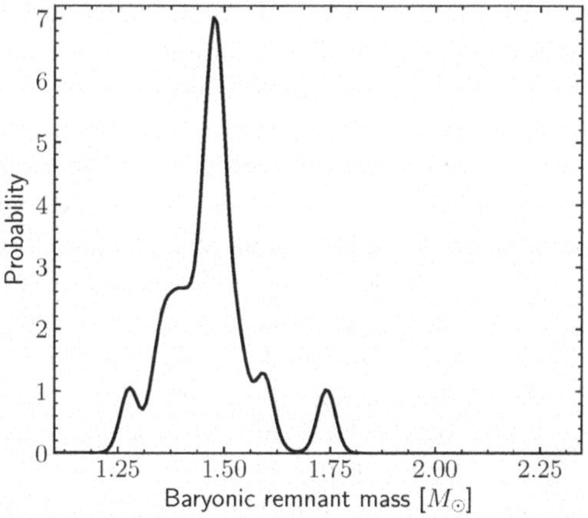


**Fig. 7** Distribution of bayronic masses of double neutron stars and non-recycled pulsars from [75]. Each measurement is approximated with a Gaussian with a width of the quoted uncertainty value or 0.03 $M_\odot$, whichever is larger. The values were converted from observed gravitational masses to baryonic masses using the equation of [107]. This figure should be compared with the theoretical predictions depicted in Fig. 4

with theoretical results in Fig. 4. A more quantitative comparison with theoretical models was done by Pejcha et al. [82] and Raithel et al. [88]. The latter work finds that single-star models have trouble explaining high-mass neutron stars. They also looked at the distribution of black hole masses.

## 4 How to Compare Observations with Theory

We see that there is a general qualitative agreement between the theoretical and observational results for normal hydrogen-rich supernovae: most theories predict explosion energies of $10^{50}$ to $10^{51}$ ergs, nickel masses of 0.01 to 0.1 $M_\odot$, and baryonic neutron star masses between 1.3 and 1.7 $M_\odot$, as is observed. This could be viewed as a success of the theory, but is not entirely unexpected. All models have some free parameters, which can be tuned to achieve the desired outcome. An example is relating one of the free theoretical parameters to the progenitor compactness. To first order, these choices affect more significantly quantities like population means or medians and less characteristics like correlations between different explosion outcomes, their slopes, and intrinsic scatters.

Observational inferences are not free from similar biases either. Analytic scaling relations always include an absolute term, which can be uncertain. As a result, works based on these relations have a considerable freedom in rescaling all of the physical quantities by a factor. Relative positions of individual supernovae are affected less. The uncertainties in fitting based on radiation hydrodynamics are more intricate, but have led to overestimates of progenitor masses when compared to inferences from pre-explosion images e.g. [57, 110]. Recent works, however, indicate a better agreement between the two independent methods [58].

Recently, an additional challenge has been recognized with inferences of explosion parameters from the light curve plateaus of Type II-P supernovae: estimates of explosion energy and ejecta mass are nearly degenerate even in radiation hydrodynamical models. Qualitatively, the degeneracy arises because higher ejecta mass leads to a longer diffusion time, which can be compensated for with higher explosion energy. If explosion energy per unit ejecta mass remains nearly constant, the observed expansion velocity, and the plateau luminosity and duration do not change much [7]. Nagy et al. [69] found explosion parameter correlations with a semianalytical code. Pejcha and Prieto [79] argued that analytic scaling relations point to nearly degenerate inferences of explosion energy and ejecta mass so that uncertainties in quantities like distance and reddening will manifest as a spread along a diagonal line in $E_{\rm exp}$–$M_{\rm ej}$ plot. They argued that this might be responsible for the claimed observational correlation between these two quantities. Goldberg et al. [34] used MESA and STELLA codes to illustrate that additional information is needed to break the degeneracy between $E_{\rm exp}$, $M_{\rm ej}$, and the progenitor radius (see especially their Fig. 26). They also argued that spectroscopic velocities secured during the shock-cooling phase in the first $\sim$15 days after the explosion could break the degeneracy. Similar conclusions were independently reached also by Dessart and Hillier [22] with multi-group radiation transfer code, which models full spectra.

The possibility of breaking the degeneracies with early observations suggested by Goldberg et al. [34] might not work for all Type II-P supernovae. There has been growing amount of evidence that some red supergiant progenitors are surrounded by circumstellar medium very close to the surface, which influences the early part of the light curve and spectra. The evidence includes modeling of early light curves [23, 31, 59–63] and "flash spectroscopy" of the progenitor surroundings (e.g. [39, 48, 123]).

Although a significant fraction of the population of stripped-envelope supernovae (spectroscopic types IIb, Ib, and Ic) likely originates from binary interactions and their ejecta masses are not representative of the progenitor initial stellar mass, explosion energies and nickel masses inferred from light curve modeling can test the explosion theories as well. Recently, [29] applied the parameterized neutrino mechanism to a population of helium stars of various masses. They found that the observed value of $M_{\rm Ni}$ are noticeably higher than the allowed range of theoretical predictions. Perhaps additional sources of energy from (magneto)-rotational processes can provide the necessary boost. It is then tempting to speculate whether the neutrino mechanism is subdominant even in normal supernovae [99].

The limitations of both theory and observations imply that moving forward with deeper and more quantitative test of supernova explosion mechanism will require investigating finer details in the observationally inferred explosion properties. There are several possibilities to move forward.

The degeneracies in inferring the explosion properties during the plateau can be put aside by focusing on the later phases dominated by radioactive decay of $^{56}$Ni. This is the approach taken by Müller et al. [67] and Anderson [4]. However, Type II-P supernovae are faint during the radioactive decay phase and the inferences of $M_{\rm Ni}$ depend on knowing well the explosion date and the distance. Future time-domain surveys could help with some of these challenges.

The alternative is to take the degeneracies explicitly into account when making the inferences and when doing comparisons with the theory. While degeneracies can be easily quantified with the covariance matrix or by directly exploring the likelihood space with Markov Chain Monte Carlo techniques, it is less straightforward how to construct the correct model. The theoretical light and velocity curves from radiation hydrodynamics never match the observations within the level of their uncertainties. As a result, the sampling of observations will then bias the inferences. Furthermore, taking into account circumstellar interaction increases the number of free parameters, where some of them might be degenerate with the explosion properties. Some of these issues might be alleviated by introducing a "transfer function" between the theoretical light and velocity curves and the observed magnitudes and velocities. This approach requires a sufficiently large training dataset to constrain the large number of additional parameters. Finally, comparisons of supernova explosion properties with theory usually implicitly assume that the observations are representative of the underlying stellar populations. Taking into account selection effects might yield new tests of the explosion mechanism, for example, by comparing relative rates of low- and high-$M_{\rm Ni}$ events.

**Acknowledgements** I am grateful to Joseph Anderson and Tuguldur Sukhbold for insightful and useful referee reports. I thank Thomas Ertl, Thomas Janka, and Bernhard Müller for sharing machine-readable versions of their results. This work has been supported by Primus grant PRIMUS/SCI/17 from Charles University, Horizon 2020 ERC Starting Grant "Cat-In-hAT" (grant agreement #803158) and INTER-EXCELLENCE grant LTAUSA18093 from the Czech Ministry of Education, Youth, and Sports.

## References

1. S.M. Adams, C.S. Kochanek, J.R. Gerke, K.Z. Stanek, The search for failed supernovae with the Large Binocular Telescope: constraints from 7 yr of data. Mon. Not. R. Astron. Soc. **469**(2), 1445–1455 (2017)
2. S.M. Adams, C.S. Kochanek, J.R. Gerke, K.Z. Stanek, X. Dai, The search for failed supernovae with the Large Binocular Telescope: confirmation of a disappearing star. Mon. Not. R. Astron. Soc. **468**(4), 4968–4981 (2017)
3. A.W. Alsabti, P. Murdin, *Handbook of Supernovae* (Springer, Berlin, 2017)
4. J.P. Anderson, A meta-analysis of core-collapse supernova $^{56}$Ni masses. Astron. Astrophys. **628**, A7 (2019)
5. J.P. Anderson, S.M. Habergham, P.A. James, M. Hamuy, Progenitor mass constraints for core-collapse supernovae from correlations with host galaxy star formation. Mon. Not. R. Astron. Soc. **424**(2), 1372–1391 (2012)
6. I. Arcavi, *Hydrogen-Rich Core-Collapse Supernovae* (Springer, Berlin, 2017), p. 239
7. W.D. Arnett, Analytic solutions for light curves of supernovae of Type II. Astrophys. J. **237**, 541–549 (1980)
8. C. Badenes, J. Harris, D. Zaritsky, J.L. Prieto, The stellar ancestry of supernovae in the magellanic clouds. I. The most recent supernovae in the large magellanic cloud. Astrophys. J. **700**(1), 727–740 (2009)
9. M.C. Bersten, O. Benvenuto, M. Hamuy, Hydrodynamical models of type II plateau supernovae. Astrophys. J. **729**(1), 61 (2011)

10. H.A. Bethe, J.R. Wilson, Revival of a stalled supernova shock by neutrino heating. Astrophys. J. **295**, 14–23 (1985)
11. S.I. Blinnikov, R. Eastman, O.S. Bartunov, V.A. Popolitov, S.E. Woosley, A comparative modeling of supernova 1993J. Astrophys. J. **496**(1), 454–472 (1998)
12. S.I. Blinnikov, F.K. Röpke, E.I. Sorokina, M. Gieseler, M. Reinecke, C. Travaglio, W. Hillebrandt, M. Stritzinger, Theoretical light curves for deflagration models of type Ia supernova. Astron. Astrophys. **453**(1), 229–240 (2006)
13. J.M. Blondin, A. Mezzacappa, C. DeMarino, Stability of standing accretion shocks, with an eye toward core-collapse supernovae. Astrophys. J. **584**(2), 971–980 (2003)
14. A. Burrows, J. Goshy, A theory of supernova explosions. Astrophys. J. **416**, L75 (1993)
15. A. Burrows, J. Hayes, B.A. Fryxell, On the nature of core-collapse supernova explosions. Astrophys. J. **450**, 830 (1995)
16. A. Burrows, E. Livne, L. Dessart, C.D. Ott, J. Murphy, A new mechanism for core-collapse supernova explosions. Astrophys. J. **640**(2), 878–890 (2006)
17. A. Burrows, L. Dessart, E. Livne, C.D. Ott, J. Murphy, Simulations of magnetically driven supernova and hypernova explosions in the context of rapid rotation. Astrophys. J. **664**(1), 416–434 (2007)
18. A. Burrows, D. Vartanyan, J.C. Dolence, M.A. Skinner, D. Radice, Crucial physical dependencies of the core-collapse supernova mechanism. Space Sci. Rev. **214**(1), 33 (2018)
19. R.A. Chevalier, N. Soker, Asymmetric envelope expansion of supernova 1987A. Astrophys. J. **341**, 867 (1989)
20. S.M. Couch, M.L. Warren, E.P. O'Connor, Simulating turbulence-aided neutrino-driven core-collapse supernova explosions in one dimension (2019). e-prints. arXiv:1902.01340
21. S. Curtis, K. Ebinger, C. Fröhlich, M. Hempel, A. Perego, M. Liebendörfer, F.-K. Thielemann,, PUSHing core-collapse supernovae to explosions in spherical symmetry. III. Nucleosynthesis yields. Astrophys. J. **870**(1), 2 (2019)
22. L. Dessart, D.J. Hillier, The difficulty of inferring progenitor masses from type-II-plateau supernova light curves. Astron. Astrophys. **625**, A9 (2019)
23. L. Dessart, D. John Hillier, E. Audit, Explosion of red-supergiant stars: influence of the atmospheric structure on shock breakout and early-time supernova radiation. Astron. Astrophys. **605**, A83 (2017)
24. K. Ebinger, S. Curtis, C. Fröhlich, M. Hempel, A. Perego, M. Liebendörfer, F.-K. Thielemann, PUSHing core-collapse supernovae to explosions in spherical symmetry. II. Explodability and remnant properties. Astrophys. J. **870**(1), 1 (2019)
25. K. Ebinger, S. Curtis, S. Ghosh, C. Fröhlich, M. Hempel, A. Perego, M. Liebendörfer, F.-K. Thielemann, PUSHing core-collapse supernovae to explosions in spherical symmetry IV: explodability, remnant properties and nucleosynthesis yields of low metallicity stars (2019). e-prints. arXiv:1910.08958
26. J.J. Eldridge, L. Xiao, E.R. Stanway, N. Rodrigues, N.Y. Guo, Supernova lightCURVE POPulation synthesis I: including interacting binaries is key to understanding the diversity of type II supernova lightcurves. Publ. Astron. Soc. Aust. **35**, 49 (2018)
27. J.J. Eldridge, N.Y. Guo, N. Rodrigues, E.R. Stanway, L. Xiao, Supernova lightCURVE POPulation synthesis II: validation against supernovae with an observed progenitor (2019). e-prints. arXiv:1908.07762
28. T. Ertl, H.T. Janka, S.E. Woosley, T. Sukhbold, M. Ugliano, A two-parameter criterion for classifying the explodability of massive stars by the neutrino-driven mechanism. Astrophys. J. **818**(2), 124 (2016)
29. T. Ertl, S.E. Woosley, T. Sukhbold, H.T. Janka, The explosion of helium stars evolved with mass loss (2019). e-prints. arXiv:1910.01641
30. T. Foglizzo, *Explosion Physics of Core-Collapse Supernovae* (Springer, Berlin, 2017), p. 1053
31. F. Förster, T.J. Moriya, J.C. Maureira, J.P. Anderson, S. Blinnikov, F. Bufano, G. Cabrera-Vives, A. Clocchiatti, T. de Jaeger, P.A. Estévez, L. Galbany, S. González-Gaitán, G. Gräfener, M. Hamuy, E.Y. Hsiao, P. Huentelemu, P. Huijse, H. Kuncarayakti, J. Martínez, G. Medina, E.F. Olivares, G. Pignata, A. Razza, I. Reyes, J. San Martín, R.C. Smith, E. Vera,

A.K. Vivas, A. de Ugarte Postigo, S.C. Yoon, C. Ashall, M. Fraser, A. Gal-Yam, E. Kankare, L. Le Guillou, P.A. Mazzali, N.A. Walton, D.R. Young, The delay of shock breakout due to circumstellar material evident in most type II supernovae. Nat. Astron. **2**, 808 (2018)
32. A. Gal-Yam, *Observational and Physical Classification of Supernovae* (Springer, Berlin, 2017), p. 195
33. J.R. Gerke, C.S. Kochanek, K.Z. Stanek, The search for failed supernovae with the Large Binocular Telescope: first candidates. Mon. Not. R. Astron. Soc. **450**(3), 3289–3305 (2015)
34. J.A. Goldberg, L. Bildsten, B. Paxton, Inferring explosion properties from type II-plateau supernova light curves. Astrophys. J. **879**(1), 3 (2019)
35. S.E. Gossan, J. Fuller, L.F. Roberts, Wave heating from proto-neutron star convection and the core-collapse supernova explosion mechanism. Mon. Not. R. Astron. Soc. **491**(4), 5376–5391 (2020)
36. M. Hamuy, Observed and physical properties of core-collapse supernovae. Astrophys. J. **582**(2), 905–914 (2003)
37. M. Herant, W. Benz, W.R. Hix, C.L. Fryer, S.A. Colgate, Inside the supernova: a powerful convective engine. Astrophys. J. **435**, 339 (1994)
38. J.J. Hester, The Crab Nebula: an astrophysical chimera. Annu. Rev. Astron. Astrophys. **46**, 127–155 (2008)
39. G. Hosseinzadeh, S. Valenti, C. McCully, D.A. Howell, I. Arcavi, A. Jerkstrand, D. Guevel, L. Tartaglia, L. Rui, J. Mo, X. Wang, F. Huang, H. Song, T. Zhang, K. Itagaki, Short-lived circumstellar interaction in the low-luminosity type IIP SN 2016bkv. Astrophys. J. **861**(1), 63 (2018)
40. F. Huang, X. Wang, J. Zhang, P.J. Brown, L. Zampieri, M.L. Pumo, T. Zhang, J. Chen, J. Mo, X. Zhao, SN 2013ej in M74: a luminous and fast-declining type II-P supernova. Astrophys. J. **807**(1), 59 (2015)
41. D. Ibeling, A. Heger, The metallicity dependence of the minimum mass for core-collapse supernovae. Astrophys. J. **765**(2), L43 (2013)
42. H.T. Janka, Conditions for shock revival by neutrino heating in core-collapse supernovae. Astron. Astrophys. **368**, 527–560 (2001)
43. H.-T. Janka, *Neutrino-Driven Explosions* (Springer, Berlin, 2017), p. 1095
44. H.-T. Janka, T. Melson, A. Summa, Physics of core-collapse supernovae in three dimensions: a sneak preview. Annu. Rev. Nucl. Part. Sci. **66**(1), 341–375 (2016)
45. Z.G. Jennings, B.F. Williams, J.W. Murphy, J.J. Dalcanton, K.M. Gilbert, A.E. Dolphin, M. Fouesneau, D.R. Weisz, Supernova remnant progenitor masses in M31. Astrophys. J. **761**(1), 26 (2012)
46. A. Jerkstrand, *Spectra of Supernovae in the Nebular Phase* (Springer, Berlin, 2017), p. 795
47. D. Kasen, S.E. Woosley, Type II supernovae: model light curves and standard candle relationships. Astrophys. J. **703**(2), 2205–2216 (2009)
48. D. Khazov, O. Yaron, A. Gal-Yam, I. Manulis, A. Rubin, S.R. Kulkarni, I. Arcavi, M.M. Kasliwal, E.O. Ofek, Y. Cao, D. Perley, J. Sollerman, A. Horesh, M. Sullivan, A.V. Filippenko, P.E. Nugent, D.A. Howell, S.B. Cenko, J.M. Silverman, H. Ebeling, F. Taddia, J. Johansson, R.R. Laher, J. Surace, U.D. Rebbapragada, P.R. Wozniak, T. Matheson, Flash spectroscopy: emission lines from the ionized circumstellar material around $<$10-day-old type II supernovae. Astrophys. J. **818**(1), 3 (2016)
49. C.S. Kochanek, J.F. Beacom, M.D. Kistler, J.L. Prieto, K.Z. Stanek, T.A. Thompson, H. Yüksel, A survey about nothing: monitoring a million supergiants for failed supernovae. Astrophys. J. **684**(2), 1336–1342 (2008)
50. J.M. LeBlanc, J.R. Wilson, A numerical example of the collapse of a rotating magnetized star. Astrophys. J. **161**, 541 (1970)
51. M. Liebendörfer, A. Mezzacappa, F.-K. Thielemann, Conservative general relativistic radiation hydrodynamics in spherical symmetry and comoving coordinates. Phys. Rev. D **63**(10), 104003 (2001)
52. M. Limongi, *Supernovae from Massive Stars* (Springer, Berlin, 2017), p. 513

53. I.Y. Litvinova, D.K. Nadezhin, Determination of integrated parameters for type-II supernovae. Sov. Astron. Lett. **11**, 145–147 (1985)
54. E. Lovegrove, S.E. Woosley, Very low energy supernovae from neutrino mass loss. Astrophys. J. **769**(2), 109 (2013)
55. Q.A. Mabanta, J.W. Murphy, How turbulence enables core-collapse supernova explosions. Astrophys. J. **856**(1), 22 (2018)
56. Q.A. Mabanta, J.W. Murphy, J.C. Dolence, Convection-aided explosions in one-dimensional core-collapse supernova simulations I: technique and validation (2019). e-prints. arXiv:1901.11234
57. K. Maguire, E. Di Carlo, S.J. Smartt, A. Pastorello, D.Y. Tsvetkov, S. Benetti, S. Spiro, A.A. Arkharov, G. Beccari, M.T. Botticella, E. Cappellaro, S. Cristallo, M. Dolci, N. Elias-Rosa, M. Fiaschi, D. Gorshanov, A. Harutyunyan, V.M. Larionov, H. Navasardyan, A. Pietrinferni, G. Raimondo, G. di Rico, S. Valenti, G. Valentini, L. Zampieri, Optical and near-infrared coverage of SN 2004et: physical parameters and comparison with other Type IIP supernovae. Mon. Not. R. Astron. Soc. **404**, 981–1004 (2010)
58. L. Martinez, M.C. Bersten, Mass discrepancy analysis for a select sample of type II-plateau supernovae. Astron. Astrophys. **629**, A124 (2019)
59. T.J. Moriya, S.-C. Yoon, G. Gräfener, S.I. Blinnikov, Immediate dense circumstellar environment of supernova progenitors caused by wind acceleration: its effect on supernova light curves. Mon. Not. R. Astron. Soc. **469**(1), L108–L112 (2017)
60. T.J. Moriya, F. Förster, S.C. Yoon, G. Gräfener, S.I. Blinnikov, Type IIP supernova light curves affected by the acceleration of red supergiant winds. Mon. Not. R. Astron. Soc. **476**(2), 2840–2851 (2018)
61. V. Morozova, A.L. Piro, M. Renzo, C.D. Ott, D. Clausen, S.M. Couch, J. Ellis, L.F. Roberts, Light curves of core-collapse supernovae with substantial mass loss using the new open-source SuperNova explosion code (SNEC). Astrophys. J. **814**(1), 63 (2015)
62. V. Morozova, A.L. Piro, S. Valenti, Unifying type II supernova light curves with dense circumstellar material. Astrophys. J. **838**(1), 28 (2017)
63. V. Morozova, A.L. Piro, S. Valenti, Measuring the progenitor masses and dense circumstellar material of type II supernovae. Astrophys. J. **858**(1), 15 (2018)
64. B. Müller, A critical assessment of turbulence models for 1D core-collapse supernova simulations. Mon. Not. R. Astron. Soc. **487**(4), 5304–5323 (2019)
65. B. Müller, H.T. Janka, Non-radial instabilities and progenitor asphericities in core-collapse supernovae. Mon. Not. R. Astron. Soc. **448**(3), 2141–2174 (2015)
66. B. Müller, A. Heger, D. Liptai, J.B. Cameron, A simple approach to the supernova progenitor-explosion connection. Mon. Not. R. Astron. Soc. **460**(1), 742–764 (2016)
67. T. Müller, J.L. Prieto, O. Pejcha, A. Clocchiatti, The nickel mass distribution of normal type II supernovae. Astrophys. J. **841**(2), 127 (2017)
68. J.W. Murphy, A. Burrows, Criteria for core-collapse supernova explosions by the neutrino mechanism. Astrophys. J. **688**(2), 1159–1175 (2008)
69. A.P. Nagy, A. Ordasi, J. Vinkó, J.C. Wheeler, A semianalytical light curve model and its application to type IIP supernovae. Astron. Astrophys. **571**, A77 (2014)
70. K. Nakamura, T. Takiwaki, T. Kuroda, K. Kotake, Systematic features of axisymmetric neutrino-driven core-collapse supernova models in multiple progenitors. Publ. Astron. Soc. Jpn. **67**(6), 107 (2015)
71. P. Nugent, M. Hamuy, *Cosmology with Type IIP Supernovae* (Springer, Berlin, 2017), p. 2671
72. E. O'Connor, *The Core-Collapse Supernova-Black Hole Connection* (Springer, Berlin, 2017), p. 1555
73. E. O'Connor, C.D. Ott, A new open-source code for spherically symmetric stellar collapse to neutron stars and black holes. Classical Quantum Gravity **27**(11), 114103 (2010)
74. E. O'Connor, C.D. Ott, Black hole formation in failing core-collapse supernovae. Astrophys. J. **730**(2), 70 (2011)
75. F. Özel, P. Freire, Masses, radii, and the equation of state of neutron stars. Annu. Rev. Astron. Astrophys. **54**, 401–440 (2016)

76. B. Paxton, P. Marchant, J. Schwab, E.B. Bauer, L. Bildsten, M. Cantiello, L. Dessart, R. Farmer, H. Hu, N. Langer, R.H.D. Townsend, D.M. Townsley, F.X. Timmes, Modules for experiments in stellar astrophysics (MESA): binaries, pulsations, and explosions. Astrophys. J. **220**(1), 15 (2015)
77. B. Paxton, J. Schwab, E.B. Bauer, L. Bildsten, S. Blinnikov, P. Duffell, R. Farmer, J.A. Goldberg, P. Marchant, E. Sorokina, A. Thoul, R.H.D. Townsend, F.X. Timmes, Modules for experiments in stellar astrophysics (MESA): convective boundaries, element diffusion, and massive star explosions. Astrophys. J. **234**(2), 34 (2018)
78. O. Pejcha, J.L. Prieto, A global model of the light curves and expansion velocities of type II-plateau supernovae. Astrophys. J. **799**(2), 215 (2015a)
79. O. Pejcha, J.L. Prieto, On the intrinsic diversity of type II-plateau supernovae. Astrophys. J. **806**(2), 225 (2015b)
80. O. Pejcha, T.A. Thompson, The physics of the neutrino mechanism of core-collapse supernovae. Astrophys. J. **746**(1), 106 (2012)
81. O. Pejcha, T.A. Thompson, The landscape of the neutrino mechanism of core-collapse supernovae: neutron star and black hole mass functions, explosion energies, and nickel yields. Astrophys. J. **801**(2), 90 (2015)
82. O. Pejcha, T.A. Thompson, C.S. Kochanek, The observed neutron star mass distribution as a probe of the supernova explosion mechanism. Mon. Not. R. Astron. Soc. **424**(2), 1570–1583 (2012)
83. A. Perego, M. Hempel, C. Fröhlich, K. Ebinger, M. Eichler, J. Casanova, M. Liebendörfer, F.K. Thielemann, PUSHing core-collapse supernovae to explosions in spherical symmetry I: the model and the case of SN 1987A. Astrophys. J. **806**(2), 275 (2015)
84. P. Podsiadlowski, *The Progenitor of SN 1987A* (Springer, Berlin, 2017), p. 635
85. D.V. Popov, An analytical model for the plateau stage of type II supernovae. Astrophys. J. **414**, 712 (1993)
86. M.L. Pumo, L. Zampieri, Radiation-hydrodynamical modeling of core-collapse supernovae: light curves and the evolution of photospheric velocity and temperature. Astrophys. J. **741**, 41 (2011)
87. M.L. Pumo, L. Zampieri, S. Spiro, A. Pastorello, S. Benetti, E. Cappellaro, G. Manicò, M. Turatto, Radiation-hydrodynamical modelling of underluminous type II plateau supernovae. Mon. Not. R. Astron. Soc. **464**(3), 3013–3020 (2017)
88. C.A. Raithel, T. Sukhbold, F. Özel, Confronting models of massive star evolution and explosions with remnant mass measurements. Astrophys. J. **856**(1), 35 (2018)
89. M.J. Raives, S.M. Couch, J.P. Greco, O. Pejcha, T.A. Thompson, The antesonic condition for the explosion of core-collapse supernovae–I. Spherically symmetric polytropic models: stability and wind emergence. Mon. Not. R. Astron. Soc. **481**(3), 3293–3304 (2018)
90. T.M. Reynolds, M. Fraser, G. Gilmore, Gone without a bang: an archival HST survey for disappearing massive stars. Mon. Not. R. Astron. Soc. **453**(3), 2885–2900 (2015)
91. W. Ricks, V.V. Dwarkadas, Excavating the explosion and progenitor properties of type IIP supernovae via modeling of their optical light curves. Astrophys. J. **880**(1), 59 (2019)
92. E.E. Salpeter, The luminosity function and stellar evolution. Astrophys. J. **121**, 161 (1955)
93. T. Shigeyama, K. Nomoto, Theoretical light curve of SN 1987A and mixing of hydrogen and nickel in the ejecta. Astrophys. J. **360**, 242 (1990)
94. S.A. Sim, *Spectra of Supernovae During the Photospheric Phase* (Springer, Berlin, 2017), p. 769
95. S.J. Smartt, Progenitors of core-collapse supernovae. Annu. Rev. Astron. Astrophys. **47**(1), 63–106 (2009)
96. S.J. Smartt, Observational constraints on the progenitors of core-collapse supernovae: the case for missing high-mass stars. Publ. Astron. Soc. Aust. **32**, e016 (2015)
97. N. Smith, The crab nebula and the class of type IIn-P supernovae caused by sub-energetic electron-capture explosions. Mon. Not. R. Astron. Soc. **434**(1), 102–113 (2013)
98. S. Spiro, A. Pastorello, M.L. Pumo, L. Zampieri, M. Turatto, S.J. Smartt, S. Benetti, E. Cappellaro, S. Valenti, I. Agnoletto, G. Altavilla, T. Aoki, E. Brocato, E.M. Corsini, A.

Di Cianno, N. Elias-Rosa, M. Hamuy, K. Enya, M. Fiaschi, G. Folatelli, S. Desidera, A. Harutyunyan, D.A. Howell, A. Kawka, Y. Kobayashi, B. Leibundgut, T. Minezaki, H. Navasardyan, K. Nomoto, S. Mattila, A. Pietrinferni, G. Pignata, G. Raimondo, M. Salvo, B.P. Schmidt, J. Sollerman, J. Spyromilio, S. Taubenberger, G. Valentini, S. Vennes, Y. Yoshii, Low luminosity type II supernovae – II. Pointing towards moderate mass precursors. Mon. Not. R. Astron. Soc. **439**(3), 2873–2892 (2014)

99. T. Sukhbold, T.A. Thompson, Magnetar-powered ordinary type IIP supernovae. Mon. Not. R. Astron. Soc. **472**(1), 224–229 (2017)
100. T. Sukhbold, S.E. Woosley, The compactness of presupernova stellar cores. Astrophys. J. **783**(1), 10 (2014)
101. T. Sukhbold, S.E. Woosley, A. Heger, A high-resolution study of presupernova core structure. Astrophys. J. **860**(2), 93 (2018)
102. T. Sukhbold, T. Ertl, S.E. Woosley, J.M. Brown, H.T. Janka, Core-collapse supernovae from 9 to 120 solar masses based on neutrino-powered explosions. Astrophys. J. **821**(1), 38 (2016)
103. K. Takáts, M.L. Pumo, N. Elias-Rosa, A. Pastorello, G. Pignata, E. Paillas, L. Zampieri, J.P. Anderson, J. Vinkó, S. Benetti, M.T. Botticella, F. Bufano, A. Campillay, R. Cartier, M. Ergon, G. Folatelli, R.J. Foley, F. Förster, M. Hamuy, V.P. Hentunen, E. Kankare, G. Leloudas, N. Morrell, M. Nissinen, M.M. Phillips, S.J. Smartt, M. Stritzinger, S. Taubenberger, S. Valenti, S.D. Van Dyk, J.B. Haislip, A.P. LaCluyze, J.P. Moore, D. Reichart, SN 2009N: linking normal and subluminous type II-P SNe. Mon. Not. R. Astron. Soc. **438**(1), 368–387 (2014)
104. K. Takáts, G. Pignata, M.L. Pumo, E. Paillas, L. Zampieri, N. Elias-Rosa, S. Benetti, F. Bufano, E. Cappellaro, M. Ergon, M. Fraser, M. Hamuy, C. Inserra, E. Kankare, S.J. Smartt, M.D. Stritzinger, S.D. Van Dyk, J.B. Haislip, A.P. LaCluyze, J.P. Moore, D. Reichart, SN 2009ib: a type II-P supernova with an unusually long plateau. Mon. Not. R. Astron. Soc. **450**(3), 3137–3154 (2015)
105. C. Thompson, Accretional heating of asymmetric supernova cores. Astrophys. J. **534**(2), 915–933 (2000)
106. T.A. Thompson, E. Quataert, A. Burrows, Viscosity and rotation in core-collapse supernovae. Astrophys. J. **620**(2), 861–877 (2005)
107. F.X. Timmes, S.E. Woosley, T.A. Weaver, The neutron star and black hole initial mass function. Astrophys. J. **457**, 834 (1996)
108. M. Ugliano, H.-T. Janka, A. Marek, A. Arcones, Progenitor-explosion connection and remnant birth masses for neutrino-driven supernovae of iron-core progenitors. Astrophys. J. **757**(1), 69 (2012)
109. V.P. Utrobin, The light curve of supernova 1987A: the structure of the presupernova and radioactive nickel mixing. Astron. Lett. **30**, 293–308 (2004)
110. V.P. Utrobin, N.N. Chugai, High mass of the type IIP supernova 2004et inferred from hydrodynamic modeling. Astron. Astrophys. **506**(2), 829–834 (2009)
111. V.P. Utrobin, N.N. Chugai, Type IIP supernova 2008 in: the explosion of a normal red supergiant. Astron. Astrophys. **555**, A145 (2013)
112. V.P. Utrobin, N.N. Chugai, Parameters of type IIP SN 2012A and clumpiness effects. Astron. Astrophys. **575**, A100 (2015)
113. V.P. Utrobin, N.N. Chugai, Luminous type IIP SN 2013ej with high-velocity $^{56}$Ni ejecta. Mon. Not. R. Astron. Soc. **472**(4), 5004–5010 (2017)
114. V.P. Utrobin, N.N. Chugai, Resolving the puzzle of type IIP SN 2016X. Mon. Not. R. Astron. Soc. **490**(2), 2042–2049 (2019)
115. B.F. Williams, T.J. Hillis, W.P. Blair, K.S. Long, J.W. Murphy, A. Dolphin, R. Khan, J.J. Dalcanton, The masses of supernova remnant progenitors in M83. Astrophys. J. **881**(1), 54 (2019)
116. S.E. Woosley, Pulsational pair-instability supernovae. Astrophys. J. **836**(2), 244 (2017)
117. S.E. Woosley, A. Heger, Nucleosynthesis and remnants in massive stars of solar metallicity. Phys. Rep. **442**(1–6), 269–283 (2007)

118. S.E. Woosley, P.A. Pinto, L. Ensman, Supernova 1987A: six weeks later. Astrophys. J. **324**, 466 (1988)
119. S.E. Woosley, A. Heger, T.A. Weaver, N. Langer, SN 1987A – presupernova evolution and the progenitor star (1997). e-prints. astro–ph/9705146
120. S.E. Woosley, A. Heger, T.A. Weaver, The evolution and explosion of massive stars. Rev. Mod. Phys. **74**(4), 1015–1071 (2002)
121. T. Yamasaki, S. Yamada, Effects of rotation on the revival of a stalled shock in supernova explosions. Astrophys. J. **623**(2), 1000–1010 (2005)
122. H. Yang, R.A. Chevalier, Evolution of the crab nebula in a low energy supernova. Astrophys. J. **806**(2), 153 (2015)
123. O. Yaron, D.A. Perley, A. Gal-Yam, J.H. Groh, A. Horesh, E.O. Ofek, S.R. Kulkarni, J. Sollerman, C. Fransson, A. Rubin, P. Szabo, N. Sapir, F. Taddia, S.B. Cenko, S. Valenti, I. Arcavi, D.A. Howell, M.M. Kasliwal, P.M. Vreeswijk, D. Khazov, O.D. Fox, Y. Cao, O. Gnat, P.L. Kelly, P.E. Nugent, A.V. Filippenko, R.R. Laher, P.R. Wozniak, W.H. Lee, U.D. Rebbapragada, K. Maguire, M. Sullivan, M.T. Soumagnac, Confined dense circumstellar material surrounding a regular type II supernova. Nat. Phys. **13**(5), 510–517 (2017)
124. L. Zampieri, *Light Curves of Type II Supernovae* (Springer, Berlin, 2017), p. 737
125. E. Zapartas, S.E. de Mink, S. Justham, N. Smith, A. de Koter, M. Renzo, I. Arcavi, R. Farmer, Y. Götberg, S. Toonen, The diverse lives of progenitors of hydrogen-rich core-collapse supernovae: the role of binary interaction. Astron. Astrophys. **631**, A5 (2019)

# Low-Mass and Sub-stellar Eclipsing Binaries in Stellar Clusters

**Nicolas Lodieu, Ernst Paunzen, and Miloslav Zejda**

## Contents



**Abstract** We highlight the importance of eclipsing double-line binaries in our understanding on star formation and evolution. We review the recent discoveries of low-mass and sub-stellar eclipsing binaries belonging to star-forming regions, open clusters, and globular clusters identified by ground-based surveys and space missions with high-resolution spectroscopic follow-up. These discoveries provide benchmark systems with known distances, metallicities, and ages to calibrate masses and radii predicted by state-of-the-art evolutionary models to a few percent. We report their density and discuss current limitations on the accuracy of the physical parameters. We discuss future opportunities and highlight future guidelines

N. Lodieu (✉)
Instituto de Astrofísica de Canarias (IAC), La Laguna, Tenerife, Spain

Departamento de Astrofísica, Universidad de La Laguna (ULL), La Laguna, Tenerife, Spain
e-mail: nlodieu@iac.es

E. Paunzen · M. Zejda
Department of Theoretical Physics and Astrophysics, Masaryk University, Brno, Czech Republic
e-mail: epaunzen@physics.muni.cz; zejda@physics.muni.cz

 
P. Kabáth et al. (eds.), *Reviews in Frontiers of Modern Astrophysics*,
https://doi.org/10.1007/978-3-030-38509-5_8

to fill gaps in age and metallicity to improve further our knowledge of low-mass stars and brown dwarfs.

# 1 The Importance of Eclipsing Binaries

## *1.1 Scientific Context*

Low-mass M dwarfs are the most common stars in the Solar neighbourhood and, more generally, in the Universe, accounting for about 70% of the entire population of hydrogen-burning stars with masses below 0.6 $M_{\odot}$ [1, 2]. Determining their physical parameters (luminosity, mass, radius) is fundamental to understand stellar evolution and constrain theoretical isochrones. At lower temperatures, the numbers of brown dwarfs, objects unable to fuse hydrogen in their interiors [3–5], with accurate mass and radius measurements remain very limited.

Eclipsing binaries (EBs) are systems with two components lying on the same plane with respect to the observer transiting each other periodically. They are fundamental probes of stellar evolution and stellar candles to measure distances of clusters [6] because the radius and mass of each component can be derived with high precision from the photometric light-curves and radial velocity monitoring, respectively [7].

The numbers of EBs has increased dramatically over the past two decades thanks to the advent of large-scale photometric and spectroscopic surveys as well as space missions. Following up on the original reviews on fundamental parameters of stars derived from EBs [8, 9], a catalogue of detached EBs with their main physical parameters including masses and radii determined to precisions better than a few percent is constantly updated [10].[1] Another public databases with physical parameters of EBs is the Binary Star Database [11],[2] which contains physical and positional parameters of the components of 120,000 stellar multiple systems compiled from a variety of published catalogues and databases.

Other unrelated projects contributed, currently supply, and will add to our knowledge of EBs. As a few example, the OGLE project principally devoted to microlensing provides huge amount (several hundred thousands) of eclipsing systems over a wide range of mass and evolutionary states towards the Galactic Bulge [12].[3] The All Sky Automated Survey (ASAS)[4] is a low cost project dedicated to constant photometric monitoring of the full sky to study variable phenomenon of any kind, including the study of EBs [13]. The Large Sky Area Multi-Object Fiber Spectroscopic Telescope (LAMOST) is a large Chinese project dedicated to

[1] The catalogue is maintained at http://www.astro.keele.ac.uk/jkt/debcat/.

[2] The Binary Star Database can be found at http://bdb.inasan.ru/.

[3] http://ogle.astrouw.edu.pl/.

[4] http://www.astrouw.edu.pl/asas/.

spectroscopy of several millions of stars with spectral classification. This program has brought several thousands of EBs over a wide range of spectral type [14], with some masses and radii determined for a few low-mass systems [15]. Other programs mainly dedicated to the discovery and tracking of minor bodies, such as the Catalina Sky Survey [16–18][5] or the Asteroid Terrestrial-impact Last Alert System (ATLAS) project [19][6] do regularly contribute to the discovery of EBs.

This review focuses on low-mass EBs with at least one M dwarf or sub-stellar companions members of star-forming regions (Sect. 2), open clusters (Sect. 3), and globular clusters (Sect. 4). We discuss the frequency of EBs in clusters and the impact of age, metallicity, and stellar variability/activity on their physical parameters (Sect. 5). This review is timely due to the most recent contribution of the *Kepler K2* mission [20, 21] to our knowledge of low-mass EBs in clusters, whose masses and radii can directly be confronted to model predictions. The study of EBs requires huge observing time investment on both photometric and spectroscopic sides needed to infer masses and radii, as demonstrated by the WIYN cluster survey [22–24], the Young Exoplanet Transit Initiative (YETI) focusing on young clusters [25], the Palomar Transient Factory survey [26, 27],[7] and the *Kepler K2* mission. We finish this review with a list of requirements and prospects to fill in gaps in the Hertzsprung-Russell (H-R) diagram (Sect. 6).

## 1.2 How Masses and Radii Are Determined Observationally

The first attempts to determine the parameters of eclipsing binaries and their components were done in the end of nineteenth century. Up to late 1960s and 1970s year a series of method were developed and used on light curves and radial velocity curves series to subtract and interpolate data from tables of different quantities (more details in [28] or [29]). The spread of computers fasten the development of many codes such as EBOP(Eclipsing Binary Orbit Program) [30], SEBM (Standard Eclipsing Binary Star Model) [31–33], WINK [34–36], LIGHT2 [37, 38], version of WUMa [39, 40] and others.

In 1971, Wilson and Devinney published the results of their code (hereafter WD) where they used for the first time the least-squares method to extract the parameters of light curves [41–44]. This WD code has been regularly upgraded up to now and could be downloaded from author's ftp.[8] Independently, users created graphical user interfaces and some minor upgrades. However, the project PHOEBE [45] is not only GUI for calculations based on WD core, nowadays it has become a more general code to models both the photometric light curve and radial velocity curves of

[5]https://catalina.lpl.arizona.edu/.

[6]https://atlas.fallingstar.com/.

[7]https://www.ptf.caltech.edu/iptf.

[8]ftp://ftp.astro.ufl.edu/pub/wilson/.

eclipsing binaries. The new version of PHOEBE2, which is still under development,[9] contains more physics and improved mathematical methods for the solutions of eclipsing binaries [46–48].

Independent codes like MECI (Method for Eclipsing Component Identification) and DEBil (Detached Eclipsing Binary Light curve fitter) [49, 50], EBAS (Eclipsing Binary Automated Solver) [51, 52], FOTEL [53], JKTEBOP,[10] ROCHE [54], NIGHTFALL,[11] BINARY MAKERS (BM) [55] are used in limited numbers of publications. Two authors of these codes also collect binary stars solutions—David H. Bradstreet, author of Binary Maker, manages the Catalog and AtLas of Eclipsing Binaries (CALEB) based only on BM solutions[12] and John Southworth the DEBCat catalogue,[13] which contains physical properties of well-studied detached eclipsing binaries where errors on the mass and radius determinations are mostly below 2%.

From the aforementioned codes, we can estimated the physical parameters of each component of a multiple system. The main parameters derived from the analysis of the light curve(s) are orbital period, (possibly) eccentricity, orbital inclination, relative ratio of the radius of the primary and secondary of the system considering the separation of the components (top panel in Fig. 1), system luminosity and photometric mass ratios. However, in some cases photometric mass ratios might be unreliable in comparison to spectroscopic mass ratios [57, 58].

The light curve solution usually requires photometric data in at least two filters. The availability of only one passband data means that some of parameters must be estimated and/or fixed. The effective temperature of primary is inferred from its spectral type or colour indices. Limb darkening coefficients are interpolated from tables e.g. [59], gravity brightening and bolometric albedo coefficients are set according to the expected type of stellar atmospheres. Then, except for the parameters mentioned above, one can determine the surface potentials, the rotational/orbital synchronicity, and the third light.

The situation improves rapidly when radial velocity measurements are available for both components (Fig. 1). In this case, it becomes possible to figure out the spectroscopic mass ratio and distance of the components, which serve as a scaling factor for the radii of each component. The combination of photometric and spectroscopic datasets leads to the determination of absolute eclipsing binary parameters in physical units, including masses, sizes, and luminosities of both components as well as distance from Earth. In this process, we can also calculate the atmosphere model and corresponding parameters (see e.g. [60–65]).

Some of aforementioned codes coupling light curve and radial velocity solutions are also capable to process additional kinds of parameters like timings of minima, interferometric measurements, and so on.

---

[9] http://phoebe-project.org.

[10] http://www.astro.keele.ac.uk/jkt/codes/jktebop.html.

[11] https://www.hs.uni-hamburg.de/DE/Ins/Per/Wichmann/Nightfall.html.

[12] http://caleb.eastern.edu/.

[13] http://www.astro.keele.ac.uk/jkt/debcat/.

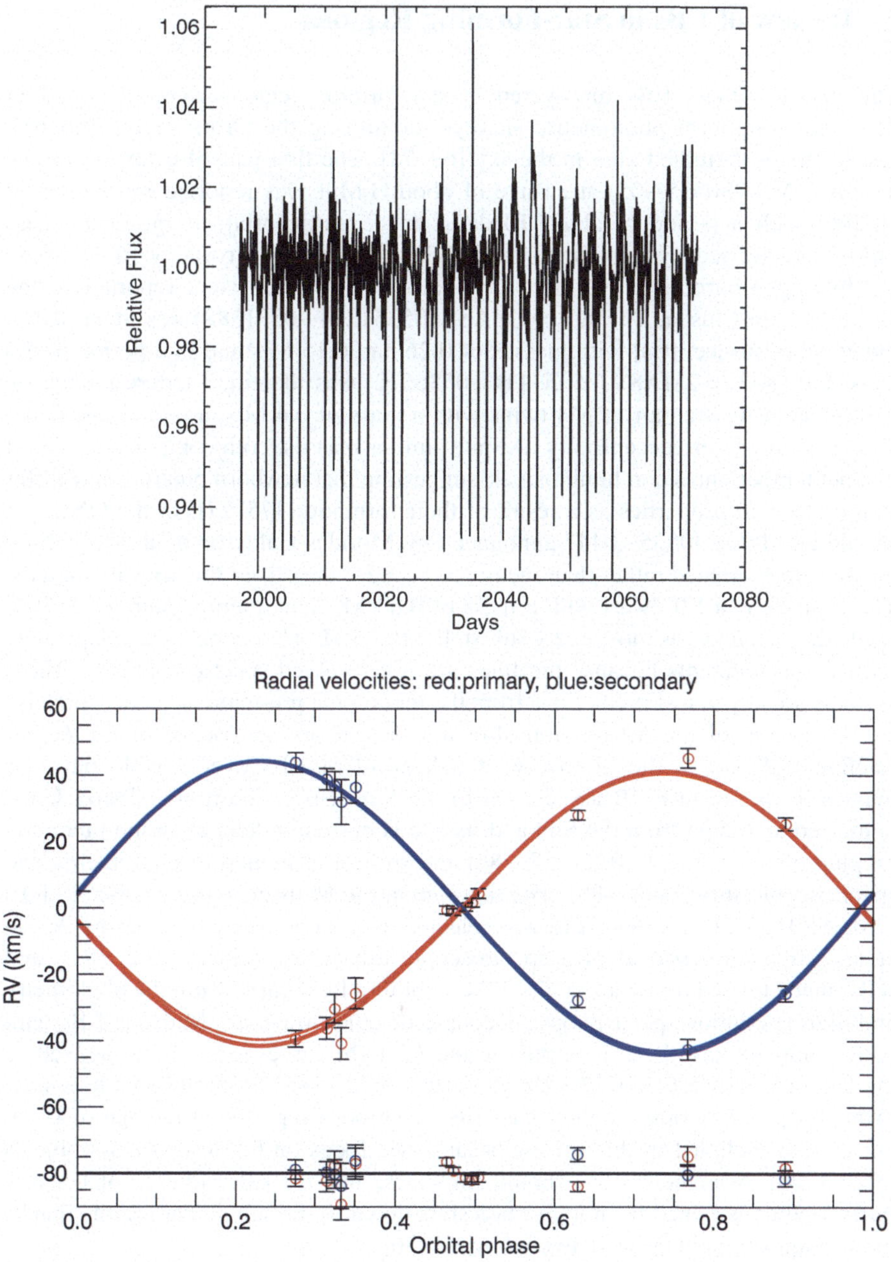


**Fig. 1** *Top: Kepler* light-curve for USco J16163068−2512201 (=EPIC 203710387) over the full ~76 days of the *K2* campaign two in Upper Scorpius. *Bottom:* Radial velocity measurements as a function of phase for the primary (blue symbols) and the secondary (red symbols) of USco J16163068−2512201. Figure taken from [56]

## 2 Review of EBs in Star-Forming Regions

The first low-mass EBs discovered in star-forming regions were identified in dedicated long-term photometric surveys monitoring the Orion region [66–68], one of the best studied area in the sky [69–74]. The first pair of eclipsing brown dwarfs (2MASS0535−05) at an age of about 1 Myr was reported by Stassun et al. [66] with a period of about 10 days, an eccentric orbit, a significant mass ratio characterised by complementary high-resolution spectroscopy [67]. Above the hydrogen-burning limit, there two pairs of low-mass stars reported in the Orion Nebula Cluster. JW 380 (= 2MASS J05351214−0531388) was identified in the Monitor project [68] with masses of 0.26 and 0.15 $M_{\odot}$ and in a period of 5.3 days. Par 1802 (= 2MASS J05351114−0536512) was identified independently by several team. It is a pair of M4 twins with a mass of 0.4 $M_{\odot}$, a period just under 5 days and non-zero eccentricity. Despite similar masses, both components exhibit distinct temperatures and luminosities, suggesting that newborn binaries may differ in the physical properties as a result of their formation [75–77]. In the ONC, we should add ISOY J0535−0447 announced by Morales-Calderón et al. [78] whose masses are estimated rather than measured because they fixed the semi-major axis. The primary is a K0 dwarf with a mass of 0.83 $M_{\odot}$ and a temperature of 5150 K while the secondary is most likely sub-stellar (0.05 $M_{\odot}$) but none have independent radius measurements because the lines are not resolved spectroscopically. These systems are key to test predictions from the theoretical pre-main-sequence models.

The advent of the *K2* mission after the loss of one gyroscope of the *Kepler* satellite [20] led to the discovery of a handful of EBs over a wide range of masses in the nearest OB association to the Sun, Upper Scorpius (USco). USco is located at 145 pc from the Sun and its age is currently debated in the literature, ranging between 5 and 10 Myr [79–86] and subject to numerous photometric and spectroscopic surveys [87–89]. The first one in the M dwarf regime, UScoCTIO 5 (2MASS J15595051−1944374), was selected as a photometric member by Ardila et al. [90] later resolved as a spectroscopic binary by Reiners et al. [91], and fully characterised by Kraus et al. [92] combining light curve from the photometry and high-resolution spectroscopy. A couple of other low-mass M dwarf EBs with EPIC numbers have been identified in the *K2* light curves [56, 86, 93] as well as the first brown dwarf (RIK72 = 2MASS J16033922−1851297) orbiting a low-mass dwarf [86]. A few other higher mass EBs have been reported at the age of USco but are not included in this review because we focus on the lowest mass objects [86, 93, 94]. Nonetheless, we should emphasise that the EB sequence of USco is fairly well constrained from high-mass stars down to the sub-stellar regime thanks to the exquisite light curve delivered by *K2* [86].

At very young ages, we should also mention the discovery of the low-mass, pre-main sequence eclipsing binary, CoRoT 223992193 (= 2MASS J06414422+0925024), whose secondary lies at the K/M border with a mass just under 0.5 $M_{\odot}$ and K dwarf primary with 0.67 $M_{\odot}$. This object belongs to the 3–6 Myr-old star-forming region NGC 2264 that was monitored continuously for 23.4 days by the CoRoT mission.

## 3 Review of EBs in Open Clusters

We can divide the stellar clusters targeted by *K2* into two groups: the intermediate-age clusters (100–200 Myr) whose main reference is the Pleiades with an age of 125±10 Myr [95–100] and the older more evolved open clusters like the Hyades (625±50 Myr) [101–108] and Praesepe (590–900 Myr) [99, 100, 102, 106, 109–112]. All these clusters are within 200 pc [113] and have metallicities close to solar or slightly super-solar [114–116].

In the Pleiades, HII 2407 known as a single-lined EB was identified eclipsing every 7.05 days [117]. The primary is well characterised with a spectral type of K1-K3, an effective temperature of 4970±95 K, a mass of 0.81±0.08 $M_{\odot}$, and a radius of 0.77±0.13 $R_{\odot}$. The secondary is a low-mass M dwarf, undetected in high-resolution spectra shortwards of 800 nm, yielding mass and radius estimates of 0.18 $M_{\odot}$ and 0.21 $R_{\odot}$, respectively. Another two EBs members of the Pleiades have been characterised photometrically and spectroscopically: HCG 76 and MHO 9 [118]. Both systems have long orbital periods with masses and radii well determined from the *K2* light curve and multiple radial velocity epochs. Two other higher mass EBs are presented in [118] as well as a possible member but not discussed in this review focusing on low-mass dwarfs. Finally, we highlight the possibility of MHO 9 being a hierarchical triple system due to its position above the Pleiades sequence in the H-R diagram (Fig. 2).

A pair of M dwarfs (2MASS J04463285+1901432) with a short period (~0.62 days) was reported by Hebb et al. [119] with masses of 0.47±0.05 and 0.19±0.02 $M_{\odot}$ in the 150 Myr-old cluster NGC 1647 [120] located at 540 pc from the Sun [121]. The system is confirmed as a photometric and spectroscopic member with a radial velocity consistent with the mean value of the cluster. This new low-mass system represent an important link between the Pleiades and older clusters discussed below.

Four low-mass EBs have been revealed in the Praesepe cluster. PTFEB 132.707+19.810 was announced by Kraus et al. [122] as a pair of 0.38+0.20 $M_{\odot}$ going around each other every 6 days and independently announced by Gillen et al. [123] as AD 3814. Another three EBs were included in the sample of low-mass EBs discussed in [123]. Two of these cluster candidates were classified as Praesepe members by four of six surveys [124–129], while the fourth one (AD 1508) is only labelled as member in two of these surveys. AD 2615 is a pair of almost equal-mass M dwarfs (0.21+0.25 $M_{\odot}$) with a period of 11.6 days and no eccentricity. The most special system, AD 3116, is composed of a M dwarf (~0.28 $M_{\odot}$) and a brown dwarf with an estimated mass of 0.052 $M_{\odot}$. The period of this system is quite short, around 2 days, and this is the only system with an significant eccentricity of 0.142. The last system, with a doubtful membership, is composed of two low-mass dwarfs close to

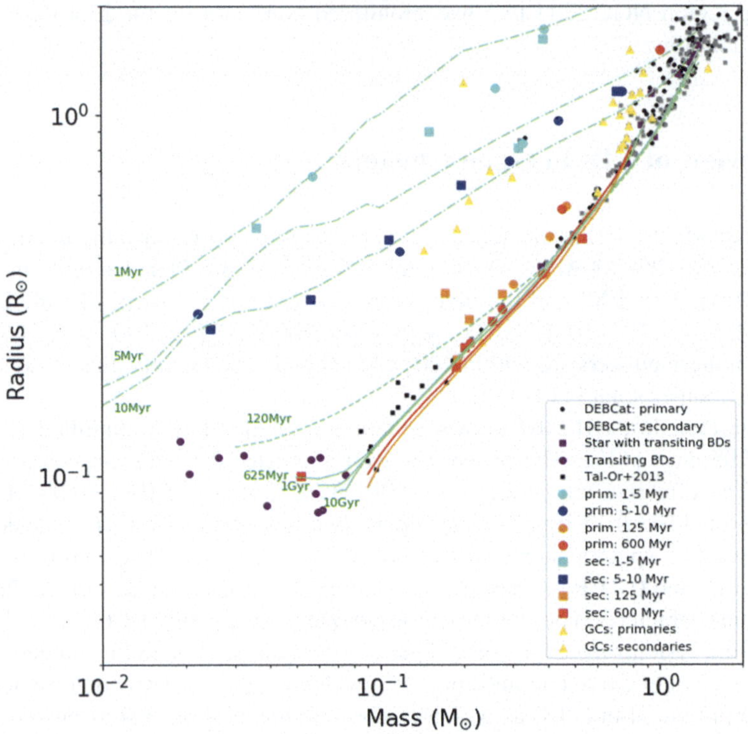


**Fig. 2** Mass-radius relation for low-mass and sub-stellar eclipsing binaries in star-forming regions, open clusters, and globular clusters. The primaries and secondaries are plotted as dots and squares, respectively. Colour scheme as follows: 1–5 Myr (cyan), 5–10 Myr (blue); 125 Myr (orange); 600 Myr (red); globular clusters (yellow). The primaries and secondaries of EBs from the DEBcat database whose accuracies are better than 2% on their masses and radii are displayed as black and grey symbols. Overplotted with green lines are the BT-Settl isochrones for ages of 1 Myr, 5 Myr, 10 Myr, 120 Myr, 625 Myr, 1 Gyr, and 10 Gyr. We added the 5 Gyr-old low-metallicity tracks at [M/H] = −2.0 and −1.0 dex as orange and red lines, respectively

the 0.5 $M_\odot$, limit set in this review, revolving every 1.55 days. For a complete census of eclipsing systems in the Beehive cluster, we should highlight the seven transiting exoplanets, including three orbiting members with masses equal or below 0.5 $M_\odot$ [130]. These four system are unambiguously confirmed as astrometric members of the Praesepe cluster from the 3D kinematic selection using the second release of *Gaia* [100].

In the Hyades, no low-mass EB was disclosed in the *K2* light curves. However, one transiting system member of the Hyades [131, 132] was announced independently by David et al. [118] and Mann et al. [133]. The primary, vA 50, has a Neptune-size planet with an upper limit on its mass of 1.1 Jupiter mass based on high-resolution spectroscopic radial velocity with an accuracy of around 0.3 km $s^{-1}$. This planet is orbiting a M dwarf member of the Hyades with a mass of 0.26 $M_\odot$ and a radius of 0.32 $R_\odot$ every ~3.5 days. This object is confirmed as a *Gaia* astrometric member located at 4.63 pc from the center of the Hyades cluster in 3D space [134].

## 4 Review of EBs in Globular Clusters

The globular clusters of our Milky Way are the oldest objects we know of. They are very massive containing up to a million of individual stars. There are two distinct populations of globular clusters, namely the classical population in the Galactic Halo [135] and the one in the Galactic Bulge [136]. The latter is younger (ages about 10 Gyr) and more metal-rich ($-0.7<$ [Fe/H] $< +0.5$ dex) than the one in the Galactic Bulge [137]. The H-R diagram for a typical globular cluster looks very different than that of an open cluster. There are no main-sequence stars of spectral types earlier than F, but there are many red giants and other objects of the late evolutionary phase (for example the horizontal branch) of low-mass stars.

Since the first discovery of different internal populations (of the main-sequence as well as the giant branches) of stars in globular clusters [138] using the *Hubble Space Telescope*, this characteristic was found for almost all known aggregates. However, there are no unique pattern or correlation with other astrophysical parameters known so far. The only possible explanation of the observations is a different (enriched) helium abundance of these internal populations [139]. However, the evolutionary mechanism behind this enrichment of helium is still unknown.

From an observational point of view, globular clusters are difficult to observe because they are very dense (up to 100 stars per arcsec$^2$), typically far away (several kpc) from the Sun, and the low-mass members have apparent magnitudes fainter than 20th magnitude. In order to resolve most of the cluster areas, large ground-based telescope and good seeing conditions or satellite measurements are needed. Especially time-series of radial velocity measurements are almost not available.

In a series of papers, the Clusters Ages Experiment (CASE; [140, 141]) investigated photometrically and spectroscopically several eclipsing binary systems in different globular clusters. Most of these systems are so-called blue stragglers which are more luminous and bluer than stars at the main-sequence turnoff point for their host cluster [142]. Therefore, these objects are the brightest main-sequence stars in the cluster and easier to observe. However, these eclipsing binary systems are peculiar in the sense that normally a significant interaction between the components took place. For example, there is a scenario in which the primary component is reborn from a former white dwarf that accreted a new envelope through mass transfer from its companion. The secondary star has lost most of its envelope while starting its ascent onto the sub-giant branch. It failed to ignite helium in its core and is currently powered by an hydrogen-burning shell [140]. The time scales of the different stages of all these processes are not known. Analysing the individual components in the mass versus radius diagram (Fig. 1) might help putting further constraints on the models.

The estimation of the binary fraction of globular cluster members is severely influenced by the above described observational constraints. Sollima et al. [143] investigated the fraction of binary systems in a sample of 13 low-density Galactic globular clusters using Hubble Space Telescope observations. They analysed the colour distribution of main-sequence stars to derive the minimum fraction of binary systems required to reproduce the observed colour-magnitude diagram morphologies. They found that all the analysed globular clusters contain a minimum binary fraction larger than 6% within the core radius. However, the estimated global fractions of binary systems range from 10 to 50% depending on the cluster. More recently, [144] determined the binary fractions for 35 globular clusters using different models including a star superposition effect. They derived a binary fraction of 6.8–10.8% depending on the assumed shape to the binary mass-ratio distribution, with the best fit occurring for a binary distribution that favours low mass ratios (and higher binary fractions). Later on, [145] presented a long-time observational campaign using FLAMES spectra of 968 red giant branch stars located around the half-light radii in a sample of ten Galactic globular clusters. From these only 21 radial velocity variables were identified as bona-fide binary stars, yielding a binary fraction of $2.2 \pm 0.5\%$. Finally, [146] found a binary fraction between 3 and 38% depending on the regions of eight globular clusters. This short overview shows already the wide range of derived values and the need for a new homogeneous analysis of all available photometric and spectroscopic data.

The newest version (November 2017) of the variable stars in Galactic globular clusters catalogue [147] was used to estimate the percentage of eclipsing binary systems in respect to all known variables. Here, we want to recall that in these old aggregates, we find mainly pulsating variables, such as Cepheids, Giants, SX Phoenicis, RR Lyrae, and RV Tauri stars. The pulsational characteristics (i.e. periods as well as amplitudes) and astrophysical driving mechanism are widely different [148]. Nevertheless, the amplitudes of these stars are comparable to those of eclipsing binaries which should not introduce a significant bias in the detection rate. The mentioned catalogue includes 5604 stars in 151 globular clusters. In total, 399 eclipsing binaries of all types (excluding field stars) are listed. The distribution of the apparent magnitudes ranges from 12 to 24 with a peak at 17.5 mag, respectively. To put this number in a broader context, we need an estimate of the total number of investigated stars per cluster and thus the overall variability ratio. This number crucially depends on the telescope used, time series characteristics, and the methods applied to analyse time series. To get a rough estimate of this number, we use five recent publications. In the following, we list the total number of observed stars, the included variable stars (known, new, and suspected), and the eclipsing binaries: 7630/40/1 [149]; 4274/59/0 [150]; 132457/359/30 [151], 31762/47/1 [152]; and 11358/13/1 [153]. The number of detected variables in globular clusters is only a few percent from which only a maximum of 10% are eclipsing binaries. Therefore, also in the future the number of known eclipsing binary systems will not significantly increase. To identify possible eclipsing binary systems with low-mass companions, available light curves have to be analysed and the best candidates for spectroscopic follow-up observations selected.

# 5 Discussion

## 5.1 Frequency of EBs

Most of the low-mass EBs identified so far in star-forming regions and open clusters come from the *CoRoT* and *Kepler* space missions, except for those members of Orion. Here in this section we provide a tentative estimate of the frequency of low-mass EBs in the regions investigated so far to spot any potential trend with age or environment.

Several studies looked at the fraction of spectroscopic binaries in the M dwarf regime. The first study f detected variables in globular clusters is only a few percent from which only a maximum of 10% are eclipsing binaries. Therefore, also in the future the number of known eclipsing binary systems will not significantly increase. To identify possible eclipsing binary systems with low-mass companions, available light curves have to be analysed and the best candidates for spectroscopic follow-up observations selected.

of M dwarf multiples revealed about 1.8±1.8% of spectroscopic binaries (0.04−0.4 au) for a small sample of a few tens of low-mass stars [154]. The CARMENES team identified nine double-line spectroscopic binaries with periods in the 1.13−8000 days interval among their sample of 342 M dwarfs, yielding a multiplicity of 2.6% [155]. The search for spectroscopic binaries in the Sloan database returned 3−4% of multiple systems with separations less than 0.4 au with a possible towards the hottest M dwarfs [156]. At later spectral types, the frequency of spectroscopic binaries among late-M dwarf (M5−M8) binaries is around 11% for separations in the 0−6 au range [157], while an independent survey of 58 M8−L6 dwarfs yielded a 0.9−11.1% multiplicity at separations closer than 1 au [158]. Lastly, we should mention the statistical occurrence of M dwarf systems in the *Kepler* field of view of 7−13% based on the fractional incidence of low-mass eclipsing binaries [159].

While the *Kepler* mission monitored a single field towards the Cygnus constellation, the *K2* mission targeted star-forming regions and open clusters in the ecliptic for periods of approximately consecutive 80 days, corresponding to semi-major axis less than a = 0.36 au. However, if we assume that a minimum of two transits are necessary to identify any eclipsing binaries with high confidence, searches in *K2* would be sensitive to periods less than about 50 days, i.e. a ≤ 0.25 au (Fig. 3).

The census of low-mass EBs in Upper Scorpius, the Pleiades, Praesepe, NGC 2264, and Ruprecht 147 is 6, 2, 5, 1, and 2, respectively. These numbers represent lower limits for several reasons intrinsic to the search for eclipsing systems: incompleteness of the samples, inhomogeneous quality of the light-curves depending on the brightness of the targets, lack of sensitivity to large mass ratios, etc. The rotation properties of M dwarf members of Upper Scorpius, the Pleiades, and Praesepe have been investigated in great details [160–163] thanks to *K2*. Using a crude selection of potential M dwarfs with effective temperature below 3800 K and $V - K_s$ colours redder than 3.8 mag, we identified 867, 566, 619 low-mass

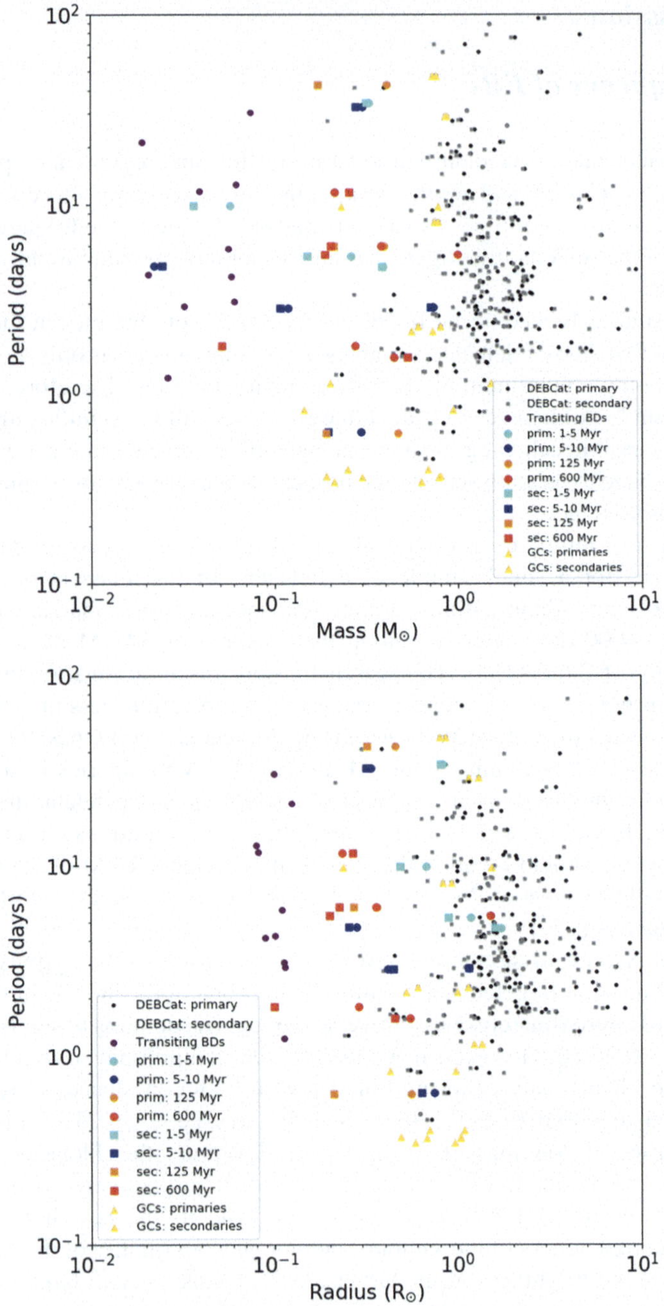


**Fig. 3** The mass-period and radius-period diagrams of low-mass EBs: the primaries and secondaries are plotted as dots and squares, respectively. Colour scheme as follows: 1–5 Myr (cyan), 5–10 Myr (blue); 125 Myr (orange); 600 Myr (red), globular clusters (yellow), field EBs from the DEBcat database (black+grey)

members in Upper Scorpius, the Pleiades, Praesepe, respectively. We derived a frequency of EBs with semi-major axis less than ~0.25 au of 0.64%, 0.35%, and 0.8% in these three regions. We estimate uncertainties up to 50% because of the low number statistics of published EBs, the rough photometric selection of M dwarf members, and the level of contamination of ground-based surveys before the advent of *Gaia*. Overall, we can argue that the frequency of EBs in clusters is below 1% for separations less than 0.25 au with no significant variation with age or environment. We also show the mass-semi-major axis and mass-eccentricity diagrams for cluster EBs in Fig. 4.

Finally, we should mention that only one brown dwarf pair is known eclipsing [66, 164], making any estimate of the frequency of sub-stellar EBs in young regions unreliable statistically. However, these types of systems should be rare although we cannot discard observational biases due to their intrinsic faintness and the lack of long-term monitoring sensitive to the sub-stellar population in star-forming regions and open clusters.

## 5.2 *The Impact of Age on Mass and Radius*

Figure 2 clearly shows that age has a strong impact on the radius of M dwarfs younger than ~500 Myr: the younger the M dwarf, the larger is its radius, as predicted by evolutionary models [165–167]. At a given mass, the radius of a Pleiades M dwarf at 120 Myr is about 10% larger than the radius of a Praesepe or a Hyades member (600–700 Myr), which is comparable to the ones of older field stars (ages > 1 Gyr). We do not see any difference at ages older than 500 Myr for low-mass M dwarfs in the H-R diagram. Models do predict differences in the sub-stellar regime but only one brown dwarf with an age larger than 500 Myr has been reported in Praesepe. The difference is small going from 600 to 120 Myr but significant moving towards much younger ages: the radius of a 0.25 $M_{\odot}$ is approximately 3 and 5 times larger at 5–10 and ~3 Myr, respectively. We observe a clear difference in radii of M dwarf members of Upper Scorpius (5–10 Myr; blue symbols) with those in Orion (<3 Myr; cyan) compared to those of the Pleiades (125 Myr; orange symbols). We note that one EB system identified in NGC2264 [168] confirms that the age of the cluster lies between the age of Orion and Upper Scorpius based on its location in the H-R diagram displayed in Fig. 2. We also remark that the system found in NGC 1647 [119] whose age is constrained to 150±10 Myr lies slightly above the evolutionary model at 120 Myr (Pleiades-like age), suggesting that a revision of the age of NGC 1647 (and possibly its distance checking the parallaxes of *Gaia* DR2) might be needed.

We also investigated the dispersion of the 11 eclipsing brown dwarfs (purple dots in Fig. 2) revealed by several missions and ground-based surveys. This dispersion might be the consequence of tides from the primary star yielding engulfment of the companions, magnetic activity, presence of cold spots on the surface of the brown dwarf, irradiation from the host star, and/or metallicity. The puzzle remains,

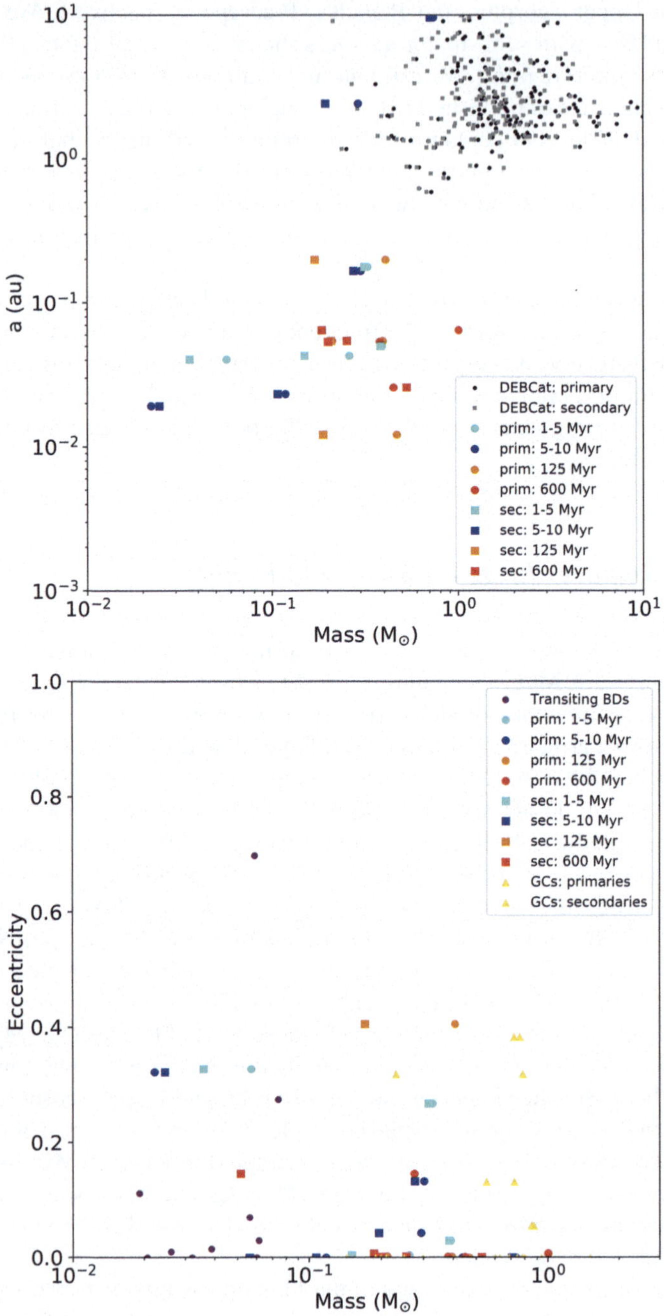


**Fig. 4** The mass-semi-major axis and mass-eccentricity diagrams of low-mass EBs in star-forming regions and clusters. The primaries and secondaries are plotted as dots and squares, respectively. Colour scheme as follows: 1–5 Myr (cyan), 5–10 Myr (blue); 125 Myr (orange); 600 Myr (red), globular clusters (yellow), field EBs from the DEBcat database (black+grey)

however, under debate. Because of the improved knowledge on the impact of the age on the radius of M dwarfs and brown dwarfs gathered over the past years mainly thanks to *Kepler K2*, we collected information on the ages of the primary stars from the discovery papers to compare the values with the inferred from the latest BT-Settl isochrones [167]. We considered the effect of age in this review keeping in mind current uncertainties on the mass determinations of the sub-stellar companions arising from the uncertainties on the mass of the primary and the grazing transits of some of the examples.

First of all, we note that two of these brown dwarfs orbit a M dwarf primary. Based on the above discussion, we can argue that the ages of these M dwarfs are older than 100 Myr although any older age is possible when taking into account the uncertainties on their masses and radii. Comparing the positions of the eclipsing brown dwarfs to the latest BT-Settl isochrones, we divided the sample into several groups. Two systems (KOI-415 and LHS 6343) appear very old, older than the others due to their small radii, consistent with the analysis of the discovery papers. We note again the low metallicity of KOI-415 ([Fe/H] = −0.24 dex) suggesting an old age. Another three systems (WASP-30, KOI-189, KOI-205) appear old, with ages around 1 Gyr or older [169–171]. The positions of CoRoT-15b and CoRoT-33b in the H-R diagram fit well the 300 Myr-old isochrone. However, we caution this point because the masses of the secondaries are ill-defined due to the intrinsic faintness of CoRoT-15 ($V \sim 16$ mag) and the grazing eclipse of CoRoT-33b [172, 173]. Nonetheless, those two systems appear as intermediate age-wise between the aforementioned systems and the (possibly) youngest ones described below. The last group of brown dwarfs exhibit inflated radii with respect to their siblings, the most extreme one being Kepler-39b [174]. Its age remains controversial depending on the method used for its determination: fit to the spectrum of the solar-type primary suggests 1.0−2.9 Gyr while the gyrochronology age infers 0.4−1.6 Gyr [175]. The brown dwarf is best fit by isochrones with ages between 50 and 120 Myr. The masses and radii of the brown dwarfs in the other systems (NLTT 41135, KELT-1b, and CoRoT-3b) are best fit by isochrones with ages bracketed by the Pleiades and Hyades isochrones [174, 176, 177]. We emphasise that three of the solar-type primaries appear over-luminous compared to the others and the BT-Settl isochrones. Overall, in spite of the current uncertainties on the masses and radii of the sub-stellar secondaries, we cannot discard age to have a significant effect on their radii due to the dependence of physical properties of brown dwarfs with gravity [178–182]. The revision of the distances of the host stars with the *Gaia* parallaxes should revise some of these discrepancies and decrease current error bars.

## *5.3 The Impact of Metallicity on Mass and Radius*

It is widely established that the fraction of stars hosting planets is larger with higher metallicity. The average metallicity of a volume-limited sample of stars with planets that have been specifically searched for planets peaks at [Fe/H] ∼ +0.1 [183, 184].

The frequency of metal-poor stars with planets is of the order of 5% or less, while more than 30% of metal-rich stars ($\geq +0.25$ dex) host planets. Moreover, there might be a trend towards low-mass planets with short periods orbiting low metallicity stars [185, 186].

We should compare the different frequencies and see whether metallicity has an impact [187].

The spectra of M dwarfs start to be affected by the dearth of metals at optical and near-infrared wavelengths for metallicities below [Fe/H] $\sim -0.5$ dex [188–191], trend extending towards at temperatures below 2500 K [192, 193]. The impact of metallicity on the sample of low-mass EBs in star-forming regions and open clusters is hard to disentangle from the effect of age because all regions have a metallicity equal or close to solar within 0.2 dex.

We also looked at the possible impact of metallicity on the dispersion of eclipsing brown dwarfs in the H-R diagram (Fig. 2). Among this sample, only two stars stand out due to their metallicity: CoRoT-33 a G9V with [Fe/H] $= +0.44 \pm 0.10$ dex [173] and KOI-415 a G0IV metal-poor solar-type analogue with [Fe/H] $= -0.24 \pm 0.10$ dex [194]. The latter is not so different from the bulk of stars with eclipsing sub-stellar companions because the difference in metallicity is less 0.2 dex but we note that its orbit is eccentric and its radius among the two lowest. CoRoT-33 is classified as a old star with an age greater than 4.6 Gyr based on a serie of indicators [173]. The radius of the brown dwarf is 40% larger than KOI-415 for an almost identical mass (59 $M_{Jup}$ vs. 62 $M_{Jup}$). Based on this comparison, we conclude that metallicity may indeed play a role in the properties of sub-stellar objects, in line with the spectral differences seen in L and T subdwarfs [192, 193].

## 5.4 The Role of Stellar Activity and Activity Cycles

The 11 year-long activity cycle on the Sun is known for a long time. The first long-term brightness changes which were interpreted as starspot cycles for M-type stars were reported by Phillips and Hartmann [195]. Chromospheric activity of F- to M-type stars can be studied using long-term Ca H&K data, for example from the Mount Wilson survey [196]. Those observations show cyclic variations yielding relations between the rotational period, the length of the activity cycle, and other stellar properties. Most important, faster rotating stars have shorter activity cycles [197], which can be explained by the classical dynamo theory. The square of the ratio of the cycle length and the rotational period can be used as a quantity to parametrise activity cycles.

In many active stars the starspots are so large that they cause brightness variations which can be few tens of percent from the mean light level [198], thus making them easily observable. The observed cycle lengths seem to converge with stellar age from a maximum dispersion around the Pleiades' age towards the solar cycle value at the Sun's age [199], and that the overall short- and long-term photometric variability increases with inverse Rossby number. The cycles of active stars are often not as regular and cyclic as their more quiet counterparts. Many active stars exhibit multiple cycle lengths simultaneously and cycle lengths in active stars are also often variable [200]. Most intriguing is the so-called flip-flop phenomenon where the activity concentrates on two permanent active longitudes, and flips between the two every few years [201].

The phenomenon are also important for the interpretation of light curves for eclipsing binary systems. The chromospherically active components of 180 low-mass pre-main sequence stars and chromospherically active binary systems have been looked at by Parihar et al. [202] and Eker et al. [203], respectively. The light curves of such systems are complicated to interpret if one or even two components show spots on different time scales and with different intensities. The effects of starspots on the light curves of eclipsing binaries, and, in particular, how they may affect the accurate measurement of eclipse timings have been investigated by Watson and Dhillon [204]. For systems containing a low-mass main-sequence star and a white dwarf, the times of primary eclipse ingress and egress can be altered by several seconds (larger effect for lower inclinations) for typical binary parameters and star-spot depressions. These effects cause a jitter in the residuals of O−C diagrams, which can also result in the false detection of spurious orbital period changes.

A nice example of how to model a light curve taking account of all the above mentioned effects is presented by Czesla et al. [205], who investigated the short-period (2.17 d) eclipsing binary CoRoT 105895502. They found a starspot with a period of about 40 days which remains quasi-stationary in the binary frame, and one starspot showing prograde motion at a rate of 2.3 degree per day, whose lifetime exceeds the duration of the observation (145 days). Only with eclipsing binary systems it is possible to study the complex correlations between chromospheric activity, spot cycles, and the astrophysical parameters in more details.

## *5.5 Flares in M Dwarfs*

Flares are mostly rapid transients lasting of the order of minutes or dozens of minutes, which are observable in different regions of the electromagnetic spectrum, from the ultra-violet to the X-ray domains. They are often observed on M dwarfs because they ideally fulfill the necessary conditions. Firstly, they are low mass stars with magnetic fields that remain active for a substantial part of their lives, and, secondly, the large difference between hot flaring regions and the cool photosphere give a higher chance of magnetic field generation. The energy of flares on M dwarfs can reach $10^{28}$–$10^{29}$ W. For example, on 2014 April 23, the *SWIFT* satellite detected a super-flare from the nearby young M-type binary DG CVn with the radiated energy about $4–9\times10^{28}$ W of energy in the 0.3–10 keV X-ray bandpass. This is about 10,000 times stronger than the most powerful solar flare on record [206]. The current available large surveys like ASAS-SN, Next Generation Transient Survey (NGTS), *Kepler/K2*, and *TESS* provide excellent photometric data to study the occurrence and

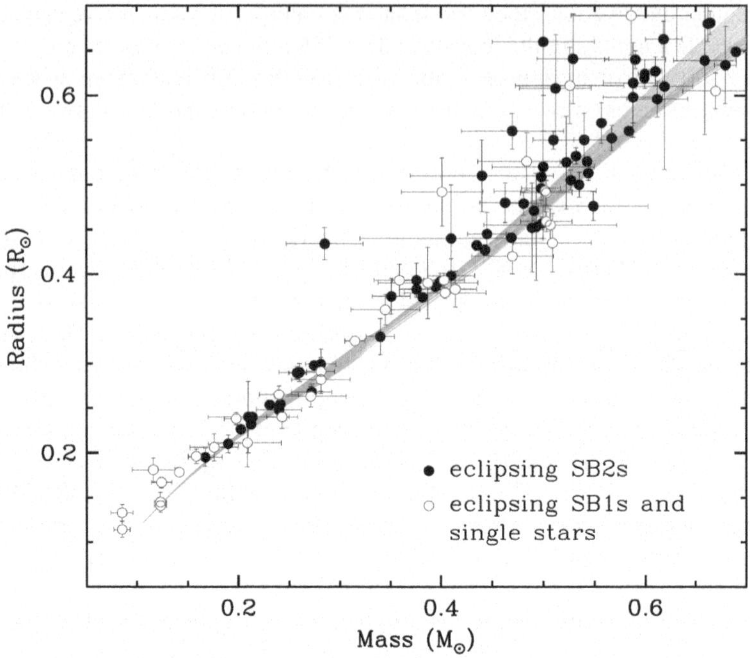


**Fig. 5** Mass-radius diagram for low-mass stars, including all measurements for double-lined eclipsing binaries (SB2; filled symbols) as well as determinations for single-lined eclipsing systems (SB1) and single stars (open symbols). Solar-metallicity Dartmouth isochrones are shown for comparison, for ages ranging from 1 to 13 Gyr (grey band) [219]

frequency of flares on M dwarfs (e.g. [207–213]). Furthermore, selected M dwarfs with observed flares were also monitored spectroscopically [214] to determine their ages and study the influence of the flare on the presence of possible exoplanets orbiting M dwarfs.

It was proposed that flares together with dark magnetic spots are responsible for the difference between the observed radii of M dwarfs and predicted theoretical values from evolutionary models. This discrepancy could be 5–10% or even more, depending on the model. Up to the beginning of the twenty-first century, only a few low-mass binary systems with M dwarf companions had measured radii with sufficient accuracy for modelling: CM Dra [215], YY Gem [216], CU Cnc [217], and GU Boo [218]. The situation in 2013 is reviewed in [219], see citations therein and Fig. 5). New solutions using more accurate photometric and spectroscopic observations as well as improved models decreased the discrepancy up to 2–3% for CM Dra [220]. The remaining difference could be caused by uncertain He abundance, for example. The calculation made for CM Dra showed that increasing the He abundance by 7% solves the remaining discrepancy in radii [220].

## 6 Future Prospects for EBs in Clusters

The numbers of photometric light curves and transiting systems available in star-forming regions and open clusters has been overwhelming thanks to the *CoRoT* mission and the (unexpected) advent of the *K2* mission focusing on the ecliptic. The future looks very bright too, with the 2-year Transiting Exoplanet Survey Satellite (*TESS*) mission [221][14] currently in space that will fully cover both the southern and northern hemispheres with a cadence of 30 min. The PLAnetary Transits and Oscillations of stars (*PLATO*) mission [222][15] is planned for a launch in 2026 aiming at targeting one million stars with two major objectives: the discovery of transiting Earth-like planets in the habitable zone of their host star and the study of stellar oscillations. However, *TESS* and *PLATO* might not contribute too much to the study of low-mass M dwarf members of star-forming regions and open clusters because they tend to avoid the ecliptic due to confusion issues and will focus on stars brighter than *K2*. We encourage the *PLATO* community to design a specific program focusing on a few clusters bracketing a large age range for several days to further constrain planetary and stellar evolutionary models.

The Young Exoplanet Transit Initiative (YETI) is an independent large ground-based involving multi-site telescopes and instruments designed to focus on nearby young regions to look for planetary transits [25]. Only one transiting planetary candidate (CVSO 30) has been identified but not yet unambiguously confirmed so far [223]. As a spin-off result, a few EBs have been discovered in several clusters, including the NGC 7243 ($\sim$250 Myr; $\sim$700 pc) and Trumpler 37 (4 Myr; 840 pc) clusters [224, 225]. The main difficulties of ground-based photometric surveys lies in the length of the nights (8 h vs. 24 h for space missions) with weather dependent conditions, the limited numbers of dedicated nights (a few per week/month vs. 27/80 days for *TESS/K2*), and the low precision of single measurements (at best a few mmag vs. less than 1 mmag from space for the same brightness). As a consequence, transiting exoplanets are tough to identify in active young stars but low-mass EBs should be easier to spot even for masses below 0.6 $M_{\odot}$.

In Fig. 2, we can clearly see a gap in age between 10 and 120 Myr. We highlight some dedicated programs to fill up that gap to investigate the evolution of masses and radii with age and constrain state-of-the-art isochrones.

- First, we should focus on the nearest (<200 pc) open clusters in this age range. The options are limited, resulting only a few regions: IC 2391 [96], IC 2602 [226], IC 4665 [227], $\alpha$ Persei [228] and NGC 2451 [229] for which *Gaia* will provide soon revised membership lists with accurate kinematics. One of the main issue though is the extension of these clusters in the sky, typically larger than most optical and infrared detectors. To reach the low-mass M dwarfs in those regions, the infrared camera VIRCAM on the VISTA telescope [230, 231]

[14] https://heasarc.gsfc.nasa.gov/docs/tess/.

[15] http://sci.esa.int/plato/.

might be the best option is the photometric accuracy can be guaranteed over several nights or weeks to look for dipping events of a few tens of magnitudes in the lowest mass members. Another alternative consists in dedicated a month of observing time with the most sensitive cameras of planet-hunter surveys with a preference to the ones most sensitive to far-red ($\geq$ 750 nm) and infrared wavelengths (1−2 μm) like the Next-Generation Transit Survey (NGTS; [232]), the TRAnsiting Planets and PlanetesImals Small Telescope (TRAPPIST; [233]), or the Search for habitable Planets EClipsing ULtra-cOOl Stars (SPECULOOS; [234]) rather than in the visible such as Trans-atlantic Exoplanet Survey (Tr-ES; [235]), Hungarian Automated Telescope Network (HATNet; [236]), the Wide Angle Search for Planets (WASP) North and South [237], the XO telescope [238], the Kilodegree Extremely Little Telescope (KELT; [239]).

- Secondly, members of young nearby moving groups younger than the Pleiades might represent ideal targets to look for exoplanets and EBs because they share the same age and metallicity[84, 179, 240–242]. Several moving groups and associations have been identified in the Solar neighbourhood (Ursa Major, TW Hya, $\beta$ Pic, AB Doradus, $\eta$ Chamaeleon, $\epsilon$ Chamaeleon, Tucana-Horologium) and hundreds of their members confirmed spectroscopically [241]. The main advantage is the closest distances of these members whose kinematics will be refined soon thanks to the releases of the *Gaia* astrometric datasets. One of the main drawback, as for all young stars, is the unknown levels of activity that might mimic the presence of planets. However, the large amplitude and RV modulation of low-mass EBs should be less affected to the intrinsic stellar activity.
- A third option is to search for low-mass companions of B- and A-type stars using X-ray data [243]. Known EBs can be located in a colour-magnitude diagram using *Gaia* data. Systems close to the zero-age-main-sequence can be easily identified because the contribution of a possible low-mass companion to the combined colour and absolute magnitude is negligible. The next steps consists in identifying those systems in the public X-ray catalogues such as *Chandra* [244], *ROSAT* [245], and *XMM-Newton* [246]. Normally, more massive stars have only weak X-ray fluxes compared to their low-mass counterparts [247]. From their spectrum and the amount of flux in X-rays, a first estimation of the astrophysical parameters of the companions can be done [248] to select systems for further spectroscopic studies.

**Acknowledgements** NL supported by the Spanish Ministry of Economy and Competitiveness (MINECO) under the grant AYA2015-69350-C3-2-P. This research has made use of the Simbad and Vizier databases, operated at the centre de Données Astronomiques de Strasbourg (CDS), of NASA's Astrophysics Data System Bibliographic Services (ADS), and the WEBDA database, operated at the Department of Theoretical Physics and Astrophysics of the Masaryk University.

## References


1. T. Henry, O. Franz, L. Wasserman, G.F. Benedict, P. Shelus, P. Ianna, J.D. Kirkpatrick, D. McCarthy, Bull. Am. Astron. Soc. **29**, 1278 (1997)
2. J.G. Winters, T.J. Henry, J.C. Lurie, N.C. Hambly, W.C. Jao, J.L. Bartlett, M.R. Boyd, S.B. Dieterich, C.T. Finch, A.D. Hosey, P.A. Ianna, A.R. Riedel, K.J. Slatten, J.P. Subasavage, Astron. J. **149**, 5 (2015). https://doi.org/10.1088/0004-6256/149/1/5
3. S.S. Kumar, E.K.L. Upton, Astron. J. **68**, 76 (1963)
4. J.C. Tarter, Bull. Am. Astron. Soc. **8**, 517 (1976)
5. G. Chabrier, I. Baraffe, Annu. Rev. Astron. Astrophys. **38**, 337 (2000)
6. J. Southworth, P.F.L. Maxted, B. Smalley, Astron. Astrophys. **429**, 645 (2005). https://doi.org/10.1051/0004-6361:20041867
7. G. Torres, J. Andersen, A. Giménez, Annu. Rev. Astron. Astrophys. **18**, 67 (2010). https://doi.org/10.1007/s00159-009-0025-1
8. D.M. Popper, Annu. Rev. Astron. Astrophys. **18**, 115 (1980). https://doi.org/10.1146/annurev.aa.18.090180.000555
9. J. Andersen, Annu. Rev. Astron. Astrophys. **3**, 91 (1991). https://doi.org/10.1007/BF00873538
10. J. Southworth (2014). e-prints. arXiv:1411.1219
11. M.A. Svechnikov, E.L. Perevozkina, VizieR Online Data Cat. **5121**, 5 (2004)
12. I. Soszyński, A. Udalski, M.K. Szymański, Ł. Wyrzykowski, K. Ulaczyk, R. Poleski, P. Pietrukowicz, S. Kozłowski, D.M. Skowron, J. Skowron, P. Mróz, M. Pawlak, K. Rybicki, A. Jacyszyn-Dobrzeniecka, Acta Astron. **67**, 297 (2017). https://doi.org/10.32023/0001-5237/67.4.1
13. K.G. Hełminiak, M. Konacki, M. RóŻyczka, J. KałuŻny, M. Ratajczak, J. Borkowski, P. Sybilski, M.W. Muterspaugh, D.E. Reichart, K.M. Ivarsen, J.B. Haislip, J.A. Crain, A.C. Foster, M.C. Nysewander, A.P. LaCluyze, Mon. Not. R. Astron. Soc. **425**, 1245 (2012). https://doi.org/10.1111/j.1365-2966.2012.21510.x
14. L. Zhang, H. Lu, X.L. Han, L. Jiang, Z. Li, Y. Zhang, Y. Hou, Y. Wang, Z. Cao, New Astron. **61**, 36 (2018). https://doi.org/10.1016/j.newast.2017.11.007
15. C.H. Lee, C.C. Lin, Res. Astron. Astrophys. **17**, 15 (2017). https://doi.org/10.1088/1674-4527/17/2/15
16. A.J. Drake, S.G. Djorgovski, A. Mahabal, E. Beshore, S. Larson, M.J. Graham, R. Williams, E. Christensen, M. Catelan, A. Boattini, A. Gibbs, R. Hill, R. Kowalski, Astrophys. J. **696**, 870 (2009). https://doi.org/10.1088/0004-637X/696/1/870
17. A.A. Mahabal, S.G. Djorgovski, A.J. Drake, C. Donalek, M.J. Graham, R.D. Williams, Y. Chen, B. Moghaddam, M. Turmon, E. Beshore, S. Larson, Bull. Astron. Soc. India **39**, 387 (2011)
18. S.G. Djorgovski, A.A. Mahabal, C. Donalek, M.J. Graham, A.J. Drake, B. Moghaddam, M. Turmon, Flashes in a star stream: automated classification of astronomical transient events. e-Prints (2012). arXiv:1209.1681. https://ui.adsabs.harvard.edu/abs/2012arXiv1209.1681D
19. J.L. Tonry, L. Denneau, A.N. Heinze, B. Stalder, K.W. Smith, S.J. Smartt, C.W. Stubbs, H.J. Weiland, A. Rest, Publ. Astron. Soc. Pac. **130**(6), 064505 (2018). https://doi.org/10.1088/1538-3873/aabadf
20. W.J. Borucki, D. Koch, G. Basri, N. Batalha, T. Brown, D. Caldwell, J. Caldwell, J. Christensen-Dalsgaard, et al., Science **327**, 977 (2010). https://doi.org/10.1126/science.1185402
21. S.B. Howell, C. Sobeck, M. Haas, M. Still, T. Barclay, F. Mullally, J. Troeltzsch, S. Aigrain, S.T. Bryson, D. Caldwell, W.J. Chaplin, W.D. Cochran, D. Huber, G.W. Marcy, A. Miglio, J.R. Najita, M. Smith, J.D. Twicken, J.J. Fortney, Publ. Astron. Soc. Pac. **126**, 398 (2014). https://doi.org/10.1086/676406
22. A.M. Geller, R.D. Mathieu, H.C. Harris, R.D. McClure, Astron. J. **137**, 3743 (2009). https://doi.org/10.1088/0004-6256/137/4/3743

23. K.E. Milliman, R.D. Mathieu, A.M. Geller, N.M. Gosnell, S. Meibom, I. Platais, Astron. J. **148**, 38 (2014). https://doi.org/10.1088/0004-6256/148/2/38
24. E.M. Leiner, R.D. Mathieu, N.M. Gosnell, A.M. Geller, Astron. J. **150**, 10 (2015). https://doi.org/10.1088/0004-6256/150/1/10
25. R. Neuhäuser, R. Errmann, A. Berndt, G. Maciejewski, H. Takahashi, W.P. Chen, D.P. Dimitrov, T. Pribulla, E.H. Nikogossian, E.L.N. Jensen, L. Marschall, Z.Y. Wu, A. Kellerer, F.M. Walter, C. Briceño, R. Chini, M. Fernandez, S. Raetz, G. Torres, D.W. Latham, S.N. Quinn, A. Niedzielski, Ł. Bukowiecki, G. Nowak, T. Tomov, K. Tachihara, S.C.L. Hu, L.W. Hung, D.P. Kjurkchieva, V.S. Radeva, B.M. Mihov, L. Slavcheva-Mihova, I.N. Bozhinova, J. Budaj, M. Vaňko, E. Kundra, Ľ. Hambálek, V. Krushevska, T. Movsessian, H. Harutyunyan, J.J. Downes, J. Hernandez, V.H. Hoffmeister, D.H. Cohen, I. Abel, R. Ahmad, S. Chapman, S. Eckert, J. Goodman, A. Guerard, H.M. Kim, A. Koontharana, J. Sokol, J. Trinh, Y. Wang, X. Zhou, R. Redmer, U. Kramm, N. Nettelmann, M. Mugrauer, J. Schmidt, M. Moualla, C. Ginski, C. Marka, C. Adam, M. Seeliger, S. Baar, T. Roell, T.O.B. Schmidt, L. Trepl, T. Eisenbeiß, S. Fiedler, N. Tetzlaff, E. Schmidt, M.M. Hohle, M. Kitze, N. Chakrova, C. Gräfe, K. Schreyer, V.V. Hambaryan, C.H. Broeg, J. Koppenhoefer, A.K. Pandey, Astron. Nachr. **332**, 547 (2011). https://doi.org/10.1002/asna.201111573
26. A. Rau, S.R. Kulkarni, N.M. Law, J.S. Bloom, D. Ciardi, G.S. Djorgovski, D.B. Fox, A. Gal-Yam, C.C. Grillmair, M.M. Kasliwal, P.E. Nugent, E.O. Ofek, R.M. Quimby, W.T. Reach, M. Shara, L. Bildsten, S.B. Cenko, A.J. Drake, A.V. Filippenko, D.J. Helfand, G. Helou, D.A. Howell, D. Poznanski, M. Sullivan, Publ. Astron. Soc. Pac. **121**, 1334 (2009). https://doi.org/10.1086/605911
27. N.M. Law, S.R. Kulkarni, R.G. Dekany, E.O. Ofek, R.M. Quimby, P.E. Nugent, J. Surace, C.C. Grillmair, J.S. Bloom, M.M. Kasliwal, L. Bildsten, T. Brown, S.B. Cenko, D. Ciardi, E. Croner, S.G. Djorgovski, J. van Eyken, A.V. Filippenko, D.B. Fox, A. Gal-Yam, D. Hale, N. Hamam, G. Helou, J. Henning, D.A. Howell, J. Jacobsen, R. Laher, S. Mattingly, D. McKenna, A. Pickles, D. Poznanski, G. Rahmer, A. Rau, W. Rosing, M. Shara, R. Smith, D. Starr, M. Sullivan, V. Velur, R. Walters, J. Zolkower, Publ. Astron. Soc. Pac. **121**, 1395 (2009). https://doi.org/10.1086/648598
28. H.N. Russell, J.E. Merrill, *The Determination of the Elements of Eclipsing Binaries* (1952). https://ui.adsabs.harvard.edu/abs/1952deeb.book.....R
29. M.S. Zverev, B.V. Kukarkin, D.Ya. Martynov, P.P. Parenago, N.F. Florya, V.P. Tsesevich, *Variable Stars*, vol. III (1947). https://ui.adsabs.harvard.edu/abs/1947pezv.book.....Z
30. P.B. Etzel, in *Photometric and Spectroscopic Binary Systems* (1981), p. 111
31. E. Budding, Astrophys. Space Sci. **22**(1), 87 (1973). https://doi.org/10.1007/BF00642825
32. E. Budding, M. Zeilik, *Some Rapidly Rotating Cool Stars and Their Activity*, vol. 254 (1986), p. 290. https://doi.org/10.1007/3-540-16763-3_200
33. E. Budding, M. Zeilik, Astrophys. J. **319**, 827 (1987). https://doi.org/10.1086/165500
34. D.B. Wood, Astron. J. **76**, 701 (1971). https://doi.org/10.1086/111187
35. D.B. Wood, Technical Report X-110-72-473, Publications of the Goddard Space Flight Center, Greenbelt (1972)
36. D.B. Wood, Publ. Astron. Soc. Pac. **85**(504), 253 (1973). https://doi.org/10.1086/129447
37. G. Hill, *Publications of the Dominion Astrophysical Observatory Victoria*, vol. 15 (Queen's Printer, Ottawa, 1979), p. 298
38. G. Hill, S. Rucinski, IAU Comm. Close Binary Stars **21**, 135 (1993)
39. S.M. Ruciński, Acta Astron. **23**, 79 (1973)
40. S.M. Rucinski, Acta Astron. **24**, 119 (1974)
41. R.E. Wilson, E.J. Devinney, Astrophys. J. **166**, 605 (1971). https://doi.org/10.1086/150986
42. R.E. Wilson, E.J. Devinney, Astrophys. J. **171**, 413 (1972). https://doi.org/10.1086/151293
43. R.E. Wilson, Astrophys. J. **672**(1), 575 (2008). https://doi.org/10.1086/523634
44. R.E. Wilson, W. Van Hamme, Astrophys. J. **780**(2), 151 (2014). https://doi.org/10.1088/0004-637X/780/2/151
45. A. Prša, T. Zwitter, Astrophys. J. **628**(1), 426 (2005). https://doi.org/10.1086/430591

46. A. Prša, K.E. Conroy, M. Horvat, H. Pablo, A. Kochoska, S. Bloemen, J. Giammarco, K.M. Hambleton, P. Degroote, Astrophys. J. Suppl. Ser. **227**(2), 29 (2016). https://doi.org/10.3847/1538-4365/227/2/29
47. M. Horvat, K.E. Conroy, H. Pablo, K.M. Hambleton, A. Kochoska, J. Giammarco, A. Prša, Astrophys. J. Suppl. Ser. **237**(2), 26 (2018). https://doi.org/10.3847/1538-4365/aacd0f
48. A. Prša, *Modeling and Analysis of Eclipsing Binary Stars: The Theory and Design Principles of PHOEBE* (IOP Publishing, Bristol, 2018). https://doi.org/10.1088/978-0-7503-1287-5
49. J. Devor, Astrophys. J. **628**(1), 411 (2005). https://doi.org/10.1086/431170
50. J. Devor, D. Charbonneau, Astrophys. J. **653**(1), 647 (2006). https://doi.org/10.1086/508609
51. O. Tamuz, T. Mazeh, P. North, Mon. Not. R. Astron. Soc. **367**(4), 1521 (2006a). https://doi.org/10.1111/j.1365-2966.2006.10049.x
52. T. Mazeh, O. Tamuz, P. North, Mon. Not. R. Astron. Soc. **367**(4), 1531 (2006b). https://doi.org/10.1111/j.1365-2966.2006.10050.x
53. P. Hadrava, Publ. Astron. Inst. Czechoslov. Acad. Sci. **92**, 1 (2004)
54. T. Pribulla, in *From Interacting Binaries to Exoplanets: Essential Modeling Tools, IAU Symposium*, vol. 282, ed. by M.T. Richards, I. Hubeny (2012), pp. 279–282. https://doi.org/10.1017/S1743921311027566
55. D.H. Bradstreet, Soc. Astron. Sci. Annu. Symp. **24**, 23 (2005)
56. N. Lodieu, R. Alonso, R. González Hernández, J. I. Sanchis-Ojeda, N. Narita, Y. Kawashima, K. Kawauchi, A. Suárez Mascareño, H. Deeg, et al., Astron. Astrophys. **584**, A128 (2015). https://doi.org/10.1051/0004-6361/201527464
57. L'. Hambálek, T. Pribulla, Contrib. Astron. Observatory Skalnate Pleso **43**(1), 27 (2013)
58. D. Terrell, R.E. Wilson, Astrophys. Space Sci. **296**(1–4), 221 (2005). https://doi.org/10.1007/s10509-005-4449-4
59. W. van Hamme, Astron. J. **106**, 2096 (1993). https://doi.org/10.1086/116788
60. A. Claret, S. Bloemen, Astron. Astrophys. **529**, A75 (2011). https://doi.org/10.1051/0004-6361/201116451
61. A. Claret, P.H. Hauschildt, S. Witte, Astron. Astrophys. **546**, A14 (2012). https://doi.org/10.1051/0004-6361/201219849
62. A. Claret, P.H. Hauschildt, S. Witte, Astron. Astrophys. **552**, A16 (2013). https://doi.org/10.1051/0004-6361/201220942
63. A. Claret, Astron. Astrophys. **600**, A30 (2017). https://doi.org/10.1051/0004-6361/201629705
64. A. Claret, Astron. Astrophys. **618**, A20 (2018). https://doi.org/10.1051/0004-6361/201833060
65. H.R. Neilson, J.B. Lester, Astron. Astrophys. **556**, A86 (2013). https://doi.org/10.1051/0004-6361/201321888
66. K.G. Stassun, R.D. Mathieu, J.A. Valenti, Nature **440**, 311 (2006). https://doi.org/10.1038/nature04570
67. K.G. Stassun, M. van den Berg, E. Feigelson, Astrophys. J. **660**, 704 (2007). https://doi.org/10.1086/513138
68. J. Irwin, S. Aigrain, S. Hodgkin, K.G. Stassun, L. Hebb, M. Irwin, E. Moraux, J. Bouvier, A. Alapini, R. Alexander, D.M. Bramich, J. Holtzman, E.L. Martín, M.J. McCaughrean, F. Pont, P.E. Verrier, M.R. Zapatero Osorio, Mon. Not. R. Astron. Soc. **380**, 541 (2007). https://doi.org/10.1111/j.1365-2966.2007.12117.x
69. M.J. McCaughrean, C.R. O'Dell, Astron. J. **111**, 1977 (1996)
70. L.A. Hillenbrand, Astron. J. **113**, 1733 (1997)
71. L.A. Hillenbrand, J.M. Carpenter, Astrophys. J. **540**, 236 (2000)
72. E.D. Feigelson, J.A. Gaffney, G. Garmire, L.A. Hillenbrand, L. Townsley, Astrophys. J. **584**, 911 (2003)
73. L.A. Hillenbrand, A.S. Hoffer, G.J. Herczeg, Astron. J. **146**, 85 (2013). https://doi.org/10.1088/0004-6256/146/4/85
74. P. Ingraham, L. Albert, R. Doyon, E. Artigau, Astrophys. J. **782**, 8 (2014). https://doi.org/10.1088/0004-637X/782/1/8

75. K.G. Stassun, R.D. Mathieu, P.A. Cargile, A.N. Aarnio, E. Stempels, A. Geller, Nature **453**, 1079 (2008). https://doi.org/10.1038/nature07069
76. P.A. Cargile, K.G. Stassun, R.D. Mathieu, Astrophys. J. **674**, 329 (2008). https://doi.org/10.1086/524346
77. Y. Gómez Maqueo Chew, K.G. Stassun, A. Prša, E. Stempels, L. Hebb, R. Barnes, R. Heller, R.D. Mathieu, Astrophys. J. **745**, 58 (2012). https://doi.org/10.1088/0004-637X/745/1/58
78. M. Morales-Calderón, J.R. Stauffer, K.G. Stassun, F.J. Vrba, L. Prato, L.A. Hillenbrand, S. Terebey, K.R. Covey, L.M. Rebull, D.M. Terndrup, R. Gutermuth, I. Song, P. Plavchan, J.M. Carpenter, F. Marchis, E.V. García, S. Margheim, K.L. Luhman, J. Angione, J.M. Irwin, Astrophys. J. **753**, 149 (2012). https://doi.org/10.1088/0004-637X/753/2/149
79. T. Preibisch, H. Zinnecker, Astron. J. **117**, 2381 (1999). https://doi.org/10.1086/300842
80. T. Preibisch, E. Guenther, H. Zinnecker, Astron. J. **121**, 1040 (2001). https://doi.org/10.1086/318774
81. C.L. Slesnick, L.A. Hillenbrand, J.M. Carpenter, Astrophys. J. **688**, 377 (2008). https://doi.org/10.1086/592265
82. M.J. Pecaut, E.E. Mamajek, E.J. Bubar, Astrophys. J. **746**, 154 (2012). https://doi.org/10.1088/0004-637X/746/2/154
83. I. Song, B. Zuckerman, M.S. Bessell, Astron. J. **144**, 8 (2012). https://doi.org/10.1088/0004-6256/144/1/8
84. M.J. Pecaut, in *IAU Symposium*, vol. 314, ed. by J.H. Kastner, B. Stelzer, S.A. Metchev (2016), pp. 85–90. https://doi.org/10.1017/S1743921315006079
85. A.C. Rizzuto, M.J. Ireland, T.J. Dupuy, A.L. Kraus, Astrophys. J. **817**, 164 (2016). https://doi.org/10.3847/0004-637X/817/2/164
86. T.J. David, L.A. Hillenbrand, E. Gillen, A.M. Cody, S.B. Howell, H.T. Isaacson, J.H. Livingston, Astrophys. J. **872**, 161 (2019). https://doi.org/10.3847/1538-4357/aafe09
87. C.L. Slesnick, J.M. Carpenter, L.A. Hillenbrand, Astron. J. **131**, 3016 (2006). https://doi.org/10.1086/503560
88. N. Lodieu, Mon. Not. R. Astron. Soc. **431**, 3222 (2013). https://doi.org/10.1093/MonthlyNoticesoftheRoyalAstronomicalSociety/stt402
89. P. Dawson, A. Scholz, T.P. Ray, K.A. Marsh, K. Wood, A. Natta, D. Padgett, M.E. Ressler, Mon. Not. R. Astron. Soc. **429**, 903 (2013). https://doi.org/10.1093/MonthlyNoticesoftheRoyalAstronomicalSociety/sts386
90. D. Ardila, E. Martín, G. Basri, Astron. J. **120**, 479 (2000)
91. A. Reiners, G. Basri, S. Mohanty, Astrophys. J. **634**, 1346 (2005). https://doi.org/10.1086/432878
92. A.L. Kraus, A.M. Cody, K.R. Covey, A.C. Rizzuto, A.W. Mann, M.J. Ireland, Astrophys. J. **807**, 3 (2015). https://doi.org/10.1088/0004-637X/807/1/3
93. T.J. David, L.A. Hillenbrand, A.M. Cody, J.M. Carpenter, A.W. Howard, Astrophys. J. **816**, 21 (2016). https://doi.org/10.3847/0004-637X/816/1/21
94. R. Alonso, H.J. Deeg, S. Hoyer, N. Lodieu, E. Palle, R. Sanchis-Ojeda, Astron. Astrophys. **584**, L8 (2015). https://doi.org/10.1051/0004-6361/201527109
95. J.R. Stauffer, G. Schultz, J.D. Kirkpatrick, Astrophys. J. Lett. **499**, 199 (1998)
96. D. Barrado y Navascués, J.R. Stauffer, R. Jayawardhana, Astrophys. J. **614**, 386 (2004)
97. P. Mazzei, L. Pigatto, Astron. Astrophys. **213**, L1 (1989)
98. S.E. Dahm, Astrophys. J. **813**, 108 (2015). https://doi.org/10.1088/0004-637X/813/2/108
99. S. Gossage, C. Conroy, A. Dotter, J. Choi, P. Rosenfield, P. Cargile, A. Dolphin, Astrophys. J. **863**, 67 (2018). https://doi.org/10.3847/1538-4357/aad0a0
100. N. Lodieu, A. Pérez-Garrido, R.L. Smart, R. Silvotti, A 5D view of the $\alpha$ Per, Pleiades, and Praesepe clusters. Astron. Astrophys. **628**, A66 (2019). https://doi.org/10.1051/0004-6361/201935533
101. A. Maeder, J.C. Mermilliod, Astron. Astrophys. **93**, 136 (1981)
102. J.C. Mermilliod, Astron. Astrophys. **97**, 235 (1981)
103. P. Mazzei, L. Pigatto, Astron. Astrophys. **193**, 148 (1988)

104. Y. Lebreton, J. Fernandes, T. Lejeune, Astron. Astrophys. **374**, 540 (2001). https://doi.org/10.1051/0004-6361:20010757
105. S. De Gennaro, T. von Hippel, W.H. Jefferys, N. Stein, D. van Dyk, E. Jeffery, Astrophys. J. **696**, 12 (2009). https://doi.org/10.1088/0004-637X/696/1/12
106. T.D. Brandt, C.X. Huang, Astrophys. J. **807**, 58 (2015). https://doi.org/10.1088/0004-637X/807/1/58
107. E.L. Martín, N. Lodieu, Y. Pavlenko, V.J.S. Béjar, Astrophys. J. **856**, 40 (2018). https://doi.org/10.3847/1538-4357/aaaeb8
108. N. Lodieu, R. Rebolo, A. Pérez-Garrido, Astron. Astrophys. **615**, L12 (2018). https://doi.org/10.1051/0004-6361/201832748
109. D.A. Vandenberg, T.J. Bridges, Astrophys. J. **278**, 679 (1984). https://doi.org/10.1086/161836
110. P. Delorme, A. Collier Cameron, L. Hebb, J. Rostron, T.A. Lister, A.J. Norton, D. Pollacco, R.G. West, Mon. Not. R. Astron. Soc. **413**, 2218 (2011). https://doi.org/10.1111/j.1365-2966.2011.18299.x
111. M. Salaris, A. Weiss, S.M. Percival, Astron. Astrophys. **414**, 163 (2004). https://doi.org/10.1051/0004-6361:20031578
112. C. Bonatto, E. Bica, L. Girardi, Astron. Astrophys. **415**, 571 (2004). https://doi.org/10.1051/0004-6361:20034638
113. Gaia Collaboration, C. Babusiaux, F. van Leeuwen, M.A. Barstow, C. Jordi, A. Vallenari, D. Bossini, A. Bressan, T. Cantat-Gaudin, M. van Leeuwen, et al., Astron. Astrophys. **616**, A10 (2018). https://doi.org/10.1051/0004-6361/201832843
114. A.M. Boesgaard, E.D. Friel, Astrophys. J. **351**, 467 (1990)
115. G. Cayrel de Strobel, F. Crifo, Y. Lebreton, in *Hipparcos - Venice '97, ESA Special Publication*, vol. 402, ed. by R.M. Bonnet, E. Høg, P.L. Bernacca, L. Emiliani, A. Blaauw, C. Turon, J. Kovalevsky, L. Lindegren, H. Hassan, M. Bouffard, B. Strim, D. Heger, M.A.C. Perryman, L. Woltjer (1997), pp. 687–688
116. M. Grenon, *IAU Joint Discussion*, vol. 13 (2000)
117. T.J. David, J. Stauffer, L.A. Hillenbrand, A.M. Cody, K. Conroy, K.G. Stassun, B. Pope, S. Aigrain, E. Gillen, A. Collier Cameron, D. Barrado, L.M. Rebull, H. Isaacson, G.W. Marcy, C. Zhang, R.L. Riddle, C. Ziegler, N.M. Law, C. Baranec, Astrophys. J. **814**, 62 (2015). https://doi.org/10.1088/0004-637X/814/1/62
118. T.J. David, K.E. Conroy, L.A. Hillenbrand, K.G. Stassun, J. Stauffer, L.M. Rebull, A.M. Cody, H. Isaacson, A.W. Howard, S. Aigrain, Astron. J. **151**, 112 (2016). https://doi.org/10.3847/0004-6256/151/5/112
119. L. Hebb, R.F.G. Wyse, G. Gilmore, J. Holtzman, Astron. J. **131**, 555 (2006). https://doi.org/10.1086/497971
120. W.S. Dias, B.S. Alessi, A. Moitinho, J.R.D. Lépine, Astron. Astrophys. **389**, 871 (2002). https://doi.org/10.1051/0004-6361:20020668
121. D.G. Turner, Astron. J. **104**, 1865 (1992). https://doi.org/10.1086/116363
122. A.L. Kraus, G.J. Herczeg, A.C. Rizzuto, A.W. Mann, C.L. Slesnick, J.M. Carpenter, L.A. Hillenbrand, E.E. Mamajek, Astrophys. J. **838**, 150 (2017). https://doi.org/10.3847/1538-4357/aa62a0
123. E. Gillen, L.A. Hillenbrand, T.J. David, S. Aigrain, L. Rebull, J. Stauffer, A.M. Cody, D. Queloz, Astrophys. J. **849**, 11 (2017). https://doi.org/10.3847/1538-4357/aa84b3
124. J.D. Adams, J.R. Stauffer, M.F. Skrutskie, D.G. Monet, S.F. Portegies Zwart, K.A. Janes, C.A. Beichman, Astron. J. **124**, 1570 (2002). https://doi.org/10.1086/342016
125. A.L. Kraus, L.A. Hillenbrand, Astron. J. **134**, 2340 (2007). https://doi.org/10.1086/522831
126. D.E.A. Baker, R.F. Jameson, S.L. Casewell, N. Deacon, N. Lodieu, N. Hambly, Mon. Not. R. Astron. Soc. **408**, 2457 (2010). https://doi.org/10.1111/j.1365-2966.2010.17302.x
127. S. Boudreault, N. Lodieu, N.R. Deacon, N.C. Hambly, Mon. Not. R. Astron. Soc. **426**, 3419 (2012). https://doi.org/10.1111/j.1365-2966.2012.21854.x
128. P. Khalaj, H. Baumgardt, Mon. Not. R. Astron. Soc. **434**, 3236 (2013). https://doi.org/10.1093/MonthlyNoticesoftheRoyalAstronomicalSociety/stt1239

129. P.F. Wang, W.P. Chen, C.C. Lin, A.K. Pandey, C.K. Huang, N. Panwar, C.H. Lee, M.F. Tsai, C.H. Tang, B. Goldman, W.S. Burgett, K.C. Chambers, P.W. Draper, H. Flewelling, T. Grav, J.N. Heasley, K.W. Hodapp, M.E. Huber, R. Jedicke, N. Kaiser, R.P. Kudritzki, G.A. Luppino, R.H. Lupton, E.A. Magnier, N. Metcalfe, D.G. Monet, J.S. Morgan, P.M. Onaka, P.A. Price, C.W. Stubbs, W. Sweeney, J.L. Tonry, R.J. Wainscoat, C. Waters, Astrophys. J. **784**, 57 (2014). https://doi.org/10.1088/0004-637X/784/1/57
130. A.W. Mann, E. Gaidos, A. Vanderburg, A.C. Rizzuto, M. Ansdell, J.V. Medina, G.N. Mace, A.L. Kraus, K.R. Sokal, Astron. J. **153**, 64 (2017). https://doi.org/10.1088/1361-6528/aa5276
131. W.F. van Altena, Astron. J. **71**, 482 (1966). https://doi.org/10.1086/109952
132. R.B. Hanson, Astron. J. **80**, 379 (1975). https://doi.org/10.1086/111753
133. A.W. Mann, E. Gaidos, G.N. Mace, M.C. Johnson, B.P. Bowler, D. LaCourse, T.L. Jacobs, A. Vanderburg, A.L. Kraus, K.F. Kaplan, D.T. Jaffe, Astrophys. J. **818**, 46 (2016). https://doi.org/10.3847/0004-637X/818/1/46
134. N. Lodieu, R.L. Smart, A. Pérez-Garrido, R. Silvotti, A 3D view of the Hyades stellar and sub-stellar population. Astron. Astrophys. **623**, A35 (2019). https://doi.org/10.1051/0004-6361/201834045
135. H. Baumgardt, M. Hilker, Mon. Not. R. Astron. Soc. **478**, 1520 (2018). https://doi.org/10.1093/mnras/sty1057
136. L.J. Rossi, S. Ortolani, B. Barbuy, E. Bica, A. Bonfanti, Mon. Not. R. Astron. Soc. **450**, 3270 (2015). https://doi.org/10.1093/mnras/stv748
137. E. Bica, C. Bonatto, B. Barbuy, S. Ortolani, Astron. Astrophys. **450**, 105 (2006). https://doi.org/10.1051/0004-6361:20054351
138. G. Piotto, L.R. Bedin, J. Anderson, I.R. King, S. Cassisi, A.P. Milone, S. Villanova, A. Pietrinferni, A. Renzini, Astrophys. J. Lett. **661**, L53 (2007). https://doi.org/10.1086/518503
139. S. Cassisi, M. Salaris, A. Pietrinferni, D. Hyder, Mon. Not. R. Astron. Soc. **464**, 2341 (2017). https://doi.org/10.1093/mnras/stw2579
140. J. Kaluzny, S.M. Rucinski, I.B. Thompson, W. Pych, W. Krzeminski, Astron. J. **133**, 2457 (2007). https://doi.org/10.1086/516637
141. J. Kaluzny, I.B. Thompson, A. Dotter, M. Rozyczka, A. Schwarzenberg-Czerny, G.S. Burley, B. Mazur, S.M. Rucinski, Astron. J. **150**, 155 (2015). https://doi.org/10.1088/0004-6256/150/5/155
142. A.R. Sandage, Astron. J. **58**, 61 (1953). https://doi.org/10.1086/106822
143. A. Sollima, G. Beccari, F.R. Ferraro, F. Fusi Pecci, A. Sarajedini, Mon. Not. R. Astron. Soc. **380**, 781 (2007). https://doi.org/10.1111/j.1365-2966.2007.12116.x
144. J. Ji, J.N. Bregman, Astrophys. J. **768**, 158 (2013). https://doi.org/10.1088/0004-637X/768/2/158
145. S. Lucatello, A. Sollima, R. Gratton, E. Vesperini, V. D'Orazi, E. Carretta, A. Bragaglia, Astron. Astrophys. **584**, A52 (2015). https://doi.org/10.1051/0004-6361/201526957
146. A.P. Milone, A.F. Marino, L.R. Bedin, A. Dotter, H. Jerjen, D. Kim, D. Nardiello, G. Piotto, J. Cong, Mon. Not. R. Astron. Soc. **455**, 3009 (2016). https://doi.org/10.1093/mnras/stv2415
147. C.M. Clement, A. Muzzin, Q. Dufton, T. Ponnampalam, J. Wang, J. Burford, A. Richardson, T. Rosebery, J. Rowe, H.S. Hogg, Astron. J. **122**, 2587 (2001). https://doi.org/10.1086/323719
148. J.R. Percy, *Understanding Variable Stars* (Cambridge University Press, Cambridge, 2011)
149. D. Deras, A. Arellano Ferro, C. Lázaro, I.H. Bustos Fierro, J.H. Calderón, S. Muneer, S. Giridhar, Mon. Not. R. Astron. Soc. **486**, 2791 (2019). https://doi.org/10.1093/mnras/stz642
150. M.A. Yepez, A. Arellano Ferro, S. Muneer, S. Giridhar, Rev. Mex. Astron. Astrofis. **54**, 15 (2018)
151. M. Rozyczka, I.B. Thompson, W. Pych, W. Narloch, R. Poleski, A. Schwarzenberg-Czerny, Acta Astronom. **67**, 203 (2017). https://doi.org/10.32023/0001-5237/67.3.1
152. Y. Tsapras, A. Arellano Ferro, D.M. Bramich, R.F. Jaimes, N. Kains, R. Street, M. Hundertmark, K. Horne, M. Dominik, C. Snodgrass, Mon. Not. R. Astron. Soc. **465**, 2489 (2017). https://doi.org/10.1093/mnras/stw2773

153. D.J. Lee, J.R. Koo, K. Hong, S.L. Kim, J.W. Lee, C.U. Lee, Y.B. Jeon, Y.H. Kim, B. Lim, Y.H. Ryu, S.M. Cha, Y. Lee, D.J. Kim, B.G. Park, C.H. Kim, J. Korean Astron. Soc. **49**, 295 (2016). https://doi.org/10.5303/JKAS.2016.49.6.295
154. D.A. Fischer, G.W. Marcy, Astrophys. J. **396**, 178 (1992)
155. D. Baroch, J.C. Morales, I. Ribas, L. Tal-Or, M. Zechmeister, A. Reiners, J.A. Caballero, A. Quirrenbach, P.J. Amado, S. Dreizler, S. Lalitha, S.V. Jeffers, M. Lafarga, V.J.S. Béjar, J. Colomé, M. Cortés-Contreras, E. Díez-Alonso, D. Galadí-Enríquez, E.W. Guenther, H.J. Hagen, T. Henning, E. Herrero, M. Kürster, D. Montes, E. Nagel, V.M. Passegger, M. Perger, A. Rosich, A. Schweitzer, W. Seifert, Astron. Astrophys. **619**, A32 (2018). https://doi.org/10.1051/0004-6361/201833440
156. B.M. Clark, C.H. Blake, G.R. Knapp, Astrophys. J. **744**, 119 (2012). https://doi.org/10.1088/0004-637X/744/2/119
157. G. Basri, A. Reiners, Astron. J. **132**, 663 (2006). https://doi.org/10.1086/505198
158. C.H. Blake, D. Charbonneau, R.J. White, Astrophys. J. **723**(1), 684 (2010). https://doi.org/10.1088/0004-637X/723/1/684
159. Y. Shan, J.A. Johnson, T.D. Morton, Astrophys. J. **813**, 75 (2015). https://doi.org/10.1088/0004-637X/813/1/75
160. L.M. Rebull, J.R. Stauffer, J. Bouvier, A.M. Cody, L.A. Hillenbrand, D.R. Soderblom, J. Valenti, D. Barrado, H. Bouy, D. Ciardi, M. Pinsonneault, K. Stassun, G. Micela, S. Aigrain, F. Vrba, G. Somers, E. Gillen, A. Collier Cameron, Astron. J. **152**, 114 (2016). https://doi.org/10.3847/0004-6256/152/5/114
161. J. Stauffer, L. Rebull, J. Bouvier, L.A. Hillenbrand, A. Collier-Cameron, M. Pinsonneault, S. Aigrain, D. Barrado, H. Bouy, D. Ciardi, A.M. Cody, T. David, G. Micela, D. Soderblom, G. Somers, K.G. Stassun, J. Valenti, F.J. Vrba, Astron. J. **152**, 115 (2016). https://doi.org/10.3847/0004-6256/152/5/115
162. L.M. Rebull, J.R. Stauffer, L.A. Hillenbrand, A.M. Cody, J. Bouvier, D.R. Soderblom, M. Pinsonneault, L. Hebb, Astrophys. J. (2017)
163. L.M. Rebull, J.R. Stauffer, A.M. Cody, L.A. Hillenbrand, T.J. David, M. Pinsonneault, Astron. J. **155**, 196 (2018). https://doi.org/10.3847/1538-3881/aab605
164. K.G. Stassun, R.D. Mathieu, J.A. Valenti, Astrophys. J. **664**, 1154 (2007). https://doi.org/10.1086/519231
165. A. Burrows, J. Liebert, Rev. Mod. Phys. **65**, 301 (1993)
166. I. Baraffe, G. Chabrier, F. Allard, P.H. Hauschildt, Astron. Astrophys. **337**, 403 (1998)
167. I. Baraffe, D. Homeier, F. Allard, G. Chabrier, Astron. Astrophys. **577**, A42 (2015). https://doi.org/10.1051/0004-6361/201425481
168. E. Gillen, S. Aigrain, A. McQuillan, J. Bouvier, S. Hodgkin, S.H.P. Alencar, C. Terquem, J. Southworth, N.P. Gibson, A. Cody, M. Lendl, M. Morales-Calderón, F. Favata, J. Stauffer, G. Micela, Astron. Astrophys. **562**, A50 (2014). https://doi.org/10.1051/0004-6361/201322493
169. D.R. Anderson, A. Collier Cameron, C. Hellier, M. Lendl, P.F.L. Maxted, D. Pollacco, D. Queloz, B. Smalley, A.M.S. Smith, I. Todd, A.H.M.J. Triaud, R.G. West, S.C.C. Barros, B. Enoch, M. Gillon, T.A. Lister, F. Pepe, D. Ségransan, R.A. Street, S. Udry, Astrophys. J. Lett. **726**, L19 (2011). https://doi.org/10.1088/2041-8205/726/2/L19
170. R.F. Díaz, G. Montagnier, J. Leconte, A.S. Bonomo, M. Deleuil, J.M. Almenara, S.C.C. Barros, F. Bouchy, G. Bruno, C. Damiani, G. Hébrard, C. Moutou, A. Santerne, Astron. Astrophys. **572**, A109 (2014). https://doi.org/10.1051/0004-6361/201424406
171. A.S. Bonomo, A. Sozzetti, A. Santerne, M. Deleuil, J.M. Almenara, G. Bruno, R.F. Díaz, G. Hébrard, C. Moutou, Astron. Astrophys. **575**, A85 (2015). https://doi.org/10.1051/0004-6361/201323042
172. F. Bouchy, M. Deleuil, T. Guillot, S. Aigrain, L. Carone, W.D. Cochran, J.M. Almenara, R. Alonso, M. Auvergne, A. Baglin, P. Barge, A.S. Bonomo, P. Bordé, S. Csizmadia, K. de Bondt, H.J. Deeg, R.F. Díaz, R. Dvorak, M. Endl, A. Erikson, S. Ferraz-Mello, M. Fridlund, D. Gandolfi, J.C. Gazzano, N. Gibson, M. Gillon, E. Guenther, A. Hatzes, M. Havel, G. Hébrard, L. Jorda, A. Léger, C. Lovis, A. Llebaria, H. Lammer, P.J. MacQueen, T. Mazeh,

C. Moutou, A. Ofir, M. Ollivier, H. Parviainen, M. Pätzold, D. Queloz, H. Rauer, D. Rouan, A. Santerne, J. Schneider, B. Tingley, G. Wuchterl, Astron. Astrophys. **525**, A68 (2011). https://doi.org/10.1051/0004-6361/201015276
173. S. Csizmadia, A. Hatzes, D. Gandolfi, M. Deleuil, F. Bouchy, M. Fridlund, L. Szabados, H. Parviainen, J. Cabrera, S. Aigrain, R. Alonso, J.M. Almenara, A. Baglin, P. Bordé, A.S. Bonomo, H.J. Deeg, R.F. Díaz, A. Erikson, S. Ferraz-Mello, M. Tadeu dos Santos, E.W. Guenther, T. Guillot, S. Grziwa, G. Hébrard, P. Klagyivik, M. Ollivier, M. Pätzold, H. Rauer, D. Rouan, A. Santerne, J. Schneider, T. Mazeh, G. Wuchterl, S. Carpano, A. Ofir, Astron. Astrophys. **584**, A13 (2015). https://doi.org/10.1051/0004-6361/201526763
174. R.J. Siverd, T.G. Beatty, J. Pepper, J.D. Eastman, K. Collins, A. Bieryla, D.W. Latham, L.A. Buchhave, E.L.N. Jensen, J.R. Crepp, R. Street, K.G. Stassun, B.S. Gaudi, P. Berlind, M.L. Calkins, D.L. DePoy, G.A. Esquerdo, B.J. Fulton, G. Fűrész, J.C. Geary, A. Gould, L. Hebb, J.F. Kielkopf, J.L. Marshall, R. Pogge, K.Z. Stanek, R.P. Stefanik, A.H. Szentgyorgyi, M. Trueblood, P. Trueblood, A.M. Stutz, J.L. van Saders, Astrophys. J. **761**, 123 (2012). https://doi.org/10.1088/0004-637X/761/2/123
175. E.E. Mamajek, L.A. Hillenbrand, Astrophys. J. **687**, 1264 (2008). https://doi.org/10.1086/591785
176. J. Irwin, L. Buchhave, Z.K. Berta, D. Charbonneau, D.W. Latham, C.J. Burke, G.A. Esquerdo, M.E. Everett, M.J. Holman, P. Nutzman, P. Berlind, M.L. Calkins, E.E. Falco, J.N. Winn, J.A. Johnson, J.Z. Gazak, Astrophys. J. **718**, 1353 (2010). https://doi.org/10.1088/0004-637X/718/2/1353
177. M. Deleuil, H.J. Deeg, R. Alonso, F. Bouchy, D. Rouan, M. Auvergne, A. Baglin, S. Aigrain, J.M. Almenara, M. Barbieri, P. Barge, H. Bruntt, P. Bordé, A. Collier Cameron, S. Csizmadia, R. de La Reza, R. Dvorak, A. Erikson, M. Fridlund, D. Gandolfi, M. Gillon, E. Guenther, T. Guillot, A. Hatzes, G. Hébrard, L. Jorda, H. Lammer, A. Léger, A. Llebaria, B. Loeillet, M. Mayor, T. Mazeh, C. Moutou, M. Ollivier, M. Pätzold, F. Pont, D. Queloz, H. Rauer, J. Schneider, A. Shporer, G. Wuchterl, S. Zucker, Astron. Astrophys. **491**, 889 (2008). https://doi.org/10.1051/0004-6361:200810625
178. N.I. Gorlova, M.R. Meyer, G.H. Rieke, J. Liebert, Astrophys. J. **593**, 1074 (2003)
179. J. Gagné, D. Lafrenière, R. Doyon, L. Malo, É. Artigau, Astrophys. J. **783**, 121 (2014). https://doi.org/10.1088/0004-637X/783/2/121
180. J.C. Filippazzo, E.L. Rice, J. Faherty, K.L. Cruz, M.M. Van Gordon, D.L. Looper, Astrophys. J. **810**, 158 (2015). https://doi.org/10.1088/0004-637X/810/2/158
181. E.C. Martin, G.N. Mace, I.S. McLean, S.E. Logsdon, E.L. Rice, J.D. Kirkpatrick, A.J. Burgasser, M.R. McGovern, L. Prato, Surface Gravities for 228 M, L, and T Dwarfs in the NIRSPEC Brown Dwarf Spectroscopic Survey. Astrophys. J. **838**(1), 73 (2017). https://doi.org/10.3847/1538-4357/aa6338
182. N. Lodieu, M.R. Zapatero Osorio, V.J.S. Béjar, K. Peña Ramírez, Mon. Not. R. Astron. Soc. **473**, 2020 (2018). https://doi.org/10.1093/MonthlyNoticesoftheRoyalAstronomicalSociety/stx2279
183. N.C. Santos, G. Israelian, M. Mayor, J.P. Bento, P.C. Almeida, S.G. Sousa, A. Ecuvillon, Astron. Astrophys. **437**, 1127 (2005). https://doi.org/10.1051/0004-6361:20052895
184. J.C. Bond, C.G. Tinney, R.P. Butler, H.R.A. Jones, G.W. Marcy, A.J. Penny, B.D. Carter, Mon. Not. R. Astron. Soc. **370**, 163 (2006). https://doi.org/10.1111/j.1365-2966.2006.10459.x
185. N.C. Santos, G. Israelian, M. Mayor, R. Rebolo, S. Udry, Astron. Astrophys. **398**, 363 (2003). https://doi.org/10.1051/0004-6361:20021637
186. R. Pinotti, L. Arany-Prado, W. Lyra, G.F. Porto de Mello, Mon. Not. R. Astron. Soc. **364**, 29 (2005). https://doi.org/10.1111/j.1365-2966.2005.09491.x
187. K. El-Badry, H.W. Rix, Mon. Not. R. Astron. Soc. **482**, L139 (2019). https://doi.org/10.1093/MonthlyNoticesoftheRoyalAstronomicalSocietyl/sly206
188. J.E. Gizis, I.N. Reid, Astron. J. **117**, 508 (1999)
189. S. Lépine, R.M. Rich, M.M. Shara, Astrophys. J. **669**, 1235 (2007). https://doi.org/10.1086/521614

190. A.J. Burgasser, K.L. Cruz, J.D. Kirkpatrick, Astrophys. J. **657**, 494 (2007). https://doi.org/10.1086/510148
191. W.C. Jao, T.J. Henry, T.D. Beaulieu, J.P. Subasavage, Astron. J. **136**, 840 (2008). https://doi.org/10.1088/0004-6256/136/2/840
192. J.D. Kirkpatrick, K. Kellogg, A.C. Schneider, S. Fajardo-Acosta, M.C. Cushing, J. Greco, G.N. Mace, C.R. Gelino, E.L. Wright, P.R.M. Eisenhardt, D. Stern, J.K. Faherty, S.S. Sheppard, G.B. Lansbury, S.E. Logsdon, E.C. Martin, I.S. McLean, S.D. Schurr, R.M. Cutri, T. Conrow, Astrophys. J. Suppl. Ser. **224**, 36 (2016). https://doi.org/10.3847/0067-0049/224/2/36
193. Z.H. Zhang, D.J. Pinfield, M.C. Gálvez-Ortiz, B. Burningham, N. Lodieu, F. Marocco, A.J. Burgasser, A.C. Day-Jones, F. Allard, H.R.A. Jones, D. Homeier, J. Gomes, R.L. Smart, Mon. Not. R. Astron. Soc. **464**, 3040 (2017). https://doi.org/10.1093/MonthlyNoticesoftheRoyalAstronomicalSociety/stw2438
194. C. Moutou, A.S. Bonomo, G. Bruno, G. Montagnier, F. Bouchy, J.M. Almenara, S.C.C. Barros, M. Deleuil, R.F. Díaz, G. Hébrard, A. Santerne, Astron. Astrophys. **558**, L6 (2013). https://doi.org/10.1051/0004-6361/201322201
195. M.J. Phillips, L. Hartmann, Astrophys. J. **224**, 182 (1978). https://doi.org/10.1086/156363
196. O.C. Wilson, Astrophys. J. **226**, 379 (1978). https://doi.org/10.1086/156618
197. S.L. Baliunas, E. Nesme-Ribes, D. Sokoloff, W.H. Soon, Astrophys. J. **460**, 848 (1996). https://doi.org/10.1086/177014
198. G.W. Lockwood, B.A. Skiff, G.W. Henry, S. Henry, R.R. Radick, S.L. Baliunas, R.A. Donahue, W. Soon, Astrophys. J. Suppl. Ser. **171**, 260 (2007). https://doi.org/10.1086/516752
199. S. Messina, E.F. Guinan, Astron. Astrophys. **393**, 225 (2002). https://doi.org/10.1051/0004-6361:20021000
200. K. Oláh, Z. Kolláth, T. Granzer, K.G. Strassmeier, A.F. Lanza, S. Järvinen, H. Korhonen, S.L. Baliunas, W. Soon, S. Messina, G. Cutispoto, Astron. Astrophys. **501**, 703 (2009). https://doi.org/10.1051/0004-6361/200811304
201. L. Jetsu, J. Pelt, I. Tuominen, Astron. Astrophys. **278**, 449 (1993)
202. P. Parihar, S. Messina, P. Bama, B.J. Medhi, S. Muneer, C. Velu, A. Ahmad, Mon. Not. R. Astron. Soc. **395**, 593 (2009). https://doi.org/10.1111/j.1365-2966.2009.14422.x
203. Z. Eker, N.F. Ak, S. Bilir, D. Doğru, M. Tüysüz, E. Soydugan, H. Bakış, B. Uğraş, F. Soydugan, A. Erdem, O. Demircan, Mon. Not. R. Astron. Soc. **389**, 1722 (2008). https://doi.org/10.1111/j.1365-2966.2008.13670.x
204. C.A. Watson, V.S. Dhillon, Mon. Not. R. Astron. Soc. **351**, 110 (2004). https://doi.org/10.1111/j.1365-2966.2004.07763.x
205. S. Czesla, S. Terzenbach, R. Wichmann, J.H.M.M. Schmitt, Astron. Astrophys. **623**, A107 (2019). https://doi.org/10.1051/0004-6361/201834516
206. R.A. Osten, A. Kowalski, S.A. Drake, H. Krimm, K. Page, K. Gazeas, J. Kennea, S. Oates, M. Page, E. de Miguel, R. Novák, T. Apeltauer, N. Gehrels, Astrophys. J. **832**, 174 (2016). https://doi.org/10.3847/0004-637X/832/2/174
207. J.R.A. Davenport, Astrophys. J. **829**(1), 23 (2016). https://doi.org/10.3847/0004-637X/829/1/23
208. E. Ilin, S.J. Schmidt, J.R.A. Davenport, K.G. Strassmeier, Astron. Astrophys. **622**, A133 (2019). https://doi.org/10.1051/0004-6361/201834400
209. T. Van Doorsselaere, H. Shariati, J. Debosscher, Astrophys. J. Suppl. Ser. **232**(2), 26 (2017). https://doi.org/10.3847/1538-4365/aa8f9a
210. S.L. Hawley, J.R.A. Davenport, A.F. Kowalski, J.P. Wisniewski, L. Hebb, R. Deitrick, E.J. Hilton, Astrophys. J. **797**(2), 121 (2014). https://doi.org/10.1088/0004-637X/797/2/121
211. M.N. Günther, Z. Zhan, S. Seager, P.B. Rimmer, S. Ranjan, K.G. Stassun, R.J. Oelkers, T. Daylan, E. Newton, E. Gillen (2019). e-prints arXiv:1901.00443
212. J.A.G. Jackman, P.J. Wheatley, C.E. Pugh, D.Y. Kolotkov, A.M. Broomhall, G.M. Kennedy, S.J. Murphy, R. Raddi, M.R. Burleigh, S.L. Casewell, Mon. Not. R. Astron. Soc. **482**(4), 5553 (2019). https://doi.org/10.1093/mnras/sty3036

213. S.J. Schmidt, B.J. Shappee, J.L. van Saders, K.Z. Stanek, J.S. Brown, C.S. Kochanek, S. Dong, M.R. Drout, S. Frank, T.W.S. Holoien, Astrophys. J. **876**(2), 115 (2019). https://doi.org/10.3847/1538-4357/ab148d
214. H.Y. Chang, Y.H. Song, A.L. Luo, L.C. Huang, W.H. Ip, J.N. Fu, Y. Zhang, Y.H. Hou, Z.H. Cao, Y.F. Wang, Astrophys. J. **834**(1), 92 (2017). https://doi.org/10.3847/1538-4357/834/1/92
215. J.C. Morales, I. Ribas, C. Jordi, G. Torres, J. Gallardo, E.F. Guinan, D. Charbonneau, M. Wolf, D.W. Latham, G. Anglada-Escudé, Astrophys. J. **691**(2), 1400 (2009). https://doi.org/10.1088/0004-637X/691/2/1400
216. G. Torres, I. Ribas, Astrophys. J. **567**(2), 1140 (2002). https://doi.org/10.1086/338587
217. I. Ribas, Astron. Astrophys. **398**, 239 (2003). https://doi.org/10.1051/0004-6361:20021609
218. M. López-Morales, I. Ribas, Astrophys. J. **631**(2), 1120 (2005). https://doi.org/10.1086/432680
219. G. Torres, Astron. Nachr. **334**(1–2), 4 (2013). https://doi.org/10.1002/asna.201211743
220. G.A. Feiden, B. Chaboyer, Astron. Astrophys. **571**, A70 (2014). https://doi.org/10.1051/0004-6361/201424288
221. G.R. Ricker, J.N. Winn, R. Vanderspek, D.W. Latham, G.Á. Bakos, J.L. Bean, Z.K. Berta-Thompson, T.M. Brown, L. Buchhave, N.R. Butler, R.P. Butler, W.J. Chaplin, D. Charbonneau, J. Christensen-Dalsgaard, M. Clampin, D. Deming, J. Doty, N. De Lee, C. Dressing, E.W. Dunham, M. Endl, F. Fressin, J. Ge, T. Henning, M.J. Holman, A.W. Howard, S. Ida, J.M. Jenkins, G. Jernigan, J.A. Johnson, L. Kaltenegger, N. Kawai, H. Kjeldsen, G. Laughlin, A.M. Levine, D. Lin, J.J. Lissauer, P. MacQueen, G. Marcy, P.R. McCullough, T.D. Morton, N. Narita, M. Paegert, E. Palle, F. Pepe, J. Pepper, A. Quirrenbach, S.A. Rinehart, D. Sasselov, B. Sato, S. Seager, A. Sozzetti, K.G. Stassun, P. Sullivan, A. Szentgyorgyi, G. Torres, S. Udry, J. Villasenor, J. Astron. Telescopes Instrum. Syst. **1**(1), 014003 (2015). https://doi.org/10.1117/1.JATIS.1.1.014003
222. I. Roxburgh, C. Catala, PLATO Consortium, Commun. Asteroseismol. **150**, 357 (2007). https://doi.org/10.1553/cia150s357
223. S. Raetz, T.O.B. Schmidt, S. Czesla, T. Klocová, L. Holmes, R. Errmann, M. Kitze, M. Fernández, A. Sota, C. Briceño, J. Hernández, J.J. Downes, D.P. Dimitrov, D. Kjurkchieva, V. Radeva, Z.Y. Wu, X. Zhou, H. Takahashi, T. Henych, M. Seeliger, M. Mugrauer, C. Adam, C. Marka, J.G. Schmidt, M.M. Hohle, C. Ginski, T. Pribulla, L. Trepl, M. Moualla, N. Pawellek, J. Gelszinnis, S. Buder, S. Masda, G. Maciejewski, R. Neuhäuser, Mon. Not. R. Astron. Soc. **460**, 2834 (2016). https://doi.org/10.1093/MonthlyNoticesoftheRoyalAstronomicalSociety/stw1159
224. R. Errmann, R. Neuhäuser, L. Marschall, G. Torres, M. Mugrauer, W.P. Chen, S.C.L. Hu, C. Briceno, R. Chini, Ł. Bukowiecki, D.P. Dimitrov, D. Kjurkchieva, E.L.N. Jensen, D.H. Cohen, Z.Y. Wu, T. Pribulla, M. Vaňko, V. Krushevska, J. Budaj, Y. Oasa, A.K. Pandey, M. Fernandez, A. Kellerer, C. Marka, Astron. Nachr. **334**, 673 (2013). https://doi.org/10.1002/asna.201311890
225. Z. Garai, T. Pribulla, L'. Hambálek, R. Errmann, C. Adam, S. Buder, T. Butterley, V.S. Dhillon, B. Dincel, H. Gilbert, C. Ginski, L.K. Hardy, A. Kellerer, M. Kitze, E. Kundra, S.P. Littlefair, M. Mugrauer, J. Nedoroščík, R. Neuhäuser, A. Pannicke, S. Raetz, J.G. Schmidt, T.O.B. Schmidt, M. Seeliger, M. Vaňko, R.W. Wilson, Astron. Nachr. **337**, 261 (2016). https://doi.org/10.1002/asna.201512310
226. P.D. Dobbie, N. Lodieu, R.G. Sharp, Mon. Not. R. Astron. Soc. **409**, 1002 (2010). https://doi.org/10.1111/j.1365-2966.2010.17355.x
227. S. Manzi, S. Randich, W.J. de Wit, F. Palla, Astron. Astrophys. **479**, 141 (2008). https://doi.org/10.1051/0004-6361:20078226
228. D. Barrado y Navascués, J. Bouvier, J.R. Stauffer, N. Lodieu, M.J. McCaughrean, Astron. Astrophys. **395**, 813 (2002)
229. M. Hünsch, S. Randich, M. Hempel, C. Weidner, J.H.M.M. Schmitt, Astron. Astrophys. **418**, 539 (2004). https://doi.org/10.1051/0004-6361:20040043

230. J.P. Emerson, in *The New Era of Wide Field Astronomy, Astronomical Society of the Pacific Conference Series*, vol. 232, ed. by R. Clowes, A. Adamson, and G. Bromage (2001), p. 339
231. G.B. Dalton, M. Caldwell, A.K. Ward, M.S. Whalley, G. Woodhouse, R.L. Edeson, P. Clark, S.M. Beard, A.M. Gallie, S.P. Todd, J.M.D. Strachan, N.N. Bezawada, W.J. Sutherland, J.P. Emerson, *Society of Photo-Optical Instrumentation Engineers (SPIE) Conference Series, Presented at the Society of Photo-Optical Instrumentation Engineers (SPIE) Conference*, vol. 6269 (2006). https://doi.org/10.1117/12.670018
232. P.J. Wheatley, D.L. Pollacco, D. Queloz, H. Rauer, C.A. Watson, R.G. West, B. Chazelas, T.M. Louden, S. Walker, N. Bannister, J. Bento, M. Burleigh, J. Cabrera, P. Eigmüller, A. Erikson, L. Genolet, M. Goad, A. Grange, A. Jordán, K. Lawrie, J. McCormac, M. Neveu, *European Physical Journal Web of Conferences, European Physical Journal Web of Conferences*, vol. 47 (2013), p. 13002. https://doi.org/10.1051/epjconf/20134713002
233. M. Gillon, E. Jehin, P. Magain, V. Chantry, D. Hutsemékers, J. Manfroid, D. Queloz, S. Udry, *European Physical Journal Web of Conferences, European Physical Journal Web of Conferences*, vol. 11 (2011), p. 06002. https://doi.org/10.1051/epjconf/20101106002
234. M. Gillon, E. Jehin, L. Delrez, P. Magain, C. Opitom, S. Sohy, *Protostars and Planets VI Posters* (2013)
235. R. Alonso, T.M. Brown, G. Torres, D.W. Latham, A. Sozzetti, G. Mandushev, J.A. Belmonte, D. Charbonneau, H.J. Deeg, E.W. Dunham, F.T. O'Donovan, R.P. Stefanik, Astrophys. J. Lett. **613**, L153 (2004). https://doi.org/10.1086/425256
236. G.Á. Bakos, J. Lázár, I. Papp, P. Sári, E.M. Green, Publ. Astron. Soc. Pac. **114**, 974 (2002). https://doi.org/10.1086/342382
237. D.L. Pollacco, I. Skillen, A. Collier Cameron, D.J. Christian, C. Hellier, J. Irwin, T.A. Lister, R.A. Street, R.G. West, D.R. Anderson, W.I. Clarkson, H. Deeg, B. Enoch, A. Evans, A. Fitzsimmons, C.A. Haswell, S. Hodgkin, K. Horne, S.R. Kane, F.P. Keenan, P.F.L. Maxted, A.J. Norton, J. Osborne, N.R. Parley, R.S.I. Ryans, B. Smalley, P.J. Wheatley, D.M. Wilson, Publ. Astron. Soc. Pac. **118**, 1407 (2006). https://doi.org/10.1086/508556
238. P.R. McCullough, J.E. Stys, J.A. Valenti, S.W. Fleming, K.A. Janes, J.N. Heasley, Publ. Astron. Soc. Pac. **117**, 783 (2005). https://doi.org/10.1086/432024
239. J. Pepper, R.W. Pogge, D.L. DePoy, J.L. Marshall, K.Z. Stanek, A.M. Stutz, S. Poindexter, R. Siverd, T.P. O'Brien, M. Trueblood, P. Trueblood, Publ. Astron. Soc. Pac. **119**, 923 (2007). https://doi.org/10.1086/521836
240. D. Montes, J. López-Santiago, M.J. Fernández-Figueroa, M.C. Gálvez, Astron. Astrophys. **379**, 976 (2001). https://doi.org/10.1051/0004-6361:20011385
241. B. Zuckerman, I. Song, Annu. Rev. Astron. Astrophys. **42**, 685 (2004)
242. T. Antoja, F. Figueras, D. Fernández, J. Torra, Astron. Astrophys. **490**, 135 (2008). https://doi.org/10.1051/0004-6361:200809519
243. S. Hubrig, O. Marco, B. Stelzer, M. Schöller, N. Huélamo, Mon. Not. R. Astron. Soc. **381**, 1569 (2007). https://doi.org/10.1111/j.1365-2966.2007.12325.x
244. I.N. Evans, F. Civano, Astron. Geophys. **59**(2), 2.17 (2018). https://doi.org/10.1093/astrogeo/aty079
245. T. Boller, M.J. Freyberg, J. Trümper, F. Haberl, W. Voges, K. Nandra, Astron. Astrophys. **588**, A103 (2016). https://doi.org/10.1051/0004-6361/201525648
246. I. Traulsen, A.D. Schwope, G. Lamer, J. Ballet, F. Carrera, M. Coriat, M.J. Freyberg, L. Michel, C. Motch, S.R. Rosen, N. Webb, M.T. Ceballos, F. Koliopanos, J. Kurpas, M.J. Page, M.G. Watson, Astron. Astrophys. **624**, A77 (2019). https://doi.org/10.1051/0004-6361/201833938
247. C. Schröder, J.H.M.M. Schmitt, Astron. Astrophys. **475**, 677 (2007). https://doi.org/10.1051/0004-6361:20077429
248. B. Stelzer, A. Marino, G. Micela, J. López-Santiago, C. Liefke, Mon. Not. R. Astron. Soc. **431**, 2063 (2013). https://doi.org/10.1093/mnras/stt225

# Globular Cluster Systems and Galaxy Formation

**Michael A. Beasley**

## Contents



M. A. Beasley (✉)
Instituto de Astrofísica de Canarias, La Laguna, Tenerife, Spain
e-mail: beasley@iac.es



P. Kabáth et al. (eds.), *Reviews in Frontiers of Modern Astrophysics*,
https://doi.org/10.1007/978-3-030-38509-5_9

**Abstract** Globular clusters (GCs) are compact, gravitationally bound systems of up to ∼1 million stars. The GCs in the Milky Way contain some of the oldest stars known, and provide important clues to the early formation and continuing evolution of our Galaxy. More generally, GCs are associated with galaxies of all types and masses, from low-mass dwarf galaxies to the most massive early-type galaxies which lie in the centres of massive galaxy clusters. GC systems show several properties which connect tightly with properties of their host galaxies. For example, the total mass of GCs in a system scales linearly with the dark matter halo mass of its host galaxy. Numerical simulations are at the point of being able to resolve globular cluster formation within a cosmological framework. Therefore, GCs link a range of scales, from the physics of star formation in turbulent gas clouds, to the large-scale properties of galaxies and their dark matter. In this chapter we review some of the basic observational approaches for GC systems, some of their key observational properties, and describe how GCs provide important clues to the formation of their parent galaxies.

## 1 Introduction

Understanding how galaxies form and evolve are major challenges in astrophysics. We presently do not have a definitive model for how nearby galaxies attain the morphologies we see, or understand fully how galaxies evolve as we look back in redshift. However, astronomers have a reasonable picture of some of the fundamental aspects of these processes. A large part of the observational and theoretical effort in understanding galaxy formation has focused on the stellar and gas content of the main components of galaxies.[1] In the case of early-type galaxies (e.g., elliptical galaxies; hereafter "ETGs") this corresponds primarily to the smoothly distributed stellar body. In the case of late-type galaxies, such as the Milky Way, these components may include a stellar and gaseous disk, and perhaps a stellar bar and bulge. However, an important additional component of galaxies is their globular cluster (GC) systems. While they may only account for <0.1% of the total stars in a galaxy [1], GCs play an out-sized role in tackling several key problems in galaxy formation.

In this chapter, some of the key properties of GC systems will be discussed, with a particular focus on their link to the formation and evolution of their host galaxies. A number of excellent reviews on GCs and GC systems exist in the literature. For the specific case of the chemical properties of Milky Way GCs see [2] and [3]. For a review on extragalactic GC systems see [4] and [1]. Forbes et al. [5] discuss some outstanding issues in GC formation and evolution. The literature focusing specifically on galaxy formation is vast and it is not possible to go into details

[1]This is often called the "baryonic" component, since stars and gas are composed of protons and neutrons which are baryons (three quarks). The poor old electrons—which are leptons—are ignored in this terminology!

here. Silk et al. [6] discuss some of the general issues in galaxy formation, while [7] focuses on the kinematics of galaxies. See [8] for details on star formation in galaxies along the Hubble sequence. Renzini [9] discusses the formation and evolution of massive ETGs. Bland-Hawthorn and Gerhard [10] go into details for the specific case of the Milky Way, while [11] discuss outstanding issues in galaxy formation from a theoretical point of view. A review of dwarf galaxies can be found in [12], and of "ultra-faint" dwarf systems here [13]. The above works are good starting points for research into GC systems and galaxy formation.

## 2 What Globular Clusters Are and Are Not

GCs are compact, centrally concentrated, gravitationally bound systems of stars. The mean mass of a GC in the Milky Way is $\sim 2 \times 10^5\,M_\odot$, and this is typical for GC systems in general. Three parameters are used to describe their physical sizes. The *core radius* ($r_c$), is the radius at which the central surface brightness of the cluster drops by half; the *half-light radius* ($r_h$), is the radius that contains half the light of the cluster, and the *tidal radius* ($r_t$), is the radius at which the density of the GC reaches zero [14]. Median values for these radii in the Milky Way are $r_c \sim 1.5$ pc, $r_h \sim 10$ pc and $r_t \sim 50$ pc [15, 16].

From the above numbers we see that GCs are also very dense stellar systems. GCs have average stellar densities of $\sim 0.2\,M_\odot\,pc^{-3}$ and central densities of up to $\sim 10^4\,M_\odot\,pc^{-3}$. These high densities and their small sizes make GCs very interesting laboratories for studies of stellar evolution and dynamics (see [17] for a review). However, more importantly—from the point of view of this chapter at least—this also means that they are relatively luminous (mean $V$-band absolute magnitudes, $M_V \sim -7.5$), concentrated stellar sources, which makes them readily observable out to large distances. Their high densities also mean that they are quite resistant to external tidal forces; the GCs in the Milky Way have survived some ten billion years of galaxy evolution (although many of the low- and high-mass clusters may well have been destroyed; [18, 19]. Within our ability to measure, all the stars in a GC are coeval (uniform age) and generally old. In addition, they are largely mono-metallic (same heavy-element abundance) to a high degree of uniformity. This makes GCs important laboratories for stellar evolution since the key parameter which dictates the evolutionary stage of any star in a cluster becomes its mass.

Extensive observational evidence suggests that galaxies are embedded in halos of "dark matter". The rotation curves of the gas in galaxies [20, 21], the motions of the stellar component (stellar velocity dispersion) [7], the hot X-ray emitting gas that surrounds massive galaxies [22, 23], gravitational lensing of background sources [24, 25] and the kinematics of GC systems [26, 27], all indicate the existence of a gravitational potential in excess of that observed for the mass of gas and stars alone.

As far as we know, GCs *at the present epoch*[2] do not have dark matter (e.g., [17, 28]). That is, they may contain dark stellar remnants from stellar evolution (e.g., brown dwarfs, black holes), but do not seem to possess any further invisible component comprising hypothetical exotic particles. This observation is important for a couple of reasons. Firstly, galaxies *do* have dark matter, so one can conclude that GCs are not galaxies! This conclusion is more profound than it first appears; a comprehensive definition of a galaxy is not so straightforward, and definitions based on size, stellar populations or stellar density can often end up including GCs (see [29]).[3] Whether or not a system has dark matter does discriminate between galaxies and stellar clusters (globular, open or young massive clusters).

Secondly, models for GC formation may be broken down into two very broad classes; "dark matter" formation models and "baryonic" formation models. In dark matter models, GCs are required to have (at least, initially) their own dark matter halo in which to form. Initial ideas included the early collapse of gas into low-mass dark halos to form GCs [30]. Alternatively, collisions of dark matter haloes may lead to GC formation at high redshift [31]. If GCs do indeed form in their own dark matter haloes, then GCs unassociated with galaxies residing in low-density environments may exist. However, so far such "free-floating" GCs have not been observed [32], although "unbound" GCs are seen to be present in galaxy clusters which are believed stripped from their parent galaxies (e.g., [33]).

The second class of models, baryonic models, make GC formation less special. They generally posit that GC formation is a natural extension of any clustered star formation, and that the main difference between other types of star cluster and GCs is one of mass, rather than formation mechanism [34, 35]. In this picture, GCs can also form in a dark matter halo, but these haloes are associated with the host galaxy, rather than individual GCs. These models also make predictions; for example in many cases they predict that GCs will obey an age–metallicity relation such that more metal-rich clusters formed in a single system may be younger than their metal-poor counterparts. Models for the formation of GC systems are further discussed in Sect. 7.

## 3 Observational Techniques

The observational techniques used to study GCs can be separated into two categories. One category is in the *spatially resolved case*. This generally (but not exclusively) applies to "nearby" clusters ($d < 1.0\,\mathrm{Mpc}$), for example, GCs found

[2]This qualifier is important to remember. Although GCs seem to have no dark matter now, this does not necessarily imply that they never had dark matter. Tidal processes could remove the vast majority of dark matter from a GC orbiting in a Milky Way-like potential over a Hubble time.

[3]Stellar systems of masses at, or below that of GCs are observed, and they typically have very high mass-to-light ratios ($M/L_V > 100$) implying high dark matter fractions. However, these are generally low-concentration, low-surface brightness objects which are typically larger ($r_\mathrm{h} > 50\,\mathrm{pc}$) than GCs and are referred to as "ultra-faint dwarf" galaxies (UFDs; [13]).

in the Milky Way and other galaxies in the Local Group. In this case, individual cluster stars can be resolved and studied, and their properties measured either via photometry or spectroscopy. The second category is the *spatially unresolved case*. By this it is meant that individual stars are not resolved, although there may still be spatially varying information. Here, the integrated light of the cluster is studied as a sum of all the stars in the cluster. Again, the both photometry and spectroscopy can be used to obtain useful information, although the analysis techniques differ from those of spatially resolved studies.

## 3.1 Spatially Resolved Methods

Spatially resolved photometry and spectroscopy are primary observational methods to obtain detailed information about the ages and chemistry of GCs. In addition, due to the proximity of Milky Way GCs, their proper motions can be measured.

### 3.1.1 Colour-Magnitude Diagrams

A primary tool used to study GCs (and resolved stellar populations in general) is the colour magnitude diagram (CMD). To construct a CMD, images of a GC are obtained in two filters (to construct a colour), and photometry is performed on these images.[4] The CMD for the Milky Way GC M13 is shown in Fig. 1. The different evolutionary phases from the tip of the asymptotic giant branch to the lower main-sequence are shown. A CMD analysis can be used to determine both the cluster age and its metallicity ([Fe/H]). Some age information is present at a number of locations in the CMD, although the key age-sensitive feature is the main-sequence turn-off (MSTO) which indicates the termination of core H-burning in stars.

The apparent magnitude of the MSTO depends on distance, so if the distance to the cluster is not well-known, distance-independent measures are generally employed, such as the difference between the position of the MSTO and the horizontal branch [36, 37]. In contrast, metallicity information comes from the horizontal location of the CMD and also from the shape of the red giant branch. Increasing metallicity makes the CMD locus shift to the red since more metals in a stellar atmosphere increases stellar atmospheric opacities, which results in lower effective temperatures (and redder colours).

To determine ages and metallicities from CMDs isochrones are used. These are theoretical stellar evolutionary tracks which predict stellar temperatures and luminosities (or observationally, magnitudes and colours) for a given age and chemical composition (see Sect. 3.2.1).

[4]The subject of photometry is a chapter in itself. Suffice to say that the standard techniques are reasonably straightforward, although a number of careful steps are required to achieve precise measurements.

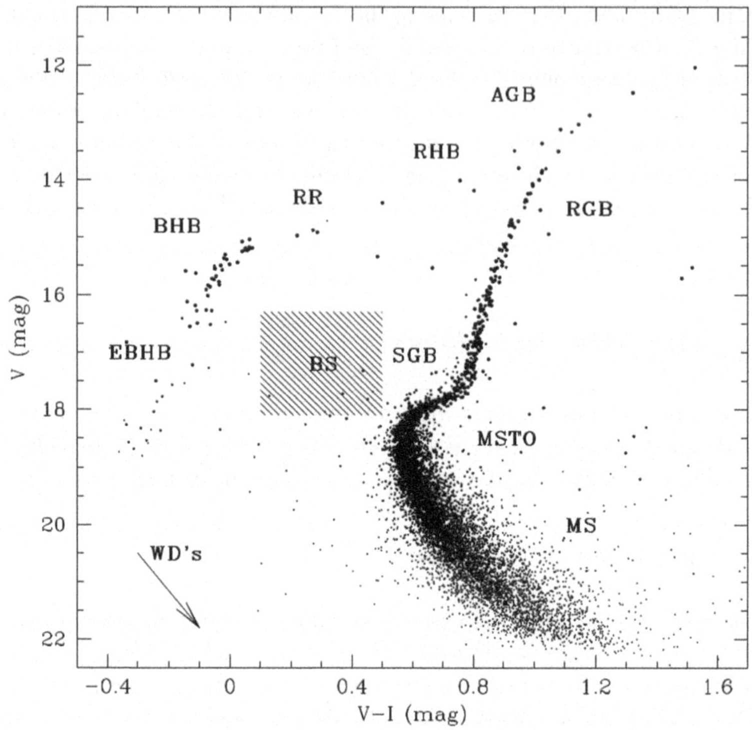


**Fig. 1** *V*, *I* colour-magnitude diagram (CMD) for the Milky Way globular cluster M13. The main features of the CMD are indicated: main-sequence (MS), main-sequence turn-off (MSTO), subgiant branch (SGB), red giant branch (RGB), asymptotic giant branch (AGB), red-side horizontal branch (RHB), RR Lyrae variables (RR), blue horizontal branch (BHB), extreme blue horizontal branch (EBHB), blue stragglers (BS) and white dwarfs (WDs)

### 3.1.2 Stellar Spectroscopy

The standard technique to study the detailed chemical abundances of stars is high resolution spectroscopy. The definition of "high resolution" is a bit vague, but resolving powers of $R > 20{,}000$ are typical. In order to interpret the spectra of stars, stellar atmosphere models are required. These are theoretical models which predict the strengths of individual atomic and molecular lines for a given effective temperature ($T_{\rm eff}$), gravity ($\log g$) and metallicity. Other factors must also be considered such as "microturbulence", which can also affect the measured line-strengths. In practice, there is a range of stellar atmosphere models which are suitable for different regimes of temperature and chemical abundance. In addition, models may assume "local thermodynamic equilibrium" (LTE) or non-LTE, which can affect the profile of some lines depending upon where they formed in the stellar atmosphere.

The most basic chemical composition measurement for a star is its metallicity. This generally refers to the ratio of iron to hydrogen ([Fe/H]), since there are many Fe lines in the optical spectra of stars and they are easily measured. However, many other chemical abundance ratios of individual elements can also be determined, and the stars in GCs show some surprises (this is discussed further in Sect. 4).

### 3.1.3 Space Motions of Globular Clusters

Stellar spectroscopy provides precise measurements of radial velocities for the stars in GCs. A radial velocity is the component of the velocity of a GC along our line of sight. However, to know the true space velocities of GCs—that is their true motions through 3-dimensional space—requires a knowledge of the cluster proper motion[5] and distance. The *Hubble Space Telescope* (*HST*) and the *GAIA* satellite have revolutionised the data available for Galactic GCs by providing proper motions and some parallaxes[6] for the Milky Way GC system [38]. The "true space velocities" for 150 GCs are now known, giving important insights into the orbits of the GCs. For example, a recent analysis for total mass of the Milky Way, based on Milky Way proper motions from the second data release of *GAIA* gives $M_{\rm tot} = 1.28^{+0.97}_{-0.48} \times 10^{12}\,{\rm M}_\odot$ [39].

## *3.2 Spatially Unresolved Methods: Integrated Light*

The observational techniques used to study spatially unresolved GCs are not too different from resolved stellar studies. However, the analysis of integrated-light information requires different approaches to that of resolved stellar studies. The integrated colours or spectra obtained for GCs generally captures the light from the entire stellar population, and the interpretation of these observations usually requires simple stellar population modelling.

### 3.2.1 Stellar Population Models

Simple stellar population (SSP) models require three primary ingredients: a set of isochrones, a stellar library and an assumption about the stellar initial mass function (IMF). The isochrones—which are invariably theoretical—predict effective temperatures, surface gravities and luminosities for a star of a given age and chemical composition. This isochrone is then populated with stars based on the adopted IMF. The stellar libraries may be empirical or theoretical and both have their

[5]The apparent movement on the sky of an object compared to a fixed background.

[6]Parallax is the apparent change of position on the sky of an object when viewed from two different positions along a given baseline. If the length of the baseline is known, by measuring the parallax angle (i.e., how much the star appears to move), the distance to the star can be determined.

advantages and disadvantages. Models made with empirical libraries are generally more successful in reproducing the colours and spectra of GCs and galaxies since they incorporate real stars present in the Milky Way or in GCs. However, this is also a disadvantage. Empirical libraries are limited to those stars observable in the Milky Way, which limits the range of metallicities available (particularly at the metal-rich end) and also imposes a fixed abundance pattern (e.g. [Mg/Fe], [Ca/Fe]) onto the stellar library. For empirical libraries, the parameter space ($T_{\rm eff}$, $\log g$, [Fe/H], [Mg/Fe]...) is generally sampled in a non-uniform manner and special interpolation techniques are implemented (e.g., [40]).

Theoretical libraries, made from stellar atmosphere models, have no such restrictions. In principle they can be constructed for a wide range of $T_{\rm eff}$, $\log g$, [Fe/H] and with arbitrary abundance patterns. However, they also suffer from several serious drawbacks. For example, no single set of stellar atmosphere models covers the full range of effective temperatures required to model a wide range of ages and metallicities in SSPs. In addition, the "linelists" which go into the stellar models are often incomplete and vary between workers. Any missing lines in the models can result in important differences between the modelled and observed stellar population. For more information on SSP modelling and its applications see [40–44].

### 3.2.2 Integrated Colours

The resulting SSPs predict colours and spectra for a simple stellar population for a given age, metallicity and chemical composition. Broad-band, optical colours (e.g., $B$, $V$, $g$, $i$...) can give useful information on ages and metallicities, but they suffer from an age–metallicity degeneracy; more metallic populations look redder, but so do older stellar populations. Disentangling the effects of age and metallicity in the optical can be aided by the use of infrared and near-UV bands for GCs. The colour distributions of GC systems is discussed in Sect. 5.1.1.

### 3.2.3 Integrated Spectra

Integrated spectra are a useful way of breaking the age–metallicity degeneracy. Spectra also provide velocity measurements for GCs, as well as potentially providing information on individual chemical abundance ratios such as the [Mg/Fe][7] ratio. Individual line-strength indices can be measured and compared to model grids, or an observed spectrum can be compared pixel-by-pixel via "full spectral fitting"

[7]Magnesium and iron are produced in short-lived, massive stars which explode as type-II supernovae, whereas iron is produced in longer-lived, lower mass stars which is released in type Ia supernova explosions. Therefore [Mg/Fe] can be used as a chemical clock—generally the higher the ratio [Mg/Fe], the shorter the timescale of star formation since low-mass stars have not had time to pollute the interstellar medium.

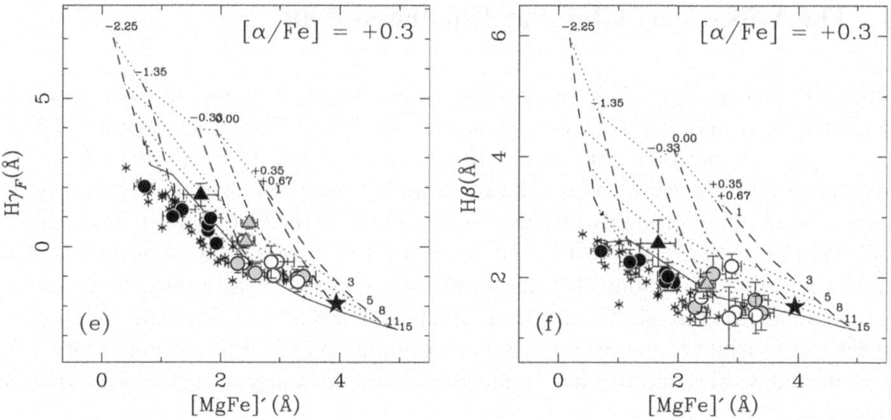


**Fig. 2** Balmer-line versus metal-line from the integrated spectra of globular clusters in the ETG NGC 1407 compared to stellar population models. Near-vertical numbers 1–15 indicate age in Gyr, while the near-horizontal numbers (−2.25 to +0.67) indicate metallicities. The figure shows that the globular clusters are very old (∼14 Gyr), and have a wide range of metallicities. Figure taken from [45]

techniques. Metallicities derived in this way, from high quality spectra, provide accuracies of 0.2–0.3 dex (decades in logarithmic spacing).

Ages can also be determined, in principle, for GCs from integrated spectra. The key age-sensitive lines in the optical integrated spectra of GCs are the Balmer hydrogen series in absorption (H$\alpha$, H$\beta$, H$\gamma$, H$\delta$ etc.) which arise from the electron transitions to the $n = 2$ quantum energy level in the hydrogen atom. The Balmer lines are sensitive to temperature and vary in strength with the location (temperature) of the main sequence turn-off. Unfortunately, there are other hot stars in GCs which are not well modeled and also contribute to the Balmer lines. Horizontal branch stars and blue stragglers are particularly problematic. These can lead to ambiguous age determinations, especially for metal-poor clusters. High-quality integrated spectra for GCs in the ETG NGC 1407 [45] are compared to SSP model "grids" in Fig. 2.

An outstanding problem with the SSP models is the "zeropoint" problem in GCs [46, 47]. When plotted in a Balmer-line, metal-line diagram, GCs tend to drop off the bottom of the grids below the oldest ages (see right-hand panel of Fig. 2). Interestingly, this is generally not seen in spectra for even the oldest galaxies! The origin of this problem is not understood, but it is possible that a combination of the peculiar abundance ratios in GCs (compared to field stars), and also atomic diffusion[8] [48] in stars near the turn-off may be responsible [47].

[8]Atomic diffusion is the collective term for processes that change the mixture of atmospheric abundances in stars due to gravity or radiation pressure. For example, heavier elements (e.g., Fe) tend to sink over long time-scales thereby lowering the observed surface abundance of the star.

# 4 The Milky Way Globular Cluster System

Since the Milky Way GC system is the closest to us, it is also the best studied. A catalogue containing the basic parameters of these GCs is maintained by W.E. Harris [4] and can be found here.[9] There are presently 158 GCs thought to be associated with the Milky Way. The true number may be closer to ∼180 clusters, since a number are probably obscured by the plane of the Galaxy [49]. The nearest GC, M4 (NGC 6121), lies just 2.2 kpc from us. The most distant—Laevens 1 (also known as Crater)—is some 145 kpc away, which puts it nearly three times as far as the Small Magellanic Cloud from which it is thought to originate. A number of GCs are visible to the naked eye. For example, M13 is a fine summer target in dark northern skies and lies just to the west from the center of the constellation of Hercules.

## *4.1 The Origin of the Milky Way Globular Clusters*

Based on their metallicities, the Milky Way GCs separate into two main populations, a metal-poor population ($\langle[\mathrm{Fe/H}]\rangle \sim -1.5$) which comprises ∼2/3 of the total system, and a metal-rich population ($\langle[\mathrm{Fe/H}]\rangle \sim -0.5$). A plot of the metallicity distribution function (a histogram of metallicities) looks "bimodal". Spatially, the metal-poor clusters are distributed roughly spherically throughout the Milky Way, reaching out to 145 kpc into the Galactic halo. The velocities of the metal-poor clusters generally show no ordered motions. However, there are some notable exceptions which might be related to the infall of dwarf galaxies bringing in their own GC populations [38, 50]. The metal-poor GC system is often referred to as the halo GC population. By contrast, most of the metal-rich clusters lie within 10 kpc of the Galactic centre. These clusters show a somewhat flattened spatial distribution, and exhibit net rotation of order 50–80 km $s^{-1}$. The general view is that the metal-rich population is associated with the old "thick" disc or bulge of the Milky Way.

CMD studies of GCs in the Milky Way indicate that the majority of clusters are older than ∼10 Gyr [37, 51]. That is, the stars in GCs are, in general, at least twice as old as our Sun. In terms of redshift ($z$), this suggests that most Milky Way globulars were formed at $z > 2$ whereas our Solar System started formation in the disc of the Milky Way somewhere in the region of $z \sim 0.45$. There are, however, a number of clusters which are a somewhat younger than the majority. Examples of these clusters are Palomar 12 and Terzan 7 which have ages ∼8 Gyr.

With increasingly precise ages for Galactic globulars, some very interesting results emerge. When combined with metallicity information, we can plot the age–metallicity relation (AMR) for GCs. This is shown in Fig. 3 and is taken from

[9] https://physwww.mcmaster.ca/~harris/mwgc.dat.

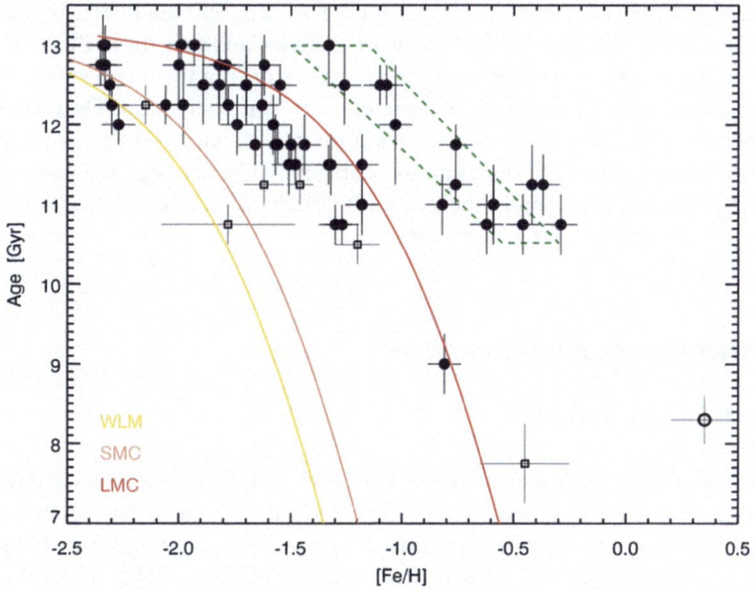


**Fig. 3** Age–metallicity relations (AMRs) for Milky Way GCs (taken from [52]). The relations split into several sequences which resemble the AMRs of dwarf galaxies (WLM, SMC, LMC) and also that of the Milky Way bulge (green dashed box). The AMRs suggest that the metal-poor Milky Way clusters may have been accreted from dwarf galaxies

[52]. The figure shows that there are at least two AMRs, and that more metal-rich clusters tend to be younger in any given relation. This is a natural consequence of chemical enrichment during star formation; stars form and the resulting energy, stellar winds and supernovae (stellar "feedback") return metals into the interstellar medium thereby enriching the subsequent generation of stars. However, what is really interesting is that the GCs split into distinct sequences. I.e., they present a bifurcated AMR. Overplotted are the expected AMRs for several dwarf galaxies (WLM, the Small Magellanic Cloud and the Large Magellanic Cloud). Also shown is the AMR for the Milky Way bulge. The figure shows that the metal-rich clusters ([Fe/H]$> -1.0$) are consistent with having formed along with the bulge of the Milky Way. The bulge is a central, old component of the Galaxy and therefore these clusters were probably formed along with the main part of the Milky Way. In the context of galaxy formation, this can be referred to as an in situ[10] GC population.

In contrast, the metal-poor clusters follow sequences that are consistent with dwarf galaxy AMRs.[11] So why do these GCs look like they formed in dwarf galaxies

[10] Formed "in-place". Astronomers like resorting to latin on occasion.

[11] This is consistent with what we know about galaxy metallicities. Galaxies follow a stellar mass–metallicity relation in that more massive galaxies are, on average, more metal-rich. This is a consequence of the fact that more massive galaxies have more stars to form metals via

when they are now in the Milky Way halo? The general conclusion is that these metal-poor "halo" clusters formed in dwarf galaxies and that these dwarf galaxies merged with the Milky Way bringing in their GCs during this process. So, we can regard many of these halo GCs as an accreted or ex-situ population. The idea that the metal-rich, central Galactic stars and GCs form in situ, while (at least an important fraction of) metal-poor halo populations (stars and GCs) are accreted and represent ex-situ populations is presently the generally accepted picture for the formation of massive galaxies and their GC systems.

## *4.2 Metallicity and Abundances*

### 4.2.1 Metallicity Spreads

In the Introduction, it was mentioned that GCs are mono-metallic. This is significant because galaxies have spreads in [Fe/H]. Indeed, the presence or absence of measurable spreads in Fe is one proposed way to distinguish galaxies from GCs, similiar to the case for the presence or absence of dark matter (Sect. 2). Fe is produced in the cores of stars during nucleosynthesis, and can only be released to the interstellar medium via supernova explosions. Therefore, a lack of spread in [Fe/H] implies only one generation of star formation has occurred in most GCs, as opposed to galaxies which may undergo multiple cycles of star formation.

However, it turns out that not all GCs are mono-metallic. A famous example is Omega Centauri (NGC 5139), which turns out to be the most massive GC in the Milky Way with $M \sim 4 \times 10^6\,M_\odot$. Spectroscopic studies show that the cluster has a spread in metallicity of up to ~1.0 dex and a complex metalicity distribution function (MDF) (e.g., [53, 54]). It is now believed that, rather than being a "true" GC, Omega Centauri is actually the nucleus of a dwarf galaxy that accreted into the Galactic halo at some point in the past [55–57]. Other GCs with spreads in [Fe/H] include M22 [58] and M54 [59].

### 4.2.2 Light Element Abundances

One of the key results to emerge in the past decade has been that GC stars show unusual patterns in some "light" elements (see [3] for a review). These elements include He, C, N, Al, Mg, Na and Ca. The picture that is emerging is that there are at least two chemically distinct populations of stars in Milky Way GCs. The first population (P1) shows a pattern of light elements which, for a given [Fe/H],

---

nucleosynthesis, and are also better able to hold onto their gas "recycled" from star formation due to their deeper potential wells.

look very similar to field[12] stars in our Galaxy. The second population (P2) of stars shows a pattern which appears unique to massive star clusters. Specifically, the abundances of the elements He, N, Na and Al are elevated, at a fixed [Fe/H], compared to those seen in P1 and field stars, while C, O and Mg are generally depressed compared to P1 and the field. The origin of these abundance variations is presently unknown, although candidates include winds from rotating massive stars, or material ejected from asymptotic giant branch stars from P1, which somehow may contaminate the (assumed to be) later P2 generation. A hard requirement here is that supernovae from high- or low-mass stars cannot contribute since this would result in a spread in [Fe/H] among the cluster stars which is only seen for a few special cases (Sect. 4.2.1).

The fact that these unusual light-element patterns are peculiar to GCs might raise the question of why this is interesting in the context of galaxy formation. It turns out the abundance pattern of the P2 generation is sufficiently different from that of Milky Way field stars that it can be used as a chemical "fingerprint" for a number of interesting processes. For example, Galactic stellar surveys suggest that, based on the fraction of "GC-like" P2 stars seen in the Galaxy, somewhere between 10–50% of Milky Way halo stars may actually come from disrupted GCs. [60, 61]. In addition, it is possible that the abundance variations seen in Milky Way GCs are also responsible for the peculiarities seen in the integrated colours of extragalactic GCs, something that must be understood in order to use the colours of extragalactic GCs to probe galaxy formation (Sect. 5.1.1).

## 5 Extragalactic Globular Cluster Systems

For all the detailed information available for the Milky Way GCs, it is only one galaxy. By studying extragalactic systems we can study a wide range of galaxy morphological types and masses in very different environments. This work has lead to important insights into the formation and co-evolution of galaxies and their GC systems.

### *5.1 Early-Type Galaxies*

The GC systems of ETGs have received most attention. This has been in part due to observational convenience. ETGs have smooth light profiles which makes GCs easily detected. ETGs also tend to have rich GC systems which also helps in their analysis. Beyond observational considerations, ETGs are of great scientific interest. They presently contain more than half of all the stars in the nearby Universe [62], and they also represent the most massive galaxies known. Analysis of the stellar

---

[12] Here the term "field" refers to stars not in star clusters.

populations of ETGs indicates that the stars are generally ancient (~10 Gyr), metal-rich (equal or higher than the solar value), and have an IMF dominated by low-mass stars (a "bottom-heavy" IMF) in their centres [9, 63, 64].

### 5.1.1 Colours and Metallicities

Early ground-based work indicated that the colour distributions of massive ETGs are quite complex and not readily fit by a single gaussian distribution [65, 66]. This lead to a major result of "bimodal" colour distributions in such galaxies, with "blue" and "red" populations. Later works, in particular with *HST*, confirmed that the colour distributions of most massive galaxies look bi- or multi-model in optical colours [67, 68]. The colour distributions for the most massive galaxies in the centres of clusters can be extremely complex and rather than showing clear bimodality, they appear broad with hints of multiple substructures [69]. Empirical colour-metallicity relations based on Milky Way GCs, or theoretical relations from SSP models that use standard horizontal-branch recipes, generally predict that such multi-modality in colours maps to multi-modality in metallicity. Detailed spectroscopic studies for a few ETGs with large sample sizes and high-quality spectra generally support the picture of metallicity multi-modality [70–73].

This result of bi-modality was both appealing, since it seems to correspond to the bi-modal metallicity distribution in the Milky Way GC system, but also problematic since producing multi-model metallicity distributions from a modelling point of view proved quite challenging (Sect. 7). Recent investigations have questioned the validity of "one-size fits all" colour-metallicity relations, and the metallicity distributions inferred from broad-band colours. It has been argued [74] that the colour-metallicity relations for GCs are non-linear in such a way that intrinsicially unimodal metallicity distributions can be multimodal in colour. This picture is in some disagreement with studies of spectroscopic metallicities (see above) and kinematics (Sect. 5.4) and is an area of ongoing research.

Recent studies have also shown that the colour-colour relations for GCs may vary in individual galaxies, and as a function of environment [75]. This is a major puzzle, since SSP models predict that for a given age and abundance ratio, any given combination of two colours should trace unique locii as a function of metallicity. Again, this puts into question whether one can use colours as proxies for metallicity for GC systems. The cause of these variations in colour are unknown, but may possibly be related to abundance variations in the GCs, perhaps similar to that seen in the Milky Way GCs (Sect. 4.2.2).

## *5.2 Late-Type Galaxies*

The GCs systems of spiral galaxies have traditionally received less observational attention than the ETGs. This mainly stems from the problems of identifying GCs in imaging with a spatially varying background (spiral arms etc.). In addition, late-type galaxies tend to have less GCs than ETGs for a given mass, and also do not reach the very high masses of the most massive ETGs. A notable exception is that of the Andromeda galaxy (M31)—the best-studied GC system of a spiral galaxy with the exception of the Milky Way. M31 has ~500 known GCs, roughly three times the size of the known Milky Way population. Most of these clusters have spectroscopic metallicity estimates, and the metallicity distribution of the clusters looks quite broad, but does not show clear sub-populations in metallicity [76]. The distance of M31 ($d = 780\,\mathrm{kpc}$) makes secure age determinations from deep CMDs impractical, but the presence of evolved stars (giant branch, horizontal branch) and spectroscopic ages suggest that the majority are old clusters. Wide-area surveys of M31 show numerous structures such as shells and tidal streams which are suggestive of past accretion events (e.g. [77]). A number of the GCs seem to be spatially and kinematically associated with some of these structures offering evidence for the accretion of GCs onto the halo of this galaxy.

Beyond the Local Group, several studies have identified what appear to be clusters associated with the discs rather than the halos of their parent spirals [78, 79]. Evidence comes both the spatial distributions of the cluster systems and also signs of ordered rotation whose rotation axis appear aligned with that of the galaxy gas or stellar disc. Since it is unlikely that merging or accretion events can give rise to such disc-like properties, the implication is that GCs can form in the discs of late-type galaxies. Similar conclusions have been reached for the GC system of the Large Magellanic Cloud (see Sect. 5.3). If the discs of spiral galaxies do contain GCs with ages comparable to Milky Way GCs, then these discs must have been formed at high redshift ($z > 2$).

## 5.3 *Dwarf Galaxies*

It is important to characterise the properties of the GC systems of dwarf galaxies for a number of reasons, not least being that the progenitors of dwarf galaxies (and their GCs) are thought to build up the halos of more massive galaxies. Most of the more massive dwarf galaxies in the Local Group have GCs and have been studied with both the resolved and unresolved methods mentioned above. There are 14 Local Group dwarfs known with GCs; and the census of these systems continues to grow with increasingly wider-field, high-resolution imaging surveys (see e.g., [80]).

The most massive Local Group dwarf with GCs is the Large Magellanic Cloud (LMC) which has 16 known clusters. CMD and spectroscopic analysis indicates that they are old and metal-poor [81, 82]—a general property of GCs in dwarfs. Interestingly, the kinematics of these clusters suggest that they might be part of a disc system which, if correct, would have formed at $z > 1$ based on the cluster ages [82]. This bears similarities to suggestions of "discy" GC systems seen in some spiral galaxies (Sect. 5.2). At the other end of the mass scale, the Pegasus dwarf has a sole GC located very close to the galaxy centre [83]. For low-mass dwarfs the properties of their GCs become particularly interesting since they offer the potential of gaining insight into the shape of the dark matter halos of these systems [84, 85].

Beyond the Local Group, GCs have been pivotal in understanding some of the properties of low surface brightness galaxies (the precise definition varies, but these are galaxies typically fainter than the night sky brightness at a dark sight of $\mu(V) \sim 22\,\mathrm{mag\,arcsec^{-2}}$). Recently, there has been a focus on "ultra-diffuse" galaxies (UDGs),[13] which are a class of low surface brightness galaxy with dwarf-like stellar masses $10^{7-8}\,\mathrm{M_{\odot}}$, but large radii (>1.5 kpc) [86]. UDGs can have rich GC systems, and velocity measurements of the GCs provided the first dynamical mass measurements for these systems [87] (see also Sect. 6).

The dwarf galaxy populations in galaxy clusters are numerically dominated by dwarf elliptical galaxies (dEs). These are spheroidal systems that look superficially like scaled-down versions of ETGs. The origin of these systems is unknown, but is thought to be related to their environment since they are rare in the field. One prominent model is that they are transformed from late-type, gas-rich dwarfs (dwarf irregulars; dIrrs) via tidal processes [88, 89]. However, studies of the properties of the GCs in late-type dwarfs and dEs suggest that this model is problematic; dEs tend to have richer GC systems that dIrrs. To solve the problem, new clusters must be formed during the transformation process and these young clusters are generally not seen in dEs [90].

## 5.4 *Kinematics*

Spectra, and therefore radial velocities, of GCs can be obtained for distances out ~20 Mpc. This includes a number of important clusters and groups such as the Virgo and Fornax clusters, the Leo group and the Centaurus group. Early work on GC kinematics focused on confirming the association of extragalactic GCs with their parent galaxies with radial velocities. It was then quickly recognised that GCs are useful tracers of the mass distributions of galaxies since they extend beyond the observable galaxy light. Studies of the velocities of GCs in ETGs, spirals and UDGs invariably show the need for dark matter to explain the observations. Typically, the observed random motions of the GCs (their velocity dispersions) is higher than would be expected if only the observed stars and gas contribute to the mass of the galaxy. Along with X-ray studies and gravitational lensing, GCs provide some of the strongest evidence for the presence of dark matter at large radii in ETGs (e.g., [26, 91–94]).

GC kinematics also brings useful insights into their connection with their host galaxies. The kinematics of the metal-rich GCs generally looks very similar to that of the central stars in galaxies [26, 53, 95–97]. On a galaxy-by-galaxy basis,

[13] Not to be confused with "ultra-faint" galaxies which have smaller sizes and significantly lower stellar masses.

the metal-rich GCs and stars have similar velocity dispersions and rotational properties. This suggests a close relation between the formation processes of the two components. The metal-poor GCs tend to show important differences from the galaxy stars, including differing velocity dispersions, rotation magnitudes (and directions) and also orbital properties. These differences reinforce the picture that the red and blue GCs—at least in part—comprise distinct populations, and that their formation pathways are also distinct.

## *5.5 Scaling Relations Between Globular Cluster Systems and Galaxies*

GC systems exhibit a number of properties that scale with those of their host galaxies. These relations not only support the idea that galaxy formation and GC formation are intimately connected, but also bring useful insight into these formation mechanisms. Some of these relations are described below.

### 5.5.1 The Total Numbers of Globular Clusters in a Galaxy

One of the most basic measures of a GC system is the total number of GCs ($N_{\mathrm{GC}}$) in the galaxy. Traditionally, the total number of GCs has been estimated by counting GCs up to the peak of the GC luminosity function (GCLF), and then doubling this number [98]. With this procedure, a symmetrical luminosity function (e.g., a normal distribution) is assumed. It turns out that this is a reasonable approximation, but in detail the GCLF shows more of a tailed distribution known as an evolved Schechter function [99] (although not directly relevant to the present discussion, it is also worth noting that the peak GCLF can be used as a "standard candle" to measure distances with an accuracy of about $\sim$10%).

Plotting $N_{\mathrm{GC}}$ versus the stellar mass ($M_*$) of the host galaxy shows that $N_{\mathrm{GC}}$ correlates positively with $M_*$. By assuming a mass-to-light ratio for the clusters, we can also plot the total GC system mass ($M_{\mathrm{GC}}$) versus $M_*$ which is useful since it compares two masses, rather than a number and a mass. This is shown in the left-hand panel of Fig. 4 taken from [100]. A positive correlation in $M_{\mathrm{GC}}$–$M_*$ may not seem so surprising; the more stars in a galaxy then the more GCs you might have (which are made of stars). However, this result is interesting since it suggests a link between the GC system (total number, or mass) and that of the host galaxy (stellar mass). Further inspection of the figure also reveals that $M_{\mathrm{GC}}$–$M_*$ is not linear. This implies that there is no one-to-one correlation between the two observables. Or, another way to look at it is that the ratio $M_{\mathrm{GC}}/M_*$ is not constant as a function of mass. It seems that low-mass galaxies (dwarfs) and some very massive galaxies are better at making GCs than stars than is the case for galaxies of intermediate masses.

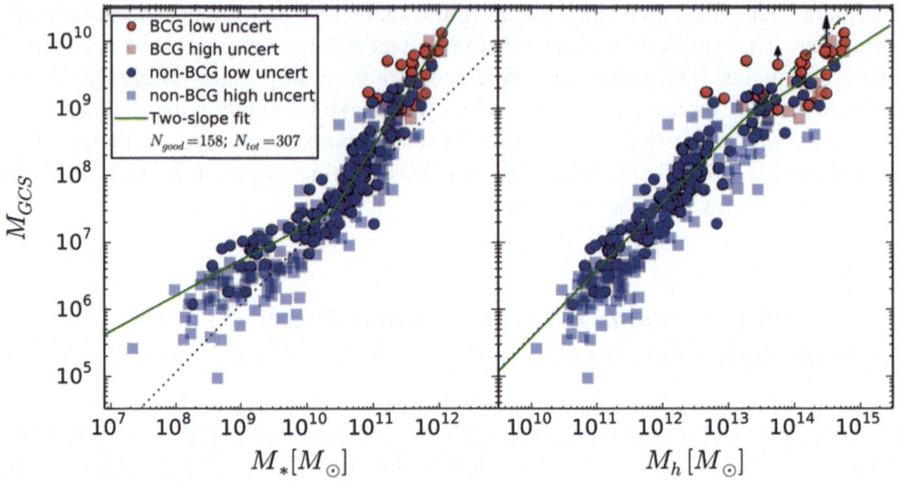


**Fig. 4** Plots of the total mass of globular cluster systems versus parent galaxy stellar mass (left panel) and dark matter halo mass (right-hand panel). The figure (taken from [100]) shows that globular cluster systems are connected to the properties of their host galaxies, and show a direct (linear) relation with the dark matter halo of their parent galaxy

The situation changes when one plots $M_{\rm GC}$ versus the dark matter halo mass ($M_{\rm halo}$) of the host galaxy. Measuring $M_{\rm halo}$ is not straightforward—we only see the gravitational effects of dark matter not the dark matter itself—however techniques such as weak lensing can provide meaningful constraints on this quantity (e.g., [100]). The right-hand panel of Fig. 4 shows $M_{\rm GC}$ plotted against $M_{\rm halo}$. Unlike the case for $M_*$, this does indicate that there is a linear relation between the two quantities which seems to hold down to at least $M_{\rm halo} \sim 10^{11}\,{\rm M}_\odot$. This implies that there is a constant ratio between the GC system mass and the host galaxy halo mass. At first sight, a linear relation between GC system *stellar mass* and host galaxy *dark matter mass* might be hard to understand. This result has been used to argue that there might exist a fundamental GC—dark matter connection. Somehow the GC system of a galaxy "knows" about what sort of dark matter halo it will end up in. The origin of this relation is not fully understood, but a relatively simple explanation may be found in [101]. Essentially, these authors argue that a linear $M_{\rm GC}$–$M_{\rm halo}$ relation emerges as a result of the central limit theorem and galaxy merging; the merging of low- and high-mass haloes and their galaxies leads to "average" halo properties and GC systems which produce a constant GC to halo mass ratio.

These results above indicate that GCs can be used to trace the properties of the dark matter of their host galaxies. One interesting application is to infer dark matter halo masses for galaxies by measuring $N_{\rm GC}$. This has been done for UDGs (Sect. 5.3). The results of several studies show that typical halo masses for these galaxies are inferred to be $10^{11}\,{\rm M}_\odot$, as derived from $N_{\rm GC}$, which is consistent with masses from GC dynamics and stellar velocity dispersions.

### 5.5.2 Sizes of Globular Cluster Systems

Another GC system property that correlates with the mass of the host galaxy is the size of the GC system [102, 103]. This is typically measured as the "half number radius" (GC $r_e$), the radius that contains half of the GC system. GC $r_e$ correlates positively with both $M_*$ and $M_{halo}$.

These correlations are similar to those seen for galaxies; galaxies obey a size–mass relation in that more massive galaxies are, on average, larger. Interestingly, galaxies at a fixed stellar mass are more compact at higher redshift. For example, at $z = 2$, galaxies with $M \sim 10^{11}\,M_\odot$ are approximately 3–4 times smaller than nearby galaxies (e.g., [104]). The origin of the "size evolution" of massive galaxies from high redshift to today is not fully understood, but it is believed that mergers play a major role by "puffing-up" compact galaxies, or by adding additional material to their outskirts to make them grow in physical size. It is possible that merging and accretion also give rise to the GC size—galaxy mass relations reported in the literature.

### 5.5.3 The Colours of Globular Cluster Systems

Studies of ETGs with a range of masses shows that the mean colour (metallicity) of the whole GC system correlates positively with mass of the host galaxy [68, 105]. More massive ETGs generally have redder GC systems. In addition, the colours of both the red and blue subpopulations scale with galaxy stellar mass. In terms of colour, the relations look a bit different, with the relation for the red clusters being about ~5 times steeper than that for the blue clusters. However, when converted into metallicity, the relations for the both the blue and red clusters appear quite similar [68].

The standard interpretation of these observations is that GCs are, on average, able to achieve higher levels of metal enrichment in more massive galaxies. This is similar to the case for the stars in galaxies themselves, which also follow a mass-metallicity relation [106, 107].

## *5.6 Extreme Globular Cluster Metallicities?*

The maximum and minimum metallicities that a GC achieves offers interesting information on the nature of the interstellar medium (ISM) and star cluster formation early in the history of a galaxy. The most metal-poor globular in the Milky Way is M15 with [Fe/H]$\sim -2.5$ dex. This is less than 1/300 of the solar value. Based on current data, it seems that a few, if any, extragalactic GCs form with metallicities much lower than this (see [108] for a recent data compilation). This "metallicity floor" of GCs is some 5 orders of magnitude higher than the metallicity of the most metal-poor stars known in the Milky Way [109]. There are a number of interpretations for these observations. Perhaps a low-metallicity ISM is unable to form massive, long-lived clusters due to different fragmentation properties of the gas at these metallicities. Alternatively, it is possible that the sites of GC formation at high-redshift are too low mass at this metallicity to be able to form massive clusters. This latter idea has been developed by Kruijssen [110]. Understanding the origin of this metallicity floor will bring useful insights into the earliest phases of star and cluster formation.

On the other end of the metallicity scale, the most metal-rich GCs in the Milky Way lie at or near solar metallicities (e.g., NGC 6528 and Pal 10) [16]. There is no strong evidence that there are significant numbers of GCs in other galaxies with [Fe/H]$>$ 0.0. This is perhaps surprising for the case of giant elliptical galaxies whose central regions tend to be dominated by super-solar metallicity stars [63, 111, 112]. This result might provide a hint that these GCs form relatively early in the star formation process in these galaxies, whose star formation timescales are inferred to be $<$1 Gyr.

## 6 Oddball Galaxies and Their Globular Clusters

In many cases in astronomy, it is the differences from the "normal" population that can give useful insights into underlying astrophysical processes. This is also the case in the study of galaxies and their GC systems. Below a few notable "oddball" systems are highlighted.

### *6.1 NGC 4365: A Trimodal Globular Cluster System*

NGC 4365 is a luminous ETG which lies about 6 Mpc behind the Virgo cluster of galaxies. Superficially it looks like a normal ETG, although it has a "kinematically decoupled core" (the galaxy centre of the galaxy rotates in the opposite sense to the main galaxy) sometimes taken to be indicative of past merger events [113]. The galaxy itself, however, appears very old [114].

It turns out that the GC system of NGC 4365 is quite unusual. NGC 4365 has been identified as a galaxy with a "trimodal" colour distribution for its GC system; it has blue, "green" and red GC subpopulations [115–118]. The origin of the central green peak has been the subject of some debate in literature, and is still uncertain. One possibility is that the green GCs may result from a gaseous galaxy merger resulting in the formation of new GCs some $\sim$4 Gyr ago. Alternatively, they may have been stripped (physically removed via gravitational interactions) from the nearby galaxy NGC 4342 [119].

## 6.2 NGC 1277: A Relic Galaxy with Only Red Globular Clusters

NGC 1277 is a particularly interesting case of an unusual GC system. NGC 1277 is an extremely old, compact and massive galaxy near the centre of the Perseus galaxy cluster. The galaxy has been identified as a candidate "relic galaxy" which is to say the remnant of the early in situ phase in massive galaxy formation without significant subsequent merging or accretion [120]. In this view, NGC 1277 is regarded as a near-pristine "core" of a normal massive ETG. Beasley et al. [121] explored the GC system with *HST* imaging and found that NGC 1277 has few, if any, blue GCs. Since the metal-poor GCs in massive galaxies are generally regarded (at least in part) as a population brought in by the accretion of lower-mass satellites (Sect. 4.1), an interpretation of the NGC 1277 observations is that the galaxy has undergone very little mass accretion since its formation at high redshift. Somehow NGC 1277 managed to avoid accreting smaller satellite galaxies during its lifetime, possibly due to the fact it is itself a satellite of the more massive galaxy NGC 1275 which may act to swallow all the material in its vicinity.

## 6.3 NGC 1052-DF2: A Galaxy with No Dark Matter?

"Dragonfly 2" (hereafter DF2; also known as [KKS2000]04) is a low surface brightness dwarf galaxy thought to lie some 20 Mpc distant in the direction of the massive ETG NGC 1052 [122]. A study of the kinematics of DF2's GCs suggests that it may have a very low dark matter fraction, compatible with no dark matter at all. This result is intriguing since galaxies generally do have dark matter, and indeed this is one of the definitions of a galaxy (see Sect. 2). More important than the definition, in modern galaxy formation theory dark matter is generally required in order to produce the galaxies we see around us so DF2 might pose a problem for cosmological models. In addition to an apparent lack of dark matter, the galaxy appears to have an unusual system of GCs in that they are extremely luminous, with a GCLF that is about a magnitude brighter than the usual value ($M_V \sim -7.5$) (GCs are generally regarded as a standard candle—see Sect. 5.5.1).

If DF2 is truly a dark matter deficient galaxy, with a rather peculiar GC system, then it challenges some of our ideas of how galaxies and their GCs can form. However, the galaxy is not without controversy. Trujillo et al. [123] have argued that DF2 is about twice as close to us as the distance reported by van Dokkum et al. [122]. This, with additional, different assumptions in their analysis, lead [123] to conclude that DF2 is an ordinary dwarf galaxy with what looks light an ordinary GC system. At the time of writing this chapter, the jury is out on this one.

# 7 Simulating Globular Cluster Systems

The improved understanding of star formation and feedback processes, and increasing computing power have allowed for increasingly sophisticated models of GC formation. "Semi-analytic" models and numerical, hydrodynamical simulations have been used to understand the connection between GCs and their host galaxies.

## *7.1 Early Models*

Some of the earliest works to make explicit connections between GC system formation and galaxy formation were largely phenomenological[14] in nature. Searle and Zinn [124] used the mono-metallicity of Milky Way GCs and the lack of a trend in this metallicity with Galactocentric radius (the lack of a metallicity gradient) to argue that the Milky Way halo and its GCs were built-up from "proto-Galactic fragments". This work contrasted quite strongly with an extremely influential paper by Eggen et al. [125] who considered a smooth, rapid collapse of an early proto-Galactic gas cloud as the precursor to the formation of the Milky Way. It turns out that to some extent these apparently contrasting models have been absorbed in the current picture of galaxy formation. The picture presented by Eggen et al. [125] looks a lot like in situ formation, while [124] has much in common with ex-situ accretion.

More specific to the formation of GC systems, [126] developed a "major merger" model whereby disk galaxies merge to form elliptical galaxies, and in the process produce new, metal-rich GCs. These newly formed metal-rich GCs comprise the "red" population, while the "blue" population is brought in as the original GC populations of the merging spirals. This model predicted multi-model colour distributions in elliptical galaxies which were subsequently observed. However, in detail the model had problems. For example it does not explain the sometimes multi-modal colour distributions of spiral galaxies, nor offers an explanation for the origin of the original blue GC populations.

Growing recognition that the galactic halo is composed (at least in part) by accreted dwarfs lead to the analytic accretion model of GC formation [127]. This modelled the build up of cluster systems in massive galaxies via the accretion of lower-mass dwarf satellites. In this model the red population is an in situ population of clusters, formed with the galaxy stars, and the blue population is brought in by the accreted satellites. Although the details of GC formation are not described in

[14] I.e., a model that describes relationships between observations, but does not stem directly from physical theory.

the model, in spirit this model is consistent with the presently favoured formation pathway for massive ETGs and their GC systems.

### 7.2 *Semi-analytical Models*

The first attempt to place GC formation in a cold dark-matter galaxy formation model was that of [128] using so-called "semi-analytic" models (SAMs). SAMs are analytical galaxy formation models that have some of their "free parameters" calibrated based on observations.[15] They model the growth and merger history of dark matter haloes as a function of redshift, following the evolution of gas, stars and galaxy formation in these haloes. Beasley et al. [128] assumed that GCs formed wherever stars formed, both in gas discs and also during gas-rich major mergers. The model had several observational successes, but in order to produce the colour bimodality a halt to star formation (a "truncation") had to be imposed in the gas discs at high redshift. Possible explanations for this truncation were the re-ionisation of the Universe, or due to the specific pressure conditions in these discs.

More recently, SAMs have been implemented which trace the merger histories of dark matter halos, galaxies and their GCs, but do not always explicitly implement GC formation [101, 129, 130]. A "merger tree", showing the growth of galaxy mass via mergers from the SAM of [101] is shown in Fig. 5. These works have had success in reproducing some of the GC–galaxy scaling relations such as the $M_{\rm GC}$–$M_{\rm halo}$ relation discussed in Sect. 5.5. An important advantage of SAMs is that they are computationally fast. Many "virtual" galaxies can be produced in a few minutes on a desktop computer. A potential limitation of SAMs is that they use approximations to calculate various physical processes such as gravitation and merging, gas cooling and star formation.

### 7.3 *Numerical Simulations*

An alternative, complementary approach to SAMs is to use hydrodynamical simulations. These are numerical simulations that follow the evolution of gas and star formation via direct simulation rather than analytic approximations. If the hydrodynamical simulations are "cosmological", then they also follow the evolution of the dark matter component of galaxies based on our understanding of the current

[15]In the case of the SAM used by Beasley et al. [128], the main calibrations were to match the galaxy luminosity function and Tully-Fisher relations.

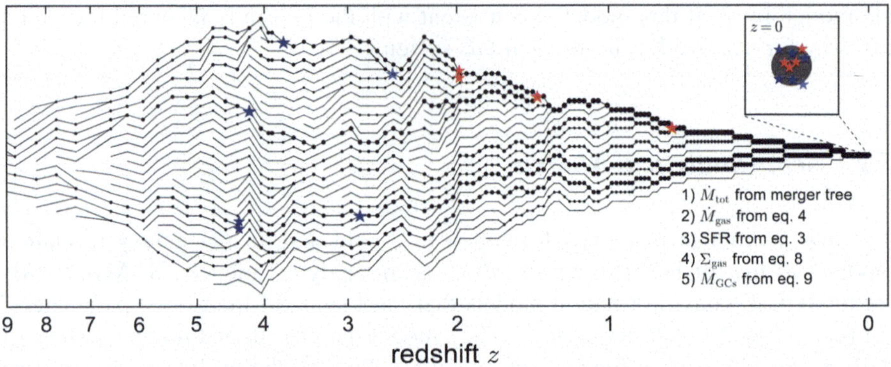


**Fig. 5** Example of a "merger tree" from the semi-analytic models of [101]. The black lines indicate the evolution of individual haloes as a function of redshift merging to become a single galaxy at $z = 0$. The blue and red stars indicate globular cluster formation events. The figure was adopted from [101]

cosmological model. For example, [131] simulated a high-redshift, disc-like galaxy and found that gaseous discs at $z \sim 3$ are plausible sites for the formation of compact objects that may go on to become present-day GCs.

Hydrodynamical simulations of galaxy and cluster formation are computationally more expensive than SAMs; a simulation may take weeks, months or even years to run on a supercomputer, depending upon the volume and sophistication of the simulation. However, they have the important advantage of being able to trace the detailed physics of gas cooling, star formation and feedback which is crucial to understanding GC formation. Unfortunately, cosmological, hydrodynamical simulations presently lack the combined spatial (sub-pc scales) and mass resolution ($\sim 10^5$ $M_\odot$) to directly model GC formation and evolution down to the present day (i.e., $z = 0$). Because of these computational limitations, "subgrid" recipes (analytic approximations similar in nature to those used in SAMs) are implemented to capture the necessary physics of cluster formation. An example of this is the E-MOSAICS project which use the EAGLE cosmological simulations with specific recipes for GC formation [132].

This said, several teams have had recent successes in directly resolving GC formation in simulations that have been limited to high redshifts [133–135]. Figure 6 shows some "virtual GCs" created in the cosmological simulations from [135].

The general picture from these and other works is that GC formation may occur anywhere that is gas-rich and turbulent, such that high pressure regions can form. These regions may be in gas discs, mergers of galaxies, in the very centres of the potential wells of massive proto-galaxies, or perhaps in cold filamentary accretion [133]. The precise conditions for cluster formation are not well understood, but high gas pressures and densities are probably a key requirement in order to create a compact, bound stellar system [34]. The above works point to a favoured epoch of

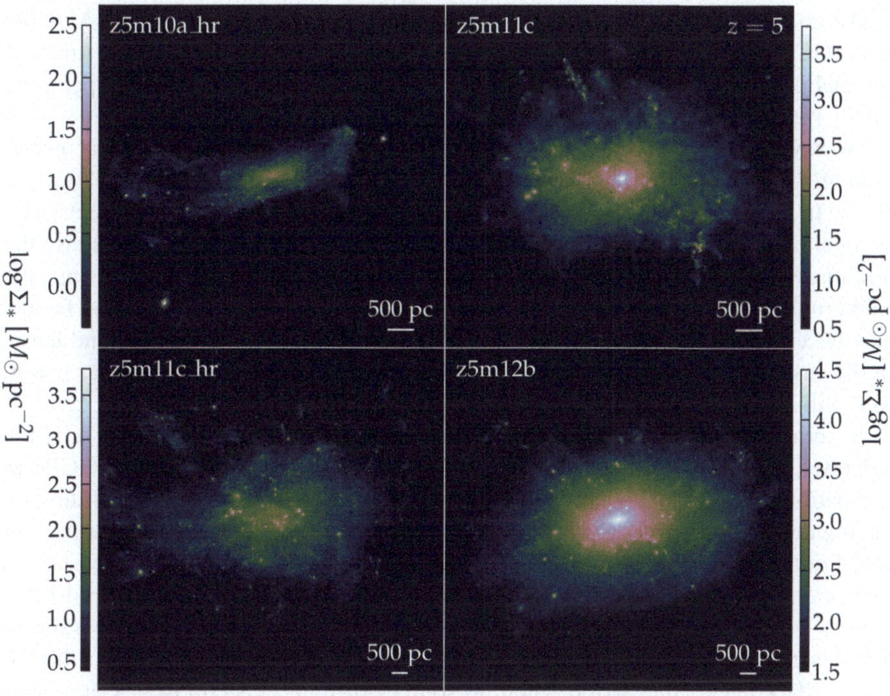


**Fig. 6** Cosmological simulations of globular cluster formation at $z = 5$ (taken from [135]). The colours indicate the density of stars, with lighter colours corresponding to higher densities. Small, point-like objects of high density are star clusters—possibly young globular clusters

GC system formation, which lies somewhere above $z > 2$. This is consistent with the ages of the vast majority of the Milky Way GCs (Sect. 4).

## 8 Globular Clusters at High Redshift

The Milky Way GCs are ancient stellar systems (Sect. 4), and cosmological simulations place the principal formation epoch of GCs at $z > 2$ (Sect. 7). Ideally, to test these models one wants to be able to directly observe the formation of GCs at high redshift, and trace this evolution across different environments and across cosmic time. The greatest challenge to identifying and studying GCs at the highest redshifts is that they are extremely compact. Using special ("PSF-deconvolution") techniques, objects with sizes ~100 pc can be resolved with *HST* at $z \sim 6$ (e.g., [136]. A typical, young GC might have $r_h < 20$ pc and so will be unresolved using standard techniques. Being able to measure the cluster size is crucial, since only a detection in itself will not distinguish a young cluster from (for example) a star forming region in a galaxy, or a compact galaxy in formation.

An additional problem is one of sensitivity. At $z = 6$, a $10^7\,M_\odot$, 10 Myr old proto-GC will have rest-frame $M_{UV} \sim 31$ ($M_V \sim 29$). This is beyond the limits of detectability of *HST*, but should be within the potential capabilities of the upcoming *James Webb Space Telescope* (*JWST*) [137, 138].

However, high redshift work studying compact objects has shown that *gravitational lensing* already has the potential to characterise "GC precursors" (GCPs; [136]. In gravitational lensing, the gravitational field of a foreground object (such as a galaxy) acts as a lense which can magnify the apparent size and increase the brightness of a background source (in this case, a GCP). In strong lensing, the background source may be stretched or show multiple images. Magnification factors may be x10–100, depending upon the precise configuration of the source and lense. In the case of *HST*, this allows for effective spatial resolutions of $\sim$10 pc at $z = 6$, well within the expected size range of GCPs.

Several studies have now claimed the possible identification GCPs at high redshift [136, 139, 140] which can be compared to the properties of GCPs in simulations (Sect. 7). This new area of research opens up the possibility of directly tracing the formation of GC systems when the Universe was a fraction of its present age. Equally as exciting is the prospect of better understanding the process of *re-ionization*. At some point early in the lifetime of the Universe (perhaps at redshifts, $z \sim 6 - 20$), neutral hydrogen distributed throughout the Universe was ionized to a plasma by source(s) of energetic (UV) photons. This process ended the so-called "dark ages", the period when no sources of light existed [141]. Candidates for the sources of energetic photons include dwarf galaxies in formation, the first "population III" stars and massive black holes in the centres of galaxies in the form of active galactic nuclei. However, it turns out that GCs, given their short star formation timescales, high redshifts of formation and sheer numbers (about 2 GCs $\mathrm{Mpc}^{-3}$) [142] may be important re-ionization sources. In fact, they may even turn out be the dominant contributors to the UV ionizing background [142, 143].

## 9 Globular Clusters and Galaxy Formation

The observational and theoretical works on GC systems, a fraction of which has been mentioned in this chapter, have allowed researchers to build a general picture of the co-evolution of galaxies and their GC systems. Many of the details still need to be worked out, but a general scenario may be described within the framework of a *two-phase* model of galaxy formation [144]. In this framework, an in situ phase builds the centres of galaxies as gas cools and forms stars in dark matter halos at $z >$ 2. Subsequently, an ex-situ phase occurs whereby lower mass galaxies are accreted over time to build massive galaxy halos. This accretion phase is still occurring at the present day.

In this context, the formation of GC systems may proceed as follows: The first generations of stars begin to form and then explode, enriching the interstellar medium to metallicities of [Fe/H]$\sim -2.5$, which is the approximate minimum metallicity seen in GCs. During the in situ phase of galaxy formation, in the densest regions of ongoing star formation, the progenitors of today's GCs proceed to form from metal-rich gas. This formation may proceed in gaseous discs or merging gas-rich systems, but preferentially occurs near the centres of what will become massive galaxies. This produces the red, metal-rich sub-populations of GC systems seen in massive galaxies. In contrast, low metallicity GCs are preferentially formed in low-mass galaxies (dwarf galaxies) which, if accreted onto a more massive galaxy, will go on to form part of the halo (blue, metal-poor) GC population of the galaxy. This represents the ex-situ phase of GC system formation. In all cases, the formation of individual GCs must occur rapidly (in a few Myr) so as to prevent a second generation of stars creating significant age or metallicity spreads in the majority of GCs, and preferentially occurs at $z > 2$.

## 10 Summary and Outlook

In this chapter, we have tried to give a taste of some of the research on GC systems and their connection to galaxy formation. In the case of the Milky Way, detailed ages and abundances of its GCs are increasingly revealing unique information on the formation of our Galaxy and the formation processes of GCs themselves. Further afield, extragalactic GC systems provide information on a range of topics, from the dark matter distributions of galaxies to the mass accretion histories of galaxies inhabiting a range of extreme environments not represented by the Local Group. Going to high redshift, astronomers are at the point where GC formation can be directly observed, and compared to increasingly sophisticated cosmological simulations. The next generation of telescopes such as *JWST* and up-coming 30-m class ground-based facilities (E-ELT, TMT, GMT) will accelerate this rapidly developing, exciting field.

So go and have a look at M13, just to the west from the center of the constellation of Hercules, and have a think about that.

**Acknowledgements** The author thanks Núria Salvador Rusiñol for proof-reading and useful feedback on the text. This work has been supported through the RAVET project by the grant AYA2016-77237-C3-1-P from the Spanish Ministry of Science, Innovation and Universities (MCIU) and through the IAC project TRACES which is partially supported through the state budget and the regional budget of the Consejería de Economía, Industria, Comercio y Conocimiento of the Canary Islands Autonomous Community.

## References

1. J.P. Brodie, J. Strader, Annu. Rev. Astron. Astrophys. **44**(1), 193 (2006). https://doi.org/10.1146/annurev.astro.44.051905.092441
2. R.G. Gratton, E. Carretta, A. Bragaglia, Astron. Astrophys. Rev. **20**, 50 (2012). https://doi.org/10.1007/s00159-012-0050-3
3. N. Bastian, C. Lardo, Annu. Rev. Astron. Astrophys. **56**, 83 (2018). https://doi.org/10.1146/annurev-astro-081817-051839
4. W.E. Harris, Annu. Rev. Astron. Astrophys. **29**, 543 (1991). https://doi.org/10.1146/annurev.aa.29.090191.002551
5. D.A. Forbes, N. Bastian, M. Gieles, R.A. Crain, J.M.D. Kruijssen, S.S. Larsen, S. Ploeckinger, O. Agertz, M. Trenti, A.M.N. Ferguson, J. Pfeffer, O.Y. Gnedin, Proc. R. Soc. London, Ser. A **474**(2210), 20170616 (2018). https://doi.org/10.1098/rspa.2017.0616
6. J. Silk, A. Di Cintio, I. Dvorkin, *Proceedings of the International School of Physics 'Enrico Fermi' Course 186 'New Horizons for Observational Cosmology'*, vol. 186 (2014), pp. 137–187. https://doi.org/10.3254/978-1-61499-476-3-137
7. M. Cappellari, Annu. Rev. Astron. Astrophys. **54**, 597 (2016). https://doi.org/10.1146/annurev-astro-082214-122432
8. J. Kennicutt, Robert C., Annu. Rev. Astron. Astrophys. **36**, 189 (1998). https://doi.org/10.1146/annurev.astro.36.1.189
9. A. Renzini, Annu. Rev. Astron. Astrophys. **44**(1), 141 (2006). https://doi.org/10.1146/annurev.astro.44.051905.092450
10. J. Bland-Hawthorn, O. Gerhard, Annu. Rev. Astron. Astrophys. **54**, 529 (2016). https://doi.org/10.1146/annurev-astro-081915-023441
11. T. Naab, J.P. Ostriker, Annu. Rev. Astron. Astrophys. **55**(1), 59 (2017). https://doi.org/10.1146/annurev-astro-081913-040019
12. A.W. McConnachie, Astron. J. **144**(1), 4 (2012). https://doi.org/10.1088/0004-6256/144/1/4
13. J.D. Simon, Annu. Rev. Astron. Astrophys. **57**, 375 (2019). https://doi.org/10.1146/annurev-astro-091918-104453
14. I. King, Astron. J. **67**, 471 (1962). https://doi.org/10.1086/108756
15. C.J. Peterson, I.R. King, Astron. J. **80**, 427 (1975). https://doi.org/10.1086/111759
16. W.E. Harris, Astron. J. **112**, 1487 (1996). https://doi.org/10.1086/118116
17. G. Meylan, D.C. Heggie, Astron. Astrophys. Rev. **8**, 1 (1997). https://doi.org/10.1007/s001590050008
18. O.Y. Gnedin, J.P. Ostriker, Astrophys. J. **474**(1), 223 (1997). https://doi.org/10.1086/303441
19. S.M. Fall, Q. Zhang, Astrophys. J. **561**(2), 751 (2001). https://doi.org/10.1086/323358
20. V.C. Rubin, J. Ford, W. Kent, Astrophys. J. **159**, 379 (1970). https://doi.org/10.1086/150317
21. Y. Sofue, V. Rubin, Annu. Rev. Astron. Astrophys. **39**, 137 (2001). https://doi.org/10.1146/annurev.astro.39.1.137
22. W. Forman, C. Jones, W. Tucker, Astrophys. J. **293**, 102 (1985). https://doi.org/10.1086/163218
23. P.J. Humphrey, D.A. Buote, F. Gastaldello, L. Zappacosta, J.S. Bullock, F. Brighenti, W.G. Mathews, Astrophys. J. **646**(2), 899 (2006). https://doi.org/10.1086/505019
24. R. Mandelbaum, C.M. Hirata, U. Seljak, J. Guzik, N. Padmanabhan, C. Blake, M.R. Blanton, R. Lupton, J. Brinkmann, Mon. Not. R. Astron. Soc. **361**(4), 1287 (2005). https://doi.org/10.1111/j.1365-2966.2005.09282.x
25. M. Cacciato, F.C. van den Bosch, S. More, R. Li, H.J. Mo, X. Yang, Mon. Not. R. Astron. Soc. **394**(2), 929 (2009). https://doi.org/10.1111/j.1365-2966.2008.14362.x
26. J. Strader, A.J. Romanowsky, J.P. Brodie, L.R. Spitler, M.A. Beasley, J.A. Arnold, N. Tamura, R.M. Sharples, N. Arimoto, Astrophys. J. Suppl. Ser. **197**(2), 33 (2011). https://doi.org/10.1088/0067-0049/197/2/33
27. J.A. Arnold, A.J. Romanowsky, J.P. Brodie, D.A. Forbes, J. Strader, L.R. Spitler, C. Foster, C. Blom, S.S. Kartha, N. Pastorello, V. Pota, C. Usher, K.A. Woodley, Astrophys. J. **791**(2), 80 (2014). https://doi.org/10.1088/0004-637X/791/2/80

28. B. Moore, Astrophys. J. Lett. **461**, L13 (1996). https://doi.org/10.1086/309998
29. B. Willman, J. Strader, Astron. J. **144**(3), 76 (2012). https://doi.org/10.1088/0004-6256/144/3/76
30. P.J.E. Peebles, Astrophys. J. **277**, 470 (1984). https://doi.org/10.1086/161714
31. P. Madau, A. Lupi, J. Diemand, A. Burkert, D.N.C. Lin (2019). e-prints arXiv:1905.08951
32. A.D. Mackey, M.A. Beasley, R. Leaman, Mon. Not. R. Astron. Soc. **460**(1), L114 (2016). https://doi.org/10.1093/MNRASl/slw076
33. E.W. Peng, H.C. Ferguson, P. Goudfrooij, D. Hammer, J.R. Lucey, R.O. Marzke, T.H. Puzia, D. Carter, M. Balcells, T. Bridges, K. Chiboucas, C. del Burgo, A.W. Graham, R. Guzmán, M.J. Hudson, A. Matković, D. Merritt, B.W. Miller, M. Mouhcine, S. Phillipps, R. Sharples, R.J. Smith, B. Tully, G. Verdoes Kleijn, Astrophys. J. **730**(1), 23 (2011). https://doi.org/10.1088/0004-637X/730/1/23
34. B.G. Elmegreen, Y.N. Efremov, Astrophys. J. **480**(1), 235 (1997). https://doi.org/10.1086/303966
35. J.M.D. Kruijssen, Mon. Not. R. Astron. Soc. **454**(2), 1658 (2015). https://doi.org/10.1093/MNRAS/stv2026
36. B. Chaboyer, P. Demarque, P.J. Kernan, L.M. Krauss, A. Sarajedini, Mon. Not. R. Astron. Soc. **283**(2), 683 (1996). https://doi.org/10.1093/MNRAS/283.2.683
37. D.A. VandenBerg, K. Brogaard, R. Leaman, L. Casagrand e, Astrophys. J. **775**(2), 134 (2013). https://doi.org/10.1088/0004-637X/775/2/134
38. E. Vasiliev, Mon. Not. R. Astron. Soc. **484**(2), 2832 (2019). https://doi.org/10.1093/MNRAS/stz171
39. L.L. Watkins, R.P. van der Marel, S.T. Sohn, N.W. Evans, Astrophys. J. **873**(2), 118 (2019). https://doi.org/10.3847/1538-4357/ab089f
40. A. Vazdekis, P. Sánchez-Blázquez, J. Falcón-Barroso, A.J. Cenarro, M.A. Beasley, N. Cardiel, J. Gorgas, R.F. Peletier, Mon. Not. R. Astron. Soc. **404**(4), 1639 (2010). https://doi.org/10.1111/j.1365-2966.2010.16407.x
41. G. Bruzual, S. Charlot, Mon. Not. R. Astron. Soc. **344**(4), 1000 (2003). https://doi.org/10.1046/j.1365-8711.2003.06897.x
42. C. Maraston, Mon. Not. R. Astron. Soc. **362**(3), 799 (2005). https://doi.org/10.1111/j.1365-2966.2005.09270.x
43. R.P. Schiavon, Astrophys. J. Suppl. Ser. **171**(1), 146 (2007). https://doi.org/10.1086/511753
44. C. Conroy, Annu. Rev. Astron. Astrophys. **51**(1), 393 (2013). https://doi.org/10.1146/annurev-astro-082812-141017
45. A.J. Cenarro, M.A. Beasley, J. Strader, J.P. Brodie, D.A. Forbes, Astron. J. **134**(1), 391 (2007). https://doi.org/10.1086/518504
46. B.K. Gibson, D.S. Madgwick, L.A. Jones, G.S. Da Costa, J.E. Norris, Astron. J. **118**(3), 1268 (1999). https://doi.org/10.1086/301013
47. A. Vazdekis, M. Salaris, N. Arimoto, J.A. Rose, Astrophys. J. **549**(1), 274 (2001). https://doi.org/10.1086/319088
48. G. Michaud, G. Alecian, J. Richer, *Atomic Diffusion in Stars* (Springer, Cham, 2015). https://doi.org/10.1007/978-3-319-19854-5
49. J. Binney, L.K. Wong, Mon. Not. R. Astron. Soc. **467**(2), 2446 (2017). https://doi.org/10.1093/MNRAS/stx234
50. D. Massari, H.H. Koppelman, A. Helmi, Astron. Astrophys. **630**, L4 (2019). https://doi.org/10.1051/0004-6361/201936135
51. A. Marín-Franch, A. Aparicio, G. Piotto, A. Rosenberg, B. Chaboyer, A. Sarajedini, M. Siegel, J. Anderson, L.R. Bedin, A. Dotter, M. Hempel, I. King, S. Majewski, A.P. Milone, N. Paust, I.N. Reid, Astrophys. J. **694**(2), 1498 (2009). https://doi.org/10.1088/0004-637X/694/2/1498
52. R. Leaman, D.A. VandenBerg, J.T. Mendel, Mon. Not. R. Astron. Soc. **436**(1), 122 (2013). https://doi.org/10.1093/MNRAS/stt1540

53. M.A. Norris, K. Gebhardt, R.M. Sharples, F.R. Faifer, T. Bridges, D.A. Forbes, J.C. Forte, S.E. Zepf, M.A. Beasley, D.A. Hanes, R. Proctor, S.J. Kannappan, Mon. Not. R. Astron. Soc. **421**(2), 1485 (2012). https://doi.org/10.1111/j.1365-2966.2012.20417.x
54. C.I. Johnson, C.A. Pilachowski, R. Michael Rich, J.P. Fulbright, Astrophys. J. **698**(2), 2048 (2009). https://doi.org/10.1088/0004-637X/698/2/2048
55. K.C. Freeman, *Globular Clusters and Nucleated Dwarf Ellipticals*. Astronomical Society of the Pacific Conference Series, vol. 48 (1993), p. 608
56. J. Hughes, G. Wallerstein, Astron. J. **119**(3), 1225 (2000). https://doi.org/10.1086/301241
57. K. Bekki, K.C. Freeman, Mon. Not. R. Astron. Soc. **346**(2), L11 (2003). https://doi.org/10.1046/j.1365-2966.2003.07275.x
58. A.F. Marino, C. Sneden, R.P. Kraft, G. Wallerstein, J.E. Norris, G. Da Costa, A.P. Milone, I.I. Ivans, G. Gonzalez, J.P. Fulbright, M. Hilker, G. Piotto, M. Zoccali, P.B. Stetson, Astron. Astrophys. **532**, A8 (2011). https://doi.org/10.1051/0004-6361/201116546
59. E. Carretta, A. Bragaglia, R.G. Gratton, S. Lucatello, M. Bellazzini, G. Catanzaro, F. Leone, Y. Momany, G. Piotto, V. D'Orazi, Astron. Astrophys. **520**, A95 (2010). https://doi.org/10.1051/0004-6361/201014924
60. S.L. Martell, E.K. Grebel, Astron. Astrophys. **519**, A14 (2010). https://doi.org/10.1051/0004-6361/201014135
61. B. Tang, C. Liu, J.G. Fernández-Trincado, D. Geisler, J. Shi, O. Zamora, G. Worthey, E. Moreno, Astrophys. J. **871**(1), 58 (2019). https://doi.org/10.3847/1538-4357/aaf6b1
62. M. Fukugita, C.J. Hogan, P.J.E. Peebles, Astrophys. J. **503**(2), 518 (1998). https://doi.org/10.1086/306025
63. S.C. Trager, S.M. Faber, G. Worthey, J.J. González, Astron. J. **120**(1), 165 (2000). https://doi.org/10.1086/301442
64. F. La Barbera, I. Ferreras, A. Vazdekis, I.G. de la Rosa, R.R. de Carvalho, M. Trevisan, J. Falcón-Barroso, E. Ricciardelli, Mon. Not. R. Astron. Soc. **433**(4), 3017 (2013). https://doi.org/10.1093/MNRAS/stt943
65. S.E. Zepf, K.M. Ashman, Mon. Not. R. Astron. Soc. **264**, 611 (1993). https://doi.org/10.1093/MNRAS/264.3.611
66. D. Geisler, M.G. Lee, E. Kim, Astron. J. **111**, 1529 (1996). https://doi.org/10.1086/117894
67. A. Kundu, B.C. Whitmore, Astron. J. **121**(6), 2950 (2001). https://doi.org/10.1086/321073
68. E.W. Peng, A. Jordán, P. Côté, J.P. Blakeslee, L. Ferrarese, S. Mei, M.J. West, D. Merritt, M. Milosavljević, J.L. Tonry, Astrophys. J. **639**(1), 95 (2006). https://doi.org/10.1086/498210
69. W.E. Harris, S.M. Ciccone, G.M. Eadie, O.Y. Gnedin, D. Geisler, B. Rothberg, J. Bailin, Astrophys. J. **835**(1), 101 (2017). https://doi.org/10.3847/1538-4357/835/1/101
70. J. Strader, M.A. Beasley, J.P. Brodie, Astron. J. **133**(5), 2015 (2007). https://doi.org/10.1086/512770
71. M.A. Beasley, T. Bridges, E. Peng, W.E. Harris, G.L.H. Harris, D.A. Forbes, G. Mackie, Mon. Not. R. Astron. Soc. **386**(3), 1443 (2008). https://doi.org/10.1111/j.1365-2966.2008.13123.x
72. A. Alves-Brito, G.K.T. Hau, D.A. Forbes, L.R. Spitler, J. Strader, J.P. Brodie, K.L. Rhode, Mon. Not. R. Astron. Soc. **417**(3), 1823 (2011). https://doi.org/10.1111/j.1365-2966.2011.19368.x
73. C. Usher, D.A. Forbes, J.P. Brodie, C. Foster, L.R. Spitler, J.A. Arnold, A.J. Romanowsky, J. Strader, V. Pota, Mon. Not. R. Astron. Soc. **426**(2), 1475 (2012). https://doi.org/10.1111/j.1365-2966.2012.21801.x
74. S.J. Yoon, S.K. Yi, Y.W. Lee, Science **311**(5764), 1129 (2006). https://doi.org/10.1126/science.1122294
75. M. Powalka, T.H. Puzia, A. Lançon, E.W. Peng, F. Schönebeck, K. Alamo-Martínez, S. Ángel, J.P. Blakeslee, P. Côté, J.C. Cuilland re, P.A. Duc, P. Durrell, L. Ferrarese, E.K. Grebel, P. Guhathakurta, S.D.J. Gwyn, H. Kuntschner, S. Lim, C. Liu, M. Lyubenova, J.C. Mihos, R.P. Muñoz, Y. Ordenes-Briceño, J. Roediger, R. Sánchez-Janssen, C. Spengler, E. Toloba, H. Zhang, Astrophys. J. Lett. **829**(1), L5 (2016). https://doi.org/10.3847/2041-8205/829/1/L5

76. N. Caldwell, R. Schiavon, H. Morrison, J.A. Rose, P. Harding, Astron. J. **141**(2), 61 (2011). https://doi.org/10.1088/0004-6256/141/2/61
77. A.W. McConnachie, M.J. Irwin, R.A. Ibata, J. Dubinski, L.M. Widrow, N.F. Martin, P. Côté, A.L. Dotter, J.F. Navarro, A.M.N. Ferguson, T.H. Puzia, G.F. Lewis, A. Babul, P. Barmby, O. Bienaymé, S.C. Chapman, R. Cockcroft, M.L.M. Collins, M.A. Fardal, W.E. Harris, A. Huxor, A.D. Mackey, J. Peñarrubia, R.M. Rich, H.B. Richer, A. Siebert, N. Tanvir, D. Valls-Gabaud, K.A. Venn, Nature **461**(7260), 66 (2009). https://doi.org/10.1038/nature08327
78. K.A.G. Olsen, B.W. Miller, N.B. Suntzeff, R.A. Schommer, J. Bright, Astron. J. **127**(5), 2674 (2004). https://doi.org/10.1086/383297
79. R.A. González-Lópezlira, Y.D. Mayya, L. Loinard, K. Álamo-Martínez, G. Heald, I.Y. Georgiev, Y. Órdenes-Briceño, A. Lançon, M.A. Lara-López, L. Lomelí-Núñez, G. Bruzual, T.H. Puzia, Astrophys. J. **876**(1), 39 (2019). https://doi.org/10.3847/1538-4357/ab113a
80. D.A. Forbes, J.I. Read, M. Gieles, M.L.M. Collins, Mon. Not. R. Astron. Soc. **481**(4), 5592 (2018). https://doi.org/10.1093/MNRAS/sty2584
81. K.A.G. Olsen, P.W. Hodge, M. Mateo, E.W. Olszewski, R.A. Schommer, N.B. Suntzeff, A.R. Walker, Mon. Not. R. Astron. Soc. **300**(3), 665 (1998). https://doi.org/10.1046/j.1365-8711.1998.01860.x
82. A.E. Piatti, E.J. Alfaro, T. Cantat-Gaudin, Mon. Not. R. Astron. Soc. **484**(1), L19 (2019). https://doi.org/10.1093/MNRASl/sly240
83. A.A. Cole, D.R. Weisz, E.D. Skillman, R. Leaman, B.F. Williams, A.E. Dolphin, L.C. Johnson, A.W. McConnachie, M. Boylan-Kolchin, J. Dalcanton, F. Governato, P. Madau, S. Shen, M. Vogelsberger, Astrophys. J. **837**(1), 54 (2017). https://doi.org/10.3847/1538-4357/aa5df6
84. D.R. Cole, W. Dehnen, J.I. Read, M.I. Wilkinson, Mon. Not. R. Astron. Soc. **426**(1), 601 (2012). https://doi.org/10.1111/j.1365-2966.2012.21885.x
85. M.D.A. Orkney, J.I. Read, J.A. Petts, M. Gieles, Mon. Not. R. Astron. Soc. **488**(3), 2977 (2019). https://doi.org/10.1093/Mon.Not.R.Astron.Soc./stz1625
86. P.G. van Dokkum, R. Abraham, A. Merritt, J. Zhang, M. Geha, C. Conroy, Astrophys. J. Lett. **798**(2), L45 (2015). https://doi.org/10.1088/2041-8205/798/2/L45
87. M.A. Beasley, A.J. Romanowsky, V. Pota, I.M. Navarro, D. Martinez Delgado, F. Neyer, A.L. Deich, Astrophys. J. Lett. **819**(2), L20 (2016). https://doi.org/10.3847/2041-8205/819/2/L20
88. B. Moore, G. Lake, N. Katz, Astrophys. J. **495**(1), 139 (1998). https://doi.org/10.1086/305264
89. R. Smith, R. Sánchez-Janssen, M.A. Beasley, G.N. Cand lish, B.K. Gibson, T.H. Puzia, J. Janz, A. Knebe, J.A.L. Aguerri, T. Lisker, G. Hensler, M. Fellhauer, L. Ferrarese, S.K. Yi, Mon. Not. R. Astron. Soc. **454**(3), 2502 (2015). https://doi.org/10.1093/MNRAS/stv2082
90. R. Sánchez-Janssen, J.A.L. Aguerri, Mon. Not. R. Astron. Soc. **424**(4), 2614 (2012). https://doi.org/10.1111/j.1365-2966.2012.21301.x
91. S.E. Zepf, M.A. Beasley, T.J. Bridges, D.A. Hanes, R.M. Sharples, K.M. Ashman, D. Geisler, Astron. J. **120**(6), 2928 (2000). https://doi.org/10.1086/316850
92. P. Côté, D.E. McLaughlin, D.A. Hanes, T.J. Bridges, D. Geisler, D. Merritt, J.E. Hesser, G.L.H. Harris, M.G. Lee, Astrophys. J. **559**(2), 828 (2001). https://doi.org/10.1086/322347
93. L. Zhu, R.J. Long, S. Mao, E.W. Peng, C. Liu, N. Caldwell, B. Li, J.P. Blakeslee, P. Côté, J.C. Cuilland re, P. Durrell, E. Emsellem, L. Ferrarese, S. Gwyn, A. Jordán, A. Lançon, S. Mei, R. Muñoz, T. Puzia, Astrophys. J. **792**(1), 59 (2014). https://doi.org/10.1088/0004-637X/792/1/59
94. A.B. Alabi, D.A. Forbes, A.J. Romanowsky, J.P. Brodie, J. Strader, J. Janz, C. Usher, L.R. Spitler, S. Bellstedt, A. Ferré-Mateu, Mon. Not. R. Astron. Soc. **468**(4), 3949 (2017). https://doi.org/10.1093/MNRAS/stx678
95. A.J. Romanowsky, J. Strader, L.R. Spitler, R. Johnson, J.P. Brodie, D.A. Forbes, T. Ponman, Astron. J. **137**(6), 4956 (2009). https://doi.org/10.1088/0004-6256/137/6/4956
96. Y. Schuberth, T. Richtler, M. Hilker, B. Dirsch, L.P. Bassino, A.J. Romanowsky, L. Infante, Astron. Astrophys. **513**, A52 (2010). https://doi.org/10.1051/0004-6361/200912482

97. V. Pota, D.A. Forbes, A.J. Romanowsky, J.P. Brodie, L.R. Spitler, J. Strader, C. Foster, J.A. Arnold, A. Benson, C. Blom, J.R. Hargis, K.L. Rhode, C. Usher, Mon. Not. R. Astron. Soc. **428**(1), 389 (2013). https://doi.org/10.1093/MNRAS/sts029
98. W.E. Harris, S. van den Bergh, Astron. J. **86**, 1627 (1981). https://doi.org/10.1086/113047
99. A. Jordán, D.E. McLaughlin, P. Côté, L. Ferrarese, E.W. Peng, S. Mei, D. Villegas, D. Merritt, J.L. Tonry, M.J. West, Astrophys. J. Suppl. Ser. **171**(1), 101 (2007). https://doi.org/10.1086/516840
100. M.J. Hudson, G.L. Harris, W.E. Harris, Astrophys. J. Lett. **787**(1), L5 (2014). https://doi.org/10.1088/2041-8205/787/1/L5
101. K. El-Badry, E. Quataert, D.R. Weisz, N. Choksi, M. Boylan-Kolchin, Mon. Not. R. Astron. Soc. **482**(4), 4528 (2019). https://doi.org/10.1093/MNRAS/sty3007
102. D.A. Forbes, Mon. Not. R. Astron. Soc. **472**(1), L104 (2017). https://doi.org/10.1093/MNRASl/slx148
103. M.J. Hudson, B. Robison, Mon. Not. R. Astron. Soc. **477**(3), 3869 (2018). https://doi.org/10.1093/mnras/sty844
104. I. Damjanov, P.J. McCarthy, R.G. Abraham, K. Glazebrook, H. Yan, E. Mentuch, D. Le Borgne, S. Savaglio, D. Crampton, R. Murowinski, S. Juneau, R.G. Carlberg, I. Jørgensen, K. Roth, H.W. Chen, R.O. Marzke, Astrophys. J. **695**(1), 101 (2009). https://doi.org/10.1088/0004-637X/695/1/101
105. J. Strader, J.P. Brodie, A.J. Cenarro, M.A. Beasley, D.A. Forbes, Astron. J. **130**(4), 1315 (2005). https://doi.org/10.1086/432717
106. C.A. Tremonti, T.M. Heckman, G. Kauffmann, J. Brinchmann, S. Charlot, S.D.M. White, M. Seibert, E.W. Peng, D.J. Schlegel, A. Uomoto, M. Fukugita, J. Brinkmann, Astrophys. J. **613**(2), 898 (2004). https://doi.org/10.1086/423264
107. E.N. Kirby, J.G. Cohen, P. Guhathakurta, L. Cheng, J.S. Bullock, A. Gallazzi, Astrophys. J. **779**(2), 102 (2013). https://doi.org/10.1088/0004-637X/779/2/102
108. M.A. Beasley, R. Leaman, C. Gallart, S.S. Larsen, G. Battaglia, M. Monelli, M.H. Pedreros, Mon. Not. R. Astron. Soc. **487**(2), 1986 (2019). https://doi.org/10.1093/MNRAS/stz1349
109. S.C. Keller, M.S. Bessell, A. Frebel, A.R. Casey, M. Asplund, H.R. Jacobson, K. Lind, J.E. Norris, D. Yong, A. Heger, Z. Magic, G.S. da Costa, B.P. Schmidt, P. Tisserand, Nature **506**(7489), 463 (2014). https://doi.org/10.1038/nature12990
110. J.M.D. Kruijssen, Mon. Not. R. Astron. Soc. **486**(1), L20 (2019). https://doi.org/10.1093/MNRASl/slz052
111. A. Gallazzi, S. Charlot, J. Brinchmann, S.D.M. White, C.A. Tremonti, Mon. Not. R. Astron. Soc. **362**(1), 41 (2005). https://doi.org/10.1111/j.1365-2966.2005.09321.x
112. I. Martín-Navarro, A. Vazdekis, J. Falcón-Barroso, F. La Barbera, A. Yıldırım, G. van de Ven, Mon. Not. R. Astron. Soc. **475**(3), 3700 (2018). https://doi.org/10.1093/MNRAS/stx3346
113. P. Surma, R. Bender, Astron. Astrophys. **298**, 405 (1995)
114. R.L. Davies, H. Kuntschner, E. Emsellem, R. Bacon, M. Bureau, C.M. Carollo, Y. Copin, B.W. Miller, G. Monnet, R.F. Peletier, E.K. Verolme, P.T. de Zeeuw, Astrophys. J. Lett. **548**(1), L33 (2001). https://doi.org/10.1086/318930
115. T.H. Puzia, S.E. Zepf, M. Kissler-Patig, M. Hilker, D. Minniti, P. Goudfrooij, Astron. Astrophys. **391**, 453 (2002). https://doi.org/10.1051/0004-6361:20020835
116. S.S. Larsen, J.P. Brodie, M.A. Beasley, D.A. Forbes, M. Kissler-Patig, H. Kuntschner, T.H. Puzia, Astrophys. J. **585**(2), 767 (2003). https://doi.org/10.1086/346219
117. C. Blom, L.R. Spitler, D.A. Forbes, Mon. Not. R. Astron. Soc. **420**(1), 37 (2012). https://doi.org/10.1111/j.1365-2966.2011.19963.x
118. A.L. Chies-Santos, S.S. Larsen, M. Kissler-Patig, Mon. Not. R. Astron. Soc. **427**(3), 2349 (2012). https://doi.org/10.1111/j.1365-2966.2012.22135.x
119. C. Blom, D.A. Forbes, C. Foster, A.J. Romanowsky, J.P. Brodie, Mon. Not. R. Astron. Soc. **439**(3), 2420 (2014). https://doi.org/10.1093/MNRAS/stu095
120. I. Trujillo, A. Ferré-Mateu, M. Balcells, A. Vazdekis, P. Sánchez-Blázquez, Astrophys. J. Lett. **780**(2), L20 (2014). https://doi.org/10.1088/2041-8205/780/2/L20

121. M.A. Beasley, I. Trujillo, R. Leaman, M. Montes, Nature **555**(7697), 483 (2018). https://doi.org/10.1038/nature25756
122. P. van Dokkum, S. Danieli, Y. Cohen, A. Merritt, A.J. Romanowsky, R. Abraham, J. Brodie, C. Conroy, D. Lokhorst, L. Mowla, E. O'Sullivan, J. Zhang, Nature **555**(7698), 629 (2018). https://doi.org/10.1038/nature25767
123. I. Trujillo, M.A. Beasley, A. Borlaff, E.R. Carrasco, A. Di Cintio, M. Filho, M. Monelli, M. Montes, J. Román, T. Ruiz-Lara, J. Sánchez Almeida, D. Valls-Gabaud, A. Vazdekis, Mon. Not. R. Astron. Soc. **486**(1), 1192 (2019). https://doi.org/10.1093/mnras/stz771
124. L. Searle, R. Zinn, Astrophys. J. **225**, 357 (1978). https://doi.org/10.1086/156499
125. O.J. Eggen, D. Lynden-Bell, A.R. Sandage, Astrophys. J. **136**, 748 (1962). https://doi.org/10.1086/147433
126. K.M. Ashman, S.E. Zepf, Astrophys. J. **384**, 50 (1992). https://doi.org/10.1086/170850
127. P. Côté, R.O. Marzke, M.J. West, Astrophys. J. **501**(2), 554 (1998). https://doi.org/10.1086/305838
128. M.A. Beasley, C.M. Baugh, D.A. Forbes, R.M. Sharples, C.S. Frenk, Mon. Not. R. Astron. Soc. **333**(2), 383 (2002). https://doi.org/10.1046/j.1365-8711.2002.05402.x
129. C. Tonini, Astrophys. J. **762**(1), 39 (2013). https://doi.org/10.1088/0004-637X/762/1/39
130. N. Choksi, O.Y. Gnedin, Mon. Not. R. Astron. Soc. **488**(4), 5409 (2019). https://doi.org/10.1093/MNRAS/stz2097
131. A.V. Kravtsov, O.Y. Gnedin, Astrophys. J. **623**(2), 650 (2005). https://doi.org/10.1086/428636
132. J. Pfeffer, J.M.D. Kruijssen, R.A. Crain, N. Bastian, Mon. Not. R. Astron. Soc. **475**(4), 4309 (2018). https://doi.org/10.1093/MNRAS/stx3124
133. N. Mandelker, P.G. van Dokkum, J.P. Brodie, F.C. van den Bosch, D. Ceverino, Astrophys. J. **861**(2), 148 (2018). https://doi.org/10.3847/1538-4357/aaca98
134. J.h. Kim, X. Ma, M.Y. Grudić, P.F. Hopkins, C.C. Hayward, A. Wetzel, C.A. Faucher-Giguère, D. Kereš, S. Garrison-Kimmel, N. Murray, Mon. Not. R. Astron. Soc. **474**(3), 4232 (2018). https://doi.org/10.1093/MNRAS/stx2994
135. X. Ma, M.Y. Grudić, E. Quataert, P.F. Hopkins, C.A. Faucher-Giguère, M. Boylan-Kolchin, A. Wetzel, J.h. Kim, N. Murray, D. Kereš (2019). e-prints arXiv:1906.11261
136. E. Vanzella, F. Calura, M. Meneghetti, M. Castellano, G.B. Caminha, A. Mercurio, G. Cupani, P. Rosati, C. Grillo, R. Gilli, M. Mignoli, G. Fiorentino, C. Arcidiacono, M. Lombini, F. Cortecchia, Mon. Not. R. Astron. Soc. **483**(3), 3618 (2019). https://doi.org/10.1093/MNRAS/sty3311
137. A. Renzini, Mon. Not. R. Astron. Soc. **469**(1), L63 (2017). https://doi.org/10.1093/MNRASl/slx057
138. L. Pozzetti, C. Maraston, A. Renzini, Mon. Not. R. Astron. Soc. **485**(4), 5861 (2019). https://doi.org/10.1093/MNRAS/stz785
139. E. Vanzella, F. Calura, M. Meneghetti, A. Mercurio, M. Castellano, G.B. Caminha, I. Balestra, P. Rosati, P. Tozzi, S. De Barros, A. Grazian, A. D'Ercole, L. Ciotti, K. Caputi, C. Grillo, E. Merlin, L. Pentericci, A. Fontana, S. Cristiani, D. Coe, Mon. Not. R. Astron. Soc. **467**(4), 4304 (2017). https://doi.org/10.1093/MNRAS/stx351
140. R.J. Bouwens, G.D. Illingworth, P.A. Oesch, M. Maseda, B. Ribeiro, M. Stefanon, D. Lam (2017). e-prints arXiv:1711.02090
141. R. Barkana, A. Loeb, Phys. Rep. **349**(2), 125 (2001). https://doi.org/10.1016/S0370-1573(01)00019-9
142. M. Boylan-Kolchin, Mon. Not. R. Astron. Soc. **479**(1), 332 (2018). https://doi.org/10.1093/MNRAS/sty1490
143. H. Katz, M. Ricotti, Mon. Not. R. Astron. Soc. **444**(3), 2377 (2014). https://doi.org/10.1093/MNRAS/stu1489
144. L. Oser, J.P. Ostriker, T. Naab, P.H. Johansson, A. Burkert, Astrophys. J. **725**(2), 2312 (2010). https://doi.org/10.1088/0004-637X/725/2/2312

# Hot Atmospheres of Galaxies, Groups, and Clusters of Galaxies

**Norbert Werner and François Mernier**

## Contents



**Abstract** Most of the ordinary matter in the local Universe has not been converted into stars but resides in a largely unexplored diffuse, hot, X-ray emitting plasma. It pervades the gravitational potentials of massive galaxies, groups and clusters of galaxies, as well as the filaments of the cosmic web. The physics of this hot medium, such as its dynamics, thermodynamics and chemical composition can be studied using X-ray spectroscopy in great detail. Here, we present an overview of the basic properties and discuss the self similarity of the hot "atmospheres" permeating the

N. Werner (✉)
MTA-Eötvös University Lendület Hot Universe Research Group, Budapest, Hungary

Department of Theoretical Physics and Astrophysics, Faculty of Science, Masaryk University, Brno, Czech Republic

School of Science, Hiroshima University, Higashi-Hiroshima, Japan

F. Mernier
MTA-Eötvös University Lendület Hot Universe Research Group, Budapest, Hungary

Eötvös University, Institute of Physics, Budapest, Hungary

SRON Netherlands Institute for Space Research, Utrecht, The Netherlands
e-mail: Francois.Mernier@esa.int

 
P. Kabáth et al. (eds.), *Reviews in Frontiers of Modern Astrophysics*,
https://doi.org/10.1007/978-3-030-38509-5_10

gravitational halos from the scale of galaxies, through groups, to massive clusters. Hot atmospheres are stabilised by the activity of supermassive black holes and, in many ways, they are of key importance for the evolution of their host galaxies. The hot plasma has been significantly enriched in heavy elements by supernovae during the period of maximum star formation activity, probably more than 10 billion years ago. High resolution X-ray spectroscopy just started to be able to probe the dynamics of atmospheric gas and future space observatories will determine the properties of the currently unseen hot diffuse medium throughout the cosmic web.

## 1 Introduction

Most of the matter in the Universe remains unseen. The invisible dark matter consists of particles that we have not yet been able to detect. Most of the ordinary matter, which we call baryons, has not been converted into stars and resides in a hot diffuse medium permeating the extended halos of galaxies, groups, and clusters of galaxies, as well as the filaments of the cosmic web.

In the innermost parts of galaxies the hot, strongly ionised, X-ray emitting plasma is often referred to as *interstellar medium (ISM)* and at large radii, well beyond the stellar component, as *circumgalactic medium (CGM)*, in groups of galaxies as *intragroup medium (IGrM)*, in clusters of galaxies as *intracluster medium (ICM)*, and in the filaments of the cosmic web as *intergalactic medium (IGM)* or *warm-hot intergalactic medium (WHIM)*. Since, the hot phases of all these media share crucial similarities—e.g. they can be well described by equations of ideal gas in a gravitational potential—we will refer to them at all scales as *hot atmospheres*.

The hot atmospheres permeating massive objects can be probed using X-ray space observatories (see the composite optical/X-ray images of a galaxy cluster and a giant elliptical galaxy in Fig. 1). They have been studied in detail in clusters of galaxies, which are particularly bright and can be observed out to redshifts $z = 1.5$–2, which corresponds to a look back time of around 10 billion years. However, the crucial role of atmospheres for the formation and evolution of individual massive galaxies is just beginning to be appreciated. They have been discovered using the *Einstein* satellite [1] and today, they are studied mainly using *Chandra* and *XMM-Newton*, which are equipped with CCD-type detectors sensitive in the approximately 0.3–10 keV band, have an energy resolution in the range of 50–150 eV, and a spatial resolution of about 1 and 15 arcsec, respectively.

However, outside the dense central regions of massive dark matter halos the atmospheric density becomes too low to be probed by current instruments. Yet, the properties of this mostly unseen diffuse medium are set by large scale shocks and cosmic feedback, therefore its observations with X-ray space telescopes could, in principle, provide key information about the physics of structure and galaxy formation. We therefore expect that the study of hot atmospheres will remain on the frontiers of astrophysics for the decades to come.

**Fig. 1** These composite optical and X-ray images show the hot, X-ray emitting atmospheres pervading the massive galaxy cluster Abell 1689 (left) and the massive elliptical galaxy NGC 5813 in the centre of a group (right). Courtesy: NASA/CXC/SAO

## *1.1 The Simple State of Hot Atmospheres*

Hot X-ray emitting atmospheres have densities in the range from $\sim 10^{-1}$ cm$^{-3}$ in the centres of massive galaxies to $\sim 10^{-4}$ cm$^{-3}$ in the outskirts. Their temperatures span from a few $10^6$ K in galaxies to $10^8$ K in the most massive clusters of galaxies. These high temperatures lead to strong ionization. The tenuous hot medium thus consists of fully ionized hydrogen and helium, enriched with highly ionized heavier elements to about a third of their Solar abundances, increasing to around Solar at the centres. This strongly ionized medium has frozen in magnetic fields with strengths of a few $\mu$G. The giroradii of thermal electrons and protons of $\sim 10^8$ cm and $\sim 10^{10}$ cm (of the order of the size of the Earth), respectively, are many orders of magnitude smaller than the particle collisional mean free paths of $10^{21}$–$10^{23}$ cm (about 1–100 kpc, comparable to the size of the Galaxy). The ratio of thermal to magnetic pressure $\beta = p/p_{\rm B} \gg 1$, which means that the magnetic pressure contribution is negligible for the hydrostatic mass determination discussed in Sect. 1.4. More generally, even though the hot atmospheres consist of weakly magnetised plasma, equations of ideal gas often provide useful approximations for their properties and in this review, we will for convenience also use the term “hot gas”.

At the densities and temperatures of hot atmospheres the ionisation equilibration time scale is short and therefore they are in a state of **collisional ionisation equilibrium** (CIE). This means that the ionisation rate is equal to the recombination rate, it is determined only by collisions and outside radiation fields do not affect the ionisation state of the plasma. The time scale of the electron-ion equilibration via Coulomb collisions is mostly shorter than $\sim$1 Gyr, therefore we can generally assume that the free electrons are in thermal equilibrium with ions. Hot atmospheres are mostly **optically thin**, which means that the density of the gas is so low that the emitted photons will never interact with other ions or electrons within the atmosphere and will thus leave the system in "straight lines" (the exception is resonant scattering at the energies of some of the strong emission lines in the dense central regions, as discussed in Sect. 5; see [2]). This, combined with the fact that the X-ray emissivity is proportional to the square of the gas density (Sect. 1.2), implies that X-ray images of hot atmospheres give us a direct view on the projected gas distribution, considerably facilitating the modelling of spectra and the interpretation of the observations.

## *1.2 X-ray Emission Processes*

As discussed above, the atmospheric gas is almost entirely ionized, made of ions and free electrons. Under these conditions, a free electron sometimes flies close to an ion and the electrostatic interaction between the two deflects the trajectory of the much less massive electron. The loss of energy due to the deceleration of the electron is then converted into a **bremsstrahlung** photon that can be detected at X-ray wavelengths. The emissivity $\epsilon$ of this process scales as

$$\epsilon \propto \bar{g} n_{\mathrm{e}}^2 (kT_{\mathrm{e}})^{-1/2} e^{-E/kT_{\mathrm{e}}}, \tag{1}$$

where $\bar{g}$ is the Gaunt-factor, which is a quantum mechanical correction factor that depends weakly on energy, $k$ is the Boltzmann constant, $T_{\mathrm{e}}$ is the electron temperature,[1] and $n_{\mathrm{e}}$ is the electron density[2] of the gas. Since such interactions can happen at a range of angles and velocities, the energy of the emitted photons can span a range of values. Bremsstrahlung thus contributes to the *continuum* of the X-ray emission. This process is dominant in rich, massive clusters permeated by gas with temperatures $\gtrsim 10^{7.5}$ K (i.e. $kT \gtrsim 3$ keV).

In addition to hydrogen and helium, hot atmospheres contain traces of highly ionized heavier elements. Like protons, these heavy ions also interact with free electrons, either by capturing them (*radiative recombination*, which also contributes to the continuum emission) or by colliding with them, which often boosts the energy of the bound electrons. These excited electrons eventually return to their lower, initial energy level (*de-excitation*) and, to conserve energy, emit an X-ray photon of an energy corresponding to the difference between the two energy levels in the ion. This produces **metal emission lines**, each of which corresponds to a specific ion of a specific chemical element.

---

[1] By convention, the temperature is often expressed in units of *energy*, as $kT_{\mathrm{e}}$, in kiloelectronvolts (keV).

[2] The ion density, $n_{\mathrm{i}}$, of the plasma with Solar metallicity can be obtained by assuming $n_{\mathrm{e}}/n_{\mathrm{i}} \simeq 1.18$.

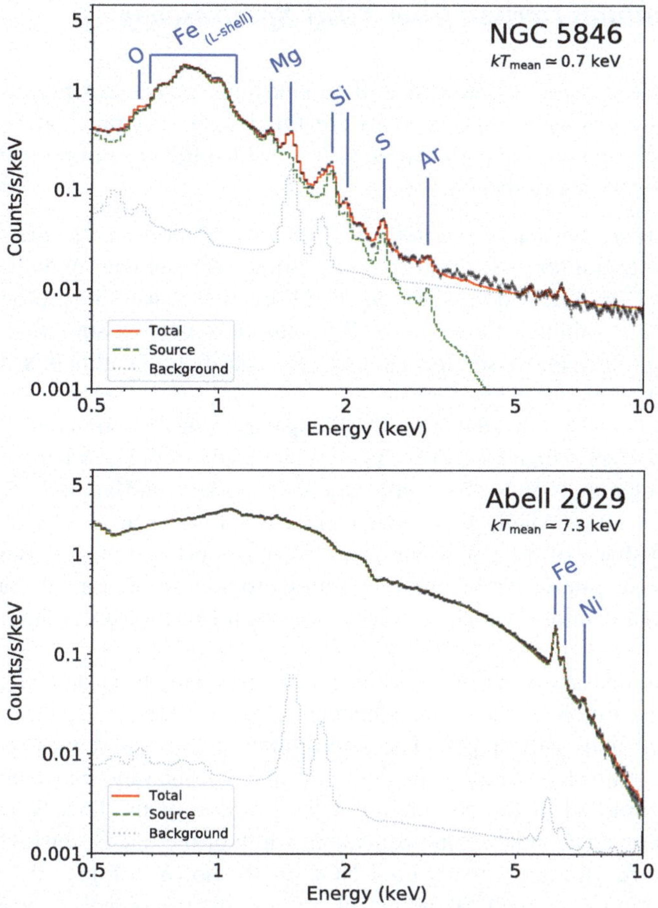


**Fig. 2** *XMM-Newton*/EPIC MOS spectra of the giant elliptical galaxy NGC 5846 (top) and the galaxy cluster Abell 2029 (bottom), both extracted within $0.1r_{500}$. When visible, metal emission lines (from their K-shell transitions, unless otherwise stated) are labeled. The MOS 1 and MOS 2 spectra have been stacked for clarity

In cooler gas pervading groups and massive ellipticals ($\sim 10^{6.5}$–$10^{7.5}$ K, i.e. $kT \sim 0.3$–3 keV), the plasma is less ionized and more of these atomic transitions can occur, giving rise to a large number of emission lines. In fact, in lower-mass systems, line emission often dominates over bremsstrahlung. Figure 2 shows a typical X-ray spectrum produced by a giant elliptical galaxy (NGC 5846, top panel) and by a massive, hot galaxy cluster (Abell 2029, bottom panel). While Abell 2029 is largely dominated by the bremsstrahlung continuum emission with only two visible Fe and Ni line complexes, most of the emission from NGC 5846 in the "bump" between $\sim$0.6 and $\sim$1.2 keV comes from a blend of Fe-L lines that are unresolved by CCD-type detectors.

### *1.3 Quantities Derived from X-ray Spectroscopy*

X-ray spectroscopy is a powerful tool to study the physics of hot atmospheres. Typically, the observed X-ray spectra are fitted with spectral models of plasma emission with several free parameters. Below, we list the most important quantities determined from X-ray spectra.

- The **electron density** of the hot gas is directly proportional to the square root of the X-ray luminosity. From spectral fitting, one can determine the emission measure $Y = \int n_e n_H dV$, where $n_H$ is the number density of protons and $V$ is the emitting volume of the source. By using reasonable assumptions about the emitting volume and assuming a smooth gas distribution within this volume, one can easily work out the gas density.
- Equation (1) shows that the bremsstrahlung spectrum will exponentially decrease at energies larger than $kT_e$. This can be seen clearly in Fig. 2, where the emission of the cooler NGC 5846 group falls below the background beyond ~3 keV, while the much hotter Abell 2029 cluster emits photons even at 10 keV. From the observed shape of the continuum, we can thus work out the gas **temperature**. The relative intensities of emission lines provide an additional constraint on the gas temperature, which is particularly useful in the cooler, line dominated systems.
- As mentioned above, each emission line corresponds to an ion of a particular element. Intuitively, the more abundant a given element is, the stronger its associated lines will appear. The **abundance** of a given element is measured from the equivalent width of its emission line, i.e. the ratio between the flux of the line and that of the continuum at the position of the line. Note, however, that, at a given abundance, the equivalent width of the line is sensitive to the gas temperature. In the relatively cool NGC 5846 shown in Fig. 2, the unresolved Fe XVII lines (i.e. with 10 bound electrons) are prominent at around ~0.7–0.9 keV. These lines do not appear in the massive hot Abell 2029, because its Fe is much more ionized and its spectrum is dominated by Fe XXV and Fe XXVI lines (with 2 and 1 bound electron, respectively).

From the temperature and density, one can determine the pressure and the entropy of the gas. Since the hot atmospheres are very tenuous and behave almost like an ideal gas, the **gas pressure** $P$ is simply defined as

$$P = \frac{\rho k T}{\mu m_p} = nkT \propto n_e kT, \tag{2}$$

where $\rho$ is the gas density, $m_p$ is the proton mass and $\mu \simeq 0.6$ is the mean molecular weight given in units of the proton mass.

Another quantity of interest is the **gas entropy**, $K$. In hot atmospheres, it is defined as[3]

$$K = \frac{kT}{n_e^{2/3}}. \tag{3}$$

In the case of a relaxed cluster in hydrostatic equilibrium, the gas with low entropy naturally sinks to the center while the gas with higher entropy floats toward higher altitudes. If a parcel of gas is moved to a different altitude (e.g. by the activity of a supermassive black hole or by a merger) its temperature and density will change due to adiabatic expansion or compression (assuming that thermal conduction is negligible in the hot plasma), but its entropy will remain unchanged. In this sense, it is a very useful quantity because it keeps memory of the thermodynamic history of the gas. The gas entropy can only be changed by shocks, radiative cooling, conduction, or mixing.

## *1.4 Hydrostatic Equilibrium*

The time that it takes for a sound wave to cross an atmosphere provides the timescale on which deviations from hydrostatic equilibrium are evened out. Therefore, if the atmosphere is undisturbed by mergers or violent active galactic nucleus (AGN) outbursts, after several sound crossing times (of order $10^9$ year for galaxy clusters) hot atmospheres should come to **hydrostatic equilibrium** with the underlying gravitational potential $\Phi$, meaning that the gravity will be balanced by gas pressure:

$$\frac{1}{\rho}\frac{\mathrm{d}P}{\mathrm{d}r} = -\frac{\mathrm{d}\Phi}{\mathrm{d}r} = -\frac{GM_{\mathrm{tot}}}{r^2}, \tag{4}$$

where $G$ is the gravitational constant and $M_{\mathrm{tot}}$ is the total (dark matter + stars + gas) mass of the system. This assumption allows us to derive the total mass distribution of the object simply by measuring the basic properties of its X-ray emitting gas. More specifically, one can write:

$$M_{\mathrm{tot}}(< r) = -\frac{kT(r)r}{G\mu m_{\mathrm{p}}}\left(\frac{\mathrm{d}\ln T}{\mathrm{d}\ln r} + \frac{\mathrm{d}\ln \rho}{\mathrm{d}\ln r}\right). \tag{5}$$

As long as the data are good enough to derive the spatial distribution of the gas density and temperature, the total mass encircled within a certain radius, $M_{\mathrm{tot}}(< r)$, can be readily obtained.

[3]This "entropy" is not exactly the same as the classical thermodynamical entropy $s$. Formally, the two quantities are related as $s = k\ \ln K^{3/2} + \text{constant}$.

The departure from hydrostatic equilibrium may be significant in clusters undergoing major mergers, where the gas is heavily disturbed. However, even in systems with prominent gas sloshing and ongoing AGN feedback, like the well known Perseus cluster, the contribution of gas motions to the total pressure support appears to be small (see Sect. 5).

## 2 Self Similarity

In the ΛCDM (cosmological constant and cold dark matter) model—favoured by cosmologists today—clusters and groups of galaxies formed at the density peaks that were born out of quantum density fluctuations in the very early Universe, which later collapsed under their own gravity. These systems grew further in a hierarchical way, i.e. larger structures formed from smaller ones, mainly via steady accretion and mergers.

This paradigm is particularly interesting in the context of the so-called "self-similarity". As shown by Kaiser [3], the assumption that the Universe has the critical density ($\Omega = 1$) implies that the wave-number (i.e. the inverse of the typical spatial scales) of the density fluctuations is distributed as a power-law. The consequence of this prediction is that, on sufficiently large scales, the Universe is expected to be self-similar. In other words, smaller dark-matter structures—containing massive galaxies or galaxy groups—appear as scaled-down versions of the largest haloes associated with massive galaxy clusters.

As shown by numerical simulations (e.g. [4]), self-similarity is expected to hold not only for the dark-matter component, but also for the hot, X-ray emitting atmospheres pervading these large scale structures. The gas falling into dark matter halos is expected to be heated by shocks. The more massive a halo is, the more shock-heated, and thus the hotter, the gas becomes. Combining the self-similar assumption with the approximation of hydrostatic equilibrium discussed earlier, one can easily show (see [5]) that, within a given overdensity limit[4] $r_\Delta$, the total mass $M_\Delta$ of a system scales with its gas temperature $T_\Delta$ as

$$M_\Delta \propto T_\Delta^{3/2}. \tag{6}$$

This simple relation shows that, in the absence of other mechanisms than gravity and shock heating, the temperature of a system reflects its gravitational potential.

[4] $r_\Delta$ (with $\Delta = 200$ or $\Delta = 500$ as commonly found in the literature) corresponds to a radius within which the total matter density reaches $\Delta$ times the critical density of the Universe at the redshift of the system. Defined this way, $r_\Delta$ can be associated with a "normalized" astrophysical radius common to each system, which takes self-similarity into account.

Another important prediction of the self-similar paradigm is that the dimensionless properties of massive galaxies, groups, and clusters are expected to be invariant. In particular, the "gas mass fraction", i.e. the fraction of hot, X-ray emitting gas mass over the total (mostly dark matter) mass is expected to be the same in all massive systems. Combining this prediction with the assumption that the total emissivity of the plasma is dominated by thermal bremsstrahlung[5] (see Sect. 1), one can also easily demonstrate that the X-ray luminosity $L_{X,\Delta}$ of a system scales with its gas temperature as

$$L_{X,\Delta} \propto T_{\Delta}^{2}. \tag{7}$$

Since $L_{X,\Delta}$ and $T_{\Delta}$ can be obtained independently from X-ray data (Sect. 1.3), Eq. (7) is extremely useful to directly test the self-similar assumption via X-ray observations of clusters, groups, and ellipticals. An example of such a study is shown in Fig. 3 (from [6]), however, we stress that many other studies are available in the literature (e.g. [7–11]).

The general conclusion of these results is that, despite a strong correlation between temperature and luminosity, the slope of the power-law is closer to 3 than to 2, which suggests a deviation from self-similarity. Moreover, as seen in Fig. 3, the relation for groups is steeper than that for clusters. In other words, groups lay on average below the expected scaling relation, which means that they must be either *hotter* or *less luminous* (or both) than initially expected. Since self-similarity assumes that gravity and shock heating are the only processes that shape the atmospheres of massive systems, the most likely cause for such deviations are additional *non-gravitational* processes, such as energy input by AGN and supernovae.

Following its definition, the entropy (Eq. (3)) scales linearly with the gas temperature. It has also been shown, using hydrodynamic simulations of gravitationally collapsed gas in hydrostatic equilibrium [14, 15], that the radial distribution of entropy is expected to increase with radius following a power-law distribution of

$$K(r) \propto r^{1.1}. \tag{8}$$

This prediction has been tested using X-ray observations of clusters, groups, and ellipticals, as shown in Fig. 4. Compared to the self-similar expectations, it appears that the radial distribution shows an excess of entropy in cluster cores (Fig. 4), which is on average stronger in low-mass systems.

[5]This assumption holds relatively well in the case of hot, massive clusters. However, as we have seen earlier, metal lines usually dominate the emissivity of cooler atmospheres pervading groups and ellipticals, which adds further complications to testing the predictions of self-similarity.

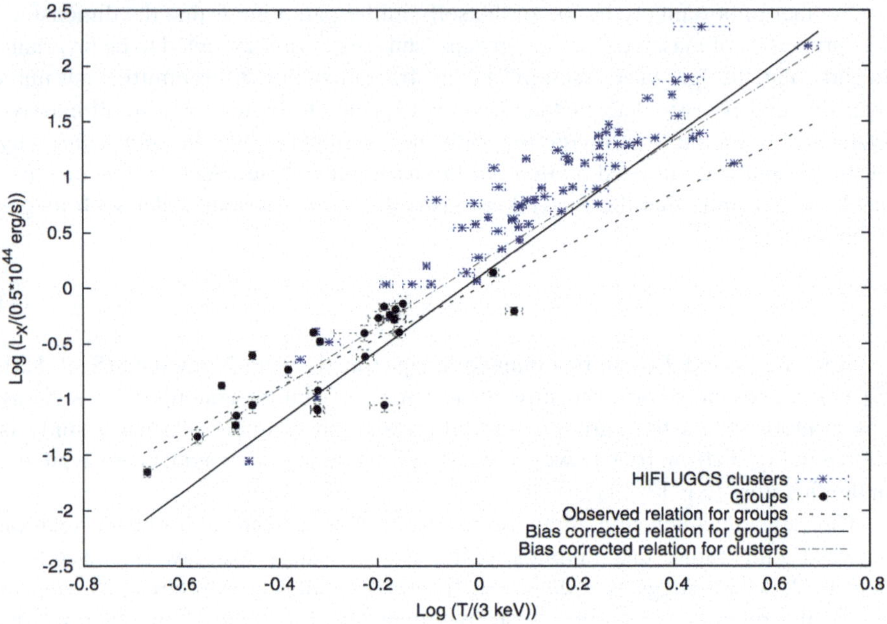


**Fig. 3** X-ray luminosity vs. temperature relation for a sample of clusters and groups [6]. The solid line and dash-dotted line show the best-fit relations for the sub-samples of groups and clusters, respectively, after corrections for biases due to selection effects. Groups appear to have a steeper relation than clusters

This entropy excess can be explained by a centrally increased temperature and/or by a central gas density drop (see Eq. (3)). In other words, there must be a mechanism at the center of these hot atmospheres that heats the gas and/or moves a significant fraction of the gas up to larger altitudes. The most likely mechanism is the feedback from the central supermassive black hole, which has more impact in the lower mass groups than in massive clusters. It has also been suggested that the entropy excess may be a signature of "pre-heating", i.e. a heating of the gas, by AGNs or stellar activity, before the formation of clusters and groups [16].

The idea that a significant fraction of gas has been lifted from the central regions of lower mass systems also agrees with the finding that, at scaled radii of $r_{2500}$ or even at $r_{500}$, the gas mass fraction is an increasing function of cluster mass up to massive clusters with virial temperatures of $kT \sim 5$ keV [8, 17–19]. In dynamically relaxed high mass clusters with $kT \gtrsim 5$ keV the gas mass fraction at $r_{2500}$ appears to be invariant [20].

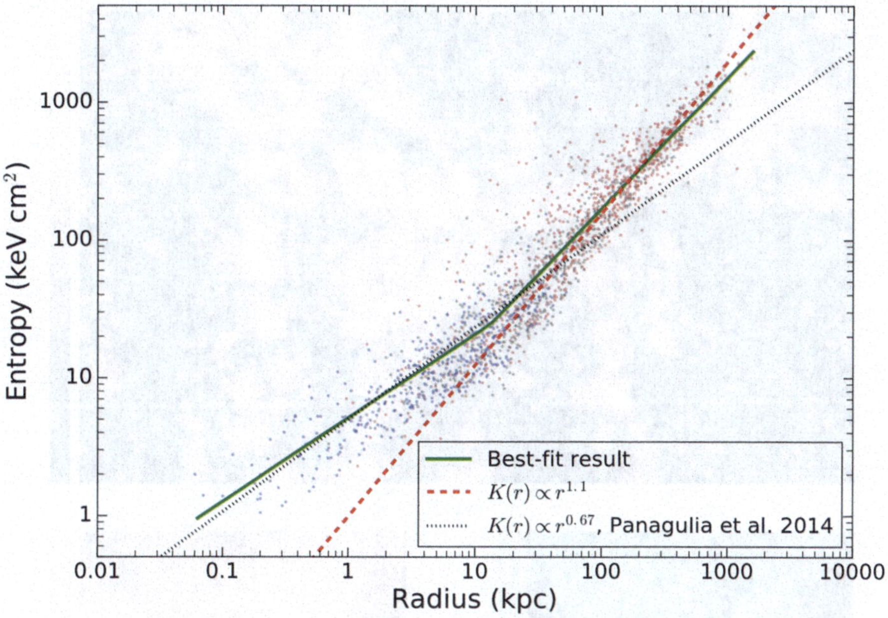


**Fig. 4** Measured radial entropy profiles for elliptical galaxies (blue data points) and for more massive systems (red and black data points). The red dashed line indicates a power-law distribution with index 1.1 predicted by gravitational collapse models and the black doted line indicates a power-law model with index 0.67 found at $r \lesssim 50$ kpc by [12]. The data and the best-fit broken power-law model (green solid line) show that the central entropy significantly flattens in the cores of these systems ([13], and references therein)

## 3 Hot Atmospheres of Galaxies and the Circumgalactic Medium

At least a part of the hot atmospheric gas of galaxies more massive than $M_{\rm crit} \approx 10^{12}\,{\rm M}_\odot$ (e.g. [21]) was accreted externally and shock heated during the process of galaxy assembly. The atmospheres were augmented significantly by stellar mass loss [22]. The contribution of stellar mass loss material increases and might actually dominate in the lower mass systems. Such atmospheric gas contains a large fraction of baryons in the local Universe (see Fig. 5). It turns out that about half of the warm-hot diffuse baryons at low redshifts may lie in galactic atmospheres (e.g. [24–26]).

Historically, the extended, volume-filling, hot atmosphere of our Milky Way was revealed by its soft X-ray emission [27–29] and further probed by X-ray absorption studies along sightliness to bright AGN [30, 31]. These studies show that it contains $2.5 \pm 1 \times 10^{10}\,{\rm M}_\odot$ of gas, which is less than its stellar mass [32]. The baryonic mass fraction within the virial radius of our Galaxy is only $6 \pm 1\%$, which falls well short of the Universal cosmic value of 16%. Studies of other spiral galaxies with *Chandra* and *XMM-Newton* find similar conclusions. When the observed density profiles are

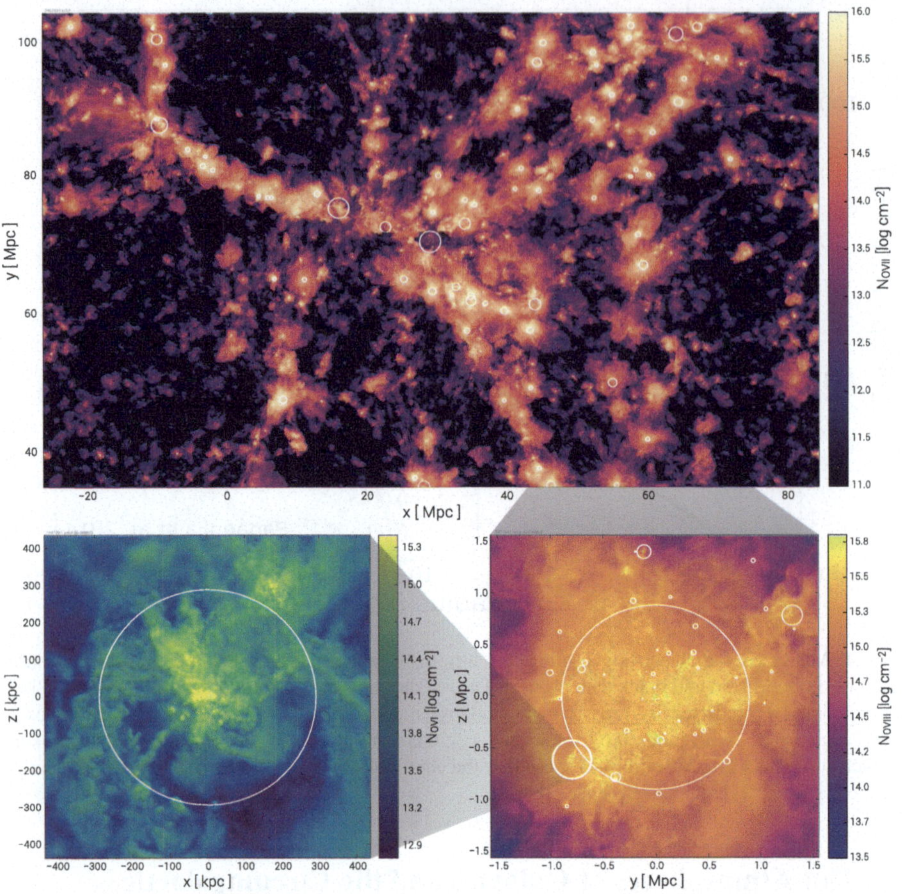


**Fig. 5** These plots show the distribution of highly ionized oxygen, which is an excellent tracer of the warm-hot diffuse gas, in the Illustris TNG100 simulations at $z = 0$. The bulk of the oxygen in the circumgalactic medium is expected to reside in O VII and O VIII ions, which have strong X-ray emission lines. The upper panel shows the column density of O VII over a depth of 15 Mpc. O VII traces the large scale filaments of the cosmic web and galaxy sized collapsed halos. The white circles show the 50 most massive halos. The lower right panel zooms in on a galaxy cluster with a halo mass of $10^{13.8}\,M_\odot$, showing its O VIII column in a box with a length of 3.5 $r_{vir}$. The large circle shows the virial radius and the small circles indicate the 50 most massive halos in the panel. The lower left panel zooms in further on a system with a halo mass of $10^{12.5}\,M_\odot$, showing the column density of O VI. From [23]

extrapolated to the virial radii of these galaxies, more than 60–70% of the baryons appear to be missing [33–35]. Stacked Sunyaev-Zeldovich (SZ) measurements of massive galaxies by the *Planck* satellite indicates that most of their baryons reside in hot diffuse atmospheres and extend beyond their virial radii [36, 37]. Bregman et al. [38] show that the atmospheric density profiles of massive spirals need to be extrapolated to 1.9–3 $R_{200}$ for their baryon to dark matter ratio to approach the cosmic value. Some kind of violent activity likely caused these galaxies to expel a large part of their hot atmospheres.

Our estimates of the baryon fractions also depend strongly on the metallicity of the gas, which was previously often assumed low. Contrary to such earlier results (e.g. [39–42]), the hot galactic atmospheres of early type galaxies are likely metal rich, with metallicities approaching the Solar value in the central regions [43] and flattening out at $\approx$0.2–0.3 Solar at larger radii (see [44] and Sect. 6). Detailed spatially resolved measurements of metallicities in the hot atmospheres of late type galaxies have not yet been performed. Such measurements are challenging due to the low number of photons.

Current measurements of the properties of galactic atmospheres are mostly limited to early type galaxies, where the presence of hot atmospheric gas in combination with radio mode AGN feedback is believed to have played a crucial role in the quenching of star formation (see Sect. 4). However, the detailed knowledge of the atmospheric gas is in many ways essential also for our understanding of the stellar components of late type, discy galaxies.

As an example, the hot atmospheric gas might also be of key importance for the evolution of spiral galaxies, feeding their discs with the fuel necessary for star formation. Milky Way mass spiral galaxies, that are forming stars at a rate similar to that of our Galaxy ($\approx$2 $M_{\odot}$ year$^{-1}$) would consume the available molecular gas in their discs in about $10^9$ years. To maintain the star forming galactic discs, they have to be fed continuously by molecular gas from outside [45]. Thermal instabilities in the hot atmospheres, leading to a "rain" onto the galactic disc (see Sect. 4), would be able to provide plenty of fuel to maintain the star formation in spirals for $\approx 10^{10}$ years. The cooling of the galactic atmospheres onto star forming discs could be stimulated by stellar feedback via powering of a galactic fountain, which produces mixing between the disc material and the atmospheric gas. The mixing reduces the cooling time of the atmosphere, making it condense and accrete onto the disc [46]. In this fountain driven accretion scenario, the key ingredient is the presence of a star-forming disc of cold gas, which helps to cool the atmosphere. If the disc gets for some reason destroyed (e.g. due to ram pressure stripping after falling into a galaxy cluster) it may not reform again and the galaxy will become red and dead.

Most X-ray observations only target the bright central parts of galaxies, groups and clusters of galaxies. Most of their hot atmospheres, which contains the dominant fraction of their baryons are still unobserved. Few systems have reliable observations all the way to their virial radii and we will have to wait for future observatories to really address the baryon content, dynamics, thermodynamics, and the chemical composition of the full extent of their atmospheres (see Sect. 5). The existing observations indicate that the gas in the cluster outskirts is clumpy and it is not entirely virialized [47]. The same might be true for the outskirts of individual galaxies, where most of the ordinary matter resides.

Even fewer observations probe the properties of the warm-hot intergalactic medium (WHIM) outside of galaxies, groups, and clusters of galaxies, which is believed to be permeating the filaments of the cosmic web. A few tentative detections in emission are likely probing the densest and hottest parts of the intergalactic medium, where only a small fraction of the WHIM resides (e.g. [48–50]). On the other hand absorption studies provided line detections only in extremely long observations toward a few bright quasars (e.g. [51, 52]). A systematic study probing the bulk of the diffuse medium in the outskirts of galaxies and permeating the cosmic web will need a substantial improvement in our observing capabilities (see Sect. 7).

## 4 Supermassive Black Hole Feedback

It is now well established that central supermassive black holes play a vital role in the evolution of their host systems. Accreting black holes are the most efficient engines in the Universe at converting rest-mass into energy, releasing $\sim 10^{20}$ erg g$^{-1}$ of accreted gas. This energy may be released in either *radiative* or *mechanical* form, depending on the accretion rate and structure of the accretion flow: in the radiative (quasar mode) AGN feedback, the radiation from vigorously accreting black holes couples with the cold gas in the host galaxy; in the so-called radio-mechanical (maintenance mode) feedback, the jets from AGN accreting at a modest rate heat or push the hot gas out. Radiatively efficient accretion and rapid black hole growth are taking place mostly early in the evolution of galaxies.

In massive galaxies, groups and clusters of galaxies hosting hot atmospheric gas, the quasar phase is followed by radiatively inefficient radio-mechanical feedback, when despite a much lower accretion rate the mechanical energy output of the jets couples efficiently to the atmospheric gas (e.g. [53]). In the absence of heating, the hot X-ray emitting gas in many clusters, groups and elliptical galaxies would cool and form stars, building much larger galaxies than are seen. However, the radio-mechanical feedback appears to be preventing the cooling of the hot gaseous atmospheres (see [54]).

In the centres of clusters of galaxies, radio-mechanical feedback is relatively well explored. X-ray studies with *Chandra* and *XMM-Newton* have shown that the expanding, AGN jet inflated radio lobes displace the hot gas, creating 'cavities' in the X-ray emitting plasma (e.g. [55] and drive weak shocks that heat the surrounding medium isotropically [56–60], preventing it from cooling (see Figs. 6 and 7). Furthermore, in their wakes, the rising bubbles filled by relativistic plasma uplift low entropy gas from the innermost regions of their host galaxies [61–66]. Current X-ray observations thus show that jets emanating from black holes accreting at modest rates are sufficiently powerful to balance the radiative cooling of hot atmospheres and limit further star-formation. The feedback is gentle and the heating rate appears to be very well tuned to the atmospheric cooling (see Fig. 7). However, no consensus has been achieved on the dominant mechanism responsible for energy transport from jets, to X-ray bubbles, and eventually into the hot plasma at large (see [67]).

There is also little consensus on how cooled or cooling gas fuels the AGN. To create a feedback loop, the thermal state of the hot atmosphere must influence the

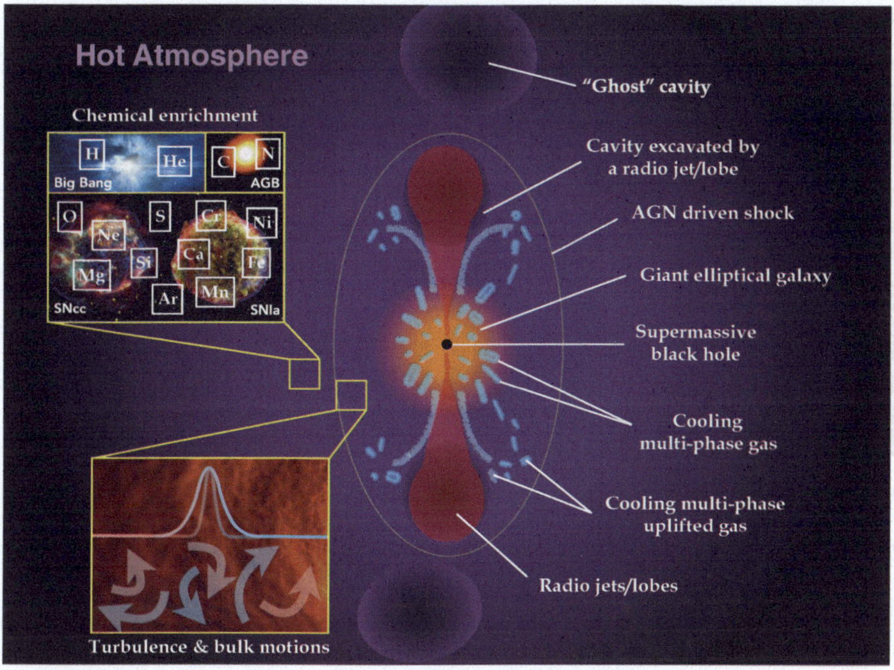


**Fig. 6** An illustration showing a chemically enriched (by SNIa, SNcc, and AGB stars), hot, X-ray emitting atmosphere, stabilized by AGN feedback. Within a radius where the cooling time of the atmospheric gas is shorter than $\sim$1 Gyr the thermally unstable atmospheric gas "precipitates" by condensing into cooler clouds that may fall towards the centre, increasing the accretion rate onto the central supermassive black hole and driving the formation of jets. The jets inflate lobes, displacing the hot atmospheric gas and creating X-ray dark cavities. Initially, the lobes/cavities expand supersonically, driving weak shocks into the surrounding medium and increasing its entropy. After detaching from the jets, the lobes/cavities rise buoyantly in the hot atmosphere, driving turbulence and uplifting low entropy gas in their updraft. The uplifted low entropy gas may cool and fall back towards the centre. As the relativistic plasma filling the bubbles looses energy it stops shining in the radio band and the cavities become "ghost" cavities. By mechanically perturbing the gas, e.g. driving shocks, uplifting the lowest entropy gas from the centre, driving turbulence, the AGN heats the atmosphere preventing its radiative cooling. AGN driven turbulent motions cause Doppler broadening of spectral lines, which can be observed by high-resolution X-ray spectrometers

power output of the AGN. A particularly important question, which still remains unanswered is: *How does AGN feedback operate across spatial scales of over 8 orders of magnitude—from the immediate vicinity of the black hole to a scale of several hundreds of kpc in clusters of galaxies?*

The most popular attempts to solve this question rely on the idea of "precipitation" [69–71]. If the ambient medium "precipitates" (cools) by condensing into cooler clouds that rain toward the centre, the accretion rate will rise by orders of magnitude and trigger a feedback response.

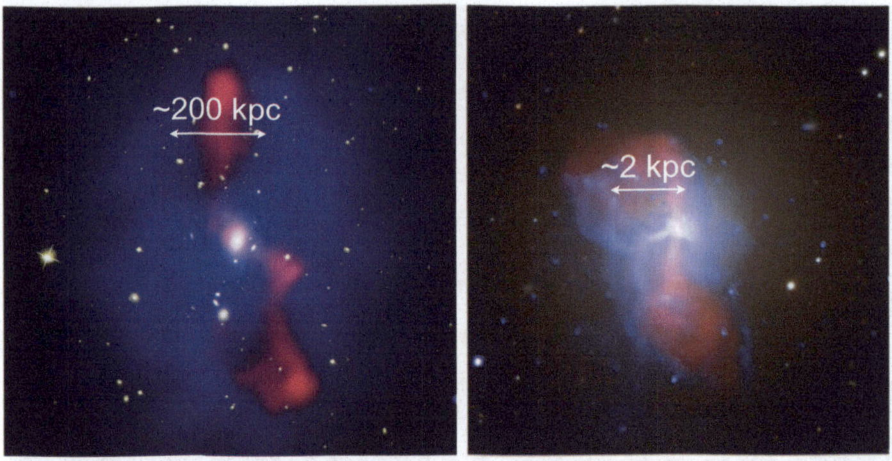


**Fig. 7** Radio-mechanical AGN feedback observed on a cluster scale (left panel) and in a galaxy (right panel). *Left panel:* Composite image of the galaxy cluster MS 0735.6+7421 [55]. *Right panel:* Composite image of M84, a massive elliptical galaxy in the Virgo cluster [68]. The *Chandra* image of the hot X-ray emitting gas is shown in blue and the VLA radio image of the jet injected relativistic plasma is shown in red on both images. A background image from the Sloan Digital Sky Survey is shown in yellow and white. The jet inflated bubbles displace the X-ray emitting plasma, forming cavities in the hot atmospheres. While the average jet power required to form the observed cavities in MS 0735.6+7421 is $1.7 \times 10^{46}$ erg $s^{-1}$ (typical for quasar luminosity but mostly released as mechanical energy), comparable to the X-ray luminosity of this massive cluster, the jet power of M84 is five orders of magnitude smaller, apparently tuned to balance the radiative cooling of its much smaller X-ray atmosphere. Courtesy: NASA/CXC/SAO

Supermassive black hole masses correlate with the luminosities and velocity dispersions of the stellar bulge components of their host galaxies indicating that they evolved together [72–76]. Recently, Lakhchaura et al. [77] found a tight correlation between the central supermassive black hole mass and the atmospheric temperatures of the brightest cluster/group galaxies (BCGs). The atmospheric temperatures are determined by the total mass of the system, implying an underlying correlation between the black hole mass and the total mass of the host galaxy. The hydrostatic analysis of BCGs confirms the existence of the underlying correlation, which turns out to be approximately linear. These results imply that the black hole masses correlate not only with the stellar components of the galaxies but also with their dark matter halos (see also the earlier results of [78, 79]). The supermassive black hole mass could be determined by the binding energy of the halo through radiative feedback during the rapid black hole growth by accretion [80, 81]. However, for the most massive galaxies, such as the BCGs, mergers are the chief channel of growth [82]. Thus the initial correlation established through quasar mode feedback could have been subsequently strengthened via numerous gas-free mergers with galaxies hosting central supermassive black holes (a natural consequence of the central limit theorem; [83]).

## 5 Atmospheric Gas Dynamics

Next to the standard X-ray observables, such as the density, temperature, and metallicity discussed above, X-ray detectors with high spectral resolution also enable direct measurements of the hot gas dynamics. While the broadening of spectral lines provides information about small scale or turbulent motions, line shifts reveal bulk, coherent, "streaming" motions on larger scales. Infall of groups and galaxies as well as accretion of surrounding matter, and the interaction of the intra-cluster medium (ICM) with the relativistic radio emitting plasma injected by the AGN are all expected to drive turbulent and bulk gas motions. Actual measurements of gas motions, however, remain sparse. The knowledge of the dynamical pressure support is also very important for the total mass measurements performed under the assumption of hydrostatic equilibrium and for cluster cosmology, which relies on accurate masses.

The first non-dispersive imaging detector with a high enough spectral resolution to determine the velocity broadening and shifts of ICM emission lines was the Soft X-ray Spectrometer (SXS) on the *Hitomi* satellite (see Fig. 8). The *Hitomi* SXS observation of the Perseus cluster revealed a 1D line-of-sight velocity dispersion reaching approximately 200 km $s^{-1}$ toward the central AGN and toward the jet inflated "ghost" bubble north of the cluster core [84, 85]. Elsewhere in the centre

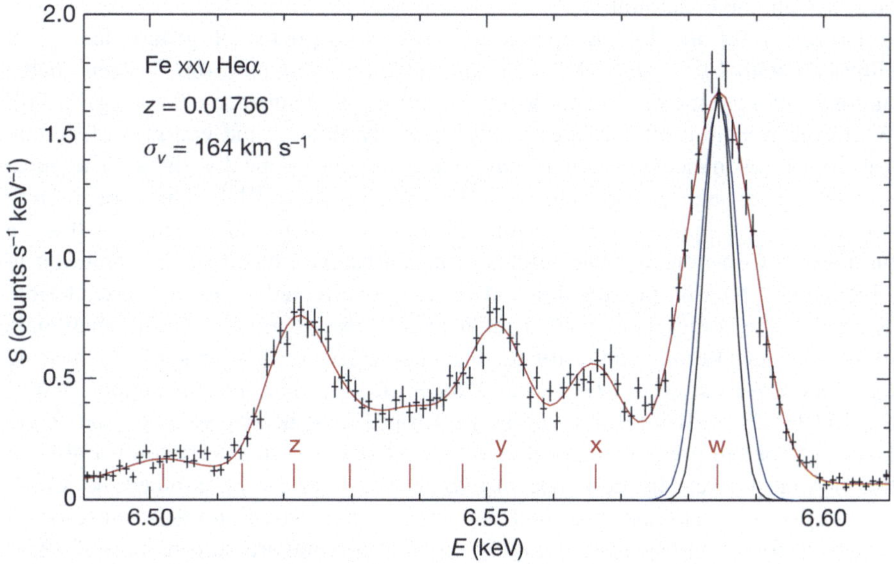


**Fig. 8** The *Hitomi* SXS spectrum of Fe XXV He$\alpha$. The black line indicates the instrumental broadening, the blue line indicates the instrumental plus thermal broadening. All the additional broadening—the difference between the blue line and the data fitted by the red model—is due to velocity broadening. The strongest resonance ('w'), intercombination ('x', 'y') and forbidden ('z') lines are indicated [84]

of the Perseus cluster probed by the *Hitomi* observations, the velocity dispersion of the ICM appears constant at around $100\,\mathrm{km\,s^{-1}}$. The observations also revealed a velocity gradient with a $100\,\mathrm{km\,s^{-1}}$ amplitude across the cluster core, consistent with a large-scale sloshing of the core gas. It turns out that if the observed gas motions are isotropic, the kinetic pressure support is less than 10% of the thermal pressure support in the cluster core. The assumption of hydrostatic equilibrium in cluster mass measurements is thus likely not violated even in the dynamically active AGN hosting cluster cores.

Based on the analysis of surface brightness fluctuations measured with *Chandra*, [86] performed indirect estimates of gas motions in the Perseus cluster even before the *Hitomi* observations. These estimates turned out to be broadly consistent with the SXS measurements. Zhuravleva et al. [86] showed that the heating rate from the dissipation of gas motions might actually be capable of balancing the radiative cooling at each radius in the Perseus cluster.

Prior to the *Hitomi* observation of the Perseus cluster, the velocity broadening of emission lines has only been observed in giant ellipticals, groups, and clusters with strongly centrally peaked surface brightness distributions, using the reflection grating spectrometers (RGS) on board *XMM-Newton* [87–89]. For more than half of the systems, the measurements of line widths provide a 68% upper limit of $200\,\mathrm{km\,s^{-1}}$. Because RGS is a slit-less dispersive spectrometer, the spectral lines observed by this instrument are also broadened by the spatial extent of the source, significantly limiting the velocity broadening measurements, plaguing them by a large systematic uncertainty.

However, for the low temperature X-ray atmospheres of groups and giant ellipticals with $kT \lesssim 0.9\,\mathrm{keV}$, the grating spectrometers on *XMM-Newton* enable indirect velocity measurements using the effect of resonance scattering [2, 90]. While the hot atmospheres are mostly optically thin, at the energies of certain transitions resonance scattering may make the gas optically thick. The most affected by resonance scattering is the Fe XVII line at 15.01 Å. The optical depth (the expected suppression of the line) decreases with the increasing small scale turbulence. Conveniently, the neighbouring unresolved blend of the same ion at 17.05 and 17.10 Å is optically thin and virtually unaffected by resonance scattering. Therefore, the comparison of their intensities allows us to measure the magnitude of resonant scattering and estimate the characteristic velocities of small scale turbulence. Werner et al. [91] and de Plaa et al. [92] measured the suppression of the 15.01 Å Fe XVII line in a sample of X-ray bright nearby galaxies and found that the turbulent pressure support could span from less than ~5% to over 40% of the thermal pressure in these systems. Ogorzalek et al. [93] combined the *XMM-Newton* RGS measurements of both line broadening (based on [89]) and resonant scattering for 13 galaxies and found a mean non-thermal pressure support of ~6%. This non-thermal pressure support is consistent with the *Hitomi* SXS measurements of the Perseus cluster and supports the picture of a quasi-continuous gentle AGN feedback.

The velocities of the cold and warm gas phases within the hot galactic, group and cluster atmospheres might also reflect the velocities in the hot ICM (e.g. [94]). Sub-mm, infrared, and optical spectra of the line emitting nebulae might thus also provide indirect estimates on the velocities in hot galactic atmospheres.

Most of these velocity measurements were performed in relaxed systems, which are currently not undergoing major mergers. Mergers produce large bulk motions as well as in strong turbulence. They also result in weak shocks and cold fronts which are discussed in detail in [95].

## 6 The Chemical Composition of Hot Atmospheres

The first X-ray spectra of clusters of galaxies, obtained in the late 1970s [96, 97], revealed the presence of K-shell Fe line emission. Few years later, emission lines of other metals were also detected by the *Einstein* satellite [98]. These observations provided the very first evidence that the ICM has been significantly enriched with metals.[6] X-ray spectroscopy is a particularly powerful tool for studying the chemical enrichment, because all important products of AGB stars (C and N) and supernovae (O, Ne, Mg, Si, S, Ar, Ca, Cr, Mn, Fe, and Ni) have their K-shell lines in the X-ray band (as well as the L-shell lines of Si, S, Ar, Ca, Fe, and Ni, with the Fe-L lines being particularly prominent). Compared to the abundances measured in other objects at longer wavelengths (e.g. stellar abundances or the chemical composition of optical emission line nebulae), the X-ray measurements of the abundances in the hot ICM are remarkably accurate. Unlike the convention in the field, the abundances measured via X-ray spectroscopy can easily be published on a linear scale and the accuracy is often mainly limited by our knowledge of the atomic data (see e.g., [99]).

The most important questions addressed by the studies of the chemical composition of atmospheric gas include:

1. How does the chemical composition of the atmospheric gas compare with the chemical composition of stars?
2. What are the astrophysical sources of the enrichment? What is the fraction of Type Ia (SNIa; i.e. a thermonuclear explosion of a white dwarf after it reached the Chandrasekhar limit of $1.4\,M_\odot$ following a mass transfer from a stellar companion or a collision with another white dwarf) to core-collapse supernovae (SNcc; i.e. a gravitational collapse of a massive star at the end of its life leading to a supernova explosion) contributing to the enrichment of the intergalactic medium?
3. What can we learn about supernova nucleosynthesis from the chemical abundance studies of hot atmospheres?

[6] All elements heavier than H and He are usually called "metals" in astrophysics.

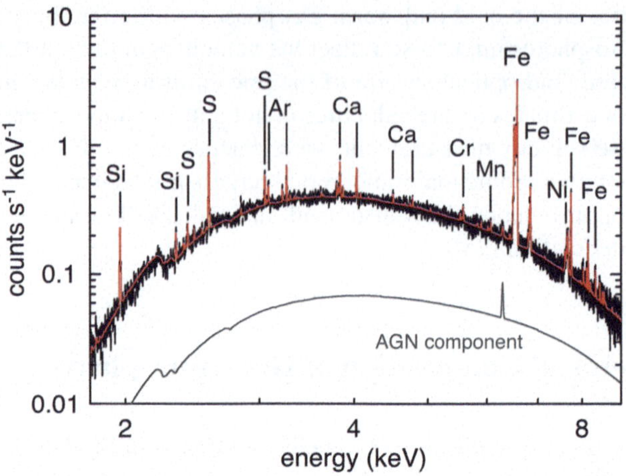


**Fig. 9** The *Hitomi* SXS spectrum of the Perseus cluster resolves the emission lines of eight different elements produced by Type Ia and core-collapse supernovae. Its analysis revealed that the relative abundances of elements produced by supernovae are very similar to the relative abundances of the same metals in the Sun [102]

4. At which cosmic epoch were the metals ejected from the stars and galaxies? Was it before or after the epoch of cluster formation?
5. What are the main transport mechanisms that drive the metals out of galaxies and spread them across the intergalactic medium?

Addressing the first three questions requires the knowledge of the detailed chemical composition of the atmospheric gas, which means measurements of the abundances of as many supernova products as possible. The unique capabilities of *XMM-Newton* and, more recently *Hitomi* (see Fig. 9), revealed that the relative abundances of the observed metals are consistent with those in the Sun [100–104]. This means that, similarly to the chemical enrichment models of our own Galaxy (for a review, see [105]), the enrichment of hot atmospheres has been dominated by only two types of supernovae: SNIa and SNcc. It seems that the contribution of population III stars (i.e. the very first stars in the Universe, made entirely of primordial hydrogen and helium) was negligible (see e.g., [106]).

Finding similar abundance ratios in our Solar neighbourhood and in the hot atmospheres of the largest objects in the Universe is surprising. The chemical composition of cluster atmospheres, which have been enriched by billions of supernovae exploding in thousands of galaxies should reflect the average chemical composition of the Universe. The result that the relative metal abundances of our Sun are so similar to those in clusters could be due to the fact that our Sun is an average star in a kind of galaxy, where most of the stars and metals in the Universe were formed. This result also indicates that the material from which our Sun formed has been a well mixed product of many supernovae.

The fourth question, about the epoch of the metal ejection from galaxies into the intergalactic gas, can be addressed in two different ways. The most direct approach is X-ray spectroscopy of galaxy clusters at various redshifts in order to look for an evolution of their chemical abundances with cosmic time. Although many studies have attempted to use this approach (e.g. [107–112]), the very large uncertainties due to the low signal-to-noise at high redshifts makes it difficult to draw firm conclusions. Moreover, for the same reason this exercise is restricted to the brightest (i.e. most massive) clusters. Despite these limitations, most studies report essentially little or no cosmic evolution in the metallicity of hot atmospheres out to $z \simeq 1$–$1.5$, when the Universe was about half of its current age.

The spatial distribution of metals in nearby systems also provides valuable information about the period when the bulk of the ICM enrichment took place (Fig. 10). In particular, the *Suzaku* X-ray satellite revealed that, instead of decreasing monotonically with radius, the Fe abundance in clusters outskirts remains remarkably uniform at $Z_{Fe} \sim 0.3$ Solar out to at least $r_{200}$ [113–115]. According to simulations (e.g. [116, 117]), this flat distribution is the key signature of metals having escaped their galaxy hosts at $z \simeq 2$–$3$ (more than 10 billion years ago), i.e. *before* these galaxies were incorporated into clusters. These results led to the so-called "early enrichment scenario", in which the bulk of metals has been ejected into the IGM before clusters started to form and grow.

As shown in Fig. 10, relaxed cool-core clusters exhibit a central increase of Fe. Because these clusters host BCGs in their centres, this Fe peak was initially

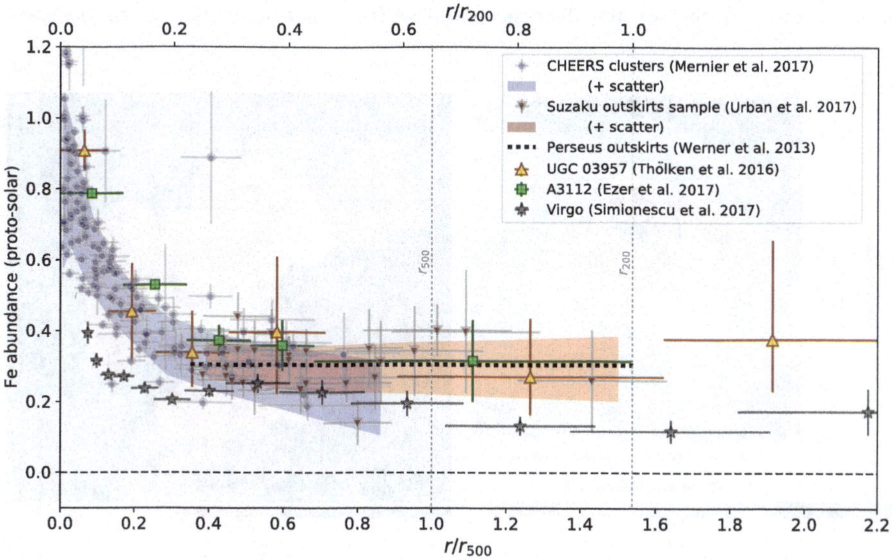


**Fig. 10** Average radial Fe abundance profiles for cool-core clusters from various measurements in the literature [123]

attributed to the stellar populations of these galaxies [118, 119]. While BCGs lack significant star formation and SNcc explosions, SNIa, due to their long delay times, may still keep enriching the BCGs with heavy metals, such as Fe and Ni. Lighter elements like O, Mg, and Si are mainly produced by SNcc and were thus not expected to be seen in large amounts within the central region. It was later found, however, that all elements, including O, Mg, and Si, exhibit a central abundance peak that follows that of Fe [44, 59, 120, 121]. This surprising discovery indicates that stars that are seen today in BCGs have little to do with the enrichment of the hot gas pervading them. In other words, the bulk of the gas enrichment must have occurred early on, *before the hot medium ended up in the BCG*, during or perhaps even before the formation of the cluster. This conclusion is also supported by the observation that while the stellar populations of BCGs have super-Solar $\alpha$/Fe abundance ratios (ratios of SNcc products over iron, which is produced mainly by SNIa; [122]), their hot atmospheres have Solar $\alpha$/Fe ratios. The different composition points to a different origin of metals.

Interestingly, the spatial distribution of metals in hot atmospheres also provides crucial clues to answer our last question. Simulations which reproduce the observed flat metal distribution in clusters outskirts (e.g. [116, 117]) indicate that the activity of supermassive black holes was necessary to eject the metals from their galactic potential before the formation of clusters. In some cases, a gentle uplift of metals by AGN feedback can be seen directly around the central dominant galaxies of nearby clusters, where metal-rich gas is observed along the radio jets of the central AGN and around their cavities [65]. The clearest observational evidence of uplift has been obtained for M87 ([61], Fig. 11 left) and for the Hydra A ([59, 124], Fig. 11 right). In these cases, however, the deep potential well of such mature systems prevents

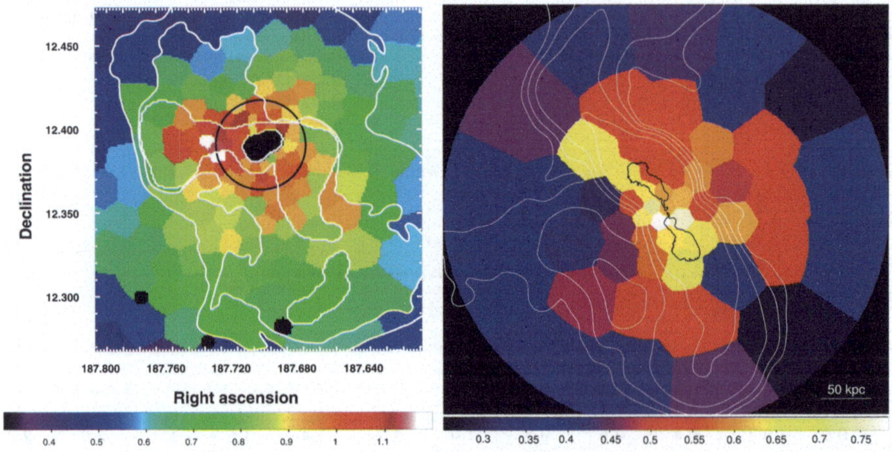


**Fig. 11** Projected Fe abundance maps for the hot atmospheres of M87 (left panel; [61]) and the Hydra A cluster (right panel; [124]). The 330 MHz and 1400 MHz radio contours, over-plotted in white and black, respectively, trace the jets from the central supermassive black hole

metals to effectively diffuse out and to enrich their surroundings at very large scales. Ejection of metals by galaxies in proto-cluster environments (i.e. at $z \gtrsim 2$) via their central supermassive black holes is yet to be observed directly.

In addition to AGN feedback, other processes may play a significant role in redistributing metals within hot atmospheres (for a review, see [125]). These are: (1) ram-pressure stripping (i.e. when a metal-rich galaxy falls into a galaxy cluster, the pressure of the pervading ICM strips its local interstellar medium and a tail of metal-rich gas forms behind the galaxy), (2) sloshing (i.e. when the ICM is slightly displaced from its potential well, circular motions start to take place and may gently transport metals on large scales), and (3) galaxy-galaxy interactions. In addition, clusters of galaxies contain a significant fraction of intracluster stars, which do not belong to any galaxy. For this reason, they can, in principle, release metals into the hot atmosphere more efficiently than stars that reside in a strong galactic potential. Currently, the exact contribution of all these mechanisms is not well known, and future dedicated observations and simulations will be needed to clarify the relative contribution of these processes.

## 7 Future Studies of Hot Atmospheres

The current generation of X-ray satellites (*Chandra*, *XMM-Newton*, and *Suzaku*) revolutionized our understanding of hot atmospheres pervading massive galaxies, groups and clusters. These missions greatly benefited from the all sky survey performed by the *ROSAT* satellite launched in 1990. The *ROSAT* survey was crucial for the mapping of the X-ray sky and discovering new systems, yet it still remains unsurpassed.

Before the launch of the major next generation X-ray observatories it will be critically important to perform a new, more sensitive sky survey. That is the objective of the eROSITA instrument onboard the recently launched Russian–German *Spectrum-Roentgen-Gamma* (*SRG*) mission, which is currently mapping the X-ray sky with a much improved spatial resolution and sensitivity [126]. This will allow the discovery of approximately 50,000–100,000 new galaxy clusters and about three millions new AGN.

Perhaps the most significant improvements in the capabilities of the future major X-ray observatories are expected in spectral resolution. They will allow us to probe directly small-scale turbulence together with large-scale gas motions. The high spectral resolution also allows us to access more metal lines, providing increasingly accurate elemental abundance measurements. Micro-calorimeters, which convert a tiny change of temperature due to the incoming X-ray photon into photon energy, have demonstrated their potential by the spectacular results from the *Hitomi* SXS observation of the Perseus cluster. A re-flight of this Japanese mission called *XRISM* (X-Ray Imaging and Spectroscopy Mission; [127]), is planned for the year 2022 and its key micro-calorimeter instrument Resolve will routinely provide X-ray spectra with an energy resolution of 5 eV.

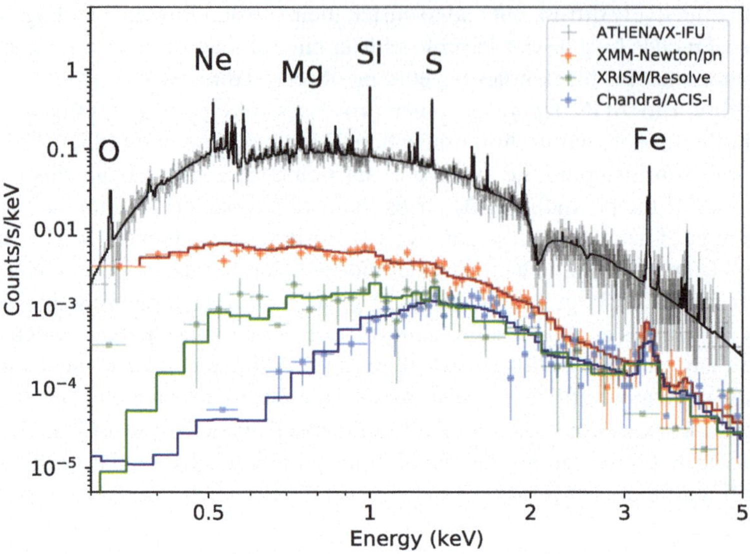


**Fig. 12** A simulated *Athena* X-IFU spectrum of a $kT = 3$ keV galaxy cluster at a redshift of $z = 1$ obtained assuming a deep 250,000 s (about 3 days) long observation. For comparison, simulated observations, assuming equally long exposure times, are shown for *XMM-Newton* EPIC/pn, *Chandra* ACIS-I, and *XRISM* Resolve

Later on around 2031, ESA's next generation X-ray observatory *Athena* (*Advanced Telescope for High-ENergy Astrophysics*) is expected to push our knowledge of hot atmospheres to yet another level. *Athena* will be equipped with a cryogenic X-ray spectrometer, based on a large array of Transition Edge Sensors (TES), offering 2.5 eV spectral resolution, with 5 arcsec per pixels, over a field of view of 5 arcmin in diameter (the X-IFU instrument; [128]). Moreover, its effective area is expected to be an order of magnitude larger than that of *XRISM* (see Fig. 12). This unprecedented sensitivity will be of vital importance as it will allow to probe high-redshift systems and provide unprecedented measurements to constrain the evolution of the thermodynamic properties and metallicities of hot atmospheres. *Athena* will be able to detect groups at early epochs (out to look-back times of ~10 Gyr), which will be essential for understanding the formation of large scale structures.

The high spectral resolution is also necessary to survey the distribution and properties of the faint circumgalactic and intergalactic medium permeating the filaments of the cosmic web. This will be the primary aim of the future Chinese *HUBS* (*Hot Universe Baryon Surveyor*) mission.[7] *HUBS* will combine a large field-of-view with an excellent spectral resolution in order to map the emission and absorption in the O VII and O VIII lines. These measurements will be crucial to

[7] http://hubs.phys.tsinghua.edu.cn/en/index.html.

detect and "map" the bulk of the ordinary matter in our nearby Universe. The yet to be approved Japanese *SuperDIOS* mission [129] will extend the capabilities of *HUBS*. Its 10 arcsec spatial resolution will allow a robust filtering of contaminating X-ray point sources and thus a more reliable census of diffuse warm-hot baryons.

The *Athena* X-ray observatory is expected to achieve a 5 arcsec spatial resolution, which is significantly better than that of *XMM-Newton*, but still worse than that of *Chandra*. In particular the studies of galaxy scale atmospheres would strongly benefit from a high spatial resolution, which would allow us to resolve out the bright point sources in these systems. But also the studies of subtle features in nearby galaxy clusters (e.g. shocks, cold fronts, ripples, etc.) or substructures in high-redshift systems (cavities, signatures of sloshing, merging sub-clusters, etc.) will benefit from combining the unprecedented spectral resolution, large effective area, and excellent *spatial* resolution. This is the ambitious goal of the proposed NASA mission *Lynx*, which aims to combine the capabilities of *Athena* with a *Chandra*-like spatial resolution and add a high resolution grating spectrometer.

The post-*Athena* (-*Lynx*) future of X-ray observatories remains far and not clear. It depends strongly on the future science questions uncovered by the upcoming facilities and on future technology that is yet to be demonstrated. However, it is never too early to "think further" and propose a very long term view on astrophysical observations (e.g. see the Voyage 2050 call by ESA). Even though the currently planned missions are expected to reveal much more about the diffuse medium outside of the gravitational potential wells of massive clusters, groups and galaxies, they will still only probe the tip of the iceberg. The physical properties of the majority of ordinary matter in the local Universe will remain unexplored even after *Athena, Lynx, HUBS*, and *Super-DIOS*. This task awaits for a dedicated mission with a large effective area, field-of-view, as well as high spatial and spectral resolution. Within about two decades, megapixel size TES detector arrays with sub-eV spectral resolution in the soft X-ray band are expected to be available. Our ambition and hope is, that such sizable high-resolution X-ray spectrometer arrays combined with large X-ray mirrors, that are currently under development, will finally enable us to build the *Cosmic Web Explorer* mission [130], which will survey the physical properties of most of the medium permeating the cosmic ocean.

**Acknowledgements** This work was supported by the Lendület LP2016-11 grant awarded by the Hungarian Academy of Sciences.

## References

1. W. Forman, C. Jones, W. Tucker, Astrophys. J. **293**, 102 (1985). https://doi.org/10.1086/163218
2. E. Churazov, I. Zhuravleva, S. Sazonov, R. Sunyaev, Space Sci. Rev. **157**, 193 (2010). https://doi.org/10.1007/s11214-010-9685-4
3. N. Kaiser, Mon. Not. R. Astron. Soc. **222**, 323 (1986). https://doi.org/10.1093/mnras/222.2.323

4. J.F. Navarro, C.S. Frenk, S.D.M. White, Mon. Not. R. Astron. Soc. **275**(3), 720 (1995). https://doi.org/10.1093/mnras/275.3.720
5. S. Giodini, L. Lovisari, E. Pointecouteau, S. Ettori, T.H. Reiprich, H. Hoekstra, Space Sci. Rev. **177**(1–4), 247 (2013). https://doi.org/10.1007/s11214-013-9994-5
6. V. Bharadwaj, T.H. Reiprich, L. Lovisari, H.J. Eckmiller, Astron. Astrophys. **573**, A75 (2015). https://doi.org/10.1051/0004-6361/201424586
7. Y.Y. Zhang, A. Finoguenov, H. Böhringer, J.P. Kneib, G.P. Smith, R. Kneissl, N. Okabe, H. Dahle, Astron. Astrophys. **482**(2), 451 (2008). https://doi.org/10.1051/0004-6361:20079103
8. G.W. Pratt, J.H. Croston, M. Arnaud, H. Böhringer, Astron. Astrophys. **498**(2), 361 (2009). https://doi.org/10.1051/0004-6361/200810994
9. H.J. Eckmiller, D.S. Hudson, T.H. Reiprich, Astron. Astrophys. **535**, A105 (2011). https://doi.org/10.1051/0004-6361/201116734
10. A. Mahdavi, H. Hoekstra, A. Babul, C. Bildfell, T. Jeltema, J.P. Henry, Astrophys. J. **767**(2), 116 (2013). https://doi.org/10.1088/0004-637X/767/2/116
11. L. Lovisari, T.H. Reiprich, G. Schellenberger, Astron. Astrophys. **573**, A118 (2015). https://doi.org/10.1051/0004-6361/201423954
12. E.K. Panagoulia, A.C. Fabian, J.S. Sanders, Mon. Not. R. Astron. Soc. **438**, 2341 (2014). https://doi.org/10.1093/mnras/stt2349
13. I.V. Babyk, B.R. McNamara, P.E.J. Nulsen, H.R. Russell, A.N. Vantyghem, M.T. Hogan, F.A. Pulido, Astrophys. J. **862**, 39 (2018). https://doi.org/10.3847/1538-4357/aacce5
14. P. Tozzi, C. Norman, Astrophys. J. **546**, 63 (2001). https://doi.org/10.1086/318237
15. G.M. Voit, S.T. Kay, G.L. Bryan, Mon. Not. R. Astron. Soc. **364**, 909 (2005). https://doi.org/10.1111/j.1365-2966.2005.09621.x
16. T.J. Ponman, D.B. Cannon, J.F. Navarro, Nature **397**(6715), 135 (1999). https://doi.org/10.1038/16410
17. A. Vikhlinin, A. Kravtsov, W. Forman, C. Jones, M. Markevitch, S.S. Murray, L. Van Speybroeck, Astrophys. J. **640**, 691 (2006). https://doi.org/10.1086/500288
18. F. Gastaldello, D.A. Buote, P.J. Humphrey, L. Zappacosta, J.S. Bullock, F. Brighenti, W.G. Mathews, Astrophys. J. **669**(1), 158 (2007). https://doi.org/10.1086/521519
19. X. Dai, J.N. Bregman, C.S. Kochanek, E. Rasia, Astrophys. J. **719**(1), 119 (2010). https://doi.org/10.1088/0004-637X/719/1/119
20. S.W. Allen, D.A. Rapetti, R.W. Schmidt, H. Ebeling, R.G. Morris, A.C. Fabian, Mon. Not. R. Astron. Soc. **383**, 879 (2008). https://doi.org/10.1111/j.1365-2966.2007.12610.x
21. C.A. Correa, J. Schaye, J.S.B. Wyithe, A.R. Duffy, T. Theuns, R.A. Crain, R.G. Bower, Mon. Not. R. Astron. Soc. **473**, 538 (2018). https://doi.org/10.1093/mnras/stx2332
22. S. Pellegrini, L. Ciotti, A. Negri, J.P. Ostriker, Astrophys. J. **856**(2), 115 (2018). https://doi.org/10.3847/1538-4357/aaae07
23. D. Nelson, G. Kauffmann, A. Pillepich, S. Genel, V. Springel, R. Pakmor, L. Hernquist, R. Weinberger, P. Torrey, M. Vogelsberger, F. Marinacci, Mon. Not. R. Astron. Soc. **477**(1), 450 (2018). https://doi.org/10.1093/mnras/sty656
24. M. Fukugita, C.J. Hogan, P.J.E. Peebles, Astrophys. J. **503**, 518 (1998). https://doi.org/10.1086/306025
25. D. Kereš, N. Katz, D.H. Weinberg, R. Davé, Mon. Not. R. Astron. Soc. **363**, 2 (2005). https://doi.org/10.1111/j.1365-2966.2005.09451.x
26. M. Fukugita, P.J.E. Peebles, Astrophys. J. **639**, 590 (2006). https://doi.org/10.1086/499556
27. S.L. Snowden, R. Egger, M.J. Freyberg, D. McCammon, P.P. Plucinsky, W.T. Sanders, J.H.M.M. Schmitt, J. Trümper, W. Voges, Astrophys. J. **485**, 125 (1997). https://doi.org/10.1086/304399
28. D.B. Henley, R.L. Shelton, Astrophys. J. Suppl. Ser. **187**, 388 (2010). https://doi.org/10.1088/0067-0049/187/2/388
29. D.B. Henley, R.L. Shelton, Astrophys. J. Suppl. Ser. **202**, 14 (2012). https://doi.org/10.1088/0067-0049/202/2/14

30. F. Paerels, A. Rasmussen, S. Kahn, J.W. Herder, C. Vries, in *XEUS – Studying the Evolution of the Hot Universe*, ed. by G. Hasinger, T. Boller, A.N. Parmer (2003), p. 57
31. A. Gupta, S. Mathur, Y. Krongold, F. Nicastro, M. Galeazzi, Astrophys. J. **756**, L8 (2012). https://doi.org/10.1088/2041-8205/756/1/L8
32. J. Bland-Hawthorn, O. Gerhard, Annu. Rev. Astron. Astrophys. **54**, 529 (2016). https://doi.org/10.1146/annurev-astro-081915-023441
33. M.E. Anderson, E. Churazov, J.N. Bregman, Mon. Not. R. Astron. Soc. **455**, 227 (2016). https://doi.org/10.1093/mnras/stv2314
34. Á. Bogdán, H. Bourdin, W.R. Forman, R.P. Kraft, M. Vogelsberger, L. Hernquist, V. Springel, Astrophys. J. **850**, 98 (2017). https://doi.org/10.3847/1538-4357/aa9523
35. J.T. Li, J.N. Bregman, Q.D. Wang, R.A. Crain, M.E. Anderson, Astrophys. J. **855**, L24 (2018). https://doi.org/10.3847/2041-8213/aab2af
36. Planck Collaboration, P.A.R. Ade, N. Aghanim, M. Arnaud, M. Ashdown, F. Atrio-Barandela, J. Aumont, C. Baccigalupi, A. Balbi, A.J. Banday et al., Astron. Astrophys. **557**, A52 (2013). https://doi.org/10.1051/0004-6361/201220941
37. J.P. Greco, J.C. Hill, D.N. Spergel, N. Battaglia, Astrophys. J. **808**, 151 (2015). https://doi.org/10.1088/0004-637X/808/2/151
38. J.N. Bregman, M.E. Anderson, M.J. Miller, E. Hodges-Kluck, X. Dai, J.T. Li, Y. Li, Z. Qu, Astrophys. J. **862**, 3 (2018). https://doi.org/10.3847/1538-4357/aacafe
39. J. Rasmussen, T.J. Ponman, Mon. Not. R. Astron. Soc. **399**, 239 (2009). https://doi.org/10.1111/j.1365-2966.2009.15244.x
40. J.N. Bregman, M.E. Anderson, X. Dai, Astrophys. J. **716**, L63 (2010). https://doi.org/10.1088/2041-8205/716/1/L63
41. M. Sun, New J. Phys. **14**(4), 045004 (2012). https://doi.org/10.1088/1367-2630/14/4/045004
42. R.M. Yates, P.A. Thomas, B.M.B. Henriques, Mon. Not. R. Astron. Soc. **464**, 3169 (2017). https://doi.org/10.1093/mnras/stw2361
43. F. Mernier, J. de Plaa, J.S. Kaastra, A.J.J. Raassen, L. Gu, J. Mao, I. Urdampilleta, N. Truong, A. Simionescu, Mon. Not. R. Astron. Soc. **478**, L116 (2018). https://doi.org/10.1093/mnrasl/sly080
44. F. Mernier, J. de Plaa, J.S. Kaastra, Y.Y. Zhang, H. Akamatsu, L. Gu, P. Kosec, J. Mao, C. Pinto, T.H. Reiprich, J.S. Sanders, A. Simionescu, N. Werner, Astron. Astrophys. **603**, A80 (2017). https://doi.org/10.1051/0004-6361/201630075
45. F. Fraternali, M. Tomassetti, Mon. Not. R. Astron. Soc. **426**(3), 2166 (2012). https://doi.org/10.1111/j.1365-2966.2012.21650.x
46. F. Fraternali, in *Gas Accretion onto Galaxies*, ed. by A. Fox, R. Davé. Astrophysics and Space Science Library, vol. 430 (2017), p. 323. https://doi.org/10.1007/978-3-319-52512-9_14
47. S. Walker, A. Simionescu, D. Nagai, N. Okabe, D. Eckert, T. Mroczkowski, H. Akamatsu, S. Ettori, V. Ghirardini, Space Sci. Rev. **215**(1), 7 (2019). https://doi.org/10.1007/s11214-018-0572-8
48. N. Werner, A. Finoguenov, J.S. Kaastra, A. Simionescu, J.P. Dietrich, J. Vink, H. Böhringer, Astron. Astrophys. **482**(3), L29 (2008). https://doi.org/10.1051/0004-6361:200809599
49. E. Bulbul, S.W. Randall, M. Bayliss, E. Miller, F. Andrade-Santos, R. Johnson, M. Bautz, E.L. Blanton, W.R. Forman, C. Jones, Astrophys. J. **818**(2), 131 (2016). https://doi.org/10.3847/0004-637X/818/2/131
50. G.E. Alvarez, S.W. Randall, H. Bourdin, C. Jones, K. Holley-Bockelmann, Astrophys. J. **858**(1), 44 (2018). https://doi.org/10.3847/1538-4357/aabad0
51. F. Nicastro, J. Kaastra, Y. Krongold, S. Borgani, E. Branchini, R. Cen, M. Dadina, C.W. Danforth, M. Elvis, F. Fiore, Nature **558**(7710), 406 (2018). https://doi.org/10.1038/s41586-018-0204-1
52. O.E. Kovács, Á. Bogdán, R.a.K. Smith, R.P. Kraft, W.R. Forman, Astrophys. J. **872**(1), 83 (2019). https://doi.org/10.3847/1538-4357/aaef78
53. E. Churazov, S. Sazonov, R. Sunyaev, W. Forman, C. Jones, H. Böhringer, Mon. Not. R. Astron. Soc. **363**(1), L91 (2005). https://doi.org/10.1111/j.1745-3933.2005.00093.x
54. B.R. McNamara, P.E.J. Nulsen, New J. Phys. **14**(5), 055023 (2012). https://doi.org/10.1088/1367-2630/14/5/055023

55. B.R. McNamara, P.E.J. Nulsen, M.W. Wise, D.A. Rafferty, C. Carilli, C.L. Sarazin, E.L. Blanton, Nature **433**, 45 (2005). https://doi.org/10.1038/nature03202
56. P.E.J. Nulsen, B.R. McNamara, M.W. Wise, L.P. David, Astrophys. J. **628**, 629 (2005). https://doi.org/10.1086/430845
57. W. Forman, P. Nulsen, S. Heinz, F. Owen, J. Eilek, A. Vikhlinin, M. Markevitch, R. Kraft, E. Churazov, C. Jones, Astrophys. J. **635**, 894 (2005). https://doi.org/10.1086/429746
58. W. Forman, C. Jones, E. Churazov, M. Markevitch, P. Nulsen, A. Vikhlinin, M. Begelman, H. Böhringer, J. Eilek, S. Heinz, R. Kraft, F. Owen, M. Pahre, Astrophys. J. **665**, 1057 (2007). https://doi.org/10.1086/519480
59. A. Simionescu, N. Werner, H. Böhringer, J.S. Kaastra, A. Finoguenov, M. Brüggen, P.E.J. Nulsen, Astron. Astrophys. **493**, 409 (2009). https://doi.org/10.1051/0004-6361:200810225
60. E.T. Million, N. Werner, A. Simionescu, S.W. Allen, P.E.J. Nulsen, A.C. Fabian, H. Böhringer, J.S. Sanders, Mon. Not. R. Astron. Soc. **407**, 2046 (2010). https://doi.org/10.1111/j.1365-2966.2010.17220.x
61. A. Simionescu, N. Werner, A. Finoguenov, H. Böhringer, M. Brüggen, Astron. Astrophys. **482**, 97 (2008). https://doi.org/10.1051/0004-6361:20078749
62. A. Simionescu, E. Roediger, P.E.J. Nulsen, M. Brüggen, W.R. Forman, H. Böhringer, N. Werner, A. Finoguenov, Astron. Astrophys. **495**, 721 (2009). https://doi.org/10.1051/0004-6361:200811071
63. N. Werner, A. Simionescu, E.T. Million, S.W. Allen, P.E.J. Nulsen, A. von der Linden, S.M. Hansen, H. Böhringer, E. Churazov, A.C. Fabian, W.R. Forman, C. Jones, J.S. Sanders, G.B. Taylor, Mon. Not. R. Astron. Soc. **407**, 2063 (2010). https://doi.org/10.1111/j.1365-2966.2010.16755.x
64. N. Werner, M. Sun, J. Bagchi, S.W. Allen, G.B. Taylor, S.K. Sirothia, A. Simionescu, E.T. Million, J. Jacob, M. Donahue, Mon. Not. R. Astron. Soc. **415**, 3369 (2011). https://doi.org/10.1111/j.1365-2966.2011.18957.x
65. C.C. Kirkpatrick, B.R. McNamara, K.W. Cavagnolo, Astrophys. J. **731**, L23 (2011). https://doi.org/10.1088/2041-8205/731/2/L23
66. C.C. Kirkpatrick, B.R. McNamara, Mon. Not. R. Astron. Soc. **452**, 4361 (2015). https://doi.org/10.1093/mnras/stv1574
67. N. Werner, B.R. McNamara, E. Churazov, E. Scannapieco, Space Sci. Rev. **215**(1), 5 (2019). https://doi.org/10.1007/s11214-018-0571-9
68. A. Finoguenov, M. Ruszkowski, C. Jones, M. Brüggen, A. Vikhlinin, E. Mandel, Astrophys. J. **686**, 911 (2008). https://doi.org/10.1086/591662
69. M. Gaspari, M. Ruszkowski, S.P. Oh, Mon. Not. R. Astron. Soc. **432**, 3401 (2013). https://doi.org/10.1093/mnras/stt692
70. G.M. Voit, M. Donahue, B.W. O'Shea, G.L. Bryan, M. Sun, N. Werner, Astrophys. J. **803**, L21 (2015). https://doi.org/10.1088/2041-8205/803/2/L21
71. B.R. McNamara, H.R. Russell, P.E.J. Nulsen, M.T. Hogan, A.C. Fabian, F. Pulido, A.C. Edge, Astrophys. J. **830**, 79 (2016). https://doi.org/10.3847/0004-637X/830/2/79
72. J. Magorrian, S. Tremaine, D. Richstone, R. Bender, G. Bower, A. Dressler, S.M. Faber, K. Gebhardt, R. Green, C. Grillmair, J. Kormendy, T. Lauer, Astron. J. **115**, 2285 (1998). https://doi.org/10.1086/300353
73. K. Gebhardt, R. Bender, G. Bower, A. Dressler, S.M. Faber, A.V. Filippenko, R. Green, C. Grillmair, L.C. Ho, J. Kormendy, T.R. Lauer, J. Magorrian, J. Pinkney, D. Richstone, S. Tremaine, Astrophys. J. **539**, L13 (2000). https://doi.org/10.1086/312840
74. L. Ferrarese, D. Merritt, Astrophys. J. **539**, L9 (2000). https://doi.org/10.1086/312838
75. J. Kormendy, L.C. Ho, Annu. Rev. Astron. Astrophys. **51**, 511 (2013). https://doi.org/10.1146/annurev-astro-082708-101811
76. R.P. Saglia, M. Opitsch, P. Erwin, J. Thomas, A. Beifiori, M. Fabricius, X. Mazzalay, N. Nowak, S.P. Rusli, R. Bender, Astrophys. J. **818**, 47 (2016). https://doi.org/10.3847/0004-637X/818/1/47
77. K. Lakhchaura, N. Truong, N. Werner (2019). e-Prints. arXiv:1904.10513

78. Á. Bogdán, A.D. Goulding, Astrophys. J. **800**, 124 (2015). https://doi.org/10.1088/0004-637X/800/2/124
79. Á. Bogdán, L. Lovisari, M. Volonteri, Y. Dubois, Astrophys. J. **852**(2), 131 (2018). https://doi.org/10.3847/1538-4357/aa9ab5
80. C.M. Booth, J. Schaye, Mon. Not. R. Astron. Soc. **405**(1), L1 (2010). https://doi.org/10.1111/j.1745-3933.2010.00832.x
81. C.M. Booth, J. Schaye, Mon. Not. R. Astron. Soc. **413**(2), 1158 (2011). https://doi.org/10.1111/j.1365-2966.2011.18203.x
82. R. Weinberger, V. Springel, R. Pakmor, D. Nelson, S. Genel, A. Pillepich, M. Vogelsberger, F. Marinacci, J. Naiman, P. Torrey, L. Hernquist, Mon. Not. R. Astron. Soc. **479**, 4056 (2018). https://doi.org/10.1093/mnras/sty1733
83. K. Jahnke, A.V. Macciò, Astrophys. J. **734**(2), 92 (2011). https://doi.org/10.1088/0004-637X/734/2/92
84. Hitomi Collaboration, F. Aharonian, H. Akamatsu, F. Akimoto, S.W. Allen, N. Anabuki, L. Angelini, K. Arnaud, M. Audard, H. Awaki, Nature **535**(7610), 117 (2016). https://doi.org/10.1038/nature18627
85. Hitomi Collaboration, F. Aharonian, H. Akamatsu, F. Akimoto, S.W. Allen, L. Angelini, M. Audard, H. Awaki, M. Axelsson, A. Bamba, Publ. Astron. Soc. Jpn. **70**(2), 9 (2018). https://doi.org/10.1093/pasj/psx138
86. I. Zhuravleva, E. Churazov, A.A. Schekochihin, S.W. Allen, P. Arévalo, A.C. Fabian, W.R. Forman, J.S. Sanders, A. Simionescu, R. Sunyaev, A. Vikhlinin, N. Werner, Nature **515**, 85 (2014). https://doi.org/10.1038/nature13830
87. J.S. Sanders, A.C. Fabian, Mon. Not. R. Astron. Soc. **412**, L35 (2011). https://doi.org/10.1111/j.1745-3933.2010.01000.x
88. J.S. Sanders, A.C. Fabian, Mon. Not. R. Astron. Soc. **429**, 2727 (2013). https://doi.org/10.1093/mnras/sts543
89. C. Pinto, J.S. Sanders, N. Werner, J. de Plaa, A.C. Fabian, Y.Y. Zhang, J.S. Kaastra, A. Finoguenov, J. Ahoranta, Astron. Astrophys. **575**, A38 (2015). https://doi.org/10.1051/0004-6361/201425278
90. M.R. Gilfanov, R.A. Sunyaev, E.M. Churazov, Sov. Astron. Lett. **13**, 3 (1987)
91. N. Werner, I. Zhuravleva, E. Churazov, A. Simionescu, S.W. Allen, W. Forman, C. Jones, J.S. Kaastra, Mon. Not. R. Astron. Soc. **398**, 23 (2009). https://doi.org/10.1111/j.1365-2966.2009.14860.x
92. J. de Plaa, I. Zhuravleva, N. Werner, J.S. Kaastra, E. Churazov, R.K. Smith, A.J.J. Raassen, Y.G. Grange, Astron. Astrophys. **539**, A34 (2012). https://doi.org/10.1051/0004-6361/201118404
93. A. Ogorzalek, I. Zhuravleva, S.W. Allen, C. Pinto, N. Werner, A.B. Mantz, R.E.A. Canning, A.C. Fabian, J.S. Kaastra, J. de Plaa, Mon. Not. R. Astron. Soc. **472**, 1659 (2017). https://doi.org/10.1093/mnras/stx2030
94. M. Gaspari, M. McDonald, S.L. Hamer, F. Brighenti, P. Temi, M. Gendron-Marsolais, J. Hlavacek-Larrondo, A.C. Edge, N. Werner, P. Tozzi, M. Sun, J.M. Stone, G.R. Tremblay, M.T. Hogan, D. Eckert, S. Ettori, H. Yu, V. Biffi, S. Planelles, Astrophys. J. **854**, 167 (2018). https://doi.org/10.3847/1538-4357/aaaa1b
95. M. Markevitch, A. Vikhlinin, Phys. Rep. **443**, 1 (2007). https://doi.org/10.1016/j.physrep.2007.01.001
96. R.J. Mitchell, J.L. Culhane, P.J.N. Davison, J.C. Ives, Mon. Not. R. Astron. Soc. **175**, 29P (1976)
97. R.F. Mushotzky, P.J. Serlemitsos, E.A. Boldt, S.S. Holt, B.W. Smith, Astrophys. J. **225**, 21 (1978). https://doi.org/10.1086/156465
98. S.M. Lea, R. Mushotzky, S.S. Holt, Astrophys. J. **262**, 24 (1982). https://doi.org/10.1086/160392
99. Hitomi Collaboration, F. Aharonian, H. Akamatsu, F. Akimoto, S.W. Allen, L. Angelini, M. Audard, H. Awaki, M. Axelsson, A. Bamba, M.W. Bautz, R. Blandford, L.W. Brenneman, G.V. Brown, E. Bulbul, E.M. Cackett, M. Chernyakova, M.P. Chiao, P.S. Coppi, E. Costantini,

J. de Plaa, C.P. de Vries, J.W. den Herder, C. Done, T. Dotani, K. Ebisawa, M.E. Eckart, T. Enoto, Y. Ezoe, A.C. Fabian, C. Ferrigno, A.R. Foster, R. Fujimoto, Y. Fukazawa, A. Furuzawa, M. Galeazzi, L.C. Gallo, P. Gandhi, M. Giustini, A. Goldwurm, L. Gu, M. Guainazzi, Y. Haba, K. Hagino, K. Hamaguchi, I.M. Harrus, I. Hatsukade, K. Hayashi, T. Hayashi, K. Hayashida, N. Hell, J.S. Hiraga, A. Hornschemeier, A. Hoshino, J.P. Hughes, Y. Ichinohe, R. Iizuka, H. Inoue, Y. Inoue, M. Ishida, K. Ishikawa, Y. Ishisaki, M. Iwai, J. Kaastra, T. Kallman, T. Kamae, J. Kataoka, S. Katsuda, N. Kawai, R.L. Kelley, C.A. Kilbourne, T. Kitaguchi, S. Kitamoto, T. Kitayama, T. Kohmura, M. Kokubun, K. Koyama, S. Koyama, P. Kretschmar, H.A. Krimm, A. Kubota, H. Kunieda, P. Laurent, S.H. Lee, M.A. Leutenegger, O. Limousin, M. Loewenstein, K.S. Long, D. Lumb, G. Madejski, Y. Maeda, D. Maier, K. Makishima, M. Markevitch, H. Matsumoto, K. Matsushita, D. McCammon, B.R. McNamara, M. Mehdipour, E.D. Miller, J.M. Miller, S. Mineshige, K. Mitsuda, I. Mitsuishi, T. Miyazawa, T. Mizuno, H. Mori, K. Mori, K. Mukai, H. Murakami, R.F. Mushotzky, T. Nakagawa, H. Nakajima, T. Nakamori, S. Nakashima, K. Nakazawa, K.K. Nobukawa, M. Nobukawa, H. Noda, H. Odaka, T. Ohashi, M. Ohno, T. Okajima, N. Ota, M. Ozaki, F. Paerels, S. Paltani, R. Petre, C. Pinto, F.S. Porter, K. Pottschmidt, C.S. Reynolds, S. Safi-Harb, S. Saito, K. Sakai, T. Sasaki, G. Sato, K. Sato, R. Sato, M. Sawada, N. Schartel, P.J. Serlemtsos, H. Seta, M. Shidatsu, A. Simionescu, R.K. Smith, Y. Soong, Ł. Stawarz, Y. Sugawara, S. Sugita, A. Szymkowiak, H. Tajima, H. Takahashi, T. Takahashi, S. Takeda, Y. Takei, T. Tamagawa, T. Tamura, T. Tanaka, Y. Tanaka, Y.T. Tanaka, M.S. Tashiro, Y. Tawara, Y. Terada, Y. Terashima, F. Tombesi, H. Tomida, Y. Tsuboi, M. Tsujimoto, H. Tsunemi, T.G. Tsuru, H. Uchida, H. Uchiyama, Y. Uchiyama, S. Ueda, Y. Ueda, S. Uno, C.M. Urry, E. Ursino, S. Watanabe, N. Werner, D.R. Wilkins, B.J. Williams, S. Yamada, H. Yamaguchi, K. Yamaoka, N.Y. Yamasaki, M. Yamauchi, S. Yamauchi, T. Yaqoob, Y. Yatsu, D. Yonetoku, I. Zhuravleva, A. Zoghbi, A.J.J. Raassen, Publ. Astron. Soc. Jpn. **70**, 12 (2018). https://doi.org/10.1093/pasj/psx156

100. J. de Plaa, N. Werner, J.A.M. Bleeker, J. Vink, J.S. Kaastra, M. Méndez, Astron. Astrophys. **465**, 345 (2007). https://doi.org/10.1051/0004-6361:20066382
101. F. Mernier, J. de Plaa, C. Pinto, J.S. Kaastra, P. Kosec, Y.Y. Zhang, J. Mao, N. Werner, Astron. Astrophys. **592**, A157 (2016). https://doi.org/10.1051/0004-6361/201527824
102. Hitomi Collaboration, F. Aharonian, H. Akamatsu, F. Akimoto, S.W. Allen, L. Angelini, M. Audard, H. Awaki, M. Axelsson, A. Bamba, M.W. Bautz, R. Blandford, L.W. Brenneman, G.V. Brown, E. Bulbul, E.M. Cackett, M. Chernyakova, M.P. Chiao, P.S. Coppi, E. Costantini, J. de Plaa, J.W. den Herder, C. Done, T. Dotani, K. Ebisawa, M.E. Eckart, T. Enoto, Y. Ezoe, A.C. Fabian, C. Ferrigno, A.R. Foster, R. Fujimoto, Y. Fukazawa, A. Furuzawa, M. Galeazzi, L.C. Gallo, P. Gandhi, M. Giustini, A. Goldwurm, L. Gu, M. Guainazzi, Y. Haba, K. Hagino, K. Hamaguchi, I.M. Harrus, I. Hatsukade, K. Hayashi, T. Hayashi, K. Hayashida, J.S. Hiraga, A. Hornschemeier, A. Hoshino, J.P. Hughes, Y. Ichinohe, R. Iizuka, H. Inoue, Y. Inoue, M. Ishida, K. Ishikawa, Y. Ishisaki, M. Iwai, J. Kaastra, T. Kallman, T. Kamae, J. Kataoka, S. Katsuda, N. Kawai, R.L. Kelley, C.A. Kilbourne, T. Kitaguchi, S. Kitamoto, T. Kitayama, T. Kohmura, M. Kokubun, K. Koyama, S. Koyama, P. Kretschmar, H.A. Krimm, A. Kubota, H. Kunieda, P. Laurent, S.H. Lee, M.A. Leutenegger, O. Limousine, M. Loewenstein, K.S. Long, D. Lumb, G. Madejski, Y. Maeda, D. Maier, K. Makishima, M. Markevitch, H. Matsumoto, K. Matsushita, D. McCammon, B.R. McNamara, M. Mehdipour, E.D. Miller, J.M. Miller, S. Mineshige, K. Mitsuda, I. Mitsuishi, T. Miyazawa, T. Mizuno, H. Mori, K. Mori, K. Mukai, H. Murakami, R.F. Mushotzky, T. Nakagawa, H. Nakajima, T. Nakamori, S. Nakashima, K. Nakazawa, K.K. Nobukawa, M. Nobukawa, H. Noda, H. Odaka, T. Ohashi, M. Ohno, T. Okajima, N. Ota, M. Ozaki, F. Paerels, S. Paltani, R. Petre, C. Pinto, F.S. Porter, K. Pottschmidt, C.S. Reynolds, S. Safi-Harb, S. Saito, K. Sakai, T. Sasaki, G. Sato, K. Sato, R. Sato, M. Sawada, N. Schartel, P.J. Serlemitsos, H. Seta, M. Shidatsu, A. Simionescu, R.K. Smith, Y. Soong, L. Stawarz, Y. Sugawara, S. Sugita, A. Szymkowiak, H. Tajima, H. Takahashi, T. Takahashi, S. Takeda, Y. Takei, T. Tamagawa, T. Tamura, T. Tanaka, Y. Tanaka, Y.T. Tanaka, M.S. Tashiro, Y. Tawara, Y. Terada, Y. Terashima, F. Tombesi, H. Tomida, Y. Tsuboi, M. Tsujimoto, H. Tsunemi, T. Go Tsuru, H. Uchida, H. Uchiyama,

Y. Uchiyama, S. Ueda, Y. Ueda, S. Uno, C.M. Urry, E. Ursino, C.P. de Vries, S. Watanabe, N. Werner, D.R. Wik, D.R. Wilkins, B.J. Williams, S. Yamada, H. Yamaguchi, K. Yamaoka, N.Y. Yamasaki, M. Yamauchi, S. Yamauchi, T. Yaqoob, Y. Yatsu, D. Yonetoku, I. Zhuravleva, A. Zoghbi, Nature **551**, 478 (2017). https://doi.org/10.1038/nature24301
103. F. Mernier, N. Werner, J. de Plaa, J.S. Kaastra, A.J.J. Raassen, L. Gu, J. Mao, I. Urdampilleta, A. Simionescu, Mon. Not. R. Astron. Soc. **480**, L95 (2018). https://doi.org/10.1093/mnrasl/sly134
104. A. Simionescu, S. Nakashima, H. Yamaguchi, K. Matsushita, F. Mernier, N. Werner, T. Tamura, K. Nomoto, J. de Plaa, S.C. Leung, Mon. Not. R. Astron. Soc. **483**(2), 1701 (2019). https://doi.org/10.1093/mnras/sty3220
105. K. Nomoto, C. Kobayashi, N. Tominaga, Annu. Rev. Astron. Astrophys. **51**, 457 (2013). https://doi.org/10.1146/annurev-astro-082812-140956
106. N. Werner, J. de Plaa, J.S. Kaastra, J. Vink, J.A.M. Bleeker, T. Tamura, J.R. Peterson, F. Verbunt, Astron. Astrophys. **449**, 475 (2006). https://doi.org/10.1051/0004-6361:20053868
107. I. Balestra, P. Tozzi, S. Ettori, P. Rosati, S. Borgani, V. Mainieri, C. Norman, M. Viola, Astron. Astrophys. **462**, 429 (2007). https://doi.org/10.1051/0004-6361:20065568
108. M.E. Anderson, J.N. Bregman, S.C. Butler, C.R. Mullis, Astrophys. J. **698**(1), 317 (2009). https://doi.org/10.1088/0004-637X/698/1/317
109. A. Baldi, S. Ettori, S. Molendi, I. Balestra, F. Gastaldello, P. Tozzi, Astron. Astrophys. **537**, A142 (2012). https://doi.org/10.1051/0004-6361/201117836
110. S. Ettori, A. Baldi, I. Balestra, F. Gastaldello, S. Molendi, P. Tozzi, Astron. Astrophys. **578**, A46 (2015). https://doi.org/10.1051/0004-6361/201425470
111. M. McDonald, E. Bulbul, T. de Haan, E.D. Miller, B.A. Benson, L.E. Bleem, M. Brodwin, J.E. Carlstrom, I. Chiu, W.R. Forman, J. Hlavacek-Larrondo, G.P. Garmire, N. Gupta, J.J. Mohr, C.L. Reichardt, A. Saro, B. Stalder, A.A. Stark, J.D. Vieira, Astrophys. J. **826**, 124 (2016). https://doi.org/10.3847/0004-637X/826/2/124
112. A.B. Mantz, S.W. Allen, R.G. Morris, A. Simionescu, O. Urban, N. Werner, I. Zhuravleva, Mon. Not. R. Astron. Soc. **472**, 2877 (2017). https://doi.org/10.1093/mnras/stx2200
113. Y. Fujita, N. Tawa, K. Hayashida, M. Takizawa, H. Matsumoto, N. Okabe, T.H. Reiprich, Publ. Astron. Soc. Jpn. **60**, S343 (2008)
114. N. Werner, J.B.R. Oonk, R.E.A. Canning, S.W. Allen, A. Simionescu, J. Kos, R.J. van Weeren, A.C. Edge, A.C. Fabian, A. von der Linden, P.E.J. Nulsen, C.S. Reynolds, M. Ruszkowski, Astrophys. J. **767**, 153 (2013). https://doi.org/10.1088/0004-637X/767/2/153
115. O. Urban, N. Werner, S.W. Allen, A. Simionescu, A. Mantz, Mon. Not. R. Astron. Soc. **470**, 4583 (2017). https://doi.org/10.1093/mnras/stx1542
116. V. Biffi, S. Planelles, S. Borgani, E. Rasia, G. Murante, D. Fabjan, M. Gaspari, Mon. Not. R. Astron. Soc. **476**(2), 2689 (2018). https://doi.org/10.1093/mnras/sty363
117. V. Biffi, F. Mernier, P. Medvedev, Space Sci. Rev. **214**(8), 123 (2018). https://doi.org/10.1007/s11214-018-0557-7
118. H. Böhringer, K. Matsushita, E. Churazov, A. Finoguenov, Y. Ikebe, Astron. Astrophys. **416**, L21 (2004). https://doi.org/10.1051/0004-6361:20040047
119. S. De Grandi, S. Ettori, M. Longhetti, S. Molendi, Astron. Astrophys. **419**, 7 (2004). https://doi.org/10.1051/0004-6361:20034228
120. J. de Plaa, N. Werner, A.M. Bykov, J.S. Kaastra, M. Méndez, J. Vink, J.A.M. Bleeker, M. Bonamente, J.R. Peterson, Astron. Astrophys. **452**, 397 (2006). https://doi.org/10.1051/0004-6361:20053864
121. E.T. Million, N. Werner, A. Simionescu, S.W. Allen, Mon. Not. R. Astron. Soc. **418**(4), 2744 (2011). https://doi.org/10.1111/j.1365-2966.2011.19664.x
122. C. Conroy, G.J. Graves, P.G. van Dokkum, Astrophys. J. **780**, 33 (2014). https://doi.org/10.1088/0004-637X/780/1/33
123. F. Mernier, V. Biffi, H. Yamaguchi, P. Medvedev, A. Simionescu, S. Ettori, N. Werner, J.S. Kaastra, J. de Plaa, L. Gu, Space Sci. Rev. **214**(8), 129 (2018). https://doi.org/10.1007/s11214-018-0565-7

124. C.C. Kirkpatrick, M. Gitti, K.W. Cavagnolo, B.R. McNamara, L.P. David, P.E.J. Nulsen, M.W. Wise, Astrophys. J. **707**, L69 (2009). https://doi.org/10.1088/0004-637X/707/1/L69
125. S. Schindler, A. Diaferio, Space Sci. Rev. **134**, 363 (2008). https://doi.org/10.1007/s11214-008-9321-8
126. A. Merloni, P. Predehl, W. Becker, H. Böhringer, T. Boller, H. Brunner, M. Brusa, K. Dennerl, M. Freyberg, P. Friedrich (2012). e-Prints. arXiv:1209.3114
127. M. Tashiro, H. Maejima, K. Toda, R. Kelley, L. Reichenthal, J. Lobell, R. Petre, M. Guainazzi, E. Costantini, M. Edison, in *Space Telescopes and Instrumentation 2018: Ultraviolet to Gamma Ray*. Society of Photo-Optical Instrumentation Engineers (SPIE) Conference Series, vol. 10699 (2018), p. 1069922. https://doi.org/10.1117/12.2309455
128. D. Barret, T. Lam Trong, J.W. den Herder, L. Piro, M. Cappi, J. Houvelin, R. Kelley, J.M. Mas-Hesse, K. Mitsuda, S. Paltani, in *Space Telescopes and Instrumentation 2018: Ultraviolet to Gamma Ray*. Society of Photo-Optical Instrumentation Engineers (SPIE) Conference Series, vol. 10699 (2018), p. 106991G. https://doi.org/10.1117/12.2312409
129. S. Yamada, T. Ohashi, Y. Ishisaki, Y. Ezoe, Y. Ichinohe, S. Kitazawa, K. Kosaka, R. Hayakawa, K. Nunomura, K. Mitsuda, N.Y. Yamasaki, T. Kikuchi, T. Hayashi, H. Muramatsu, Y. Nakashima, Y. Tawara, I. Mitsuishi, Y. Babazaki, D. Seki, K. Otsuka, M. Ishihara, K. Osato, N. Ota, M. Tomariguchi, D. Nagai, E. Lau, K. Sato, the DIOS team, J. Low Temp. Phys. **193**(5), 1016 (2018). https://doi.org/10.1007/s10909-018-1918-z
130. A. Simionescu, S. Ettori, N. Werner, D. Nagai, F. Vazza, H. Akamatsu, C. Pinto, J. de Plaa, N. Wijers, D. Nelson, E. Pointecouteau, G.W. Pratt, D. Spiga, E. Lau, M. Rossetti, F. Gastaldello, V. Biffi, E. Bulbul, J.W. den Herder, D. Eckert, F. Fraternali, B. Mingo, G. Pareschi, G. Pezzulli, T.H. Reiprich, J. Schaye, S.A. Walker, J. Werk (2019). e-Prints. arXiv:1908.01778

# The Establishment of the Standard Cosmological Model Through Observations

**Ricardo Tanausú Génova-Santos**

## Contents



**Abstract** Over the last decades, observations with increasing quality have revolutionized our understanding of the general properties of the Universe. Questions posed for millenia by mankind about the origin, evolution and structure of the cosmos have found an answer. This has been possible mainly thanks to observations of the Cosmic Microwave Background, of the large-scale distribution of matter structure in the local Universe, and of type Ia supernovae that have revealed the accelerated expansion of the Universe. All these observations have successfully converged into the so-called "concordance model". In spite of all these observational successes, there are still some important open problems, the most obvious of which are what generated the initial matter inhomogeneities that led to the structure observable in today's Universe, and what is the nature of dark matter, and of the

R. T. Génova-Santos (✉)
Instituto de Astrofísica de Canarias, La Laguna, Tenerife, Spain

Departamento de Astrofísica, Universidad de La Laguna (ULL), La Laguna, Tenerife, Spain
e-mail: rgs@iac.es



P. Kabáth et al. (eds.), *Reviews in Frontiers of Modern Astrophysics*,
https://doi.org/10.1007/978-3-030-38509-5_11

dark energy that drives the accelerated expansion. In this chapter I will expand on the previous aspects. I will present a general description of the Standard Cosmological Model of the Universe, with special emphasis on the most recent observations that have allowed to establish this model. I will also discuss the shortfalls of this model, its most pressing open questions, and will briefly describe the observational programmes that are being planned to tackle these issues.

# 1 Cosmological Models

## *1.1 First Cosmological Views and Models*

Since its beginnings, mankind has questioned about the origin, history and evolution of the Cosmos. During the fifth century B.C. classical philosophers and thinkers like Anaxagoras, Leucippus of Mileto or Democritus hypothesised about the main constituents of the Universe. After that, the first models attempting to describe the closest observable Universe (the orbits of the Solar System planets) appeared. Aristotle (fourth century B.C.) proposed the geocentric model and a finite and static Universe built up of fire, air, water and earth. In the second century B.C. Ptolemy further developed this model by introducing the concept of epicycles. Regrettably this was the prevalent model during more than 17 centuries, and throughout the full Middle Ages. In 1543 Nicolaus Copernicus caused a big breakthrough with the publication of his book *De revolutionibus orbium coelestium*, where he suggested that the Sun, rather than the Earth, was the centre of the Solar System. This idea had initially been put forward by Aristarchus of Samos in the third century B.C., although it had been totally ignored. Copernicus theory marked a major event in the history of science, triggered the Copernican Revolution, and made a pioneering contribution to the so-called Scientific Revolution, which together with other events and developments in other fields in turn marked the emergence of the modern science during the early modern period. Later in 1584 Giordano Bruno went further and dared to suggest that not even the Solar System was the centre of the Universe, with the Sun being just one more star amongst many others. This had well-known terribly bad consequences for him. In 1605 Johannes Kepler introduced refinements on the Copernican model, such as proposing elliptical instead of circular planet orbits, based on observations carried out by Galileo Galilei. We can probably say that these developments marked the transition of Cosmology from a philosophical and speculative to a more scientific discipline, which involved an attempt to perform the first mathematical description of the Cosmos based on the observations.

It is often considered that the Scientific Revolution that started with the ideas of Copernicus was culminated with the publication by Isaac Newton of his "Mathematical Principles of Natural Philosophy" in 1687, where he formulated the laws of motion and universal gravitation, thence lying the foundation of classical mechanics. Newton proposed a finite and static Universe, in which matter is uniformly distributed. He thought that gravitational forces between planets should lead to an unstable Solar System and, sticking to his religious beliefs, he stated that God's intervention made it an stable system. This idea was challenged 100 years later by Pierre-Simon Laplace, who thought that mathematics should be able to explain the motion of planets without the need of God. In the same way as Newton did, Albert Einstein naturally attempted to apply his theory to cosmology, this time the theory of General Relativity, published in 1915. Einstein also embraced the prevailing idea at that time of a dynamically static Universe. Instead of resorting to God, in this case he modified his original field equations by the introduction of the cosmological constant, $\Lambda$, which he viewed as a repulsive form of gravity that kept the Universe stable. This changed dramatically with the discovery of the Universe expansion by Edwin Hubble in 1929, which led Einstein to recognise as his biggest blunder the assumption of a static universe and the introduction of the cosmological constant.

## 1.2 Big Bang Cosmology

Although confirmed observationally in 1929 by Edwin Hubble, the idea of an expanding universe had been proposed in 1922 by Alexander Friedmann, based on Einstein's fundamental equations. In 1927 Georges Lemaître, backed by the first observations of Edwin Hubble, concluded that the Universe was in fact expanding, and this naturally drove him to propose for the first time the idea of a Big Bang (he initially coined his idea "primeval atom"). Later in 1946 George Gamow [1] proposed a hot Big Bang model to explain the primeval build-up of atoms heavier than hydrogen. He realised that reaching the binding energies of those atoms required temperatures of the order of $\sim 10^9$–$10^{10}$ K. According to this model the Universe was sufficiently hot and dense in its beginnings, and later expanded out and cooled down, in such a way that now it should have an average non-zero temperature. This remnant temperature should show up as a background radiation with a temperature of $\sim 5$ K, the so-called "Cosmic Microwave Background" (CMB), according to the prediction by Ralph A. Alpher and Robert Herman [2].

At that time, a competing alternative was the "Steady State Model", which was proposed by three prominent physicists: Hermann Bondi, Thomas Gold and Fred Hoyle. This model proposed that the observed expansion of the Universe was associated with a spontaneous and continuous creation of matter, which led to a Universe with a fixed average density [3]. The existence of the Cosmic Microwave Background could not fit by any means in this theory, and therefore its discovery in 1964 by Arno Penzias and Robert Wilson[1] [5] led to the refutal of this model, and to the final recognition of the Big Bang Model. Ever since this has been the prevailing model to explain the beginning and later expansion of our Universe. The three main,

[1]Note that, even before this, the CMB had been indirectly detected by Adams [4] through the local excitation of CN molecules in our Galaxy. This excitation was attributed to some kind of "unknown" radiation with temperature $\sim 2.3$ K.

and independent, pillars supporting nowadays this theory are: (1) the expansion of the Universe; (2) the abundance of light elements; and (3) the CMB.

### 1.3 The Standard Model of Cosmology

All the previous developments and discoveries led to the establishment of the what we nowadays call the "Standard Model of Cosmology" (SMC). This model is supported mainly on General Relativity and by the validity of the "Cosmological Principle", which states that the Universe is homogeneous and isotropic on large scales ($\gtrsim$100 Mpc). The geometry of a homogeneous and isotropic Universe in a four-dimensional (space and time) system is described by the Robertson-Walker metric,

$$ds^2 = c^2dt^2 - a^2(t)\left[\frac{dr^2}{1-kr^2} + r^2\left(d\theta^2 + \sin^2\theta d\phi^2\right)\right], \tag{1}$$

where $a(t)$ is the scale factor and $k = -1, 0, +1$ respectively for an open, flat or closed Universe. Alexander Friedmann used this metric to solve the Einstein's field equations of General Relativity, leading to the so-called Friedmann equations:

$$\left(\frac{\dot{a}}{a}\right)^2 = \frac{8\pi G}{3}\rho - \frac{kc^2}{a^2} + \frac{\Lambda c^2}{3}, \tag{2}$$

$$\frac{\ddot{a}}{a} = -\frac{4\pi G}{3}\left(\rho + \frac{3p}{c^2}\right) + \frac{\Lambda}{3}c^2, \tag{3}$$

where $p$ and $\rho$ are respectively pressure and density, $G$ is the gravitational constant, $c$ the speed of light and $\Lambda$ is the "cosmological constant". Equation (2) gives the expansion rate, whereas Eq. (3) gives the acceleration of the expansion. By defining the following dimensionless mass, curvature and dark-energy (or vacuum energy) density parameters:

$$\Omega_{\rm m} = \frac{8\pi G}{3H_0^2}\rho_0, \quad \Omega_{\rm k} = -\frac{kc^2}{a_0^2H_0^2} \quad \text{y} \quad \Omega_\Lambda = \frac{\Lambda c^2}{3H_0^2}, \tag{4}$$

where $H_0 = (\dot{a}_0/a_0)$ denotes the present value of the Hubble parameter, the first Friedmann equation (2) can be written at $t_0$ (current time) as[2]

$$\Omega_{\rm m} + \Omega_{\rm k} + \Omega_\Lambda = 1. \tag{5}$$

[2] A more detailed explanation of the derivation of the Friedmann equations can be found in classical references like [6].

The total density of the Universe can be written as $\Omega_{\rm tot} = \Omega_{\rm m} + \Omega_\Lambda = 1 - \Omega_{\rm k}$, and the critical density is the value required for a flat geometry ($\Omega_{\rm k} = 0 \Rightarrow \Omega_{\rm tot} = 1$), $\rho_c = 3H_0^2/(8\pi G) = 1.879\,h^2 \times 10^{-29}\,{\rm g\,cm^{-3}}$, in such a way that $\Omega_{\rm m} = \rho_0/\rho_c$. This flat geometry is confirmed by observations, as we will see in Sect. 2.

We now have evidences for a non-zero dark energy component (see Sect. 2), with $\Omega_\Lambda \approx 0.69$. The other dark component of the Universe is coined "dark matter", the first evidences of which date from 1933 when Fritz Zwicky [7] encountered a discrepancy on the total mass of the Coma cluster inferred through the velocity dispersion of individual member galaxies and through its emitted light. Another well-known evidence came later through the analysis of the rotation curves of galaxies [8], while nowadays we have even further compelling evidences of this component derived from gravitational lensing, the large-scale structure of the Universe, or the CMB, as we will see in Sect. 2. These observations also agree on that this component should be "cold". Therefore, the matter density is decomposed into a cold-dark-matter component and a baryonic component, $\Omega_{\rm m} = \Omega_{\rm b} + \Omega_{\rm c}$, and we know that $\Omega_{\rm c} \approx 0.27$ (see Sect. 2). Therefore $\Omega_\Lambda + \Omega_{\rm c} \approx 0.95$, and then we can say that only 5% of the total matter-energy budget of the Universe is ordinary matter, the rest being dark matter and dark energy. For this reason the current SMC has been coined $\Lambda$CDM.

The first Friedmann equation (5) describes the composition and the geometry of the Universe. Based on compelling observational evidences, in the $\Lambda$CDM model the universe is considered to be flat, so $\Omega_{\rm k} = 0$ is fixed. Then, given the existence of the two matter components, we end up with two independent parameters, which are usually considered to be $\Omega_{\rm b}$ and $\Omega_{\rm m}$, while the density of dark energy is derived from Eq. (5) as $\Omega_\Lambda = 1 - \Omega_{\rm b} - \Omega_{\rm c}$. A third important free parameter in the model, which describes the evolution of the Universe, is the Hubble constant, $H_0$. Nowadays it is common practice to use as a free parameter the angular acoustic scale at recombination, $\theta_{\rm MC}$, instead of $H_0$, because it has less correlations with other cosmological parameters.

The previous three parameters accounting for the composition and evolution of the Universe must be complemented with three additional parameters. Two of them describe the initial conditions of the primordial density perturbations, whose later evolution leads to structure formation. More specifically these parameters are the amplitude $A_{\rm s}$ and the spectral index $n_{\rm s}$ of the primordial power spectrum of scalar perturbations,

$$P(k) = A_{\rm s}\left(\frac{k}{k_0}\right)^{n_{\rm s}-1}, \tag{6}$$

where $k_0$ is an arbitrary pivot scale.

Astrophysical parameters, related to different processes like radiation transfer, ionisation or recombination, must also be considered. The most important of these, which becomes the sixth parameter in the ΛCDM model, is the optical depth of Thomson scattering to recombination, $\tau$, which is defined as

$$\tau = \sigma_{\rm T} \int_0^{z_{\rm rec}} n_{\rm e}(z')\, dz', \qquad (7)$$

where $\sigma_{\rm T}$ is the Thomson cross section and $n_{\rm e}(z)$ the electron density at redshift $z$.

We commented earlier that the geometry parameter, $\Omega_{\rm k}$, is usually considered as a fixed parameter in this parametrisation. Another important parameter that is usually fixed is the CMB temperature (or equivalently the radiation density), thanks to the outstanding precision with which this parameter can be measured by fitting the CMB frequency spectrum, $T_0 = 2.7260 \pm 0.0013$ K [9].

To sum up, we have a 6-parameter model, $\{\Omega_{\rm b}, \Omega_{\rm c}, \theta_{\rm MC}, A_{\rm s}, n_{\rm s}, \tau\}$, which describes the global properties of our Universe, and which provides an astonishingly good description of the available data, as we will see in Sect. 2.

## 1.4 Extensions to the ΛCDM Model

As we have seen before in Sect. 1.3, the ΛCDM model has six independent parameters, and provides an excellent fit to all kind of available data. Two other parameters, curvature $\Omega_{\rm k}$ and the radiation density $T_0$, are very well constrained by observations and are left fixed. Different extensions to this model, involving new parameters, can be considered. We will now briefly describe some of these extensions.

### 1.4.1 Early-Universe Physics and Initial Conditions

A complete cosmological model must include an statistical description of the tiny initial perturbations that gave rise to the structure we see in today's Universe. The physical framework that provides this is inflation [10, 11], according to which the early Universe underwent a brief period of accelerated expansion during which quantum fluctuations were stretched out to become the classical fluctuations that we see today. Inflationary cosmology provides an elegant explanation of the key features of our Universe: flat geometry ($\Omega_{\rm k} = 0$), adiabatic, nearly scale-invariant ($n_{\rm s} \lesssim 1$, very close to unity), and Gaussian perturbations in all scales. However, some other phenomenological models of inflation (see [12] for a review), or even alternatives to inflation, predict departures from these conditions, or distinctive signatures that may be seen in the data.

**Scale-Dependent Primordial Fluctuations** In the base ΛCDM model primordial fluctuations are modelled by a pure power-law with spectral index $n_{\rm s}$ (Eq. (6)). The simplest inflationary models predict a nearly scale-invariant spectrum, $n_{\rm s} \approx 1$.

However, there are inflationary models allowing for a running of the spectral index, $dn_s/d\ln k$:

$$P(k) = A_s \left(\frac{k}{k_0}\right)^{n_s - 1 + (1/2)(dn_s/d\ln k)\ln(k/k_0)} . \tag{8}$$

**Tensor Perturbations** Although in the base ΛCDM model perturbations are assumed to be purely scalar modes, primordial tensor perturbations (gravitational waves) should also be originated during inflation. The tensor-mode spectrum is usually described as

$$P_t(k) = A_t \left(\frac{k}{k_0}\right)^{n_t} . \tag{9}$$

The ratio between the amplitude of the tensor to the scalar modes, at the reference scale $k_0$, is denoted as $r = A_t/A_s$. This parameter is proportional to the fourth power of the energy scale of inflation. Slow-roll inflation models satisfy the consistency relation $n_t = -r/8$. As it is shown in Fig. 1, tensor modes contribute to the temperature anisotropies in the large scales (low multipoles, $\ell$), but are completely overshadowed by the scalar modes. On the other hand, a confirmation of the existence of tensor modes could be unambiguously achieved by detecting the so-called B-mode anisotropy at large angular scales (see Sect. 2.1).

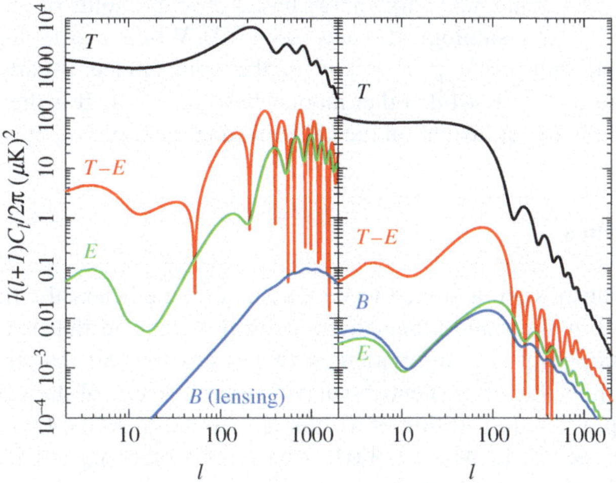


**Fig. 1** Temperature (black), E-mode (green), B-mode (blue) and TE cross-correlation (red) CMB power spectra produced by scalar perturbations (left) and tensor perturbations (gravitational waves; right). The amplitude of the tensor perturbations has been fixed at $r = 0.22$. In the left-hand panel the B-mode spectrum induced by weak gravitational lensing is also depicted. Figure taken from [13]

**Spatial Curvature** The base ΛCDM model assumes Friedmann-Lemaître-Robertson-Walker metric with a flat geometry. As we will see in Sect. 2 all current observations are consistent with $\Omega_k = 0$ to very high accuracy. In fact, providing an explanation for the flatness of our observed Universe was one of the main motivations of the inflationary cosmology. However, there are specific inflationary models that generate open [14] or closed [15] universes. The detection of a deviation from $\Omega_k = 0$ would have profound implications both for inflationary models and for fundamental physics.

**Isocurvature Perturbations** As it was said above, in ΛCDM the primordial perturbations are considered to be adiabatic. This is what observations currently show, and in fact it is a key prediction of single-field inflation. On the other hand, isocurvature fluctuations (fluctuations in the relative energy densities of different species at a fixed total density) could be produced by sub-dominant fields during the inflation era. Detecting them could then yield important insights on physics at the energy scale of inflation.

### 1.4.2 Dark Energy

The late-time accelerated expansion of the Universe is still considered one of the most mysterious aspects of standard cosmology. In the base ΛCDM model the acceleration is provided by a cosmological constant Λ that is introduced in Einstein equations of General Relativity (see Sect. 1.3). However there are many other alternatives, like a dynamical dark energy based in scalar fields or modifications of general relativity on cosmological scales (see [16]). While a cosmological constant has equation of state $w = p/\rho = -1$, scalar field models usually have a time varying $w$ with $w \geq -1$, while other models have $w < -1$. It is then important to try to constrain $w$ to get insight on the nature of dark energy.

### 1.4.3 Neutrinos

In ΛCDM neutrinos are assumed to be massless. Nonetheless the flavour oscillations observed in solar and atmospheric neutrinos indicate that neutrinos should be massive, implying striking evidences of physics beyond the standard model. Constraining the value of the neutrino mass is therefore one of the key questions in fundamental physics. In addition of assuming zero mass, ΛCDM fixes the number of relativistic species to $N_{\rm eff} = 3.046$ (the reason of being slightly larger than three is that the three standard model neutrinos were not completely decoupled at electron-positron annihilation). Exploring different $N_{\rm eff}$ values would be useful to test extensions of the standard model that predict the existence of new light particles.

### 1.4.4 Variation of Fundamental Constants

The ΛCDM model assumes the validity of General Relativity in cosmological scales, and of the standard model of particle physics in small scales. One possible extension to this model, which may have important implications in fundamental physics, is to consider a possible variation of dimensionless constants. These include the fine-structure constant, $\alpha$, the electron-to-proton mass ratio, $\mu$, and the gravitational constant, $\alpha_g = Gm_p^2/\hbar c$. While a variation of $G$ can affect Friedmann equations, and also raise the issue of consistency in the overall theory of gravity, variations of the non-gravitational constants such as $\alpha$ or $\mu$ could be produced for instance if there exists a new interaction between light and atoms mediated by a new massless scalar field [17]. Another important prediction of the ΛCDM scenario is the redshift evolution of the CMB temperature (or radiation density) which, under the assumption of adiabatic expansion of the Universe and photon-number conservation, should be $T_{\mathrm{CMB}}(z) = T_0(1 + z)$. However, there are non-standard scenarios like a non perfectly transparent Universe, decaying vacuum cosmologies or modified gravity models, which lead to photon mixing and/or violation of photon number conservation, and therefore to a deviation from the previous standard redshift evolution.

# 2 Observational Probes

In the previous section we have described the basics of the ΛCDM model. Of course, the establishment of this model is not only a product of human thinking, but of the analysis and interpretation of huge amount of data coming from different kind of observations. Over the last decades, technological improvements in the design and development of astronomical instrumentation have led to data with increasing quality and sensitivity, allowing to determine cosmological parameters with unprecedented accuracy. This is why we usually say that we live in the era of "precision cosmology". As a result of the beautiful consensus between completely different and independent kind of observations, the current ΛCDM model has been coined the "concordance model". We will now describe the three most important cosmological probes.

## *2.1 The Cosmic Microwave Background*

The CMB currently stands as the observational probe giving the tightest constraints on cosmological parameters. Since the discovery by the *COBE* satellite of the temperature anisotropies [18], different experiments (from the ground, balloon-borne or space satellites) have been specifically designed to measure and characterise these anisotropies with gradually finer angular resolution and sensitivity. These anisotropies (with amplitude $\sim 10^{-5}$ with respect to the mean CMB temperature) are produced by fluctuations of the baryon-photon fluid at the last-scattering surface ($z \sim 1100$) in the gravitational potential wells produced during the inflationary period. All the statistical information of these anisotropies can be neatly encoded in the so-called "power spectrum", which is defined in terms of a spherical harmonics expansion:

$$\frac{\Delta T}{T_0}(\theta, \phi) = \sum_{\ell=2}^{\infty} \sum_{m=-\ell}^{\ell} a_{\ell m} Y_{\ell m}(\theta, \phi), \tag{10}$$

where $\Delta T$ represents the CMB temperature spatial variations with respect to their mean value $T_0$. Given that under $\Lambda$CDM temperature fluctuations are Gaussian, the $a_{\ell m}$ coefficients must be random and statistically independent variables. The value of $a_{\ell m} a^*_{\ell m}$ averaged over the full sky gives an estimate of the power associated with each multipole $\ell$, and for this reason we define the power spectrum of the temperature variation as the variance of this quantity,

$$C_\ell = \frac{1}{2\ell + 1} \sum_{m=-\ell}^{\ell} a_{\ell m} a^*_{\ell m} = \langle |a_{\ell m}|^2 \rangle. \tag{11}$$

It is common practice to plot the quantity $\mathcal{D}_\ell = \ell(\ell + 1)C_\ell/(2\pi)$.

The peaks in the power spectrum (see Fig. 3) are associated with the spatial distribution of the CMB anisotropies, and reflect the acoustic oscillations in the baryon-fluid plasma in the early Universe, which are frozen during the matter-radiation decoupling. The relative positions and amplitudes of these peaks are tightly sensitive to a number of cosmological parameters, in particular to the total energy density and to the curvature of the Universe. This motivates the interest to achieve an accurate characterisation of the CMB power spectrum.

The BOOMERanG experiment was the first that clearly delineated the first peak of the CMB power spectrum [19], at $\ell \approx 200$ (angular scales of $\approx 1°$ on the sky), inferring the flatness of the Universe. These results were confirmed soon after that by other balloons like MAXIMA [20] or ARCHEOPS [21]. In the following $\sim$5 years, ground-based experiments like VSA [22], CBI [23] or ACBAR [24] were able to measure higher-order peaks out to $\ell \approx 1500$. NASA's *WMAP* satellite observed the full sky with angular resolution of 15 arcmin [25], comparable to previous ground-based experiments, resulting in a significant improvement in the measurement of the first peak, and also covering the second and third peaks out to $\ell \approx 1000$. Slightly later on time, ACT [26] and SPT [27], a new generation of ground-based experiments, reached angular resolutions of $\sim$1 arcmin on small sky regions ($\sim$200 deg$^2$), allowing to measure the power spectrum out to $\ell \sim 4000$ (see Fig. 3).

The third generation of CMB space missions, after *COBE* and *WMAP*, was ESA's *Planck* satellite. *Planck* observed the full sky with an angular resolution of 5 arcmin

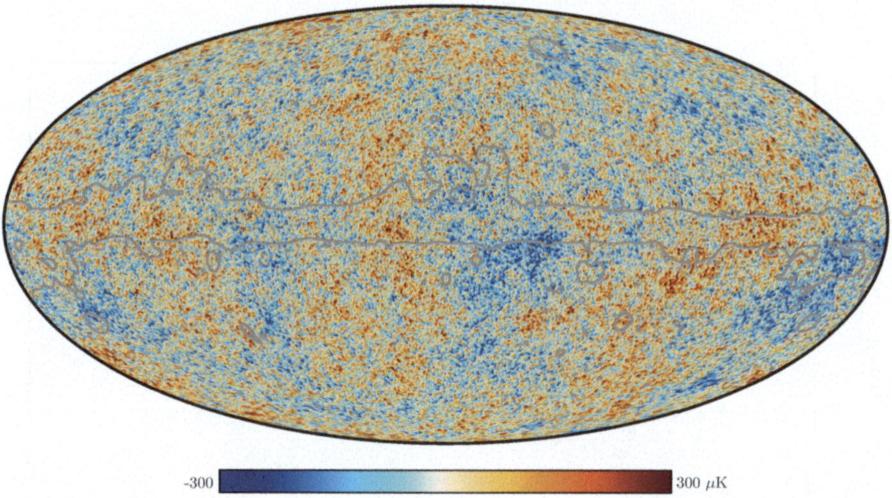


**Fig. 2** Full-sky CMB map derived from *Planck* multi-frequency data [28], projected into Galactic coordinates. The grey lines indicate the masked regions (mostly encompassing the Galactic plane) that have been excluded in the cosmological analyses. Figure taken from the *Planck* Legacy Archive (https://www.cosmos.esa.int/web/*planck*/)

[28]. In addition to its improved angular resolution and sensitivity, another important improvement of *Planck* was its wide frequency coverage, with nine independent frequency channels between 30 and 857 GHz. This allowed a more efficient subtraction of Galactic foregrounds thanks to a more accurate measurement of the spectra of the different emission components, mainly free-free and synchrotron emissions, which are important at low frequencies $\lesssim$50 GHz, and thermal dust emission, which is important at higher frequencies $\gtrsim$150 GHz (a concise review on Galactic foregrounds can be found in [29]). Figure 2 shows the final *Planck* full-sky map of the primordial CMB anisotropies, resulting from a combination of the different frequency bands to minimise foreground contamination. Grey lines on this map enclose regions with potential residual foreground contamination, mostly along the Galactic plane, that are masked out for cosmological analyses.

The final *Planck* power spectrum of these anisotropies ($\mathcal{D}_\ell = \ell(\ell+1)C_\ell/(2\pi)$, where $C_\ell$ is given by Eq. (11)), calculated in regions not affected by this mask, is shown by the blue points in Fig. 3, in comparison with recent measurements by other experiments. It can be seen that these data allowed to clearly measure the first seven acoustic peaks in the temperature (TT) power spectrum, and to trace the damping tail of the anisotropies out to $\ell \approx 2500$. The CMB temperature angular power spectrum is complemented by the polarisation power spectrum, that is usually decomposed into the gradient-like E-mode (even parity) pattern and the curl-like B-mode (odd parity) pattern [30, 31]. While E-modes are generated by both scalar and tensor perturbations, B-modes can only be generated by tensor

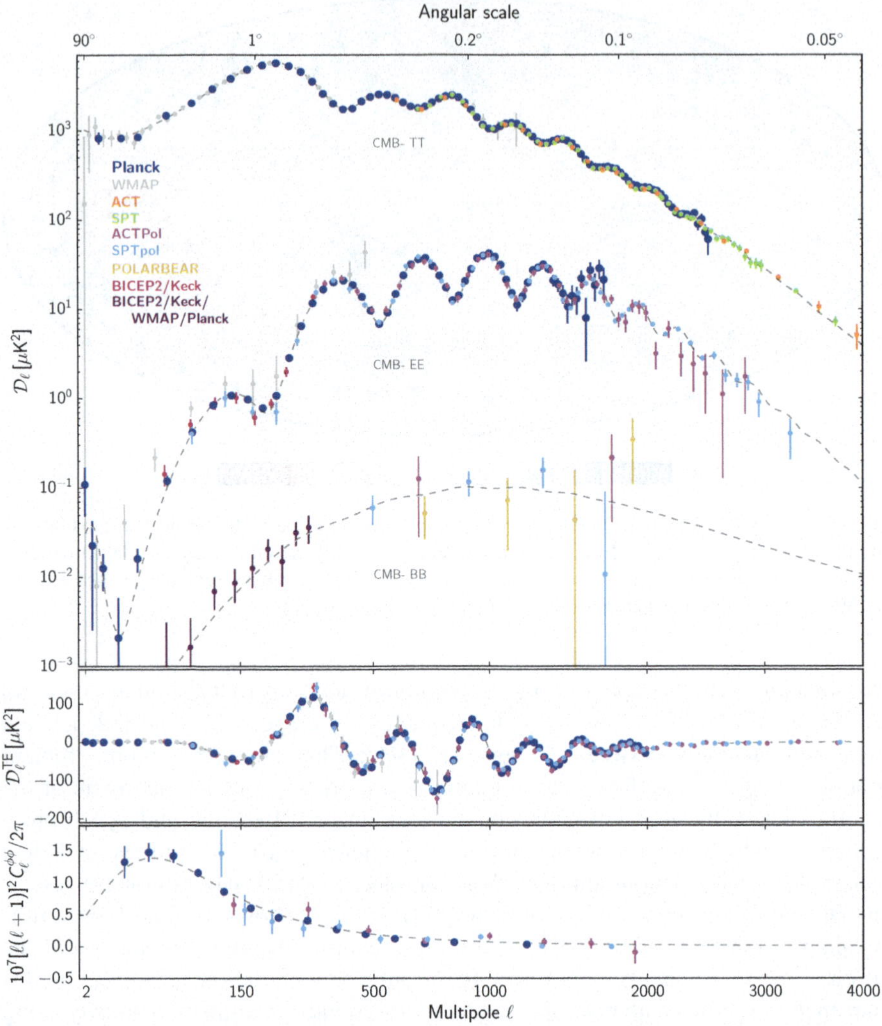


**Fig. 3** CMB power spectra showing the most recent measurements coming from different experiments [28]. The top panel shows the auto-power spectra of CMB temperature, E-mode and B-mode anisotropies, while the middle panel shows the TE cross-correlation. The bottom panel shows the lensing power spectra. The dashed lines show the best-fit ΛCDM model to *Planck* temperature, polarisation and lensing data. *Planck* data cover all the angular scales up to $\ell \approx 2500$, while data from ACT and SPT with finer angular resolution help to reach $\ell \approx 4000$. Figure taken from the *Planck* Legacy Archive

perturbations[3] (see Fig. 1). The EE power spectrum as well as the TE cross-power spectrum have been nowadays measured to high accuracy, while only upper limits on the BB power do exist. The *Planck* TE and EE power spectra successfully provide high-sensitivity measurements of other 11 peaks (see Fig. 3). *Planck* surpassed all previous experiments in both angular resolution and sensitivity, except for the higher angular resolution measurements of ACT and SPT that allowed to measure the damping tail out to even higher multipoles.

Apart from the improved frequency coverage, sensitivity and angular resolution, *Planck* data brought two other important improvements over previous missions and experiments. On the one hand, the lower systematics of *Planck* polarisation data (in particular on the TE cross-correlation and on the EE auto-correlation) allowed for the first time to obtain constraints on cosmological parameters with comparable accuracies to those derived from temperature data only. On the other hand, *Planck* data allowed an accurate measurement in the full-sky of the lensing of the CMB photons as they traverse the large-scale structure of the Universe, which essentially leads to a smoothing of the acoustic peaks. This effect was detected for the first time by the ACT telescope [33], and now *Planck* has measured it with sufficient accuracy as to allow reconstructing a full-sky map of the lensing potential (see Figure 6 of [28]), and making it useful for the first time to improve the precision in the determination of cosmological parameters. This also allowed breaking internal parameter degeneracies in such a way that CMB data can now be used without the need for external cosmological data to derive precise constraints on all 6 $\Lambda$CDM parameters simultaneously. In particular, while the amplitude of the TT power spectrum is proportional to $A_s e^{-2\tau}$, *Planck* low-$\ell$ polarisation data allow an independent measurement of $\tau$, therefore breaking the $A_s - \tau$ degeneracy associated with temperature data. On the other hand, *Planck* CMB lensing data contribute to break the well-known CMB "geometric degeneracy" in the $\Omega_m - \Omega_\Lambda$ plane (see Fig. 4). While previously, the combination of *WMAP*, ACT and SPT data allowed to measure dark energy with 3.5% precision ($\Omega_\Lambda = 0.721 \pm 0.025$ [25]), now *Planck* data alone lead to a precision of 0.8% ($\Omega_\Lambda = 0.6889 \pm 0.0056$ [34]). This has contributed to consolidate the CMB as the cosmological probe giving the tightest constraints on cosmological parameters nowadays.

The final *Planck* constraints on the base 6-parameter $\Lambda$CDM model are shown in Table 1. The first column corresponds to constraints coming from the TT power spectrum and the low multipoles ($\ell < 30$) of the polarisation EE power spectrum, which is used to determine $\tau$ after breaking its degeneracy with $A_s$. The results shown in the second and third columns come from the TE and EE power spectra, respectively, while those in the fourth column correspond to a combination of *Planck* temperature, polarisation and lensing data. The perfect agreement between temperature and polarisation data is a significant achievement of *Planck*, and a powerful consistency check of the underlying model, and of *Planck* data themselves. All parameters except $\tau$ are measured with precisions better than 1%, with the best-

[3] A didactic review on CMB polarisation theory can be found in [32].

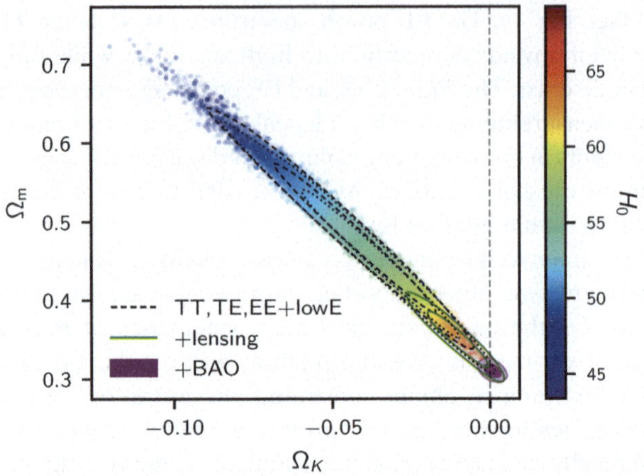


**Fig. 4** Constraints on the $\Omega_k - \Omega_m$ plane for a non-flat universe, coming from *Planck* temperature and polarisation data (dashed and dotted black contours, for two different likelihood implementations), and in combination with Planck lensing data (green contours) and with BAO data (violet regions). In each case 68% and the 95% confidence levels are shown. Points show samples from the chains, and their colour indicate the value of the Hubble constant. This shows the well-known "geometric degeneracy", which is partially alleviated thanks to Planck lensing data and, more significantly, after the inclusion of BAO data. Figure taken from [34]

**Table 1** Best-fit constraints from CMB-only Planck data [34] on the base 6-parameter ΛCDM model

| Parameter | TT + LowEE | TE + LowEE | EE | TT,TE,EE + lensing |
|---|---|---|---|---|
| $\Omega_b h^2$ | $0.02212 \pm 0.00022$ | $0.02249 \pm 0.00025$ | $0.0240 \pm 0.0012$ | $0.02237 \pm 0.00015$ |
| $\Omega_c h^2$ | $0.1206 \pm 0.0021$ | $0.1177 \pm 0.0020$ | $0.1158 \pm 0.0046$ | $0.1200 \pm 0.0012$ |
| $100\theta_{MC}$ | $1.04077 \pm 0.00047$ | $1.04139 \pm 0.00049$ | $1.03999 \pm 0.00089$ | $1.04092 \pm 0.00031$ |
| $\tau$ | $0.0522 \pm 0.0080$ | $0.0496 \pm 0.0085$ | $0.0527 \pm 0.0090$ | $0.0544 \pm 0.0073$ |
| $\ln(10^{10}\,A_s)$ | $3.040 \pm 0.016$ | $3.018^{+0.020}_{-0.018}$ | $3.052 \pm 0.022$ | $3.044 \pm 0.014$ |
| $n_s$ | $0.9626 \pm 0.0057$ | $0.967 \pm 0.011$ | $0.980 \pm 0.015$ | $0.9649 \pm 0.0042$ |

Different columns correspond to different combinations of temperature, polarisation and lensing CMB data

determined parameter being the acoustic scale, $\theta_{MC}$, which is directly related to the Hubble constant, $H_0$, and has precision as good as 0.3%.

While in Table 1 we show the constraints on the six independent parameters that form the base ΛCDM model, Table 2 of [28] also lists the constraints on a number of derived parameters. Some of the most interesting ones are the Hubble parameter, $H_0 = 67.66 \pm 0.42\,\mathrm{km\,s^{-1}\,Mpc^{-1}}$, the dark-energy density parameter, $\Omega_\Lambda = 0.6889 \pm 0.0056$, and the age of the Universe, $13.787 \pm 0.020\,\mathrm{Gyr}$.

Overall the results from *Planck* data were consistent with previous results from *WMAP* and other observatories, although with a significant tightening on the precision of most of the measured cosmological parameters. One example is the measured red tilt on $n_s$ with respect to scale invariance ($n_s = 1$), which is now measured at $8.4\sigma$, while *WMAP* measured it at $2.2\sigma$. This is a great success of the $\Lambda$CDM model, but at the same time made feel a bit disappointed people eager to find new issues pointing to new physics beyond $\Lambda$CDM that could have been previously overlooked. There are however some slight differences on some of the previous parameters with respect to *WMAP*. One important change brought in by *Planck* data is the measurement of the reionisation optical depth, which is found to be $\tau = 0.0544 \pm 0.0073$, while *WMAP* had measured $\tau = 0.089 \pm 0.014$ [25]. The smaller value given by *Planck* is driven by improved cleaning of the Galactic dust emission in polarisation. Actually, when *WMAP* polarisation maps are cleaned using the *Planck* 353 GHz channel, they are fully consistent with *Planck* low-frequency polarisation maps. This is an important achievement of *Planck*, as it is the fact that the value of $\tau$ coming from *Planck* low-$\ell$ polarisation data is fully consistent with the value coming from a completely independent dataset as it is the combination of *Planck* TT, *Planck* CMB lensing and Baryon Acoustic Oscillation (BAO) data [35]. On the other hand, *Planck's* value of $\Omega_\Lambda$ is smaller by $1.3\sigma$ than the one derived by *WMAP* and, as a consequence, in a flat-geometry $\Omega_m$ is larger by $1.3\sigma$. *Planck's* value of $H_0$ is smaller by $1.0\sigma$, and the age of the universe is bigger by $0.42\sigma$. This is why sometimes it has been said that *Planck* found that the Universe is slightly fatter and older. The lower value of $H_0$ found by *Planck* led to some tension with cosmic-distance ladder measurements [36], at the $\sim$3–$4\sigma$ level, and this is nowadays being broadly discussed in the community. This will be commented in more detail in Sect. 3, together with some other possible tensions or anomalies in the data.

The *Planck* team also explored extensions on the $\Lambda$CDM base 6-parameter model (like the ones commented in Sect. 1.4), finding in general good agreement with previous analyses based on other CMB datasets. For instance, no evidence was found of a running of the spectral index ($dn_s/d\ln k$ in Eq. (8)), this being consistent with the predictions of the simplest slow-roll inflation models. When the curvature density parameter is not fixed at $\Omega_k = 0$, from temperature and polarisation data it is found $\Omega_k = -0.044^{+0.018}_{-0.015}$, which is an apparent detection of curvature well above $2\sigma$ (see black contours of Fig. 4). However, the inclusion of CMB lensing data pushes $\Omega_k$ back into consistency with a flat geometry within $2\sigma$ (green contours in Fig. 4), while including BAO data convincingly breaks the geometric degeneracy (violet regions in Fig. 4) leading to a perfectly flat geometry with $\Omega_k = 0.0007 \pm 0.0019$. The low-$\ell$ temperature power spectrum allows constraining the amplitude of tensor modes (see black solid line in the right-hand panel of Fig. 1), giving $r_{0.002} < 0.10$ (95% C.L.) after combining with TE, EE and lensing, at a pivot scale of $0.002\,\mathrm{Mpc}^{-1}$. Combination with polarisation BB power spectra from the BICEP2 and Keck Array experiments [37] of course tightens this constraint to $r_{0.002} < 0.065$ (95% C.L.). This is nowadays the best constraint on a component of tensor modes in the early Universe. *Planck* data also allowed to study the nature of dark energy by constraining a time-varying equation of state using the parametrisation $w(a) = w_0 + (1 + a)w_a$ (see Sect. 1.4.2). *Planck* data alone allow a very wide volume of dynamical dark-energy parameter space, although with unrealistically high $H_0$ values. The addition of external data (BAO and SNe) narrows the constraints to the $\Lambda$CDM values, $w_0 = -1$, $w_a = 0$. In what concerns neutrino masses, while the default *Planck* analysis assumed the minimal mass $\sum m_\nu = 0.06\,\text{eV}$ allowed by neutrino flavour oscillation experiments, leaving it as a free parameter results in a 95% CL upper limit of $\sum m_\nu < 0.54\,\text{eV}$, when *Planck* TT and low-$\ell$ EE power spectra are used. Adding *Planck* polarisation and lensing data tightens this constraint to $\sum m_\nu < 0.24\,\text{eV}$. Interestingly, increasing the neutrino masses leads to lower $H_0$ values, aggravating the tension with the distance-ladder determinations of [36] (see Sect. 3). When the number of neutrino relativistic species is left as a free parameter, the resulting constraints are also fully compatible with the standard value of $N_{\text{eff}} = 3.046$. A detailed discussion on all these analyses, as well as on other extensions to the base $\Lambda$CDM model can be found in [34].

## 2.2 *The Large Scale Structure of the Universe*

The density perturbations created during inflation got frozen on the CMB at $z \approx 1100$ (400,000 years after the big bang), and continued evolving to the present day, shaping the large-scale matter distribution visible in the nearby Universe. The three-dimensional spatial distribution of the Large Scale Structure (LSS) of the Universe at low redshifts is very sensitive to several cosmological parameters, and in particular to dark energy. Low-redshift observations of the LSS measure the Universe over the last several billion years, when the dark energy dominates. Comparing the constraints coming from these observations with those derived from the CMB then requires extrapolating predictions to the present-day Universe starting from initial conditions over 13 billion years ago. This is in fact a quite strong test of our cosmological model.

Nowadays the two main LSS observables providing cosmological information are the galaxy clustering and the characterisation of number counts of galaxy clusters.

### 2.2.1 Galaxy Clustering

The same acoustic oscillations in the baryon-fluid plasma in the early Universe that were discussed in Sect. 2.1, and which lead to the series of peaks and troughs that are seen in the CMB power spectrum (see Fig. 3), create a specific anisotropy in the distribution of galaxies (and of larger structures like clusters of galaxies) in the local Universe. This anisotropy, usually called "clustering of galaxies", is mathematically encoded in the two-point correlation function, $\xi(r)$, which gives the expected number of pairs of galaxies with one galaxy in a volume $\delta V_1$ and another galaxy in the volume $\delta V_2$, through the following formula:

$$\langle n_{\text{pair}} \rangle = \bar{n}^2 \left[1 + \xi(r)\right] \delta V_1 \delta V_2, \qquad (12)$$

where $\bar{n}$ is the mean number density of galaxies and $r$ is the separation of the two volume elements. If the galaxies were unclustered then $\xi(r) = 0$, so in fact $\xi(r)$ measures the excess clustering of galaxies at separation $r$. It can be shown that $\xi(r)$ is the Fourier transform of the matter power spectrum (see Eq. (6)),

$$\xi(r) = \int P(k)e^{-i\mathbf{k}.\mathbf{r}} \frac{d^3k}{(2\pi)^3}. \tag{13}$$

Therefore, by measuring the galaxy two-point correlation function we can extract information about the matter power spectrum

$$P(k) = \int \xi(r)e^{i\mathbf{k}.\mathbf{r}} d^3r, \tag{14}$$

which is the quantity more directly related to theory.

The wiggles visible in the power spectrum (see Fig. 5) reflect the oscillations of the radiation-baryon plasma, and are coined the Baryon Acoustic Oscillation. They essentially reflect the imprint of the primordial perturbations in the distribution of galaxies in the local Universe, and are equivalent to the acoustic oscillations seen in the CMB power spectrum that were imprinted by the same perturbations at redshift $z \sim 1100$. The BAO wiggles are actually standing sound waves in the

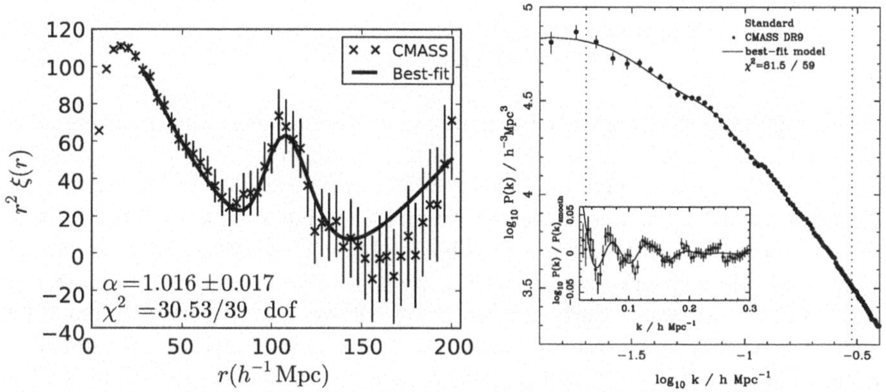


**Fig. 5** LSS correlation function (left) and power spectra (right). Data points come from the BOSS/CMASS galaxy sample, and the solid lines represent the best-fitting models. The left-hand plot shows the best-fitting dilation scale, $\alpha$, which measures the relative position of the BAO peak in the data with respect to a fiducial model, and also the $\chi^2$ statistic giving the goodness of the fit. In the right-hand plot the vertical lines delimit the range of scales ($0.02 < k < 0.3$ $h$Mpc$^{-1}$) that are used to fit the data, and the inset shows both the model and the data divided by the best-fitting model with no BAO in the same $k$-range. Figure taken from [38] (by permission of Oxford University Press)

pre-recombination universe that imprint a characteristic scale on the late-time matter clustering of galaxies at the radius of the sound horizon,

$$r_d = \int_{z_\mathrm{d}}^{\infty} \frac{c_s(z)}{H(z)} dz, \tag{15}$$

evaluated at the drag epoch $z_\mathrm{d}$, during the decoupling of photons and baryons shortly after recombination. This scale actually provides a "standard ruler", that can be measured both in the CMB anisotropies and in the LSS maps at low redshifts. It is seen as a localised peak in the correlation function, or as a series of damped oscillations in the power spectrum (see Fig. 5). An anisotropic BAO analysis measuring the BAO feature along the line of sight allows to constrain the expansion rate $H(z)$, in a highly complementary and independent way to SNe measurements. On the other hand, measuring the BAO feature in the transverse direction allows constraining the comoving angular diameter distance[4] $D_M(z)$, which is related to the physical angular diameter distance by $D_M(z) = (1+z)D_A(z)$. BAO measurements in fact constrain the combinations $H(z)r_d$ and $D_M(z)/r_d$. The cosmological model has then to be adjusted in such a way that radial and angular clustering match, then constraining the product $H(z)D_A(z)$. This was first proposed as a cosmological test by Alcock and Paczynski in 1979 [40], and has ever since been coined the Alcock-Paczynski test. If instead we average the clustering in 3D over all directions, then matching the clustering scale to the expected comoving clustering is sensitive to[5]

$$D_V(z) = [czD_M^2(z)/H(z)]^{1/3}. \tag{16}$$

This highlights the importance of mapping the three-dimensional distribution of galaxies in the Universe, in order to be able to reconstruct the matter correlation function and its power spectrum. This requires of course not only measuring the angular positions of galaxies on the sky, which is easy, but also their radial distances, which is not so easy. Although it is possible to obtain photometric redshifts through broad-band colour observations, spectroscopy is clearly the most precise method to obtain redshift estimates. Early spectroscopic galaxy redshift surveys include the Center for Astrophysics Redshift Survey [42], the Third Reference Catalogue of Bright Galaxies (RC3; [43]) or Las Campanas Redshift Survey (LCRS, [44]). The advent of multi-object spectrographs (MOS) now allows obtaining multiple galaxy spectra simultaneously. Key facilities that have used this technique to undertake wide-field redshift surveys include the AAΩ instrument on the Anglo-Australian Telescope, which has been used to conduct the 2-degree Field Galaxy Redshift Survey (2dF-GRS [45]) and WiggleZ [46], and the Sloan Telescope, which has been used for the Sloan Digital Sky Survey (SDSS; [47]). The Baryon Oscillation

[4] For a brief review of cosmology distance definitions see [39].

[5] A comprehensive review on galaxy clustering and BAO theory can be found in [41].

Spectroscopic Survey (BOSS), part of SDSS-III [48], was designed to obtain spectroscopic redshifts of 1.5 million galaxies to at $z = 0.2$–$0.7$ over a sky area of 10,000 square degrees.

The first clear detections at low redshift of the BAO signal came from galaxy clustering analyses of the 2dF [49] and from the luminous red galaxy (LRG) sample of the SDSS [50], at redshift $z \approx 0.3$. The WiggleZ survey allowed to increase the redshift coverage to $0.4 < z < 1.0$, with a precision of 3.8% [46]. The initial goals of the BOSS survey were to improve the precision of the measurement down to 1% in the redshift range $z = 0.2$–$0.7$, and to achieve for the first time a detection of the BAO peak at $z > 2$ using the 3D structure in the Ly$\alpha$ forest absorption towards 160,000 high-redshift quasars. Both goals were successfully fulfilled. Figure 5 shows the galaxy correlation function $\xi(r)$ and power spectrum $P(k)$ derived from the CMASS sample of the BOSS survey [38], corresponding to data release 9 (DR9). The CMASS DR9 samples comprises 264,283 galaxies distributed over 3275 square degrees, and with redshifts $0.43 < z < 0.7$. Further data releases have resulted in an increased number of galaxies distributed over progressively larger sky areas. A catalogue of positions and redshifts of 147,000 QSOs at $0.8 < z < 2.2$ extracted from the extended BOSS (eBOSS) data [51], as well as the absorption by the Ly$\alpha$ forest towards the line of sight of 157,783 QSOs at $2.1 < z < 3.5$ [52], have also been used to reconstruct the BAO signature. Figure 6 shows a summary of the BAO distance measurements ($D_V$, which is a combination of $H(z)$ and $D_M(z)$, see Eq. (16)) resulting from these measurements, as well as from previous surveys at lower-redshifts, normalised by the expected value coming from the best-fit $\Lambda$CDM model to the *Planck* CMB observations. The excellent agreement between independent BAO measurements at different redshifts, as well as with the CMB measurements at $z = 1100$, is a remarkable success of both the observations and the theoretical model.

As it was explained before, these BAO measurements are crucial to set constraints on cosmological parameters, in particular on the expansion rate and thus on dark energy. Also, combination with the CMB crucially breaks some parameter degeneracies, allowing to obtain even tighter cosmological constraints (see Fig. 4). As an example, in Fig. 7, we show constraints on $\Omega_{\mathrm{m}}$ and $\Omega_\Lambda$ coming from BAO measurements alone, and from the combination of BAO and CMB. Nowadays BAO data alone can actually allow to measure dark energy at more than $3\sigma$, while combination with CMB measurements from *Planck* increases the significance to $21\sigma$ [54].

### 2.2.2 Galaxy Clusters

Different observables related to galaxy clusters have important applications in cosmology. The clustering of galaxy clusters, and their BAO signature, can be used to constrain cosmological parameters in the same way as the clustering of galaxies [55], although typically with lower constraining power due to the smaller number of objects. The combination of X-ray and Sunyaev-Zel'dovich (SZ) effect (this is the

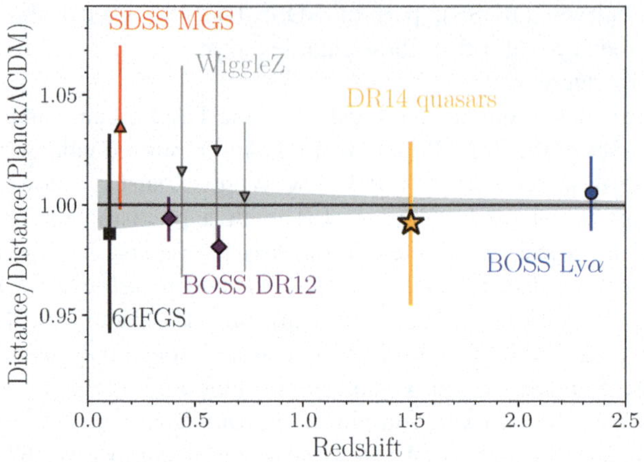


**Fig. 6** BAO distance measurements ($D_V$) coming from BOSS DR12 galaxies [53], BOSS DR14 QSOs [51], and from the Lyα forest [52], in comparison with measurements from previous surveys at lower redshifts. The results have been normalised to the expected value coming from the best-fit ΛCDM cosmology from *Planck* CMB data. The shaded grey area delimits the extrapolated 68% C.L. region derived from this best-fit CMB model. Figure taken from [51] (by permission of Oxford University Press)

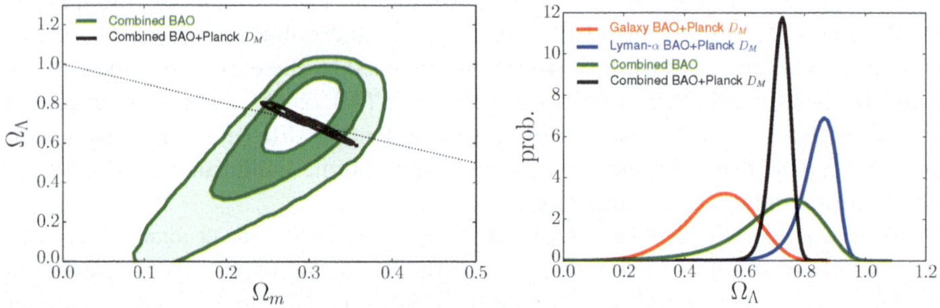


**Fig. 7** Constraints derived from BOSS BAO measurements on open ΛCDM models. The left panel shows constraints on the $\Omega_\Lambda - \Omega_\mathrm{m}$ plane (dotted line corresponds to a flat geometry), while the right panel shows the $\Omega_\Lambda$ one-dimensional probability density function. Green contours and lines show constraints using only BAO measurements, derived from the combination of galaxy and Lyα forest measurements, while black represents the combination of BAO and *Planck* CMB measurements. The results derived from the combination of the CMB with either BAO galaxy measurements or BAO Lyα measurements are shown in red and blue, respectively. Figure taken from [54]

inverse Compton scattering of CMB photons as they transverse the hot gas inside galaxy clusters) observations can be directly used to measure cosmic distances, and in turn to estimate the Hubble constant at the redshift of the cluster [56], although this method is prone to different kind of systematic effects and is not very popular nowadays. Peculiar velocities of galaxy clusters, which can be measured through the kinetic SZ effect, probe directly the gravitational potential. Estimates of the gas mass fraction through X-rays or through the SZ effect can be used to infer the matter density parameter, $\Omega_m$, if the baryonic density $\Omega_b$ is fixed from any other method. The power spectrum of the temperature anisotropies induced by the SZ effect can be used to constrain a combination of $\Omega_m$ and of the amplitude of matter fluctuations in scales of 8 $h^{-1}$ Mpc, $\sigma_8$; more specifically it allows to constrain the product $\sigma_8^8\Omega_m^3$ [57].

Apart from all these observables and applications, the most popular application of galaxy clusters to cosmology nowadays comes from the characterisation of their number counts, which essentially gives the number of clusters per redshift and mass bin, in a given solid angle element $\Delta\Omega_i$:

$$\bar{N}(z_i, M_j) = \frac{\Delta\Omega_i}{4\pi} \int_{z_i}^{z_{i+1}} dz \frac{dV}{dz} \int_{\ln M_j}^{\ln M_{j+1}} d\ln M \frac{dn}{d\ln M}. \tag{17}$$

Cosmology enters here through the volume element $dV/dz$, and through the mass function, $dn/d\ln M$. While the amplitude of the mass function is proportional to $\sigma_8$, its shape is proportional to $\Omega_m$. There is a degeneracy between these two parameters, in such a way that these analyses constrain the product $\sigma_8\Omega_m^{0.3}$. This degeneracy is broken by using external priors on other parameters. In addition to the shape of the local mass function, its redshift evolution, which is sensitive to the growth of linear density perturbations, can be used to constrain the dark energy density parameter and its equation of state. This was done by Vikhlinin et al. [58] using a sample of just 86 galaxy clusters, from which they obtained a 5$\sigma$ detection of dark energy (see Fig. 8).

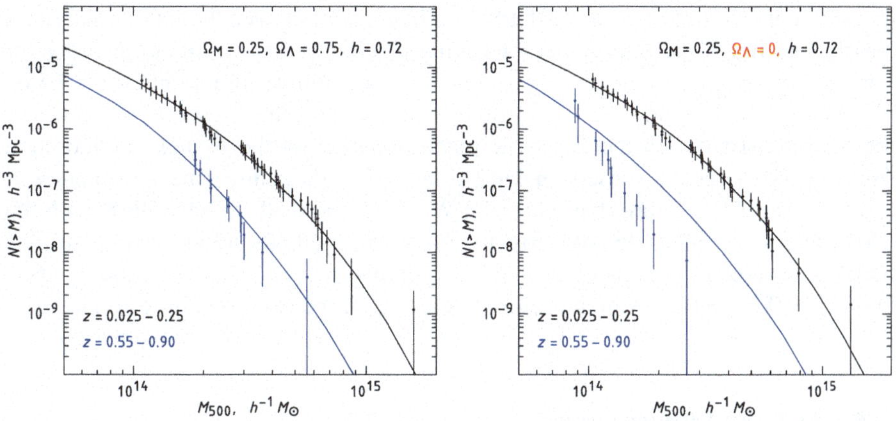


**Fig. 8** Illustration of the sensitivity of the mass function of clusters of galaxies to the cosmological model. The data and the model have been computed for a model with (left panel) and without (right panel) dark energy, considering a low- and a high-redshift galaxy sample. It becomes immediately clear that a model without dark energy provides a very poor fit to the data in the high redshift bin. Figure taken from [58] (© AAS. Reproduced with permission)

Obtaining meaningful cosmological constraints from the galaxy clusters mass function requires statistically-large cluster samples together with accurate estimates of redshifts and masses. While nowadays X-ray and SZ surveys provide large samples and redshifts can be estimated to high accuracy by follow-up spectroscopic observations, the mass estimates are subject to systematics that are likely to introduce biases on the inferred cosmological parameters. Different mass proxies can be used, like the X-ray luminosity, the SZ flux, the optical richness, the velocity dispersion of individual galaxies, or weak lensing measurements. Of course, a good proxy should have a small scatter between the estimated and the real cluster mass, and a low bias. All cosmological analyses assume a certain bias between the estimated and the real mass, which must be fixed or fitted for, and is denoted by $(1-b)$. This bias can result either from cluster physics that are not properly accounted for or from instrumental effects.

Thanks to its large and well characterised catalogue of galaxy clusters detected via the SZ effect, the results from the *Planck* collaboration led to an important step forward in this kind of analyses. The 2013 *Planck* cosmological analysis using cluster number counts [59] relied on the $Y_X - M_{500}$ relation (relation between X-ray luminosity and the mass enclosed inside $r_{500}$, the radius at which the density is 500 times the critical density of the Universe) as a mass proxy, and either fixed the mass bias parameter at $(1-b) = 0.8$, or used a flat prior inside the range [0.7, 1.0]. On the other hand, the 2015 *Planck* analysis [60] relied on improved mass proxies based on weak lensing measurements, and also on the use of the lensing of the CMB photons on the cluster position, a new technique that is starting to be exploited. Using two different cluster samples with measurements of the gravitational shear they obtained respectively $(1-b) = 0.688 \pm 0.072$ (WtG in Fig. 9) and $(1-b) = 0.780 \pm 0.092$ (CCCP in Fig. 9), while the CMB lensing measurement led to $(1-b) = 1.01 \pm 0.19$. The scatter between these three estimates clearly highlights the importance of having a reliable mass estimate in order for this method to produce accurate cosmological constraints, in particular on the $\sigma_8$ parameter, which is the one more strongly affected by this issue. Figure 9 shows the final constraints on the $\Omega_m - \sigma_8$ plane obtained by the *Planck* collaboration from the 2015 results (based on DR2), and using their three mass estimates. It can be seen in this figure that while the WtG method leads to agreement with the CMB, the $\sigma_8$ estimates derived from the other two methods are in tension with the CMB. The value of the mass bias that is required to reconcile both measurements is $(1-b) = 0.58 \pm 0.04$, which is close to the first of the two previous estimates.

## 2.3 Type Ia Supernovae

Type Ia supernova (hereafter SNIa) are thought to arise from thermonuclear explosions of white dwarfs in binary systems, and are known to produce rather uniform brightness at their maximum, therefore being considered as standard candles as their visual magnitude depends primarily on their distance to the observer. Their V-band

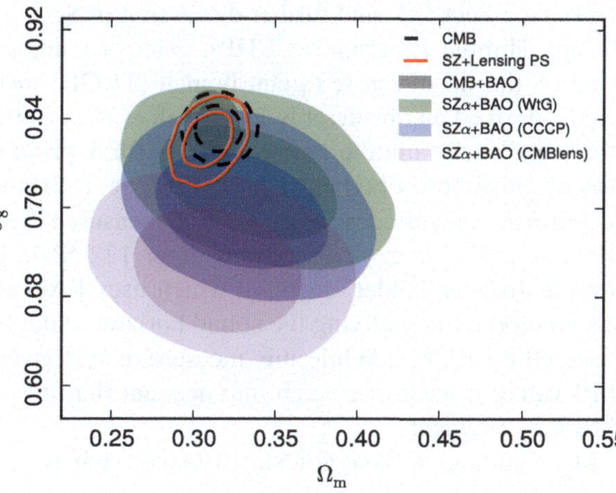


**Fig. 9** Constraints on the $\Omega_m - \sigma_8$ plane derived from SZ cluster number counts by the *Planck* collaboration [60], using different mass estimates (WtG, CCCP and CMBlens), compared with those coming from the primary CMB (black contours), from the primary CMB and BAO (black shaded area) and from the combination of primary CMB and the lensing power spectrum (red contours). Credit: [60], reproduced with permission © ESO

peak luminosities present a scatter of approximately 0.4 mag [61]. This scatter can be notably reduced by introducing some corrections, the most important of which has to do with the correlation between the peak luminosity and the light curve shape (LCS), first introduced by Phillips [62]. Two other important refinements are the correction for the correlation between SN colour and extinction and the application of $K$-corrections for the redshifting effects. After these corrections are introduced, the dispersion in well-measured optical band peak magnitudes is reduced to only $\sim$0.12 magnitudes, allowing each measured SN to provide a luminosity-distance estimate with precision of $\sim$6%.

Their ability to measure distances to such a good precision bestows SNIa important cosmological applications, as the luminosity-distance is sensitive to the density parameters $\Omega_m$, $\Omega_k$, $\Omega_\Lambda$ and to the expansion rate $H(z)$ (see relevant equations in [39]). At low redshifts, where the cosmic expansion is still linear, and therefore the effects of curvature and dark-energy are negligible (see Fig. 10), SNIa observations can be used to measure this linearity and thence to obtain estimates of the Hubble constant [63]. This requires knowledge of the absolute SNIa magnitude, which is usually calibrated through cepheid distances, using the period-luminosity relation (Leavitt law). The most-commonly used anchors are geometric distances to Milky Way Cepheids, eclipsing binaries in the Large Magellanic Clouds or in M31, and the water megamasers in NGC4258. This was the approach that was followed by the *HST* Key Project, to get an estimate with 11% precision, $H_0 = 72 \pm 8\,\mathrm{km\,s^{-1}\,Mpc^{-1}}$ [64]. The error of this measurement is driven by systematics related to the absolute calibration, and therefore recent efforts have focused on improving this calibration rather than on decreasing the statistical errors by increasing the number of objects. This has been the goal of the SH0ES ("Supernova and $H_0$ for the Equation of State") programme, whose most recent measurement is $H_0 = 74.03 \pm 1.42\,\mathrm{km\,s^{-1}\,Mpc^{-1}}$, which corresponds to 1.9% precision [36] and is in tension at a level of 4.3$\sigma$ with the CMB measurement from

*Planck* (see Sect. 2.1, and further discussion in Sect. 3.1). Recently, the "Carnegie-Chicago Hubble Program" (CCHP), using a completely independent calibration based on the tip of the red giant branch (TRGB) method, which uses population II stars, derived a considerably lower value, $H_0 = 69.8 \pm 1.9\,\mathrm{km\,s^{-1}\,Mpc^{-1}}$ [65], which sits in the middle between the SH0ES programme and the CMB values. Also recently, [66] obtained $H_0 = 67.77 \pm 1.30\,\mathrm{km\,s^{-1}\,Mpc^{-1}}$ after calibrating the intrinsic magnitudes of 207 SNIa measured by the "Dark Energy Survey" (DES) at $0.018 < z < 0.85$, plus other 122 SNIa at lower redshifts, using the "inverse distance ladder" method, which uses BAO measurements as a reference. This method requires fixing the sound horizon scale, for which they used the value measured by *Planck*. While this measurement derived from SNIa is closer to the CMB value, it has to be taken into account that it is not fully independent of the CMB measurements.

At cosmologically-significant distances, where the effects of the matter and density content of the Universe become important, the luminosity distance is obtained through an integration of the expansion rate over the redshift, which depends on the cosmological model. Therefore, the dimming of the standard candles at high redshifts can be used to constrain cosmological parameters. Importantly, contrary to the local measurement of $H_0$, the measurement of the expansion rate at high redshifts is independent of the absolute luminosity of the SNIa, which however is considered to be constant with redshift. In the late 1990s two independent teams, the "High-z Supernova Research Team" [67] and the "Supernova Cosmology Project" [68], unexpectedly found that distant supernova (out to $z = 0.8$) are $\sim$0.25 mag dimmer than they would be in a decelerating universe, indicating that the Universe is currently undergoing an accelerated expansion. When analysed assuming a universe with matter and a cosmological constant, these results provided evidence of $\Omega_\Lambda > 0$ at more than 99%. This important discovery was recognised with the 2011 Physics Nobel Prize. Since then, new observations have compellingly confirmed this result. The largest high-redshift ($z \approx 0.4$–$1.0$) samples to date come from the ESSENCE survey [69] and from the CFHT Supernova Legacy Survey (SNLS [70]). The left-hand panel of Fig. 10 shows the Hubble diagram resulting from the SNLS high-z SNIa sample, in combination with other observations at lower redshifts. Following what has become common practice in this field, instead of distances this figure represents magnitudes versus redshift. It is clearly seen that a model with $\Omega_\Lambda = 0$ is strongly disfavoured by the data. At intermediate distances ($0.1 < z < 0.4$) the SDSS-II supernova survey [72] has resulted in 500 spectroscopically confirmed SNIa. At very high redshifts, *HST* surveys (see e.g. [73]) have yielded $\sim$25 SNIa at $z > 1$ suitable for cosmological analyses.

The greatest cosmological utility of SNIa comes from the combination of different data sets spanning a wide redshift range. However, combination and homogenisation of data from different instruments and telescopes in such a way that they are useful for cosmological analyses entails a major difficulty, specially in what concerns survey-to-survey relative flux calibration, joint light curve fitting, and consistent use of $K$-corrections. The joint light-curve analysis (JLA; [71]) contains 740 spectroscopically confirmed SNIa with high-quality light curves, coming from

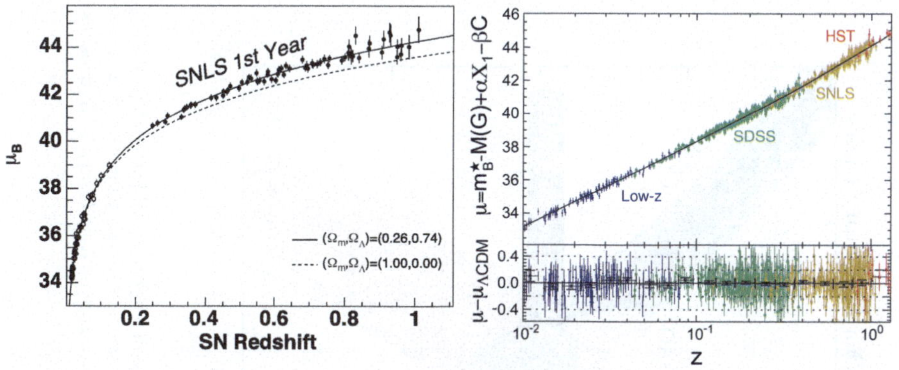


**Fig. 10** Hubble diagrams showing data points obtained from Type Ia supernova observations. The left panel shows measurements from the Supernova Legacy Survey ($z > 0.2$) in combination with nearby SNIa measurements. Two cosmological models are shown superimposed, one corresponding to the best-fit cosmological model with $\Omega_\Lambda = 0.74$, and another one without dark energy, clearly highlighting that the data favour $\Omega_\Lambda > 0$. The right panel shows the results from the joint light-curve analysis (JLA), with combined measurements from four different surveys, providing uniform redshift coverage out to $z = 1$, together with the distance modulus redshift relation of the best-fit ΛCDM cosmology. Credit: [70] (left) and [71] (right), reproduced with permission © ESO

the combination of two major surveys, the SDSS-II supernova survey and the SNLS, in addition to low-$z$ observations, and very high-$z$ data from the *HST* (see right-hand panel of Fig. 10). This was the default sample for cosmological analyses, until the recent advent of the "Pantheon" sample [74], which contains 1048 SNe spanning the redshift range $0.01 < z < 2.3$, thanks to the addition of Pan-STARRS1 Medium Deep Survey and various low-redshift and *HST* samples.

The left panel of Fig. 11 clearly shows that these new data have resulted in a significant improvement of the cosmological constraints, when compared with the original data of the High-z Supernova Research Team [67]. The Pantheon data [74] alone allow a high-significance detection of dark energy. When both statistical and systematic uncertainties are combined together, the result for a non-flat universe is $\Omega_\Lambda = 0.733 \pm 0.113$ ($6\sigma$ detection), while for $\Omega_k = 0$ the result is $\Omega_\Lambda = 0.702 \pm 0.022$ ($32\sigma$ detection). This demonstrates the ability of SNIa observations to constrain cosmology on their own. However, much tighter cosmological constraints are achieved through the combination with other cosmological probes like the CMB or the BAO. The great advantage is that SNIa likelihoods are typically orthogonal to other measurements of cosmological parameters, the reason for this being the lower mean redshift coverage compared to most of other methods. This is crucial to break the geometric degeneracy of the CMB in the $\Omega_\Lambda - \Omega_m$ plane (see in Fig. 4 this effect in the $\Omega_k - \Omega_m$ plane). The combination of the CMB and SNIa also allows exploring cosmological models with equation of state of dark energy $w \neq 1$ ($w = 1$ corresponds to a cosmological constant), by breaking the degeneracy between $w$ and $\Omega_m$ (see right-panel of Fig. 11).

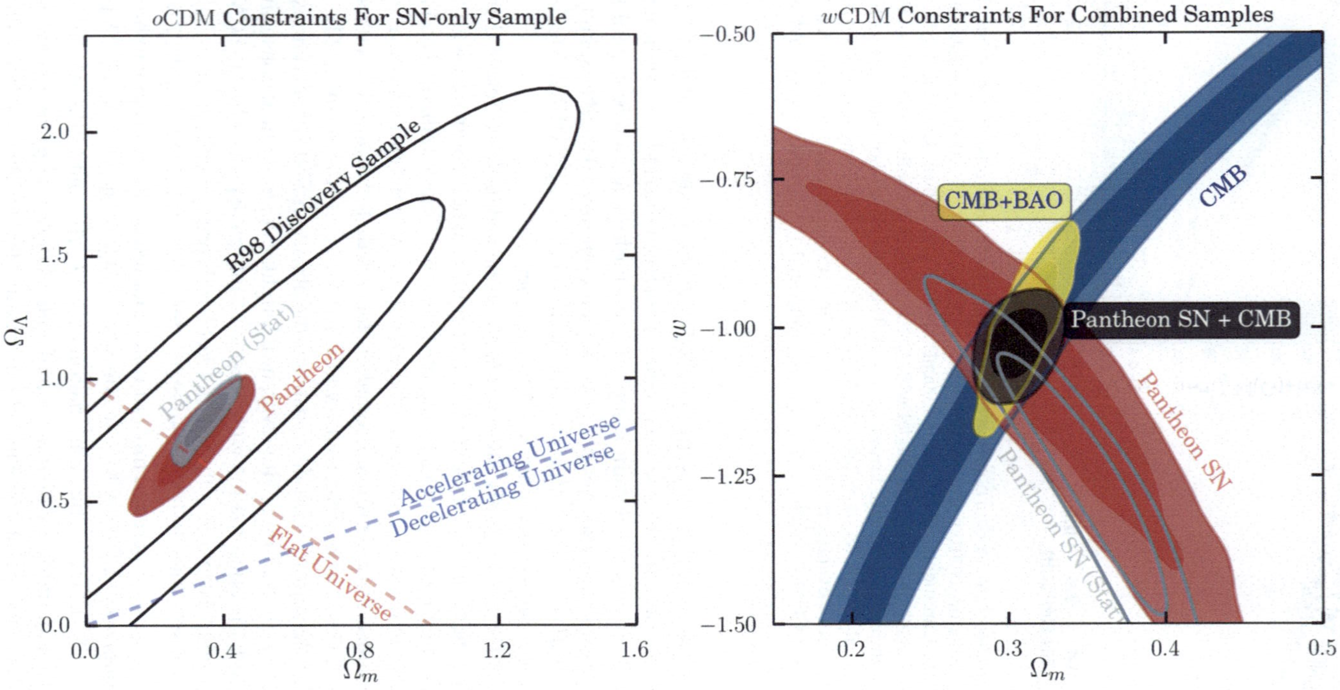


**Fig. 11** Cosmological constraints derived from the Pantheon SNIa sample. Left: constraints on the $\Omega_\Lambda - \Omega_\mathrm{m}$ plane for a non-flat $\Lambda$CDM cosmology derived from the Pantheon sample in comparison with those coming from the R98 sample of [67]. Right: constraints derived from the Pantheon sample on the $\Omega_\mathrm{m} - w$ plane for a CDM cosmology with a dark energy component with equation of state parameter $w$, in comparison with those coming from *Planck* CMB data, and with the combination of the two, and with BAO. Figure extracted from [74] (© AAS. Reproduced with permission)

# 3 Tensions, Anomalies and Open Problems in the Standard Model

The previous two sections have clearly shown that we nowadays have a very solid and well-established model that is able to describe the general properties of the Universe with just six parameters. It is quite remarkable that completely different and independent observations, spanning a wide range of redshifts (from the early to the present universe) agree to great precision on the values of these parameters. In the CMB itself, which is currently the probe giving the most stringent constraints on the model, there are three different observables (temperature anisotropies, polarisation anisotropies, and CMB lensing) giving fully consistent constraints. While the CMB temperature or polarisation anisotropies were not able to jointly fit the mass, curvature and dark-energy parameters (this is the well-known geometric degeneracy; see Fig. 4), the CMB lensing, which has recently started to be exploited to constrain cosmological parameters, helps to alleviate this degeneracy. Combination with either BAO or SNIa observations totally breaks this degeneracy and, importantly, the result is the same independently on what of these two probes is used.

This outstanding agreement between observations in the framework of a relatively simple 6-parameter model is a remarkable achievement, and has probably incited cosmologists to further scrutinise the data in the search for possible tensions or discrepancies that could point to new physics or yet undiscovered phenomena beyond the standard model. This search has led to some anomalies and tensions between some specific parameters, which lie at a not sufficiently high significance level as to be called "discrepancies". Apart from this, of course, it has to be born in mind that, while the model is well established, it points to the existence of two entities, dark matter and dark energy, which we currently do not understand. Neither we do properly understand the physics of inflation, which is a basic ingredient of the current model. These are open key problems that should drive the research in cosmology in the coming decades. In the following sections we will briefly describe these aspects.

## *3.1 Tensions*

### 3.1.1 $H_0$ Estimates in the Local and in the Early Universe

The tension that is currently being more hotly debated concerns the value of the Hubble constant, for which standard distance-estimation techniques in the local universe using SNIa (see Sect. 2.3) seem to give higher values than the one inferred from the parameters that best-fit the CMB anisotropies at $z = 1100$ (see Sect. 2.1). As it was commented before, the value derived by the SH0ES team from SNIa observations, $H_0 = 74.03 \pm 1.42\,\mathrm{km\,s^{-1}\,Mpc^{-1}}$ [36], differs at the $4.3\sigma$ level with the value derived from the *Planck* data, $H_0 = 67.66 \pm 0.42\,\mathrm{km\,s^{-1}\,Mpc^{-1}}$ [34]. Independent local estimates rely on measuring time delays on strong-lensed QSOs. Applying this technique on six lensed systems the H0LiCOW team has recently reached a similar precision to other techniques on $H_0$, obtaining $H_0 = 73.3^{+1.7}_{-1.8}\,\mathrm{km\,s^{-1}\,Mpc^{-1}}$ [75], a value that is in tension at the $3.1\sigma$ level with the CMB. When the two local estimates (time-delay cosmography [75] and distance ladder [36]) are combined, the difference with respect to the CMB increases to $5.3\sigma$. However, recently the CCHP team, calibrating SNIa distances using the TRGB method, came up with a lower value, $H_0 = 69.8 \pm 1.9\,\mathrm{km\,s^{-1}\,Mpc^{-1}}$ [65], which alleviates the tension with the CMB.

The fact that different CMB experiments (*Planck*, ACT, SPT), on the one hand, and different local estimates (time-delays and SNIa), on the other hand, agree, is telling, and therefore the question of whether there is new physics beyond $\Lambda$CDM seems justified. Importantly, any proposed extension to the current model must alleviate the current tension but not at the cost of increasing the differences on other parameters, something that is not always easy to achieve. There are some recently proposed extensions to the model, like the existence of a scalar field acting as an early dark energy component [76], or a model with self-interacting neutrinos [77]. However, despite it is important to keep eyes open in the search for extensions to the current model that could lead to fresh discoveries, before extracting more firm conclusions it seems convenient that the data are examined in more detail in the search for possible systematics that could explain the differences. Also, future observations with higher precision will no doubt help to better establish the significance of the tension. More detailed discussions about this aspect can be found on the notes of a workshop recently held at the Kavli Institute for Theoretical Physics that was specifically dedicated to this topic [78].

### 3.1.2 $\sigma_8$ from Galaxy Cluster Number Counts and from the CMB

Another apparent tension that has fostered interest on the last years concerns the joint $\Omega_\mathrm{m} - \sigma_8$ estimates coming from cluster number counts and from the CMB, as illustrated in Fig. 9 and previously referred to in Sect. 2.2.2. The tension initially pointed out in the *Planck* 2013 cosmological results (using DR1) is partially relieved in the *Planck* 2015 results (DR2) thanks to the lower value of $\sigma_8$ derived from the CMB, which in turn is due to the lower value of $\tau$. Using the CCCP value for the mass bias $(1-b)$, which is the default method in the *Planck* analysis [60], the tension with the CMB is at $1.5\sigma$. This tension is not unique of *Planck*, but is also present in the cosmological analyses extracted from the number counts of other SZ surveys like ACT [79] or SPT [80], although typically at a lower significance because they are based on smaller cluster samples which leads to larger error bars. Currently the major uncertainty on these analyses is the mass bias estimate. According to the *Planck* 2015 analysis the value of the mass bias that is required to reconcile the CMB and number counts estimates is $(1-b) = 0.58 \pm 0.04$ [60], which means that clusters should have mass $\approx 40\%$ lower than derived from hydrostatic equilibrium estimates. While current numerical simulations and weak lensing estimates agree that the mass bias should instead be $(1 - b) \approx 0.8$, there could be aspects related to incorrect modelling of cluster physics or systematics on the data that could explain this difference.

If these possible systematics are overlooked, we might start considering extensions to the $\Lambda$CDM model that could explain the differences between the low- and high-redshifts determinations of $\sigma_8$. The most obvious one would be a non-minimal sum of neutrino masses. This scenario was considered in [60], and the conclusion was that in fact while this led to a reduced tension on $\sigma_8$ it was at the cost of increasing the tension between other parameters. Another possibility discussed in [60] is that baryonic physics may influence the late-time evolution of density perturbations in such a way that feedback from active galactic nuclei could damp growth and therefore reduce $\sigma_8$. In any case, before developing any further these or other theoretical interpretations, it may be advisable to deepen our understanding of possible systematics associated with this analysis. As an example, recently [81] pointed out that neglecting the relativistic corrections to the SZ frequency spectrum could lead to significant biases on the determination of $\sigma_8$ from the SZ power spectrum. And, more importantly, a more precise determination of the mass bias parameter will give the definitive answer about the real tension between the CMB and cluster number counts. Improved cluster mass estimates from the CMB lensing, in the short term, or from future lensing surveys like *Euclid*, *WFIRST* and LSST that may allow to reach $1\%$ precision in the determination of the mass bias parameter, will definitely contribute to this.

### 3.1.3 *Planck* Low- and High-Multipole Data

Although overall there is a quite remarkable consistency between the results of different CMB experiments, when they are analysed in closer detail some slight differences show up. Compared to *WMAP* results, *Planck* prefers a somewhat lower expansion rate, higher dark matter density and higher power spectrum amplitude, as it has been discussed in several *Planck* papers [35], and also in [82]. It seems that these differences arise from the different multipole ranges that are sampled by *Planck* and *WMAP*. In fact [82] noted that there were some parameter shifts, in some cases at around $2 - 3\sigma$, between the $\Lambda$CDM parameters derived using *Planck* multipoles $\ell < 1000$ or $\ell > 1000$. They also pointed out some differences between high-$\ell$ *Planck* and SPT data, concluding that the previous tensions may not be due to new physics but rather to systematics not accounted for in the data. Furthermore, none of the extensions that have been discussed in the literature seem to be preferred by the data. These issues were revisited in detail by Aghanim et al. [83], and the conclusion was that, given the dimensionality of the model, the differences found between parameters are not statistically significant, and *Planck* low-$\ell$ and high-$\ell$ data are consistent with each other within around $10\%$ PTE.

### 3.1.4 High Amplitude of the Lensing Potential in the *Planck* TT Power Spectrum

As it was commented in Sect. 2.1, thanks to the increased quality of current datasets, the lensing of CMB photons by the LSS of the universe can now be used to constrain cosmological parameters. The lensing causes different effects in the CMB: a smoothing of the acoustic peaks and troughs in the temperature and polarisation power spectra, a conversion of E-modes into B-modes, and a generation of significant non-Gaussianity that can be measured through the CMB 4-point functions. *Planck* data show some differences in the amplitude of the lensing potential derived from the former and from the latter of these estimates, which are at around the $2\sigma$ level [35]. The difference in the combination of parameters $\sigma_8\Omega_{\rm m}^{0.25}$, which is a good proxy for the lensing amplitude, is just $1.3\sigma$ [83]. However, as initially pointed out by Addison et al. [82], this difference increases to $2.2\sigma$ [83] if only multipoles $\ell > 1000$ are used to quantify the effect of the smoothing of the CMB peaks.

## 3.2 *Anomalies*

Although there is an excellent consistency between the CMB data and the $\Lambda$CDM model, several features or anomalies in the CMB maps and power spectra have extensively been discussed in the literature, with the aim to precisely assess their significance and understand if there are really hints for extensions of the $\Lambda$CDM model. One of the CMB anomalies that has been more hotly debated is a lack of power and correlation on large angular scales, which shows up both in the power spectrum at multipoles $\lesssim$40 (see Fig. 3) and in the two-point angular correlation function. This lack of power is actually the main driver of the difference between the best-fit parameters from the low-$\ell$ and high-$\ell$ power spectra that were previously commented in Sect. 3.1. Particularly low is the quadrupole amplitude, as first hinted at in *COBE* data. Bennett et al. [84] analysed this effect in *WMAP*-7yr data and found that its statistical significance was quite sensitive to the assumed foreground mask, to the statistical estimator, and also to the impact of the Integrated Sachs-Wolfe (ISW) on large angular scales. They also pointed out that the *WMAP* angular correlation function is consistent with the $\Lambda$CDM expectation within $2\sigma$ for all angular scales, and then downplayed the significance of this effect. Its significance is however larger (between $2.5\sigma$ and $3\sigma$, depending on the estimator used) in *Planck* data, due to the fact that *Planck* power spectrum is lower than the one from *WMAP* [85]. Remarkably, artificially scaling up the low-$\ell$ data to better match the best-fit model leads to an even lower value of $H_0$, which aggravates the tension with SNIa data (see Sect. 3.1).

Another well-known anomaly is the "cold spot", a non-Gaussian negative feature seen in the CMB around $l = 209°$, $b = -57°$ with a size of $\sim$5°, first identified in *WMAP*-1yr data using a spherical Mexican hat wavelet analysis [86], and with a significance of around 1.4–2.3$\sigma$ [84]. This feature was confirmed by *Planck* data, with a probability $\lesssim$1%, the exact value depending on the exact size of the filter used [87]. Other two anomalies, also detected in *WMAP*-1yr data, and that have subsequently been extensively scrutinised are: a remarkable alignment between the directions of the quadrupole and the octopole, which are supposed to be randomly oriented; and a hemispherical asymmetry between the low-$\ell$ power spectra calculated in two halves of the sky separated with respect to the position $l = 237°$, $b = -20°$, with significance between 95% and 99%. A more detailed overview of these an other anomalies can be found in different reviews [88, 89] or in the relevant *WMAP* [84] and *Planck* [87] papers.

Currently there is no discussion that these anomalies do exist, both in *WMAP* and *Planck* data, this in fact being a good consistency cross-check of two different independent datasets, nor there is debate about their statistical significance. The question is whether these anomalies in the data may be connected or not with cosmic anomalies in the model. While some groups emphasise that we still lack an understanding of these large-scale features that seem to violate statistical isotropy and the scale invariance of inflationary perturbations [88], there is general consensus that they do not provide sufficiently high statistical significance as to justify the endeavour of exploring extensions of the current $\Lambda$CDM model [84]. The key issue has to do with the interpretation of the statistical significance of the different tests that are applied to the data. Scott [89] points out that the statistical significance of all these features is usually assessed after they were actually discovered, disregarding other anomalies that could have been found and actually were not (this is what statisticians refer to as "multiplicity of tests"). As an example [89] mentions that the statistical significance of the cold spot is usually assessed by looking to the probability of finding features with just the same angular size and amplitude, while we should instead marginalise over the scale as well as over the potential filter shapes. Of course, there is not a fully objective way to look for these effects. Therefore, in order to account for them, [89] recommends focusing on 5$\sigma$ anomalies, and this would inevitably downplay the significance of the previous effects that is typically 2–3$\sigma$. Also, it has to be taken into account the correlation between different anomalies when assessing their combined significance.

## *3.3 Open Problems*

In the previous sections we have described our current understanding of the Universe, which essentially is that it evolved from an early hot and dense phase, leading to the later hierarchical assembly of galaxies, clusters and superclusters at our epoch. However, the successes of the current $\Lambda$CDM model must inevitably be balanced by the fact that they rest upon three unknown or yet not well understood entities beyond the Standard Model of particle physics: inflation, dark matter and dark energy. These are the main mysteries in contemporary cosmology and, arguably, in all of physics.

The theory of inflation is supported by various observational indirect evidences. It was initially proposed in the 1980s to resolve a number of puzzles of standard big bang cosmology, such as the entropy, flatness, horizon, smoothness and monopole problems [90]. More importantly, inflation provides the seeds for structure formation. During inflation initial quantum fluctuations were stretched, thence leading to the cosmological fluctuations that later grew and, by gravitational instability, generated the structure of today's universe. Several predictions of inflation have been successfully confirmed by observations, namely that the Universe should have a flat geometry, nearly scale-invariant perturbations, and nearly Gaussian perturbations in all scales. However, at present we can not assert that any of the available observations prove that inflation is actually correct, and for this reason it remains being a theoretical framework rather than a model. We currently lack a precise understanding of how exactly inflation occurred, of the physics of extremely high energies that are required to drive this period, nor we have a unique scenario of inflation. It is recognised that nowadays the best way to attain advances in this field is through the detection of the faint B-mode pattern in the CMB polarisation which should have been created by the gravitational waves generated during inflation (see Sect. 2.1 and Fig. 1). The detection of this signal would help constraining the energy scale of inflation—a key discriminant between different inflation models—and its dynamics, would in fact confirm the fourth prediction of inflation, which is the generation of a spectrum of tensor modes (gravitational waves), and would help measuring the exact shape of this spectrum. This is the main goal of the ongoing and planned CMB experiments.

As it was explained in the previous sections, there are compelling evidences that we have knowledge of just 5% of the total mass-energy content of the universe, while 26% is in the form of dark matter and 69% in the form of dark energy. Current observations provide strong evidences for a non-baryonic nature of dark matter. Current evidences for the existence of non-baryonic matter come from oscillation experiments, which have shown that at least two of the three neutrino species have non-zero mass. However, CMB and LSS observations indicate that they only make up a small fraction of the dark matter in the Universe. In parallel to direct dark-matter searches in experiments like LHC, astrophysics and cosmology can provide insightful information about the nature of dark matter. Measuring the exact value of the total absolute neutrino mass, which can be done by measuring the tiny suppression of structure formation produced by neutrinos, is important to understand the mechanisms that gave them their mass. On the other hand, measuring the density profiles of dark matter halos and the matter power spectrum with very high precision can help to set constraints on the dark matter mass. This will be the goal of future LSS surveys.

The origin of cosmic acceleration, which implies the existence of some distributed component of energy density not associated with matter concentrations and which exerts negative pressure, also remains a deep conundrum. One of the key issues to advance our understanding of the nature of dark energy concerns the measurement of its equation of state, as it was commented in Sect. 1.4.2. As it was described in Sect. 2, all current measurements are consistent with a cosmological constant, i.e. vacuum energy with $w = -1$, and do not show any hint for a time-variation of this value. However, particle theories suggest that the dark energy density should not be constant and may have varied by many orders of magnitude over the history of the Universe. This would explain why its current value is so small, while the generally accepted theory of strong interactions (quantum chromodynamics) makes a contribution to the vacuum energy that is over 50 orders of magnitude larger than the observed value. Measuring the value of $w$ to higher precision, and constraining a possible time-variation of this parameter, is one of the main goals of future LSS surveys like *Euclid*, DESI or LSST.

## 4 Concluding Remarks

In this chapter we have shown that we currently have a consistent model describing the basic properties of the Universe we live in, its origin, evolution and matter-density content. This model is supported on Einstein's General Relativity, and states that the universe originated in a hot-dense phase called Big Bang, it underwent an early phase of expansion that created light elements via the big bang nucleosynthesis and the CMB, and a much earlier phase of accelerated expansion, called inflation, during which the initial quantum fluctuations were stretched to cosmic sizes, leaving an imprint on the CMB in the form of its anisotropies, and also leading to the formation of LSS due to gravitational instability. This model is strongly supported by a variety of observations, of which here we have described the most important ones: CMB, the LSS and Type Ia supernovae. Technological advances over the last decades have allowed to exploit these cosmological probes to unprecedented precision, allowing to infer cosmological parameters with precision better than 1%, and leading to what we have come to call the "era of precision cosmology". It is also an outstanding achievement that totally different and independent cosmological probes, like the CMB, the LSS or SNIa, give consistent constraints on the cosmological parameters. In fact, even using only the CMB, the data from the *Planck* satellite have allowed to obtain consistent constraints from three different observables: temperature anisotropies, polarisation anisotropies, and the lensing of CMB photons.

Current observations are nicely described by a set of just six parameters, three of which describe the composition and evolution of the Universe, two the initial conditions of the density perturbations, while the sixth is an astrophysical parameter related to the optical depth of Thomson scattering to recombination. This model has been coined ΛCDM, to emphasise that it is dominated by a dark energy component, which causes the observed accelerated expansion that seems to be associated with a cosmological constant Λ (i.e. vacuum energy with equation-of-state parameter $w = -1$) with a density parameter $\Omega_\Lambda = 0.69$, and by a cold dark matter component with $\Omega_c = 0.26$, while ordinary baryonic matter contributes to just $\Omega_b = 0.05$. The spectral index of the density perturbations is found to be $n_s = 0.9646 \pm 0.0042$, which is a small deviation, at the $8\sigma$ level, from perfect scale invariance ($n_s = 1$), in agreement with the simplest inflationary models. Extensions from this model have been explored using existing data, but finding no compelling evidences. For instance, when different probes are combined no hints have been found of deviations from a perfectly flat geometry, with the best constraint $\Omega_k = 0.0007 \pm 0.0019$, coming from a combination of CMB and BAO, being astonishingly consistent with $\Omega_k = 0$. No evidences have been found either for a running of the spectral index accounting for a scale-dependency of the primordial fluctuations, as is predicted by some inflationary models. The equation of state of dark energy is found to be fully consistent with $w(a) = -1$, with no evidence of any time-variation of this value. The upper limit on the neutrino mass, as well as the number of relativistic species, are also fully compatible with the standard values. These results do not necessarily imply that these deviations do not exist, but rather that the existing data do not favour any of them within their error bars, and for this reason we can firmly assert that $\Lambda$CDM is the model that best describe these data.

Despite the overall stunning consistency between different and independent datasets, there are a few claimed tensions in some specific cosmological parameters. The most significant one, and currently more hotly discussed, concerns the expansion rate. SNIa-based distance-estimation techniques in the local Universe give a value for $H_0$ that differs with the CMB measurement at $z = 1100$ at $4.3\sigma$. If the SNIa measurements are combined with other local estimates based on time delays on strong-lensed QSOs, then the discrepancy with the CMB increases to $5.3\sigma$. Several ideas about extensions of $\Lambda$CDM that could restore concordance between the two measurements have been recently explored in the literature, but none of them seem to provide a convincing solution, either because in some cases a better agreement is achieved thanks to enlarging the uncertainties, or in some other cases at the cost of increasing the discrepancies in other parameters. Other important tension concerns the value of the amplitude of the scalar fluctuations, $\sigma_8$, derived from the CMB and from the number counts of galaxy clusters. In this case the discrepancy is at $\sim 1.5\sigma$, but the exact value is strongly dependent on the value assumed for the mass bias parameter. A more firm assessment of the real discrepancy requires a more reliable estimate of this bias parameter, for which currently different mass estimates give a considerable scatter. Therefore, it seems clear that before continuing to explore extensions to the current $\Lambda$CDM model, it may be better to wait until systematics affecting current analyses are reduced with the help of additional datasets with better sensitivity. An example are the new cluster mass estimates extracted from CMB lensing, that have just started to be exploited.

Finally, in spite of having a well consolidated model describing all current observations, we may not forget that two of its main ingredients, dark matter and dark energy, are currently unknown entities. Future LSS surveys will be key to shed new light on the nature of these entities, particularly important being the characterisation of the dark energy equation of state, and a possible time-variation of this parameter. This will be crucial to understand what is causing the observed accelerated expansion of the universe, or if alternatively our model requires modifications of General Relativity on cosmological scales. Equally important is understanding the generation of the primordial density fluctuations, for which the preferred model is inflation. Understanding the physics that drove this process could lead to important discoveries at energy scales not previously explored, and in this aspect ongoing and future CMB polarisation experiments will play a major role.

**Acknowledgements** Most of this work was written during a 5-week visit of the author to the University of Cambridge, in summer 2019. The author thanks the hospitality of the Cavendish Astrophysics group during this visit. The author also thanks Francisco-Shu Kitaura for reading parts of the text, and the referee for a careful reading of the text and useful comments. Some of the figures presented here have been taken from the "Planck Image Gallery" (ESA and Planck Collaboration).

## References

1. G. Gamow, Phys. Rev. **70**, 572 (1946)
2. R.A. Alpher, R. Herman, Nature **62**, 774 (1948)
3. H. Bondi, T. Gold, Mon. Not. R. Astron. Soc. **108**, 252 (1948)
4. W.S. Adams, Astrophys. J. **93**, 11 (1941)
5. A.A. Penzias, R.W. Wilson, Astrophys. J. **142**, 419 (1965)
6. S. Weinberg, *Gravitation and Cosmology: Principles and Applications of the General Theory of Relativity* (Wiley, New York, 1972)
7. F. Zwicky, Helv. Phys. Acta **6**, 110 (1933)
8. V.C. Rubin, W.K. Ford Jr., N. Thonnard, Astrophys. J. **238**, 471 (1980)
9. D.J. Fixsen, Astrophys. J. **707**, 916 (2009)
10. A.A. Starobinsky, ZhETF Pisma Redaktsiiu **30**, 719 (1979)
11. A.H. Guth, Phys. Rev. D **23**, 347 (1981)
12. A. Linde, *Inflationary Cosmology*, vol. 738 (Springer, Berlin, 2008), p. 1
13. A. Challinor, *Astrophysics from Antarctica*, vol. 288 (2013), p. 42
14. A. Linde, Phys. Rev. D **59**, 023503 (1999)
15. A. Linde, J. Cosmol. Astropart. Phys. **5**, 002 (2003)
16. S. Tsujikawa, Modified gravity models of dark energy, in *Lectures on Cosmology*. Lecture Notes in Physics, vol. 800 (Springer, Berlin, 2010), p. 99
17. J.-P. Uzan, Rev. Mod. Phys. **75**, 403 (2003)
18. G.F. Smoot, C.L. Bennett, A. Kogut et al., Astrophys. J. Lett. **396**, L1 (1992)
19. P. de Bernardis, P.A.R. Ade, J.J. Bock et al., Nature **404**, 955 (2000)
20. R. Stompor et al., Astrophys. J. Lett. **561**, L7 (2001)
21. A. Benoît et al., Astron. Astrophys. **399**, L19 (2003)
22. C. Dickinson et al., Mon. Not. R. Astron. Soc. **353**, 732 (2004)
23. J.L. Sievers et al., Astrophys. J. **591**, 599 (2003)
24. C.L. Kuo et al., Astrophys. J. **600**, 32 (2004)
25. C.L. Bennett, D. Larson, J.L. Weiland et al., Astrophys. J. Suppl. Ser. **208**, 20 (2013)
26. J.W. Fowler, V. Acquaviva, P.A.R. Ade et al., Astrophys. J. **722**, 1148 (2010)
27. K.T. Story, C.L. Reichardt, Z. Hou et al., Astrophys. J. **779**, 86 (2013)
28. Planck 2018 Results I (2018). arXiv: 1807.06205
29. C. Dickinson, in *Draft Proceedings for the Conference Rencontres de Moriond 2016 on Cosmology* (2016). arXiv: astro-ph/1606.03606
30. M. Zaldarriaga, U. Seljak, Phys. Rev. D **55**, 1830 (1997)
31. M. Kamionkowski, A. Kosowsky, A. Stebbins, Phys. Rev. D **55**, 7368 (1997)
32. W. Hu, M. White, New Astron. **2**, 323 (1997)
33. S. Das, B.D. Sherwin, P. Aguirre et al., Phys. Rev. Lett. **107**, 021301 (2011)
34. Planck 2018 Results VI (2018). arXiv: 1807.06209

35. Planck 2013 Results XIII, Astron. Astrophys. **594**, A13 (2016)
36. A.G. Riess, S. Casertano, W. Yuan, L.M. Macri, D. Scolnic, Astrophys. J. **876**, 85 (2019)
37. BICEP2 Collaboration, Keck Array Collaboration, P.A.R. Ade et al., Phys. Rev. Lett. **121**, 221301 (2018)
38. L. Anderson, E. Aubourg, S. Bailey et al., Mon. Not. R. Astron. Soc. **427**, 3435 (2012)
39. D.W. Hogg, Distance measures in cosmology (2000). arXiv:astro-ph/9905116
40. C. Alcock, B. Paczynski, Nature **281**, 358 (1979)
41. W.J. Percival, Lectures given at Post-Planck Cosmology, Ecole de Physique des Houches, and New Horizons for Observational Cosmology, International School of Physics Enrico Fermi, Varenna (2013). arXiv:astro-ph/1312.5490
42. J. Huchra, M. Davis, D. Latham, J. Tonry, Astrophys. J. Suppl. Ser. **52**, 89 (1983)
43. G. de Vaucouleurs, A. de Vaucouleurs, H.G. Corwin Jr. et al., Sky Telescope **82**, 621 (1991)
44. S.A. Shectman, S.D. Landy, A. Oemler et al., Astrophys. J. **470**, 172 (1996)
45. M. Colless, B.A. Peterson, C. Jackson et al. (2003). arXiv:astro-ph/0306581
46. M.J. Drinkwater, R.J. Jurek, C. Blake et al., Mon. Not. R. Astron. Soc. **401**, 1429 (2010)
47. D.G. York, J. Adelman, J.E. Anderson Jr. et al., Astron. J. **120**, 1579 (2000)
48. D.J. Eisenstein, D.H. Weinberg, E. Agol et al., Astron. J. **142**, 72 (2011)
49. S. Cole, W.J. Percival, J.A. Peacock et al., Mon. Not. R. Astron. Soc. **362**, 505 (2005)
50. D.J. Eisenstein, I. Zehavi, D.W. Hogg et al., Astrophys. J. **633**, 560 (2005)
51. M. Ata, F. Baumgarten, J. Baumgarten et al., Mon. Not. R. Astron. Soc. **473**, 4773 (2018)
52. J.E. Bautista, N.G. Busca, J. Guy et al., Astron. Astrophys. **603**, A12 (2017)
53. S. Alam, M. Ata, S. Bailey et al., Mon. Not. R. Astron. Soc. **470**, 2617 (2017)
54. É. Aubourg, S. Bailey, J.E. Bautista et al., Phys. Rev. D **92**, 123516 (2015)
55. F. Marulli, A. Veropalumbo, M. Sereno et al., Astron. Astrophys. **620**, A1 (2018)
56. M. Bonamente, M.K. Joy, S.J. LaRoque et al., Astrophys. J. **647**, 25 (2006)
57. Planck 2015 Results XXII, Astron. Astrophys. **594**, A22 (2016)
58. A. Vikhlinin, A.V. Kravtsov, R.A. Burenin et al., Astrophys. J. **692**, 1060 (2009)
59. Planck 2013 Results XX, Astron. Astrophys. **571**, A20 (2013)
60. Planck 2015 Results XXIV, Astron. Astrophys. **594**, A24 (2016)
61. A.G. Riess, W.H. Press, R.P. Kirshner, Astrophys. J. **473**, 88 (1996)
62. M.M. Phillips, Astrophys. J. Lett. **413**, L105 (1993)
63. A. Sandage, G.A. Tammann, Astrophys. J. **256**, 339 (1982)
64. W.L. Freedman, B.F. Madore, B.K. Gibson et al., Astrophys. J. **553**, 47 (2001)
65. W.L. Freedman, B.F. Madore, D. Hatt et al. (2019). arXiv: 1907.05922
66. E. Macaulay, R.C. Nichol, D. Bacon et al., Mon. Not. R. Astron. Soc. **486**, 2184 (2018)
67. A.G. Riess, A.V. Filippenko, P. Challis et al., Astron. J. **116**, 1009 (1998)
68. S. Perlmutter, G. Aldering, G. Goldhaber et al., Astrophys. J. **517**, 565 (1999)
69. W.M. Wood-Vasey, G. Miknaitis, C.W. Stubbs et al., Astrophys. J. **666**, 694 (2007)
70. P. Astier, J. Guy, N. Regnault et al., Astron. Astrophys. **447**, 31 (2006)
71. M. Betoule, R. Kessler, J. Guy et al., Astron. Astrophys. **568**, A22 (2014)
72. M. Sako, B. Bassett, A.C. Becker et al., Publ. Astron. Soc. Pac. **130**, 064002 (2018)
73. A.G. Riess, L.-G. Strolger, S. Casertano et al., Astrophys. J. **659**, 98 (2007)
74. D.M. Scolnic, D.O. Jones, A. Rest et al., Astrophys. J. **859**, 101 (2018)
75. K.C. Wong, S.H. Suyu, G.C.-F. Chen et al. (2019). arXiv: 1907.04869
76. V. Poulin, T.L. Smith, T. Karwal, M. Kamionkowski, Phys. Rev. Lett. **122**, 221301 (2019)
77. C.D. Kreisch, F.-Y. Cyr-Racine, O. Doré (2019). arXiv: 1902.00534
78. L. Verde, T. Treu, A.G. Riess (2019). arXiv: 1907.10625
79. M. Hasselfield, M. Hilton, T.A. Marriage et al., J. Cosmol. Astropart. Phys. **7**, 008 (2013)
80. S. Bocquet, A. Saro, J.J. Mohr et al., Astrophys. J. **799**, 214 (2015)
81. M. Remazeilles, B. Bolliet, A. Rotti, J. Chluba, Mon. Not. R. Astron. Soc. **483**, 3459 (2019)
82. G.E. Addison, Y. Huang, D.J. Watts et al., Astrophys. J. **818**, 132 (2016)
83. Planck Intermediate Results LI, N. Aghanim, Y. Akrami et al., Astron. Astrophys. **607**, A95 (2017)
84. C.L. Bennett, R.S. Hill, G. Hinshaw et al., Astrophys. J. Suppl. Ser. **192**, 17 (2011)

85. Planck 2013 Results XV, Astron. Astrophys. **571**, A15 (2014)
86. P. Vielva, E. Martínez-González, R.B. Barreiro, J.L. Sanz, L. Cayón, Astrophys. J. **609**, 22 (2004)
87. Planck 2013 Results XXIII, Astron. Astrophys. **571**, A23 (2014)
88. D.J. Schwarz, C.J. Copi, D. Huterer, G.D. Starkman, Classical Quantum Gravity **33**, 184001 (2016)
89. D. Scott (2018). arXiv: 1804.01318
90. Planck 2013 Results XXII, Astron. Astrophys. **571**, A22 (2014)

# Exploiting Solar Visible-Range Observations by Inversion Techniques: From Flows in the Solar Subsurface to a Flaring Atmosphere

**Michal Švanda, Jan Jurčák, David Korda, and Jana Kašparová**

## Contents



**Abstract** Observations of the Sun in the visible spectral range belong to standard measurements obtained by instruments both on the ground and in the space. Nowadays, both nearly continuous full-disc observations with medium resolution and dedicated campaigns of high spatial, spectral and/or temporal resolution constitute a holy grail for studies that can capture (both) the long- and short-term changes in the dynamics and energetics of the solar atmosphere. Observations of photospheric spectral lines allow us to estimate not only the intensity at small regions, but also various derived data products, such as the Doppler velocity and/or the components of the magnetic field vector. We show that these measurements contain not only


M. Švanda (✉)
Astronomical Institute, Charles University, Praha, Czech Republic

Astronomical Institute of the Czech Academy of Sciences, Ondřejov, Czech Republic
e-mail: michal@astronomie.cz

J. Jurčák · J. Kašparová
Astronomical Institute of the Czech Academy of Sciences, Ondřejov, Czech Republic

D. Korda
Astronomical Institute, Charles University, Praha, Czech Republic


 
P. Kabáth et al. (eds.), *Reviews in Frontiers of Modern Astrophysics*,
https://doi.org/10.1007/978-3-030-38509-5_12

direct information about the dynamics of solar plasmas at the surface of the Sun but also imprints of regions below and above it. Here, we discuss two examples: First, the local time-distance helioseismology as a tool for plasma dynamic diagnostics in the near subsurface and second, the determination of the solar atmosphere structure during flares. The methodology in both cases involves the technique of inverse modelling.

## 1 Introduction

The Sun is our closest star and as such it serves as a prototype for the whole class of such cosmic objects [1]. As compared to the other stars, the Sun is routinely observed from a luxurious proximity 24 h a day, 7 days a week. The availability of the long-term near-continuous synoptic observations together with short-term high-resolution campaign observations consists a holy grail for highly detailed studies of the solar structure and dynamics. The theories and hypotheses based on the observations of the Sun serve as constraints for theories of stellar structure and evolution.

The Sun is an active and variable star with various phenomena observed in its atmosphere termed the *solar activity*. These phenomena exist due to the solar interior being interwoven with the magnetic field, which takes various forms. The origin of it is still being debated in the literature [2]. Overall, the solar magnetic field depicts cyclic behaviour on a large range of time scales from seconds to millennia [3]. There is a dynamo process running deep inside the Sun, possibly at the base of the convection zone, which recycles and strengthens the global magnetic field. Emergence of the magnetic flux tubes through the convection zone towards the surface and above forms complexes of strong magnetic field localisations, the *active regions* [4]. In active regions, we typically find the most prominent phenomena of solar activity—sunspots, prominences, and flares. All these processes are only partially understood, thus much of the on-going and future research is dedicated to them.

The proper understanding of the violent phenomena of solar activity, namely flares and coronal mass ejections, is necessary as these phenomena may in effect interfere with human technology, which is crucial for the quality of life as we know it [5]. There are many open questions and without answering those we will not be able to successfully predict the solar activity, which is one of the main goals of the solar research. For instance, triggers for flares may lie in the dynamics of the plasmas at the solar surface and below and the flaring processes may strongly depend on the parameters of the atmosphere above the surface. Visible-range observations may in principle bring important information about both solar interior and atmosphere.

## 2 Solar Photosphere in the Visible Range

In the visible-range of electromagnetic radiation, the photosphere is the dominant source of solar radiation. The term photosphere is often replaced by "solar surface", but this is not very precise. Following a text-book definition, the photosphere describes the thin near-spherical shell where the visible and infrared solar continua originate [6]. Its width is several hundreds km and the lower boundary of the photosphere is defined at the optical depth unity ($\tau_{500} = 1$)—level at which the intensity at $\lambda = 500$ nm is approximately given by the continuum source function. The opacity is dominated by bound-free $H^-$ transitions. The formation height of the continuum radiation is wavelength dependent. For instance it is deeper in the near-infrared part of the spectrum, where free-free contributions of $H^-$ dominate.

The photosphere is only several hundreds kilometers thick. One of the definitions adopts a thickness of about 300 km, which corresponds to the scale over which the gas pressure decreases by an order of magnitude. Even though a realistic description of the vertical stratification of the photosphere have been achieved already by 1-D models [7], observations are very convincing in the fact that the evaluations of the structure of the photosphere needs to be considered in all three dimensions [8]. Already the observations in the visible continuum by amateur telescopes reveal the pattern of granulation, the distinct convection pattern that corresponds to the thermal dissipation scale [9]. The granular cells have a typical diameter of about 1 Mm and a life time of about 10 min. Both these parameters depict very wide statistical distributions. According to the current paradigm, the granules are the final products of the turbulent-convection cascade emerging from the bottom of the convection zone [10]. Other convection-like features are present and measured in the photosphere: the mesogranules, the supergranules and possibly the giant cells.

High-resolution observations reveal not only the granular cells, but also features of smaller size, such as magnetic bright points and other indications for the small-scale magnetic field [11]. In these magnetic-field concentrations the $\tau = 1$ level lies even deeper, about 200 km as compared to the surrounding "quiet" photosphere. In these regions, the magnetic pressure contribution to hydrostatic balance reduces the gas density. Such thin flux tubes form a magnetic network that is preferentially located at supergranular boundaries. Even larger magnetic field concentrations create anomalous granular patterns (flux-emergence regions, "plage" regions) and strong enough magnetic field inhibits the convection completely by creating pores and sunspots [12].

Thus the transition from the 1-D vertically stratified models towards a fully 3-D time-dependent solution of (magneto)hydrodynamical equations with realistic equation of state was eminent. And it was successful [13]. The physics behind the processes active in the quiet photosphere and closely below and above are believed to be understood quite well. That is demonstrated by the fact that the numerical simulations reproduce the observable photospheric properties remarkably well [9]. Detailed simulations reproduce even specialties such as elongated granules in the regions with emerging magnetic fields or particular flows around spots and pores [14, 15].

## 3 Sub-surface Dynamics

From the analyses of solar observations together with proper modelling based on the theory of stellar structure and evolution it turns out that the Sun is a main-sequence star with an effective temperature of $5778 \pm 3$ K, radius of $(6.960 \pm 0.001) \times 10^8$ m and a mass of $(1.9889 \pm 0.0003) \times 10^{30}$ kg [16]. Because of the action of gravity, such a star has a significant radial stratification, generally showing three main layers. The central region up to some 25% of radius is occupied by the *core*, where the physical conditions allow the thermonuclear fusion to run mostly by means of the proton-proton chain. This is the region where the solar luminosity, equivalent of almost $4 \times 10^{26}$ W, originates. The central temperature reaches 15.7 MK and the pressure of $2.5 \times 10^{16}$ Pa according to the state-of-the-art models [17].

At around 25% of solar radius the temperature of the ambient plasma decreases below the value of around 7 MK. The efficiency of the thermonuclear reactions is very low henceforth, however the state parameters of the plasma are such that the plasma is fully ionised and the opacity is very low for the high-energy photons originating from the proton-proton chain. Due to the temperature gradient the energy is transported towards the surface via radiation, photons diffuse by free-free radiative scattering. This layer is called *the radiative zone*.

At 71% of the solar radius [18, 19] the mean plasma temperature decreases under ~2 MK and recombination of ions starts to occur locally. The opacity rapidly increases due to the bound-free radiative absorption of photons by heavy elements. The energy transport via radiation is no longer possible. The convection sets in and begins to be the main energy transport agent until the surface, through the whole *convection zone*. The ever-changing granulation is a consequence of the convective overshoot from the convection zone underlying the photosphere. The convection zone is a principal layer, where the dynamo is seated somewhere and from where the flux tubes later forming sunspots and other phenomena rise up.

Studying the convection zone is not straightforward. The sub-surface of the Sun is optically thick, preventing us from directly observing the interior layers. Understanding the properties of the plasma in these regions has consequences for theories of convection, stability of sunspots, the dynamics of stratified convection, and others. Most of current knowledge about convection comes primarily from computational work, e.g. [20–22]. Helioseismic inversions of the sub-surface flows play an important role in constraining these theories.

The physics of convection in the low-viscosity, large-density and large-temperature gradient regime associated with the convection zone is not well known [23]. Most of the convection zone is moderately vertically stratified, but as the plasma approaches the photosphere, it undergoes a rapid expansion because of the steep near-surface density gradient. With the rapid decrease of temperature in the near-surface layers ionisation zones of several elements (e.g. helium and hydrogen) form, and the release of the latent heat by the recombination is thought to power the various scales of turbulent convection.

## 3.1 Oscillations

A powerful way of imaging the solar interior is via inferences gathered from studying the statistics of the acoustic and surface gravity waves detectable at the surface. Solar pressure and surface gravity modes are generated randomly by the vigorous turbulence in the upper convection zone. These oscillations are best observed in the solar photosphere by measuring Doppler shifts of photospheric absorption lines.

The spatio-temporal power spectrum of solar oscillations provides the best evidence for the existence of the resonant standing waves. These resonant modes form the structure of non-intersecting ridges in a $k - \omega$ diagram, where $k$ stands for the wave number and $\omega = 2\pi\nu$ for the cyclic frequency of the waves. The position of the ridges is a very sensitive indicator of the internal structure of a star. An example of the $k - \nu$ diagram is in Fig. 1. One can see obvious signatures of resonant modes depicted by ridges of increased spectral power.

A vast majority of the modes seen in the solar $k - \omega$ diagram belong to the group of $p$-modes, which are acoustic modes where the pressure is a restoring force. The $p$-modes propagate mostly in a convectively unstable layers, that is in the near-surface convection zone. The ridge positioned lowest on the $k - \omega$ diagram is a $f$-mode ridge. $f$-modes are surface gravity modes, similar to the oscillations of the free surface. Frequencies and wave numbers (which relate to the wavelengths) are examples of helioseismic observables.

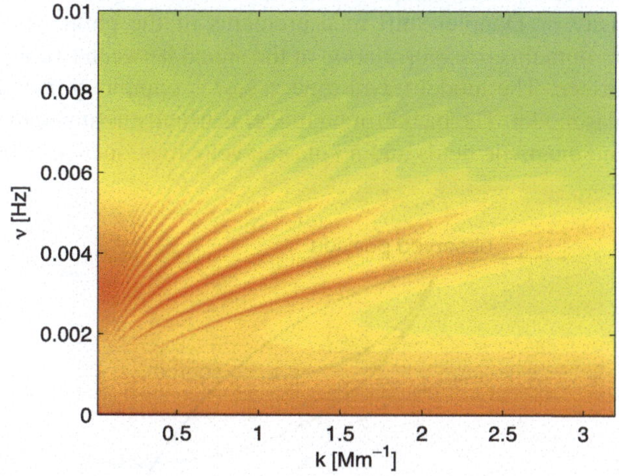


**Fig. 1** An example $k - \nu$ diagram as constructed from 24-h Dopplergram series recorded by Helioseismic and Magnetic Imager (HMI) on-board of Solar Dynamics Observatory (*SDO*). Signal of low frequency corresponds to the convection zone. The lowest positioned ridge is the $f$-mode ridge, others are $p$-mode ridges. "Five-minute oscillations" with frequency around 3.3 mHz are result of interference of $p$-modes. A lack of signal above 5.3 mHz is related to an acoustic cut-off frequency. The colour scale in arbitrary units indicates the spectral power with greenish colours having a low power and reddish colours having a large power

## 3.2 Helioseismology

Changes in the internal structure of the Sun, including localised changes, induce shifts in frequencies of the modes. Forward modelling allows us to relate anomalies like flows, thermal hot/cold spots etc. to changes in helioseismic observables. It is not straightforward to infer the characteristics of these anomalies from observables. Sophisticated techniques of inverse modelling—or "inversions" in short—must be adopted.

The aim of helioseismic inversions is to reveal the structure of the subsurface flows (rotation, meridional circulation, convection), magnetic fields, and to measure deviations in the plasma state parameters (temperature, density, pressure) from a quiet Sun average.

Localised perturbations in the parameters of solar plasmas may be investigated using the approach of local helioseismology. The time–distance helioseismology [24], one of the local helioseismic methods, has proven to be very useful. In recent years, time–distance helioseismology has been used to determine near-surface flows [25–28], flows beneath sunspots [29–34] and flows in their vicinity [25], to study the rotational gradient at the base of the convection zone [35], etc.

Time–distance inversions (see a review [36] and a scheme in Fig. 2) are based on a concept of a perturbed travel time $\delta\tau$,

$$\delta\tau = \tau_{\rm obs} - \tau_{\rm model}. \quad (1)$$

The observed travel time $\tau_{\rm obs}$ is measured from observations (usually from a series of intensity or Doppler-shift measurements of the photospheric lines) by means of a time-domain cross-correlation of the signal between two different points on the solar surface. The model travel time $\tau_{\rm model}$ is computed from the assumed background solar model. The background model is usually assumed to be vertically stratified with no magnetic fields and no plasma velocities, such as Model S [17].

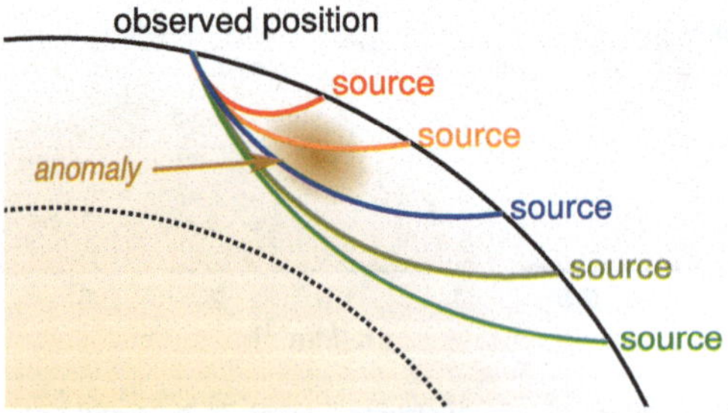


**Fig. 2** Schematic representation of the time–distance technique. The seismic waves are excited at various sources and travel through the solar interior to the observed position. Different colour indicate different modes (with different wavelengths). Some of these ray travel through an anomaly in the interior, which affects the total travel time of the affected waves. From combining the travel-time measurement of various modes one can learn about the nature of the anomaly

The actual state of the solar interior at a certain point may depart from the background model, i.e., state quantities may be perturbed or flows may be present. The usual assumption used in a vast majority of helioseismic studies is that these perturbations of the plasma state are small compared to the background model. Thus we may assume a linear perturbation and solve the linearised equation of the wave propagation.

The perturbed travel time $\delta\tau$ can theoretically be computed from the model as

$$\delta\tau(\boldsymbol{r}) = \int_{\odot} \mathrm{d}^3\boldsymbol{x}\,\boldsymbol{K}(\boldsymbol{x}) \cdot \delta\boldsymbol{q}(\boldsymbol{x}). \tag{2}$$

where $\boldsymbol{r}$ indicates the horizontal position vector and $z$ the vertical coordinate in a plane-parallel approximation of the 3-D position vector $\boldsymbol{x} = (\boldsymbol{r}, z)$. Here $\boldsymbol{K}$ is a vector of *sensitivity kernels* which represents the linear sensitivity of the travel-time measurements to the perturbations $\delta\boldsymbol{q}$ [37, 38]. The sensitivity kernels are defined by the background model and the level of approximation (ray approximation, Born approximation, etc.). The vector of perturbations may contain quantities such as flows, speed of sound, density, etc. Since the solar oscillations are excited by vigorous surface convection, in reality the measured travel time contains a large random-noise realisation component, which needs to be taken into account in further analyses.

The goal of an inverse problem is to invert Eq. (2) to determine $\delta\boldsymbol{q}$ from the measurements of $\delta\tau$. Such an inversion is usually not possible due to the presence of random noise.

In the Sun, a full spectrum of waves with different properties is excited. Individual modes propagate more-or-less independently throughout the solar interior, thus by measuring the travel times of different modes we obtain independent information about the nature of the solar interior. By filtering, it is possible to separate distinct modes of the waves and measure their travel times. Consequently, for each filtered wave mode there is a separate Eq. (2), therefore also the travel times may symbolically be written in the vector form as $\delta\boldsymbol{\tau}$. On the other hand, all these equations share the same vector of perturbers $\delta\boldsymbol{q}$. In helioseismic inversions one combines all independently measured travel times to learn about the perturbers.

Helioseismic inversions are usually performed by using two principal methods: The regularised least squares (RLS) and optimally localised averaging (OLA). The RLS method [39] seeks to find the models of the solar interior, which provide the best least-squares fit to the measured travel-time maps, while regularising the solution (e.g., by requiring the smooth solution). The OLA method was developed for geoseismology [40, 41]. Subtractive-OLA (SOLA) method [42], a form suitable for use in helioseismology, is based on explicitly constructed spatially confined averaging kernels by taking linear combination of sensitivity kernels, while simultaneously keeping the error magnification small. The resulting coefficients are then used to linearly combine the travel-time maps and obtain an estimate for structure and magnitude of solar plasma perturbations. A SOLA-type inversion is the principal method demonstrated here. The SOLA has been used in time–distance local helioseismology in the past [28, 43], where the ability of SOLA inversions to reveal the structure of 3-D internal flows was demonstrated. An efficient approach to solve fully consistent SOLA inversions was has been also introduced [44].

## 3.3 SOLA Time–Distance Inversions

### 3.3.1 The Basic Principle

Unlike the RLS approach, the SOLA methodology does not minimise the difference between the measured travel times and those modelled from the modified solar model. SOLA assumes that the inverted estimates of the quantities may be obtained using a linear combination of the travel-time maps, symbolically written for $\alpha$-th component $\delta q_\alpha$ of the vector $\delta \boldsymbol{q}$ as

$$\delta q_\alpha^{\rm inv}(\boldsymbol{r}_0; z_0) = \boldsymbol{w}(\boldsymbol{r} - \boldsymbol{r}_0; z_0) \cdot \delta \boldsymbol{\tau}(\boldsymbol{r}), \tag{3}$$

where $\boldsymbol{w}$ are the inversion weight functions. From an oversimplified point of view, SOLA performs the deconvolution of the travel-time maps.

Taking into account a noise term (the measured travel times are noisy) and using Eqs. (2) and (3), we obtain

$$\delta q_\alpha^{\rm inv}(\boldsymbol{r}_0; z_0) = \int_\odot {\rm d}^3\boldsymbol{x}' \mathcal{K}(\boldsymbol{r}' - \boldsymbol{r}_0, z') \cdot \delta \boldsymbol{q}(\boldsymbol{x}') + \text{noise}\,, \tag{4}$$

where the object $\mathcal{K}$ is referred to as the (vector) *averaging kernel*. It is defined as

$$\mathcal{K}\left(\boldsymbol{r}', z; z_0\right) = \int {\rm d}^2\boldsymbol{r}\, \boldsymbol{w}\left(\boldsymbol{r} - \boldsymbol{r}_0; z_0\right) \cdot \boldsymbol{K}\left(\boldsymbol{r}' - \boldsymbol{r}, z\right) \tag{5}$$

and quantifies the level of spatial smearing of the real quantity $\delta \boldsymbol{q}$.

In the SOLA method we search for the weights $\boldsymbol{w}$ so that the vector averaging kernel $\mathcal{K}$ resembles the target function $\mathcal{T}$. The target function is an user-selected initial estimate for the averaging kernel which indicates the desired localisation in the Sun. An example of commonly used target function is a 3-D Gaussian with a peak at a chosen target depth $z_0$.

In practise the SOLA cost function $\chi^2_{\rm SOLA}$ is minimised, which can be written in a simplified form

$$\chi^2_{\rm SOLA} = \left|\left|\mathcal{K}(\boldsymbol{x}') - \mathcal{T}(\boldsymbol{x}')\right|\right|^2 + \mu\sigma_\alpha^2 + \nu \sum_{\beta \neq \alpha} \left|\left|\mathcal{K}_\beta(\boldsymbol{x}')\right|\right|^2 + \chi^2_{\rm other}. \tag{6}$$

The first term $\left|\left|\mathcal{K}(\boldsymbol{x}') - \mathcal{T}(\boldsymbol{x}')\right|\right|^2$ of the right-hand side of Eq. (6) then evaluates the misfit between the averaging kernel and a target function $\mathcal{T}$.

The second term of Eq. (6) is the regularisation of the random-noise level in the inverted quantity,

$$\sigma_\alpha^2 = \boldsymbol{w}^{\mathrm{T}} \boldsymbol{\Lambda} \boldsymbol{w}, \tag{7}$$

where $\boldsymbol{\Lambda}$ is the travel-time noise covariance matrix. The matrix $\boldsymbol{\Lambda}$ may be obtained either from the reference model or from the travel-time measurements.

### 3.3.2 The Cross-Talk

The third term $\sum_{\beta \neq \alpha} \left|\left|\mathcal{K}_\beta(\boldsymbol{x}')\right|\right|^2$ in Eq. (6) is related to the cross-talk and regularises the non-diagonal components (that is all components except for the inverted $\alpha$-th) of the averaging kernel. The meaning of the cross-talk is illustrated by Eq. (4). There, the inverted estimate of the $\alpha$-th component is composed not only from the appropriate $\delta q_\alpha$ smoothed with the respective averaging kernel $\mathcal{K}_\alpha$, but also from the "irrelevant" perturbers $\delta q_\beta$ for $\beta \neq \alpha$ smoothed with their respective averaging kernels $\mathcal{K}_\beta$.

The physical meaning of the averaging kernel is demonstrated in Figs. 3 and 4. According to Eq. (4) the inverted quantity ($v_x^{\mathrm{inv}}$ in this case) is obtained by cross-correlation of the averaging-kernel vector and vector $\delta \boldsymbol{q}$ of the real physical quantities. For each position $\boldsymbol{x}_0 = (\boldsymbol{r}_0, z_0)$, the weighted average of real quantity $\delta q_\beta$ is computed where the weights equal to $\mathcal{K}_\beta$. The final $v_x^{\mathrm{inv}}$ is then a sum of the noise term and individual weighted averages for all $\beta$s (columns one to four in Fig. 3). In the case of $v_x$ inversion in Fig. 4 one can see that $v_x^{\mathrm{inv}}$ is mostly given by real $v_x$ averaged over around 15 Mm in horizontal and 4 Mm in vertical direction (see contours in Fig. 4). The other contributions are small because $\mathcal{K}_y$, $\mathcal{K}_s$, $v_z$, and

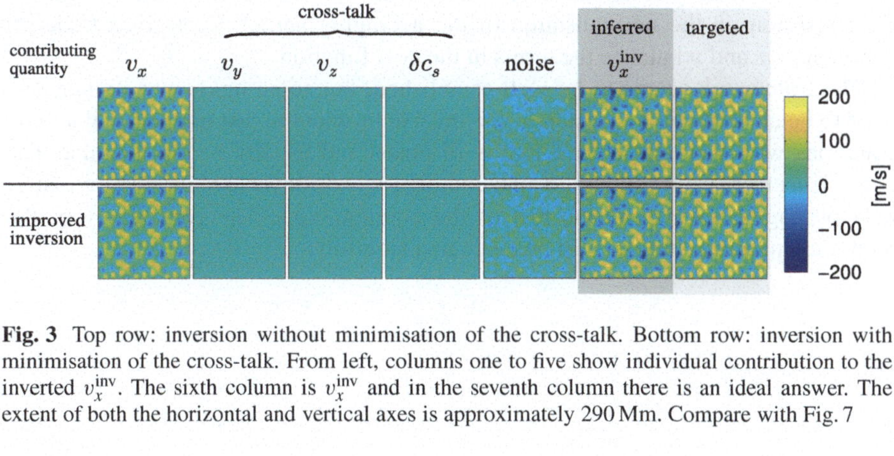


**Fig. 3** Top row: inversion without minimisation of the cross-talk. Bottom row: inversion with minimisation of the cross-talk. From left, columns one to five show individual contribution to the inverted $v_x^{\mathrm{inv}}$. The sixth column is $v_x^{\mathrm{inv}}$ and in the seventh column there is an ideal answer. The extent of both the horizontal and vertical axes is approximately 290 Mm. Compare with Fig. 7

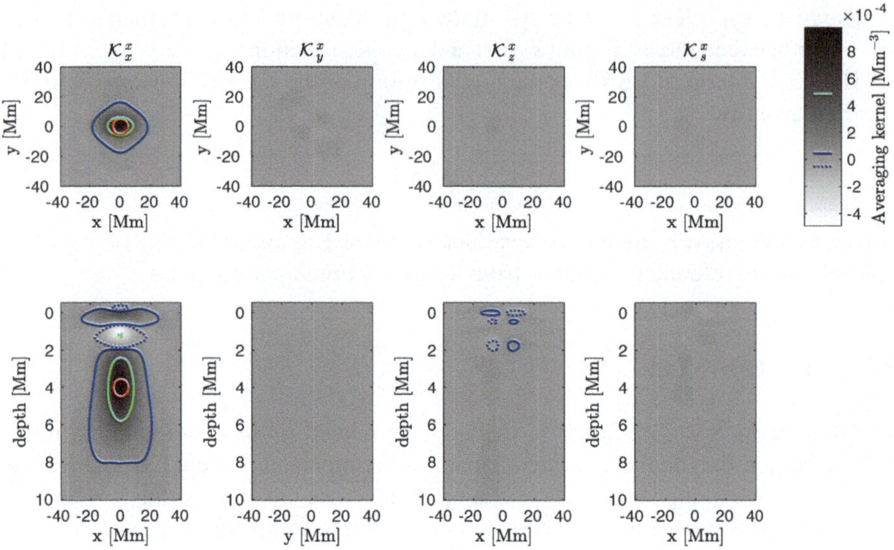


**Fig. 4** The averaging kernel for $v_x$ inversion at 4.0 Mm depth. The columns indicate the contributions from the individual quantities considered in the inversion, the terms not in the direction of the inversion indicate the leakage of the other quantities—the cross-talk. The red curve corresponds to the half-maximum of the target function at the target depth. The solid green and dotted green curves correspond to plus and minus of the half-maximum of the averaging kernel at the target depth, the blue solid and blue dotted lines correspond to +5 and −5% of the maximum of the averaging kernel at the target depth, respectively. In the top row, there are horizontal slices of the averaging kernel at the target depth and in the bottom row, there are vertical slices perpendicular to the symmetries

$\delta c_s$ are small, thus products of $\mathcal{K}_\beta\, \delta q_\beta$ are small for these off-diagonal components (cf. individual contributions in Fig. 3).

To complete the discussion of the terms in the inversion cost function we add that $\chi^2_{\mathrm{other}}$ contains additional terms such as the regularisation of the weights or the constraint on the normalisation of the averaging kernel. User-given trade-off parameters $\mu$ and $\nu$ balance the terms in the cost function.

The framework seems to be well established, but some unknowns remain that need to be cleared out. For instance, the inverse problem is cast in a set of algebraic equations with a system matrix that is ill-posed and usually nearly singular. The numerical errors from calling the mathematical pseudo-inverse of the system matrix may propagate throughout the rest of the computations. The consistency of the numerical implementation must be evaluated carefully.

### 3.3.3 Interpretation of the Inversion

The description above gives an indication that there are three different quantities that are needed to be determined in order to interpret the time–distance inversion results properly.

- *Inverted map* $\delta q_\alpha^{\mathrm{inv}}$, which gives the inverted estimates for the physical quantity $\delta q_\alpha$ at positions $\boldsymbol{r}_0$ and at target depth $z_0$.
- *Noise level* $\sigma_\alpha$ gives the root-mean-square value of the random-noise component in the inverted map. The particular noise-component realisation in the inversion is not known and is an inseparable part of the inverted map.
- Vector *averaging kernel* $\mathcal{K}$, defined by Eq. (5), gives an important information about the smearing of the inverted quantity. It also gives the real depth sensitivity (whereas the value of $z_0$ is the target depth which only serves as a parameter). In Fig. 5 one can see the depth dependence of averaging kernels of four inversions with different target depths $z_0 = (0.5, 2, 4, 6)$ Mm. A careful reader can see that the peak sensitivity of these kernels is different from their indicated target depths.

  The averaging kernel also contains information about the presence of the sidelobes. Sidelobes are places with non-negligible values of the averaging kernel which are at larger distances from the central point. Such an averaging kernel is not well localised (see an example in Fig. 6). Thus, the inverted estimate in a given point is spoiled by the properties of plasma which is far away and which possibly has different properties.

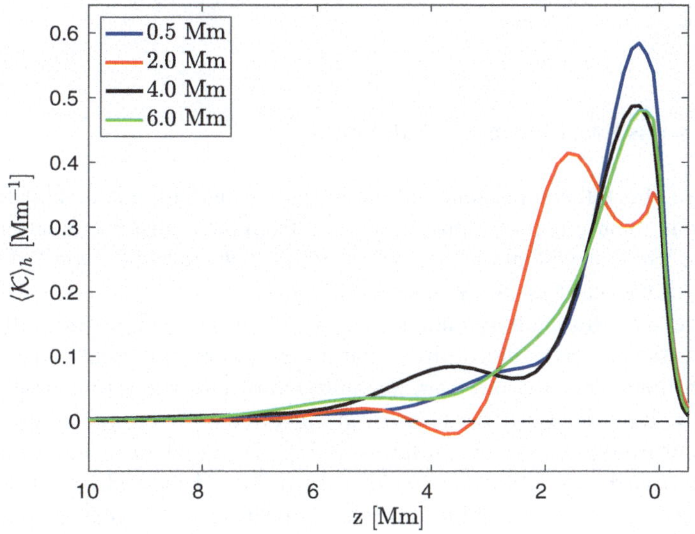


**Fig. 5** Examples of the horizontal integrals of the averaging kernels for $v_z$ inversion as a function of depth

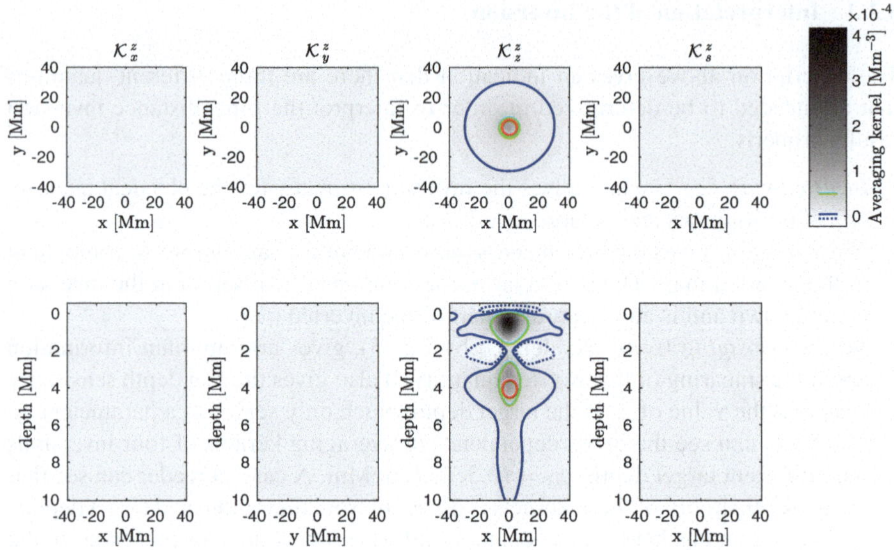


**Fig. 6** The averaging kernel for $v_z$ inversion at 4.0 Mm depth; however, the sensitivity of the inversion is mostly around 1 Mm below the surface. See Fig. 3 for the definition of contours

The off-diagonal components for $\alpha \neq \beta$ indicate the leakage of unwanted perturbations to the inverted map, the cross-talk. Ideally these cross-talk components of the averaging kernel should be negligible, so that the inverted map is not polluted by the cross-talk.

### 3.3.4 Lessons Learnt from the Validation

One way to check the consistency of the inversion modelling is a validation using known inputs. These known inputs may come from the realistic simulation of solar convection. Such a validation then gives limits that could be expected from the inversions of the real-Sun observables [45].

The inversions for the horizontal flow components ($v_x$ and $v_y$) are well-posed in the near-surface layers. The results of the inversion are in a great agreement with the known flows which come from a simulation of the convection zone. In Fig. 3 one can see individual contributions to the inverted $v_x^{\rm inv}$. From the left, first five columns correspond to individual terms in Eq. (4), the sixth column is a sum of individual contributions (left-hand side of Eq. (4)) and the right-most box shows an ideal answer of our inversion (noiseless smearing of the real quantity by the inversion target function). In the first row, minimisation of the cross-talk was not applied, while in the second row, it was applied. The differences in corresponding columns are negligible. Thus, horizontal flow is a strong perturber and possible contributions from the remaining weak perturbers (the vertical flow and the sound-speed perturbations) are negligible even when the cross-talk is not minimised.

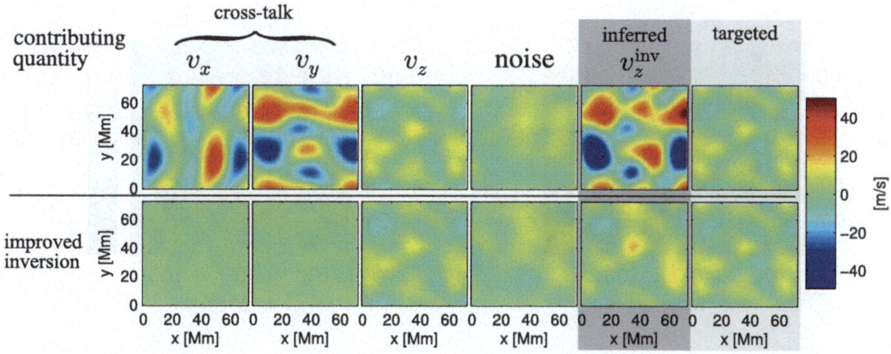


**Fig. 7** All components of an example $v_z$ inversion at 1 Mm depth. Top row with the cross-talk ignored, bottom row with the cross-talk minimised. We demonstrate that if cross-talk is not addressed, horizontal components will leak into the inverted $v_z$ and cause a bias. Obviously, inferred estimate is then in the anticorrelation with the expectations

With the travel times averaged over 24 h or so, it is possible to invert for the horizontal flow snapshot on supergranular scales roughly in the upper 5 Mm of the convection zone. An example of the averaging kernel of such an inversion is shown in Fig. 4. For deeper inversions different approaches must be taken. One should either increase the level of averaging (by increasing the extent of the averaging kernel) or invert for a statistical representation. In the latter case the ensemble averaging over a set of representatives of the same kind (e.g. set of individual supergranules) is a suitable approach [46, 47].

For the successful inversion for the vertical flow component ($v_z$) it is essential to minimise the cross-talk, otherwise the results are significantly distorted by the leakage mainly from the horizontal components of the flow; the inverted vertical flow is almost in anticorrelation with the expected output [48, 49], see Fig. 7. The leakage from the sound-speed perturbations is acceptably small. For the travel times averaged over 24 h or so it is not possible to invert for the flow snapshot deeper than around 1 Mm. In Fig. 6 there is the averaging kernel for $v_z$ inversion at 4.0 Mm depth. It is evident that the averaging kernel is localised around 1 Mm below the surface, not around targeted 4 Mm (compare to equivalent averaging kernel for the horizontal flow component in Fig. 4). The depth sensitivity of several inversions is nicely seen in Fig. 5, where horizontal integrals of averaging kernels for $v_z$ inversions in different depths are plotted. Even though the target depth is changing from 0.5 to 6.0 Mm, the depth sensitivity of the inversion is nearly the same. Attempts to perform deeper inversions lead either to a noise level much larger that the signal, or to a strong departure from the target function (as in this case) [50]. Attempt to invert for the snapshot of the vertical flow on supergranular scales with a signal-to-noise ratio larger than one naturally converges towards the shallow "surface" inversion.

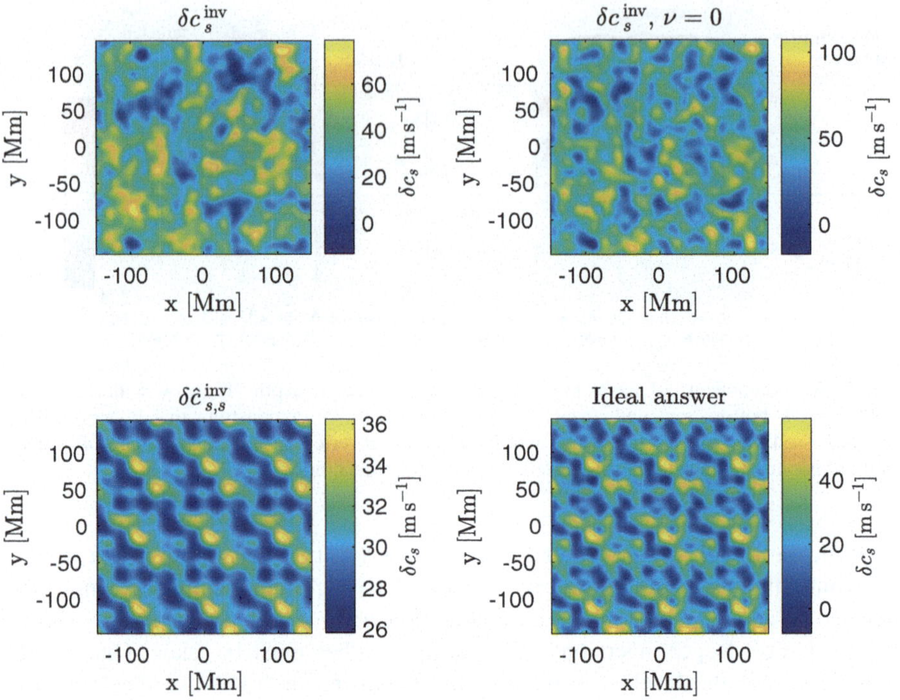


**Fig. 8** Top row: $\delta c_s^{\rm inv}$ with minimisation of the cross-talk and $\delta c_s^{\rm inv}$ without minimisation of the cross-talk. Bottom row: contribution in the direction of inversion and the ideal answer

The realistic perturbation of the speed of sound is also a weak perturber. Thus the cross-talk minimisation is essential. Unlike in the case of the vertical flow, in the case of the sound-speed perturbations the cross-talk is positively correlated. Thus the cross-talk "helps" to "measure" $\delta c_s$. Such an approach leads to an overestimation of the magnitude of the sound-speed variations by almost a factor of two in the near-surface layers. Such an inversion, however, is not the inversion for $\delta c_s$. An example of inverted sound-speed perturbation in near-surface layers is plotted in Fig. 8. The first row shows the inverted maps with and without cross-talk minimisation (similar to the fourth column in Fig. 3). In the second row there are, from left to right, individual contribution to inverted result with the cross-talk minimisation (similar to the fourth column in the bottom row in Fig. 3) and the ideal answer.

By using the local helioseismology, one can learn about the dynamics of the solar subsurface. The knowledge of this dynamics is important for assessment of the evolution of the magnetic field. Unfortunately, helioseismic inversions do not seem to help in constraining the subsurface structure of the magnetic field in the active regions due to a very strong interaction of various kinds of waves in the magnetised plasma [34].

# 4 Information About the Higher Atmosphere

The coupling between the solar sub-surface and the structures of the magnetised atmosphere is striking [51]. For instance, a particular flow topology may contribute in triggering the solar flares. In a sudden eruption of a quiet-Sun filament followed by a solar flare, the zonal shear flow possibly contributed to triggering the filament eruption and the consequent flare ignition [52, 53].

The upper atmospheric layers have a very low density and it is difficult to speak about them in terms of the spherical shells. The chromosphere is a warped and highly dynamic surface that is thin around kilogauss concentrations of the magnetic field and thick in regions where acoustic shocks appear in otherwise cool internetwork gas. Higher up the chromospheric fibrils are structured by the magnetic field that extend from the network and from plage regions. The chromosphere is very dynamic, where we find evidence for magnetic reconnections, Alfvén waves, magnetically guided converted acoustic waves and many more which form the chromospheric structures. The chromosphere loses much of the energy by radiation in strong resonance lines of hydrogen, helium, calcium and magnesium.

These higher layers do not contribute much in the visible range to the total solar irradiance, because the photosphere overshines them. However, locally there are phenomena, which may outshine even the background photosphere.

## *4.1 White-Light Flares*

Solar flares are widely believed to be a consequence of reconnection of the coronal magnetic field in a peculiar configuration. The magnetic energy stored in the entangled coronal loops is released suddenly during the flare and a large portion of the flare energy is radiated away in a wide range of wavelengths emerging from the intensively heated flare atmosphere.

In the visible range of wavelengths the usual line emission is often accompanied by enhancement of continuum radiation and such flares are called *white-light flares* (WLF, [54, 55]). There are various mechanisms proposed for enhancement of the optical continuum: hydrogen bound-free and free-free transitions, Thomson scattering, and $H^-$ emission. Furthermore, each mechanism may dominate in different atmospheric layers spanning from the photosphere through the temperature minimum region to the chromosphere, and all require an increase in temperature and electron density in those layers.

However, it is still debated how these layers are heated (see Fig. 9). Several processes have been proposed: electron and/or proton bombardment, XEUV heating, Alfvén wave dissipation, etc. [56]. Moreover, it has been shown that the photosphere and the chromosphere can be radiatively coupled via photospheric heating by $H^-$ absorption of the hydrogen Balmer continuum, which originates in the chromosphere. This backwarming then can lead to increased photospheric ($H^-$) radiation [57].

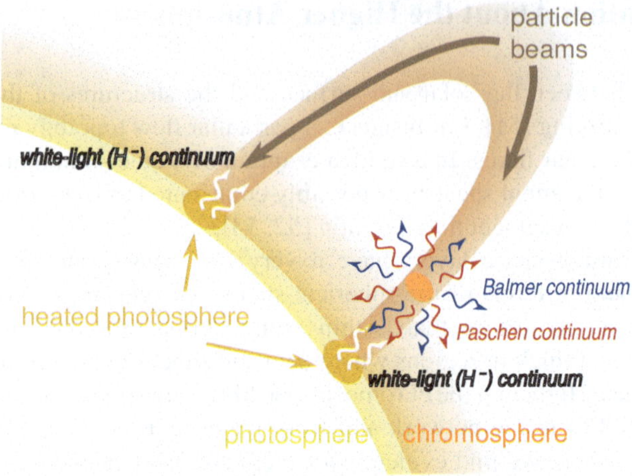


**Fig. 9** Schematic model of two possible origins of the white-light flare. In the left leg of the coronal loop the model of heating of the photosphere by particle beams is considered. In the right leg the schematic representation of the backwarming model is considered

To disentangle the contributions to the visible-light continuum, specific observations and dedicated models are needed. Combining flare observations of several photospheric and chromospheric lines together with visible-light continuum and non-LTE modelling, i.e. allowing departures from local thermodynamical equilibrium (LTE), a semi-empirical model of a WLF was constructed [58]. Recently, using off-limb flare observations in the *SDO*/HMI pseudo-continuum and non-LTE RHD approach [59], the presence of the hydrogen Paschen continuum originating in the chromosphere was reported. Other HMI off-limb sources related to a flare were found to be of two kinds (chromospheric and coronal), and interpreted as free-bound continuum (and possible line emission) and Thomson scattering, respectively [60, 61]. Furthermore, the hydrogen Balmer continuum was observed during flares by *IRIS* [62–64]. Additionally, near-infrared emission at 1.56$\mu$m was detected during several X-class WLFs [65, 66]. In an undisturbed solar atmosphere this emission is considered to originate at the opacity minimum located below $\tau_{500} = 1$.

## 4.2 Spectral Lines and Their Inversion

It is the goal of solar spectroscopy to determine reliably various physical parameters in the solar atmosphere. Spectral profiles of polarized light, i.e., the four Stokes profiles $I$, $Q$, $U$, and $V$, contain a lot of information about the conditions in the line-forming region. In the solar photosphere, the shape of polarized light profiles of magnetically sensitive lines is typically governed by Zeeman effect. Thus observations of such lines allow us to determine also the properties of the magnetic field, a physical parameter that plays an important role in many processes of solar activity.

To obtain estimates of the physical parameters in the solar photosphere, one of the fundamental steps is to solve the radiative transfer equation (RTE) for polarized light

$$\frac{\mathrm{d}\mathbf{I}}{\mathrm{d}\tau} = \mathbf{K}(\mathbf{I} - \mathbf{S}), \tag{8}$$

where $\mathbf{I} = [I, Q, U, V]^T$ is the Stokes vector of polarized light, $\mathbf{S}$ is the source function, and $\mathbf{K}$ is the so-called propagation matrix. In the solar photosphere, which is sufficiently dense and collisional rates are high enough, we can assume LTE. This implies that the source function is not a source of polarized light and is used in a form $\mathbf{S} = [B_\nu(T), 0, 0, 0]^T$, where $B_\nu$ represents the Planck function. Propagation matrix $\mathbf{K}$ contains absorption, dichroism, and dispersion coefficients that are dependent on the physical parameters of the atmosphere.

The formal solution of Eq. (8) can be written in the form

$$\mathbf{I}(0) = \int_0^\infty \mathrm{d}\tau\, \mathbf{O}(0, \tau)\mathbf{K}(\tau)\mathbf{S}(\tau), \tag{9}$$

where $\mathbf{I}(0)$ is the outgoing intensity of polarized light, $\tau$ is the optical depth, and $\mathbf{O}$ is the evolution operator. It can be shown that the evolution operator must fulfill the following equation [67]:

$$\frac{\mathrm{d}\mathbf{O}(\tau, \tau')}{\mathrm{d}\tau} = \mathbf{K}(\tau)\mathbf{O}(\tau, \tau'). \tag{10}$$

Equation (10) is the reason why there is no general analytical solution of RTE for polarized light. Solution in the form of attenuation exponential is valid only in specific cases because the matrices do not commute in general. The specific case is called Milne-Eddington (ME) atmosphere and it fulfills two necessary conditions. First, the propagation matrix $\mathbf{K}$ is independent on optical depth, i.e., the physical parameters defining the matrix elements are constant with optical depth. Second, the source function depends linearly on optical depth. As the temperature determines the shape of the source function via the Planck function, the assumption of the linear dependence of source function on the optical depth determines the temperature stratification, which is not realistic. If more realistic and thus more complicated models of atmosphere are used, the RTE is solved numerically.

For any given model atmosphere the RTE (8) is solved and the resulting intensities of polarized light are hereafter denoted as $\mathbf{I}^{\mathrm{syn}}$. The goal is to estimate a model atmosphere that produces $\mathbf{I}^{\mathrm{syn}}$ as similar as possible to the observed intensities of polarized light $\mathbf{I}^{\mathrm{obs}}$, i.e., to minimize the cost function

$$\chi^2 = \frac{1}{\nu} \sum_{k=1}^{4} \sum_{i=1}^{M} \left[ I_k^{\mathrm{obs}}(\lambda_i) - I_k^{\mathrm{syn}}(\lambda_i) \right]^2 , \tag{11}$$

where the internal sum is performed over $M$ wavelength samples indicated by index $i$, the external sum is performed over individual Stokes profiles, and the value of $\nu$ corresponds to the number of degrees of freedom, that is the difference between the number of observables and the number of free parameters of the model.

There are two principal methods to minimise $\chi^2$. First, compute $\mathbf{I}^{\mathrm{syn}}$ from a huge grid of atmospheric models and search for the best match with the $\mathbf{I}^{\mathrm{obs}}$ within the grid. Examples how to do it quickly using the method of the principal component analysis were published [68, 69]. Second, use a model atmosphere and alter the physical parameters of the model until the $\chi^2$ is minimized.

The second method is widely used for spectral lines formed in the solar photosphere where LTE conditions apply. There the inversion can be computed in a reasonable time even if RTE (8) is solved numerically. Apart from the LTE assumption, the inversion codes also assume that the model atmosphere is in a hydrostatic equilibrium. Although this assumption is only rarely fulfilled in solar atmosphere, it does not influence directly the resulting values of free parameters of the inversion (magnetic field strength, inclination and azimuth, line-of-sight velocity). The assumption of hydrostatic equilibrium allows us to compute the pressure from the temperature stratification directly. The pressure is further used to derive parameters like density and electron pressure that is a necessary parameter for the line synthesis in the SIR code (Stokes Inversion based on Response functions). These parameters influence the opacity of the atmosphere and thus the height scale of the obtained model atmosphere, but the Stokes profiles of photospheric lines are typically not sensitive to these physical parameters.

For huge datasets and automatic data reduction, the inversion methods usually assume Milne-Eddington atmosphere, the synthesis of Stokes profiles is computed analytically, and therefore these codes are very fast. Spectropolarimetric observations from satellites are automatically inverted using ME codes, e.g., VFISV code designed for SDO/HMI data [70] or MERLIN code[1] used to *Hinode*/SOT data. The disadvantage of ME codes is a very simplified model atmosphere that cannot produce asymmetric Stokes profiles. However, with increasing spatial resolution of the telescopes, the asymmetries of the Stokes profiles are commonly observed and more complicated atmospheric models have to be assumed to explain them.

A very powerful method of spectral-line inversion was introduced in [71]. This inversion code SIR allows for realistic temperature stratification and also the magnetic field vector and the macroscopic line-of-sight velocity can change with height in the atmosphere. The methodology is very flexible so it allows to

[1] https://www2.hao.ucar.edu/csac/csac-data/sp-data-description.

invert many spectral-line profiles at once, thereby improving the resulting inverted estimates of the physical parameters.

The method is based on the concept of *response functions*, which come from the linearisation of the RTE (8). The response function $\boldsymbol{R}(\lambda, \tau)$, where $\lambda$ is the wavelength and $\tau$ is the optical depth, describes a linear relation between the atmospheric parameters and the emergent intensity.

$$\delta \boldsymbol{I}(\lambda) = \int_0^\infty \boldsymbol{R}(\lambda, \tau) \delta x(\tau) \mathrm{d}\tau, \tag{12}$$

where $\delta \boldsymbol{I}$ represents the modification of the emergent Stokes spectrum $\boldsymbol{I}$ which is caused by the perturbation $\delta x$ in a single physical parameter for simplicity. Vector $\boldsymbol{R}$ is a vector of response functions corresponding to the Stokes-vector components and tells us how the observed spectrum responds to the modifications of the physical conditions in the model. Response functions behave the same way as partial derivatives of the spectrum with respect to the physical quantities. Within linear approximation, response functions give the sensitivities of the emerging Stokes profiles to perturbations of plasma parameters in the given model atmosphere. Example of response function of the Stokes $I$ profile to the temperature perturbation in a model atmosphere corresponding to the sunspot umbra is shown in Fig. 10. Here one can see the sensitivity of the pair of iron lines to temperature at different atmospheric levels.

In an iterative process the atmosphere is modified in such a way that the cost function (11) is minimized. In a linear approximation the cost function of the altered atmospheric model $\chi^2(\mathbf{x} + \delta\mathbf{x})$ can be approximated by the Taylor series of the cost function of the original atmospheric model $\chi^2(\mathbf{x})$

$$\chi^2(\mathbf{x} + \delta\mathbf{x}) \simeq \chi^2(\mathbf{x}) + \delta\mathbf{x}^T(\nabla\chi^2 + \mathbf{H}'\delta\mathbf{x}), \tag{13}$$

where $\nabla\chi^2$ are partial derivatives of the cost function and can be derived from the response functions, $\mathbf{H}'$ is so-called curvature matrix containing the second partial derivatives of $\chi^2$, i.e., $H'_{ij} = 1/2\ \partial^2\chi^2/\partial x_i \partial x_j$. The elements of the curvature matrix are non-trivial functionals of the response functions.

The inversion code SIR [71] uses the Marquardt algorithm to minimize the cost function. Following Eq. (13), the goal is to solve the equation

$$\nabla\chi^2 + \mathbf{H}'\delta\mathbf{x} = 0, \tag{14}$$

i.e., the model atmosphere condensed in the vector $\mathbf{x}$ is altered in a way that its modification $\delta\mathbf{x}$ fulfills the equation

$$\delta\mathbf{x} = -\mathbf{H}'^{-1}\nabla\chi^2. \tag{15}$$

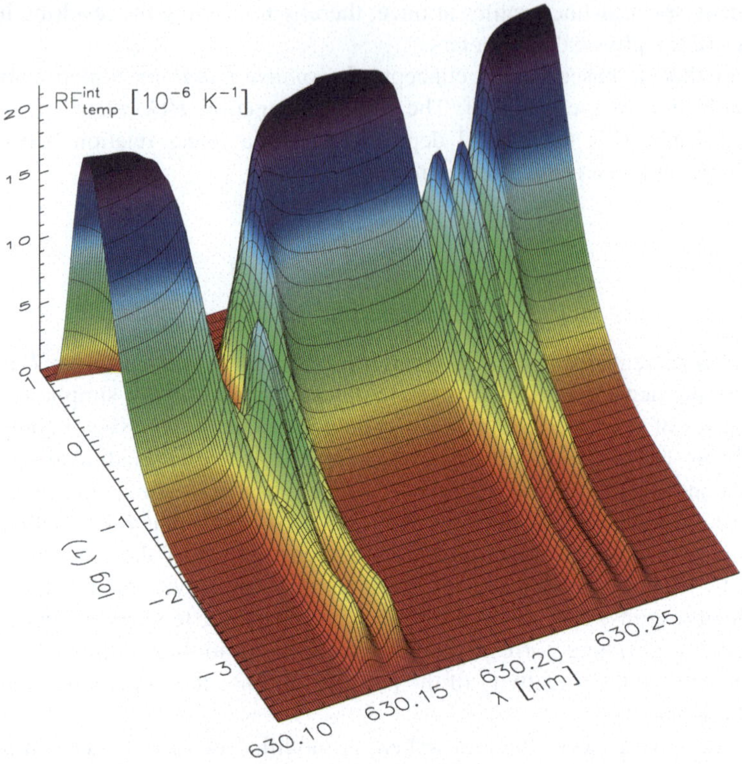


**Fig. 10** Response function of the Stokes *I* profile to the temperature perturbations for the pair of neutral Fe lines around 630 nm. It illustrates that the continuum intensity is determined by temperature around $\log \tau = 0$ and the line wings and cores are more sensitive to temperature changes at higher atmospheric layers

Note that the change of the atmospheric model is further controlled by an additional parameter in the code, with its value based on the evolution of the cost function during the iterative process of the spectral line fitting. In each iteration the response functions need to be calculated again to ensure that the final solution in the last step fits well to the linear approximation from the previous step. The problem of radiative transfer of polarised light and the spectral line inversions is described in detail in [72].

The inversion code SIR is widely used by the community but there are other inversion codes that allow for stratification of plasma parameters with height in the atmosphere and are also based on linearisation of the RTE (8), see e.g. [73, 74].

### 4.3 Atmosphere in the Flare

The SIR code [71] is designed to invert spectral lines for situations where non-LTE effects are not important, thus its use is usually limited to the investigation of the structure of the magnetised photosphere. However, there are case studies, including the example below [75], where some information about the upper atmospheric levels may also be inferred. In the following we describe a special application of SIR code to spectropolarimetric measurements of photospheric lines obtained during a solar flare.

The Solar Optical Telescope (SOT, [76]) aboard the *Hinode* satellite [77] has provided observations in broad-band filters as well as spectropolarimetry of a pair of photospheric Fe I lines. Its continuum broad-band filters have been used to detect and analyse WLF emission [78]. Here, we focus on spectropolarimetric data. The spectropolarimeter attached to the SOT has been measuring the Stokes profiles of the Fe I 630.15 and 630.25 nm lines since 2006. Analyses of these data have given us great insight into the structure of the magnetic field in the solar photosphere, both in quiet-Sun and active regions. In all cases, these Fe I lines were observed in absorption on the solar disk. This is caused by a temperature decrease with height, and thus a decrease in the source function in the layers of the solar photosphere where these lines are formed. These lines were detected in emission only in the case of observations at the extreme solar limb [79].

In September 2017 the raster scan of *Hinode* spectropolarimeter (SP) crossed a WLF ribbon and captured a unique set of emission profiles that allow us to study in detail the response of the solar photosphere to an X9.3-class flare (SOL2017-09-06T11:53, see also [80]). Continuum intensity maps of *Hinode* rasters are shown in Fig. 11 and the blue contour indicates a region where the intensity enhancement is caused by the WLF ribbon and not by the evolution of the sunspot fine structure.

The above mentioned code SIR was used to determine the physical properties of the solar photosphere. To account for the complex emission profiles observed in the flare ribbon, the temperature was allowed to change at several (five) optical depths $\tau$.

As shown in Fig. 12, the inversion code explains the flare intensity difference by enhanced temperatures at all optical depths, i.e. predicts also the heating of the solar photosphere around $\log \tau = 0$, where the photospheric continuum forms. For the pixel shown in Fig. 12, the continuum enhancement of $0.3 I^c_{QS}$ (continuum intensity of the surrounding quiet Sun) is achieved by an increase in temperature at $\log \tau = 0$ by 840 K compared to the post-WLF phase.

Such an increase in temperature at the deepest photospheric layers is unlikely for several reasons. First, the minimum of temperature stratification in any of the investigated pixels is not below $\log \tau = -0.5$. Second, the observed emission profiles are never in pure emission. There always is a slight decrease in intensity in the far wings of the Fe I 630.15 and 630.25 nm lines that indicates the temperature decrease above $\log \tau = 0$. Furthermore, if the temperature increase below the solar surface is real, the heating will have to be of a specific type, e.g. increase with depth because the atmospheric density significantly rises at those layers.

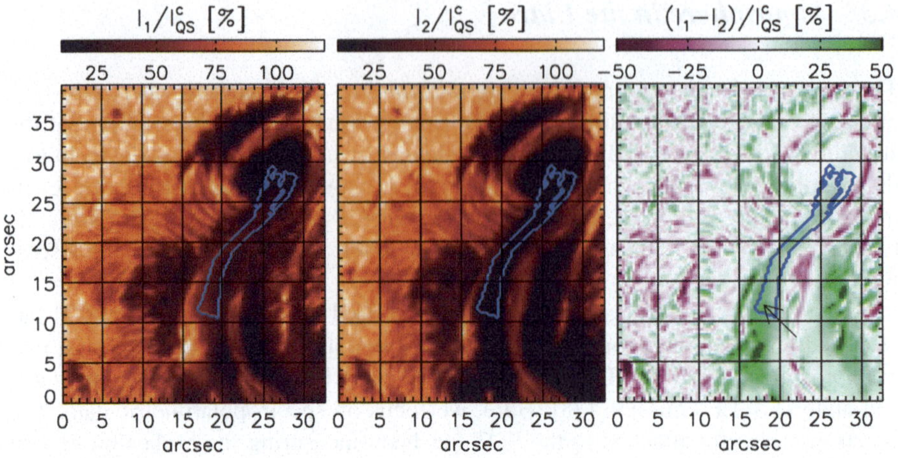


**Fig. 11** Continuum intensity maps reconstructed from the two *Hinode* raster scans. The left map was scanned between 11:57 UT and 12:04 UT and captured a WLF ribbon; the middle map was scanned between 12:19 UT and 12:42 UT. On the right is the intensity difference between these two scans. The blue contour indicates the region where we ascribe the intensity difference to the WLF. The arrow points to a pixel where we observed the Stokes profiles displayed in Figs. 12 and 14. The labels $I_1$ and $I_2$ correspond to the continuum intensities observed during the first and second *Hinode*/SP scan, respectively

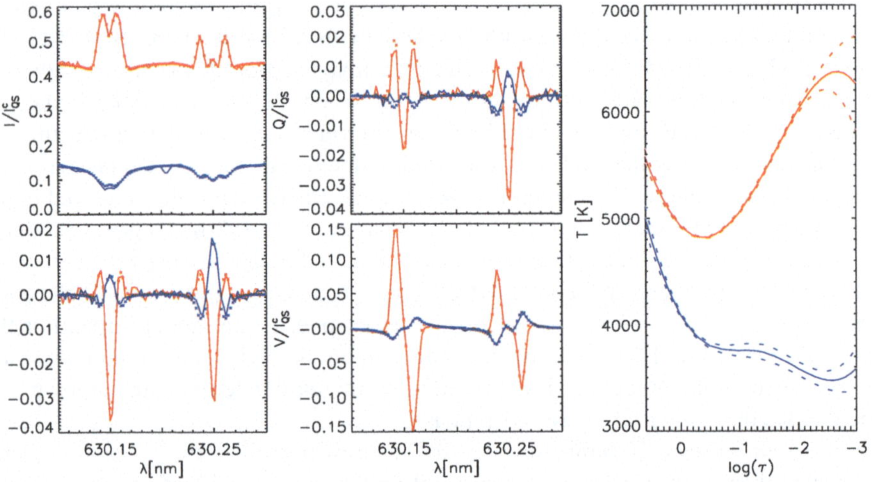


**Fig. 12** Left, small plots: Comparison of the Stokes profiles observed during the WLF (red lines) and in the post-WLF phase (blue lines); the asterisk symbols in the respective colours indicate the best fit of these profiles achieved with the inversion code. Right: Temperature stratifications obtained by the inversion code for the flare (red) and post-WLF (blue) phase; the dashed lines show the error margin determined by the inversion code SIR

The assumption that the flare atmosphere is probably not altered significantly around $\log \tau = 0$ is further supported by the semiempirical flare models F1, F2, F3, F1* [57, 81] and the FLA and FLB models for WLF [58]. These coincide with the quiet-Sun model C of [82] at heights $z < 0$ km, i.e. at $\log \tau > 0$ [57, Fig. 1]. Also, models of continuum emission in a sunspot atmosphere heated by non-thermal electron beams [83] indicate that the temperature below $z = 0$ km increases not more than by 100 K for one of the strongest beam heating they used. Little or no photospheric heating was observed also in other models [84, 85].

Moreover, the observations of off-limb visible continuum sources during a M7.7 and a M1.7 flare were recently analysed [59]. Using analytical formulae and detailed radiation-hydrodynamical simulations, the authors conclude that the dominant source of the off-limb visible continuum radiation is the Paschen recombination continuum and there are also smaller contributions by Thomson scattering and by the hydrogen free-free emission, see Fig. 13. Furthermore, both HMI observations

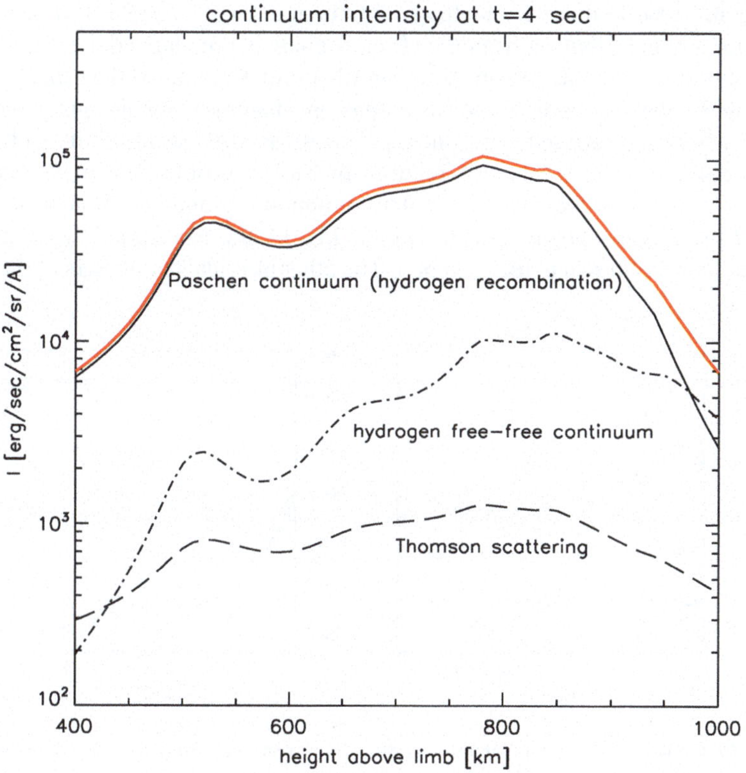


**Fig. 13** Vertical variations of the line-of-sight contributions to the continuum intensity from FLARIX radiation-hydrodynamical simulation at time $t = 4$ s. The total continuum emission (red line) is dominated by the Paschen continuum (full black line) with minor contributions from the hydrogen free-free (dotted-dashed line) and Thomson scattering (dashed line) components. See also [59]. Courtesy of P. Heinzel

and numerical simulations show, see Fig. 1 in [59] and Fig. 13, respectively, that the continuum radiation from these sources is located in the solar chromosphere, i.e. well above the line-forming region of the Fe I 630.15 and 630.25 nm lines. In the studied cases of M-class flares, the off-limb continuum intensity was around $0.1 I^c_{QS}$.

Naturally, the inversion code SIR cannot account for such potential sources of chromospheric continuum emission as it is applied to a photospheric line. Instead, SIR compensates for the continuum rise by increasing the temperature around the photospheric continuum-formation layer. To investigate whether such a contribution of continuum intensity from the solar chromosphere is realistic, a set of inversions was performed on the Stokes profiles observed in the region encircled by the blue contour in Fig. 11. In order to mimic such a chromospheric optically-thin contribution, the Stokes $I$ intensity was artificially decreased by a flat continuum, where for each pixel the observed WLF continuum was decreased to the continuum intensity observed in the post-WLF phase (the step of the decrease was $0.01 I^c_{QS}$). This fine step of decreasing the $I^c_{QS}$ was used to find the best match of temperature stratifications for the flare and post-WLF phases at each pixel, and does not necessarily mean a matching continuum intensity. Also, it is assumed that the continuum contribution from the chromosphere is unpolarised, i.e. the Paschen continuum; therefore, the Stokes $Q$, $U$, and $V$ profiles remained the same.

In Fig. 14 the results of these inversions are displayed for the pixel shown in Fig. 12, where the red and blue lines are identical in these plots. It is clear that the inversion code SIR can also reliably fit Stokes profiles when the Stokes $I$ profile is artificially decreased by a flat continuum. Examples of such decreased Stokes $I$ profiles are shown in the left plot in Fig. 14; the corresponding temperature stratifications are shown in the right plot. The other physical parameters of the model

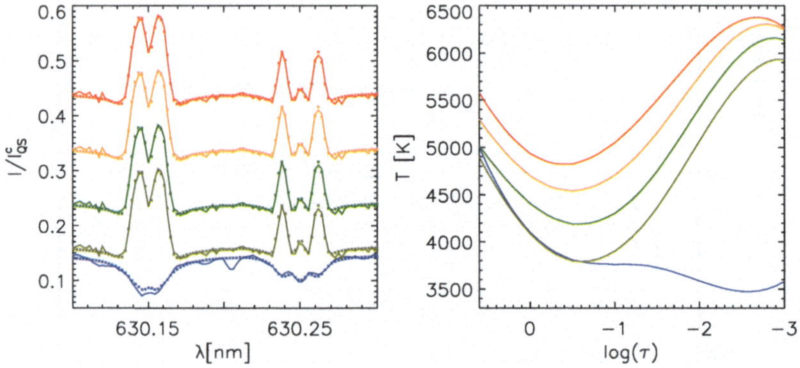


**Fig. 14** Comparison of the temperature stratifications resulting from the inversion of Stokes profiles, where the Stokes $I$ profile was reduced by a flat continuum of $0 I^c_{QS}$ (red), $0.1 I^c_{QS}$ (orange), $0.2 I^c_{QS}$ (green), and $0.28 I^c_{QS}$ (olive). The blue Stokes $I$ profile corresponds to the post-WLF phase at the same location and the temperature stratification indicated by the blue line corresponds to this profile. Left: Lines correspond to the observed profiles, and the asterisk symbols to the best fits obtained by the inversion code SIR

atmosphere (such as magnetic field structure) are not affected significantly by the modification of the Stokes $I$ profile.

Next, the best match is sought between model temperatures obtained from the inversions of the reduced Stokes $I$ profiles and the post-WLF temperature stratification in the same pixel in the interval of $\log\tau = [0.3, -0.3]$. For the case of the pixel shown in Fig. 14, the best match is displayed by olive lines, and the temperature stratifications match within their uncertainties in the range of optical depths from $\log\tau = 0.7$ to $-0.7$. This is the range of optical depths where most of the photospheric continuum forms.

As implied before, the reduction of the Stokes $I$ profile by a flat continuum is optimised to achieve the match of the temperature value at $\log\tau = 0$ for the flare and post-WLF phases. This does not necessarily mean that the continuum intensity would be the same for the reduced Stokes $I$ profile and the profile observed during the post-WLF phase. In the case of the pixel shown in Fig. 14, a discrepancy of $0.02 I^{c}_{QS}$ is found between the continuum intensity of the olive-coloured emission profile and the blue post-WLF profile. This is caused by a small contribution to the continuum intensity from the upper layer of the solar photosphere that is heated during the flare phase.

This is clearly seen in Fig. 15, where the response functions of the two compared Stokes $I$ profiles to the changes in their respective temperature stratifications are shown. The difference between these response functions (shown in the right panel of Fig. 15) indicates that in the case of the emission profile, the photospheric continuum also forms in higher photospheric layers, mostly between $\log\tau = -1$ and $-2$.

The spectral-line inversions of the photospheric line thus allowed us to infer some information about the higher levels of the atmosphere. First of all, it is clear that the

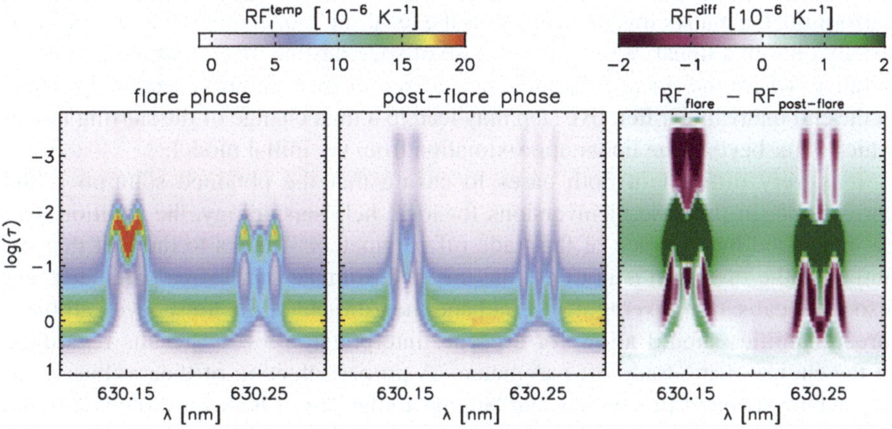


**Fig. 15** Response functions of the Stokes $I$ profile to the temperature for the model atmosphere fitting the emission profile shown in Fig. 14 (left panel, in olive green) and for the model atmosphere fitting the post-WLF profile in Fig. 14 (middle panel, in blue). Their difference is shown in the right panel

flare modifies the atmosphere significantly. Second, from the presented modelling of the *photosphere* it turned out that there very likely is a layer in the *chromosphere* or above, where a spectral continuum contribution forms during the flare event.

## 5 Summary

In the two examples we demonstrated that the visible-range electromagnetic radiation originating in the thin layer of the solar photosphere actually may contain also information about the adjacent layers of both the solar interior and the solar atmosphere. This information must however be sought out by using sophisticated methods. In the two examples we showed the main methodology was that of the mathematical inversion.

The framework of the two is very similar. Both methods rely on existence of functions having a meaning of Fréchet derivatives of the functional with respect to the free parameters of the model. In the case of local-helioseismology inversions these functions are termed sensitivity kernels, in the case of spectral-line inversion they are called response functions. Their meaning is very much the same. They represent a linear response of the perturbation in the free parameters of the model (flow and speed of sound in case of local helioseismology, atmospheric stratification in the case of spectral-line inversions) in the observables (travel times of the waves in the case of local helioseismology, the Stokes vector in the case of the spectral-line inversion).

Both problems end up with construction of the cost function which is minimised with respect to the free parameters of the model. The exact mathematical method how this is solved is different. Whereas the local helioseismology relies on performing the matrix inversion and thus the resulting model modifications are valid strictly only in a linear regime, in the case of spectral-line inversions the process is iterative, where the linear relation is considered in each iteration separately. Thus, in the end many linear iterative step may lead to a final change of the starting model which is far beyond the linear approximation from the initial model.

It is very difficult in both cases to ensure that the obtained solution is the correct one. In the case of inversions for local helioseismology, the solution may be influenced by a choice of the trade-off parameters. Thanks to the fact that the output of the inversion is not only the inverted estimate of the free parameter, but also the localisation (averaging) kernel and the estimate of the noise level, all these three quantities should allow for a proper interpretation of the results regardless of the choice of the trade-off parameters. A proper selection of their values is far beyond the scope of this review and belongs to the “art of helioseismology”. In the case of the spectral-line inversion, the goodness of the fit is described purely by the $\chi^2$ value, from which the uncertainties of the estimates of the free parameters may be evaluated. It is not possible to ensure that the iterative process reaches the global minimum. It is thus important to select a proper starting atmospheric model which helps the convergence process. A selection of the best starting model is very peculiar and might also be termed with a word “art” and is influence also by the experience of the user knowing the code and its behaviour.

**Acknowledgements** The authors were supported by the Czech Science Foundation under grants 18-06319S (M.Š., J.J. and D.K.) and 19-09489S (J.K.). J.J., J.K., and M.Š. acknowledge the support from the project RVO:67985815. D.K. is supported by the Grant Agency of Charles University under grant No. 532217. We thank Juraj Lörinčík and Petr Heinzel for useful comments on an early draft of the manuscript and the referee for constructive comments.

## References

1. O. Engvold, J.C. Vial, A. Skumanich, *The Sun as a Guide to Stellar Physics* (Elsevier, Amsterdam, 2019). https://doi.org/10.1016/C2017-0-01365-4
2. P. Charbonneau, Living Rev. Sol. Phys. **7**(1), 3 (2010). https://doi.org/10.12942/lrsp-2010-3
3. I.G. Usoskin, Living Rev. Sol. Phys. **14**(1), 3 (2017). https://doi.org/10.1007/s41116-017-0006-9
4. Y. Fan, Living Rev. Sol. Phys. **6**(1), 4 (2009). https://doi.org/10.12942/lrsp-2009-4
5. C.J. Schrijver, Space Weather **13**(9), 524 (2015). https://doi.org/10.1002/2015SW001252
6. A.I. Shapiro, H. Peter, S.K. Solanki, *Chapter 3 – The Sun's Atmosphere* (Elsevier, Amsterdam, 2019), pp. 59–85. https://doi.org/10.1016/B978-0-12-814334-6.00003-0
7. J.E. Vernazza, E.H. Avrett, R. Loeser, Astrophys. J. **30**, 1 (1976). https://doi.org/10.1086/190356
8. G. Cauzzi, A. Asensio Ramos, K. Reardon, K. Janssen, *The Dynamic Sun: Challenges for Theory and Observations*, vol. 600 (ESA Special Publication, Paris, 2005), p. 12.1
9. Å. Nordlund, R.F. Stein, M. Asplund, Living Rev. Sol. Phys. **6**(1), 2 (2009). https://doi.org/10.12942/lrsp-2009-2
10. H. Hotta, H. Iijima, K. Kusano, Sci. Adv. **5**(1), 2307 (2019)
11. J. Sánchez Almeida, J.A. Bonet, B. Viticchié, D. Del Moro, Astrophys. J. **715**(1), L26 (2010). https://doi.org/10.1088/2041-8205/715/1/L26
12. R.F. Stein, Living Rev. Sol. Phys. **9**(1), 4 (2012). https://doi.org/10.12942/lrsp-2012-4
13. T.M.D. Pereira, M. Asplund, R. Collet, I. Thaler, R. Trampedach, J. Leenaarts, Astron. Astrophys. **554**, A118 (2013). https://doi.org/10.1051/0004-6361/201321227
14. M. Rempel, M. Schüssler, R.H. Cameron, M. Knölker, Science **325**(5937), 171 (2009). https://doi.org/10.1126/science.1173798
15. M.C.M. Cheung, M. Rempel, A.M. Title, M. Schüssler, Astron. Astrophys. **720**(1), 233 (2010). https://doi.org/10.1088/0004-637X/720/1/233
16. M. Stix, *The Sun: An Introduction* (Springer, Berlin, 2004)
17. J. Christensen-Dalsgaard, W. Dappen, S.V. Ajukov, E.R. Anderson, H.M. Antia, S. Basu, V.A. Baturin, G. Berthomieu, B. Chaboyer, S.M. Chitre, A.N. Cox, P. Demarque, J. Donatowicz, W.A. Dziembowski, M. Gabriel, D.O. Gough, D.B. Guenther, J.A. Guzik, J.W. Harvey, F. Hill, G. Houdek, C.A. Iglesias, A.G. Kosovichev, J.W. Leibacher, P. Morel, C.R. Proffitt, J. Provost, J. Reiter, E.J. Rhodes, Jr., F.J. Rogers, I.W. Roxburgh, M.J. Thompson, R.K. Ulrich, Science **272**, 1286 (1996). https://doi.org/10.1126/science.272.5266.1286
18. S. Basu, H.M. Antia, Mon. Not. R. Astron. Soc. **287**, 189 (1997). https://doi.org/10.1093/mnras/287.1.189
19. J.N. Bahcall, A.M. Serenelli, M. Pinsonneault, Astrophys. J. **614**, 464 (2004). https://doi.org/10.1086/423027
20. A. Vögler, S. Shelyag, M. Schüssler, F. Cattaneo, T. Emonet, T. Linde, Astron. Astrophys. **429**, 335 (2005). https://doi.org/10.1051/0004-6361:20041507

21. D. Benson, R. Stein, Å. Nordlund, in *Solar MHD Theory and Observations: A High Spatial Resolution Perspective*, ed. by J. Leibacher, R.F. Stein, H. Uitenbroek. Astronomical Society of the Pacific Conference Series, vol. 354 (2006), p. 92
22. M. Rempel, M. Schüssler, M. Knölker, Astrophys. J. **691**, 640 (2009). https://doi.org/10.1088/0004-637X/691/1/640
23. S.M. Hanasoge, K.R. Sreenivasan, Sol. Phys. **289**, 3403 (2014). https://doi.org/10.1007/s11207-014-0471-4
24. T.L. Duvall, Jr., S.M. Jefferies, J.W. Harvey, M.A. Pomerantz, Nature **362**, 430 (1993). https://doi.org/10.1038/362430a0
25. L. Gizon, T.L. Duvall, Jr., R.M. Larsen, J. Astrophys. Astron. **21**, 339 (2000)
26. T.L. Duvall, Jr., L. Gizon, Sol. Phys. **192**, 177 (2000)
27. J. Zhao, A.G. Kosovichev, Astrophys. J. **603**, 776 (2004). https://doi.org/10.1086/381489
28. J. Jackiewicz, L. Gizon, A.C. Birch, Sol. Phys. **251**, 381 (2008). https://doi.org/10.1007/s11207-008-9158-z
29. T.L.J. Duvall, S. D'Silva, S.M. Jefferies, J.W. Harvey, J. Schou, Nature **379**, 235 (1996). https://doi.org/10.1038/379235a0
30. J. Zhao, A.G. Kosovichev, T.L. Duvall, Jr., Astrophys. J. **557**, 384 (2001). https://doi.org/10.1086/321491
31. S. Couvidat, A.C. Birch, A.G. Kosovichev, Astrophys. J. **640**, 516 (2006). https://doi.org/10.1086/500103
32. R. Cameron, L. Gizon, T.L. Duvall, Jr., Sol. Phys. **251**, 291 (2008). https://doi.org/10.1007/s11207-008-9148-1
33. L. Gizon, H. Schunker, C.S. Baldner, S. Basu, A.C. Birch, R.S. Bogart, D.C. Braun, R. Cameron, T.L. Duvall, S.M. Hanasoge, J. Jackiewicz, M. Roth, T. Stahn, M.J. Thompson, S. Zharkov, Space Sci. Rev. **144**, 249 (2009). https://doi.org/10.1007/s11214-008-9466-5
34. H. Moradi, C. Baldner, A.C. Birch, D.C. Braun, R.H. Cameron, T.L. Duvall, L. Gizon, D. Haber, S.M. Hanasoge, B.W. Hindman, J. Jackiewicz, E. Khomenko, R. Komm, P. Rajaguru, M. Rempel, M. Roth, R. Schlichenmaier, H. Schunker, H.C. Spruit, K.G. Strassmeier, M.J. Thompson, S. Zharkov, Sol. Phys. **267**, 1 (2010). https://doi.org/10.1007/s11207-010-9630-4
35. S.M. Hanasoge, T.L. Duvall, Astrophys. J. **693**, 1678 (2009). https://doi.org/10.1088/0004-637X/693/2/1678
36. L. Gizon, A.C. Birch, H.C. Spruit, Annu. Rev. Astron. Astrophys. **48**(1), 289 (2010). https://doi.org/10.1146/annurev-astro-082708-101722
37. L. Gizon, A.C. Birch, Astrophys. J. **614**, 472 (2004). https://doi.org/10.1086/423367
38. A.C. Birch, L. Gizon, Astron. Nachr. **328**, 228 (2007). https://doi.org/10.1002/asna.200610724
39. A.G. Kosovichev, Astrophys. J. Lett. **461**, L55 (1996). https://doi.org/10.1086/309989
40. G.E. Backus, J.F. Gilbert, Geogr. J. **16**, 169 (1968)
41. G.E. Backus, J.F. Gilbert, R. Soc. Lond. Philos. Trans. Ser. A **266**, 123 (1970)
42. F.P. Pijpers, M.J. Thompson, Astron. Astrophys. **262**, L33 (1992)
43. J. Jackiewicz, L. Gizon, A.C. Birch, M.J. Thompson, Astron. Nachr. **328**, 234 (2007). https://doi.org/10.1002/asna.200610725
44. J. Jackiewicz, A.C. Birch, L. Gizon, S.M. Hanasoge, T. Hohage, J.B. Ruffio, M. Švanda, Sol. Phys. **276**(1–2), 19 (2012). https://doi.org/10.1007/s11207-011-9873-8
45. D. Korda, M. Švanda, Astron. Astrophys. **622**, A163 (2019). https://doi.org/10.1051/0004-6361/201833000
46. M. Švanda, Astrophys. J. Lett. **759**, L29 (2012). https://doi.org/10.1088/2041-8205/759/2/L29
47. T.L. Duvall, S.M. Hanasoge, S. Chakraborty, Sol. Phys. **289**, 3421 (2014). https://doi.org/10.1007/s11207-014-0537-3
48. J. Zhao, D. Georgobiani, A.G. Kosovichev, D. Benson, R.F. Stein, Å. Nordlund, Astrophys. J. **659**, 848 (2007). https://doi.org/10.1086/512009
49. M. Švanda, L. Gizon, S.M. Hanasoge, S.D. Ustyugov, Astron. Astrophys. **530**, A148 (2011). https://doi.org/10.1051/0004-6361/201016426
50. D. Korda, M. Švanda, J. Zhao, Astron. Astrophys. **629**, A55 (2019). https://doi.org/10.1051/0004-6361/201936268

51. S. Wedemeyer-Böhm, A. Lagg, Å. Nordlund, Space Sci. Rev. **144**(1–4), 317 (2009). https://doi.org/10.1007/s11214-008-9447-8
52. T. Roudier, M. Švanda, N. Meunier, S. Keil, M. Rieutord, J.M. Malherbe, S. Rondi, G. Molodij, V. Bommier, B. Schmieder, Astron. Astrophys. **480**(1), 255 (2008). https://doi.org/10.1051/0004-6361:20077973
53. T. Roudier, B. Schmieder, B. Filippov, R. Chandra, J.M. Malherbe, Astron. Astrophys. **618**, A43 (2018). https://doi.org/10.1051/0004-6361/201832937
54. Z. Švestka, Space Sci. Rev. **5**, 388 (1966). https://doi.org/10.1007/BF02653250
55. D.F. Neidig, Sol. Phys. **121**, 261 (1989). https://doi.org/10.1007/BF00161699
56. T.R. Metcalf, R.C. Canfield, J.L.R. Saba, Astrophys. J. **365**, 391 (1990). https://doi.org/10.1086/169494
57. M.E. Machado, A.G. Emslie, E.H. Avrett, Sol. Phys. **124**, 303 (1989). https://doi.org/10.1007/BF00156272
58. P.J.D. Mauas, M.E. Machado, E.H. Avrett, Astrophys. J. **360**, 715 (1990). https://doi.org/10.1086/169157
59. P. Heinzel, L. Kleint, J. Kašparová, S. Krucker, Astrophys. J. **847**, 48 (2017). https://doi.org/10.3847/1538-4357/aa86ef
60. J.C. Martínez Oliveros, S. Krucker, H.S. Hudson, P. Saint-Hilaire, H. Bain, C. Lindsey, R. Bogart, S. Couvidat, P. Scherrer, J. Schou, Astrophys. J. Lett. **780**, L28 (2014). https://doi.org/10.1088/2041-8205/780/2/L28
61. P. Saint-Hilaire, J. Schou, J.C. Martínez Oliveros, H.S. Hudson, S. Krucker, H. Bain, S. Couvidat, Astrophys. J. Lett. **786**, L19 (2014). https://doi.org/10.1088/2041-8205/786/2/L19
62. P. Heinzel, L. Kleint, Astrophys. J. Lett. **794**, L23 (2014). https://doi.org/10.1088/2041-8205/794/2/L23
63. L. Kleint, P. Heinzel, P. Judge, S. Krucker, Astrophys. J. **816**, 88 (2016). https://doi.org/10.3847/0004-637X/816/2/88
64. A.F. Kowalski, J.C. Allred, A. Daw, G. Cauzzi, M. Carlsson, Astrophys. J. **836**, 12 (2017). https://doi.org/10.3847/1538-4357/836/1/12
65. Y. Xu, W. Cao, C. Liu, G. Yang, J. Jing, C. Denker, A.G. Emslie, H. Wang, Astrophys. J. **641**, 1210 (2006). https://doi.org/10.1086/500632
66. Y. Xu, W. Cao, J. Jing, H. Wang, Astrophys. J. Lett. **750**, L7 (2012). https://doi.org/10.1088/2041-8205/750/1/L7
67. E. Landi Degl'Innocenti, M. Landi Degl'Innocenti, Sol. Phys. **97**, 239 (1985). https://doi.org/10.1007/BF00165988
68. D.E. Rees, A. López Ariste, J. Thatcher, M. Semel, Astron. Astrophys. **355**, 759 (2000)
69. H. Socas-Navarro, A. López Ariste, B.W. Lites, Astrophys. J. **553**(2), 949 (2001). https://doi.org/10.1086/320984
70. J.M. Borrero, S. Tomczyk, M. Kubo, H. Socas-Navarro, J. Schou, S. Couvidat, R. Bogart, Sol. Phys. **273**(1), 267 (2011). https://doi.org/10.1007/s11207-010-9515-6
71. B. Ruiz Cobo, J.C. del Toro Iniesta, Astrophys. J. **398**, 375 (1992)
72. J.C. del Toro Iniesta, *Introduction to Spectropolarimetry*. Introduction to Spectropolarimetry, by Jose Carlos del Toro Iniesta (Cambridge University Press, Cambridge, 2003), pp. 244, ISBN 0521818273
73. C. Frutiger, S.K. Solanki, M. Fligge, J.H.M.J. Bruls, Astron. Astrophys. **358**, 1109 (2000)
74. M. van Noort, Astron. Astrophys. **548**, A5 (2012). https://doi.org/10.1051/0004-6361/201220220
75. J. Jurčák, J. Kašparová, M. Švanda, L. Kleint, Astron. Astrophys. **620**, A183 (2018). https://doi.org/10.1051/0004-6361/201833946
76. S. Tsuneta, K. Ichimoto, Y. Katsukawa, S. Nagata, M. Otsubo, T. Shimizu, Y. Suematsu, M. Nakagiri, M. Noguchi, T. Tarbell, A. Title, R. Shine, W. Rosenberg, C. Hoffmann, B. Jurcevich, G. Kushner, M. Levay, B. Lites, D. Elmore, T. Matsushita, N. Kawaguchi, H. Saito, I. Mikami, L.D. Hill, J.K. Owens, Sol. Phys. **249**, 167 (2008). https://doi.org/10.1007/s11207-008-9174-z

77. T. Kosugi, K. Matsuzaki, T. Sakao, T. Shimizu, Y. Sone, S. Tachikawa, T. Hashimoto, K. Minesugi, A. Ohnishi, T. Yamada, S. Tsuneta, H. Hara, K. Ichimoto, Y. Suematsu, M. Shimojo, T. Watanabe, S. Shimada, J.M. Davis, L.D. Hill, J.K. Owens, A.M. Title, J.L. Culhane, L.K. Harra, G.A. Doschek, L. Golub, Sol. Phys. **243**, 3 (2007). https://doi.org/10.1007/s11207-007-9014-6
78. G.S. Kerr, L. Fletcher, Astrophys. J. **783**, 98 (2014). https://doi.org/10.1088/0004-637X/783/2/98
79. B.W. Lites, R. Casini, R. Manso Sainz, J. Jurčák, K. Ichimoto, R. Ishikawa, T.J. Okamoto, S. Tsuneta, L. Bellot Rubio, Astrophys. J. **713**, 450 (2010). https://doi.org/10.1088/0004-637X/713/1/450
80. C.G. Giménez de Castro, J.P. Raulin, J.F. Valle Silva, P.J.A. Simões, A.S. Kudaka, A. Valio, Space Weather **16**, 1261 (2018). https://doi.org/10.1029/2018SW001969
81. M.E. Machado, E.H. Avrett, J.E. Vernazza, R.W. Noyes, Astrophys. J. **242**, 336 (1980). https://doi.org/10.1086/158467
82. J.E. Vernazza, E.H. Avrett, R. Loeser, Astrophys. J. Suppl. Ser. **45**, 635 (1981). https://doi.org/10.1086/190731
83. J.X. Cheng, M.D. Ding, M. Carlsson, Astrophys. J. **711**, 185 (2010). https://doi.org/10.1088/0004-637X/711/1/185
84. P.J.A. Simões, G.S. Kerr, L. Fletcher, H.S. Hudson, C.G. Giménez de Castro, M. Penn, Astron. Astrophys. **605**, A125 (2017). https://doi.org/10.1051/0004-6361/201730856
85. G.S. Kerr, L. Fletcher, A.J.B. Russell, J.C. Allred, Astrophys. J. **827**, 101 (2016). https://doi.org/10.3847/0004-637X/827/2/101

# Starburst Galaxies

**Ivana Orlitova**

## Contents



**Abstract** The rate of star formation varies between galaxy types and evolves with redshift. Most stars in the universe have formed in episodes of an exceptionally high star-forming activity, commonly called a starburst. We here summarize basic definitions and general properties of starbursts, together with their observational signatures. We overview the main types of starburst galaxies both in the local universe and at high redshift, where they were much more common. We specify

I. Orlitova (✉)
Astronomical Institute of the Czech Academy of Sciences, Ondřejov, Czechia
e-mail: orlitova@asu.cas.cz



P. Kabáth et al. (eds.), *Reviews in Frontiers of Modern Astrophysics*,
https://doi.org/10.1007/978-3-030-38509-5_13

similarities and differences between the local and distant samples and specify the possible evolutionary links. We describe the role of starburst galaxies in the era of cosmic reionization, relying on the most recent observational results.

## 1 Star Formation and Starburst

The rate of star formation is one of the basic characteristics of a galaxy. The stellar content and stellar age provide observational signatures that allow us to derive the history of galaxy evolution. The star formation rate (SFR), defined as the stellar mass increase per unit time, $dM_*/dt$, can range from zero, such as in gas-poor elliptical galaxies, to hundreds or thousands of solar masses per year ($M_\odot$ year$^{-1}$) in the most vigorously star-forming galaxies. For illustration, the Milky Way (MW) forms stars at a rate of less than 1 $M_\odot$ year$^{-1}$. Star formation (SF) can be continuous and more-or-less regular, or it can consist of short and intense bursts separated by long intervals of quiescence. Understanding how galaxies make stars and what drives the differences between them is an active field of research. We will here overview the main observational methods and results, focusing specifically on galaxies that undergo a burst of star formation.

The SFR alone may not be a sufficient parameter to characterize how important is the SF in a given galaxy. In principle, the more massive the galaxy, the larger the SFR can be, if enough cold gas ($\sim$10 K) is available. It is convenient to define relative quantities that allow a better comparison between galaxies. By factoring out the galaxy stellar mass, we obtain the specific star formation rate (sSFR), the SFR per unit mass. The sSFR is proportional to the birthrate parameter $b$, the ratio of the current SFR to the average past SFR [1]. Another useful quantity is the surface density of star formation, $\Sigma_{SFR}$, which measures the SFR per unit area and which correlates with the gas surface density $\Sigma_{gas}$ (combined atomic H I and molecular $H_2$). This relation is commonly called 'Schmidt-Kennicutt' and has been discussed in detail in Kennicutt's review [1]. An updated picture, using recent multi-wavelength data, clarifies that the relation is mainly driven by molecular $\Sigma_{H_2}$ [2, 3].

Star formation proceeds through the assembly of cold gas into dense clouds which eventually undergo gravitational collapse once they attain masses of $\sim 10^6\,M_\odot$ and sizes up to 100 pc. What drives this gas compression and what triggers the SF is a subject of many discussions as we will present in the following sections (shock waves, stellar density waves such as bars or tidal interactions). How is the available gas converted into stars is then described by the SFR efficiency $\epsilon = SFR/M_{gas}$ (definition used in extragalactic studies, different from that for individual clouds in the MW). The inverse of the efficiency is called the depletion time, which states how long can the galaxy continue forming stars at the current rate with the given mass of atomic H I and molecular $H_2$. A large spread of efficiencies is observed in local galaxies, with an average of $\sim$5% per $10^8$ years [1–3], corresponding to a depletion time of 2 Gyr (which can in reality be a factor of two longer due to recycling of gas from stellar winds).

**Fig. 1** Starburst region 30 Doradus (also known as Tarantula nebula) in the Large Magellanic Cloud. The multi-colour UV and optical image was obtained using the Hubble Space Telescope and shows the central concentration of young stars (called R136) in the NGC 2070 star cluster, which is situated at the core of the nebula. R136 contains several O stars and Wolf-Rayet stars and produces most of the energy that makes the 30 Doradus visible. Credit: NASA/ESA

When a galaxy undergoes an exceptionally intense phase of SF, we speak of a starburst (Fig. 1). A starburst galaxy is thus not a separate class and it is rather an evolutionary phase in the life of a galaxy. The starburst activity may not be uniformly spread across the galaxy. While global starbursts preferentially occur in dwarfs, the starbursts in massive galaxies are usually localized in a small volume, most often in the circumnuclear region, where $10^8$–$10^{10}\,M_\odot$ of gas is confined to a radius of $< 1$ kpc. The reasons behind the vigorous star formation are not yet fully understood and we will mention here the main scenarios. The most appealing is the effect of gravitational interaction between galaxies, ranging from a close passage to a complete merger (from pioneering papers such as [4, 5] to recent observations [6] and simulations [7]). The tidal forces perturb the orbits of gas and stars and enable the gas flow toward the galaxy center. However, it is not yet clear whether such an encounter is a necessary condition (some starburst galaxies seem to be isolated), nor whether the encounter is a sufficient condition (tidal forces between dwarf galaxies are modest). A similar transfer of gas is also possible by the action of gravitational instabilities such as stellar bars [8, 9]. Subsequently, local dynamical processes including the pressure, the local velocities, the starlight and self-gravitation decide on the future of the gas clouds (as reviewed by [9, 10]). The accumulation of gas in the central parts is essential for yet another phenomenon: feeding of the active galactic nucleus (AGN), i.e. the central galactic region dominated by a supermassive black hole. Indeed, a simultaneous AGN and starburst activity exists in numerous galaxies, as we will describe in the following sections. It is still a matter of debate if the coexistence is coincidental or if there is a causal relationship [11]. If no apparent causes of starburst are visible, we can speculate about the time delays between different phenomena or invoke the role of shock waves due to preceding star formation and supernova explosions [9, 12].

There is no unique and rigorous definition of a starburst, and a good definition is being discussed in the literature still today [3, 13, 14]. The classical definition is related to the amount of gas in the galaxy and the maximum lifetime of the starburst: the galaxy would consume all of its remaining gas in a limited time were the SFR constant. If this consumption (depletion) time is much shorter than the age of the universe, then we define the galaxy as a starburst. In other words, the efficiency of SF must be at least an order higher than in an average local galaxy [1]. However, the problem with this definition is the uneven gas fraction in different galaxy types. The depletion time in a massive, gas-poor galaxy can be short, yet without any sensible reason to call it a starburst. An alternative definition is based on the birthrate parameter: one can estimate the time that the galaxy would need in order to create all of its existing stars at the present SFR. This approach represents a disadvantage for massive galaxies where the timescales to reproduce their stellar mass are excessively long. Moreover, both of these approaches have a built-in redshift dependence—the varying age of the universe changes the quantitative meaning of starburst parameters at each redshift [13]. One therefore has to be careful what is meant by a starburst in each study. Heckman [13] proposed to define a starburst based on $\Sigma_{\mathrm{SFR}}$, which treats equivalently both nearby and high-redshift galaxies. High $\Sigma_{\mathrm{SFR}}$ implies large surface densities of both gas and stellar mass [1]. As a consequence, such regions will have a high gas pressure, a high rate of mechanical energy deposition and a high density of radiant energy. $\Sigma_{\mathrm{SFR}}$ is hence directly linked to the galaxy physical properties.

The definition of a starburst may have to stay loose, we can nevertheless require a parallel application of several criteria, and, in addition, add a qualitative requirement that the starburst must have a global impact on its host galaxy evolution—powerful outflows, significant mass fraction transformed to stars, disturbed structure, enhanced luminosity, etc. Typically, authors require $b \geq 3$ or $b \geq 10$ to select starburst galaxies [15, 16]. The undisputed starbursts have durations of $10^7$–$10^8$ years and $\Sigma_{\mathrm{SFR}} \sim 1$–$100\,\mathrm{M}_\odot\,\mathrm{year}^{-1}\,\mathrm{kpc}^{-2}$, which exceeds $\Sigma_{\mathrm{SFR}}$ of the MW by several orders of magnitude [13]. The most extreme of them form stars with efficiencies close to 100% per $10^8$ year [1]. What is certain is that the upper bound for efficiency must be set by causality—conversion of the entire self-gravitating gas reservoir into stars in one dynamical time.

Starbursts (defined by $b \geq 3$) currently form 20% of the present-day massive stars [13], representing a non-negligible baryonic constituent of the local universe (if $b \geq 10$ is considered, the fraction drops to 3%). Because of their young age, they contain large numbers of massive stars and X-ray binaries, and they thus offer unique opportunities for studying the high-mass objects and their feedback in the form of mechanical energy (winds, jets) and energetic photons (ultraviolet, X-rays).

The collective action of stellar winds, supernovae and jets drives powerful outflows of gas on the galactic scale. Also known under the name superwinds, the large-scale outflows transport material, including heavy elements, to the intergalactic medium (IGM). The localized starburst hence directly affects a large volume of space. Starburst galaxies were much more frequent in the early universe than they are today and they played an important role in shaping the IGM throughout the history of the universe. Besides, they were responsible for forming most of the stars in the universe as they were the building blocks of present-day galaxies. SF then sharply dropped after redshift $z \sim 2$ (i.e. age of the universe ~3 Gyr). The first starbursts also played yet another role in cosmology: the ultraviolet radiation of young massive stars was probably responsible for converting the IGM from neutral to ionized in the first billion years after the Big Bang (so called Cosmic Reionization).

The goal of the following sections is to provide a basic description of how are starbursts observed and how the observational methods lead to their "classification" both at low and high redshift. Section 2 describes the starburst signatures at various spectral wavelengths. Section 3 overviews the major starburst galaxy types in the local universe. Section 4 presents the high-redshift galaxy classes, all detected thanks to the starburst signatures. Section 5 focuses on the role of starburst galaxies in the cosmic reionization. Finally, Sect. 6 brings concluding remarks and future prospects, including the observational facilities that are in preparation.

## 2 Observational Signatures of a Starburst

Star formation manifests itself by a variety of features across a large portion of the electromagnetic spectrum. We observe the direct starlight of young stars as well as the light reprocessed by interstellar gas and dust. Detailed and insightful reviews of SFR indicators were provided by Kennicutt [3] and Calzetti [17].

### *2.1 Ultraviolet and Optical Continuum*

The UV and optical light of young stellar populations is dominated by massive stars with the lifetimes of $10^6$–$10^8$ years. The O and B stars produce most of their energy in the far-ultraviolet (FUV) band, including a large fraction at Lyman continuum wavelengths ($\lambda < 912$ Å), capable of ionizing hydrogen atoms. In parallel, massive stars drive powerful stellar winds that affect kinematics of the interstellar medium (ISM) and produce specific spectral features. The importance of the UV observations in starbursts was demonstrated by the International Ultraviolet Explorer (*IUE*) satellite, operating for almost 20 years before the end of the twentieth century. The *IUE* was followed by the Far Ultraviolet Spectroscopic Explorer (*FUSE*), the Galaxy Evolution Explorer (*GALEX*), and the Hubble Space Telescope (*HST*). Thanks to the *IUE*, the first UV catalogue of stellar types was issued in 1985 [18].

In the dust-poor case, the bright UV continuum is a prominent signature of the young massive stars, i.e. of a starburst. The bright UV is also observable for high-redshift galaxies and represents one of the major methods of galaxy detection (Sect. 4.1, Fig. 6). Furthermore, empirical relations exist between the SFR and the UV luminosity, and therefore both FUV and near-UV (NUV) luminosities serve as SFR estimators [17]. Nevertheless, the total SFR estimation must ideally take into account the effect of dust. The dust efficiently reprocesses the UV light into infrared (IR) radiation (see Sect. 2.5). Therefore, combined UV and IR measurements provide a more precise determination of the galaxy SFR [17].

The FUV light short of the Lyman edge ($\lambda = 912$ Å) ionizes neutral hydrogen. As a result, most of this radiation is absorbed in the ISM or IGM and does not reach our telescopes. Instead, a sharp drop, known as the Lyman break, appears in the FUV spectrum. This property is conveniently used for determining the galaxy redshift at various epochs of cosmic history (Sect. 4.1, Fig. 6). Part of the absorbed ionizing radiation is reprocessed into emission lines which are formed by recombination and which become another SF signature (Sects. 2.3 and 2.4). If, on the other hand, the column densities of interstellar H I are low enough not to remove all of the Lyman continuum, the remaining radiation will escape to the surrounding IGM and ionize it—which is of great interest mainly to the high-$z$ studies and the era of cosmic reionization ($z > 6$, Sect. 5.2).

## 2.2 UV Absorption Lines

UV-bright galaxies allow observation of starburst properties through UV absorption lines formed in the foreground ISM (Fig. 2). The lines are especially well detectable in starbursts producing copious amounts of young stars and thus UV photons. The lines inform us about chemical composition of the ISM gas and its ionization state (e.g. Si II, Si III, Si IV), about kinematics and about optical depth of the gas. The *FUSE* satellite opened the way to studying the UV lines in the MW and nearby galaxies, as it was the first UV facility with sufficient spectral resolution. The *HST* has then revolutionized the field with progressively more sensitive spectrographs

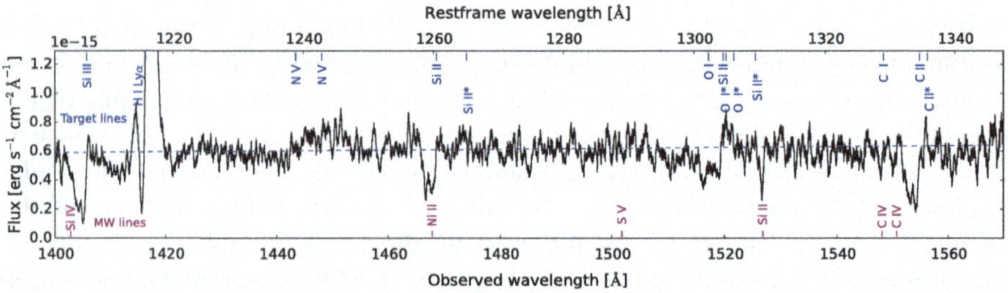


**Fig. 2** UV absorption lines of the galaxy SDSS-J030321.41-075923.2. The major lines originating both in the galaxy and in the Milky Way have been labeled. Source: *HST* archive

and improving resolution: GHRS, STIS, COS. Conversely, at high redshift, UV lines are reachable with ground-based optical telescopes, which need to have a sufficient collecting area and a spectrograph of sufficient resolution.

The UV absorption lines probe gas kinematics in nearby starbursts, showing that their properties may range from static to high-velocity ($\sim$1000 km s$^{-1}$) outflows [19, 20], and may vary between different gas phases. Outflows were also found to be ubiquitous at high redshift [21]. The UV absorption lines thus offer useful tools for answering questions about stellar feedback, ISM enrichment, IGM enrichment, gas flows, and galaxy evolution. We describe in Sect. 2.3 that the UV lines also represent valuable complementary information for interpreting the Lyman-alpha line. Eventually, the optical depth of the UV lines has been explored as a tool for studying the escape of ionizing UV continuum from galaxies [19]. If the neutral gas situated along the line of sight does not completely cover the ionizing source, we should observe residual flux in the absorption lines. This can in turn be translated to the Lyman continuum escape fraction [22]. However, in reality, the interpretation of residual fluxes is not straightforward, is dependent on the gas geometry, its chemistry, on spectral resolution, and on the applied model. Observational testing of the method against the directly measured Lyman continuum escape has only started recently, after the first successful Lyman continuum detections have been confirmed [23, 24].

## 2.3 *Lyman-Alpha Line of Hydrogen (UV)*

Lyman-alpha (Ly$\alpha$), at wavelength $\lambda = 1215.6$ Å, is the first transition of hydrogen and in principle its brightest spectral line. In galaxies, it is produced by the recombination process in the regions of ionized gas mostly around hot stars and AGN. The probability that the recombination cascade passes through the first transition is $\sim$70% and Ly$\alpha$ thus reprocesses a major part of the stellar ionizing radiation. It may hence appear as the ideal signature of star-forming galaxies and it was indeed predicted to be the beacon of high-$z$ galaxies as early as the 1960s [25]. However, the reality is less straightforward: Ly$\alpha$ undergoes a complex radiative transfer in the neutral ISM of the galaxy and the IGM. By symmetry, its cross-section for interaction with neutral hydrogen is large and the line becomes optically thick at H I column densities as low as $10^{13}$ cm$^{-2}$, i.e. basically in all galaxies (see a useful review of Ly$\alpha$ physics by M. Dijkstra [26]). Observational confirmation of Ly$\alpha$ only came almost 20 years after the prediction and the line was far weaker than expected ([27] in low redshift; [28] in $z \sim 3$; first *HST* Ly$\alpha$ images in [29]).

The Ly$\alpha$ physics started to emerge with the growing observational samples both at low and high redshift, and, in parallel, with the development of numerical models. Building on analytical solutions that were only possible for a few extreme cases [30], the numerical codes explore a range of geometries and conditions, from homogeneous spherical set-ups [31, 32], through clumpy media [33] to full hydrodynamic simulations [34]. The resonant scattering off hydrogen atoms

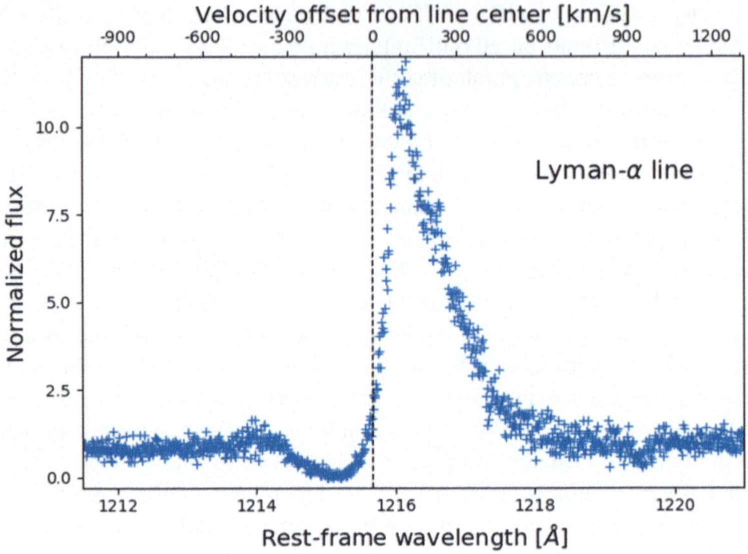


**Fig. 3** Lyα spectrum of Mrk 259, a nearby irregular starburst galaxy from the LARS sample. The P-Cygni profile with its asymmetric red peak and the absorption trough blueward from the line center (1215.67 Å, dashed vertical line) result from radiative transfer. Credit: *HST* archive

increases the probability of Lyα destruction by dust, therefore the dust role is enhanced with respect to other UV wavelengths. The Lyα line is further sensitive to the geometry of the medium, its clumpiness, and the macroscopic and microscopic kinematics. To escape from the ISM, the Lyα photons must either shift to regions where the H I density is low, or shift in frequency to the line wings (by interaction with hydrogen atoms) where the scattering cross-section is smaller. As a result, Lyα galaxy images commonly show a "halo", which here means a low surface brightness Lyα emission extending beyond the stellar continuum image [35–37]. This feature is likely due to photons produced in the starburst regions and transferred to the outskirts of the galaxy. Conversely, the Lyα escape through a frequency shift is detected in spectral line profiles, which range from asymmetric P-Cygni (absorption in the blue, emission in the red, see Fig. 3), through single red peaks (mostly in high $z$), double- or multiple-peaks, to damped absorption profiles (with no escaping photons). All of the profiles result from radiative transfer of photons from the line core to the line wings.

Despite its complex interpretation, Lyα is one of the primary tools for galaxy detection at high redshift (Sect. 4.2). Now that we are aware of its limitations and we search in the correct luminosity range (much fainter than the theoretical one), Lyα stays detectable out to redshifts beyond $z \sim 10$. As demonstrated in the local universe (from the earliest studies such as [27, 38]), the Lyα emission is only present in a subset of star-forming galaxies and therefore introduces selection biases that may not be fully understood today. Lyα is mostly bright at low-mass, low-metallicity, dust-poor starburst galaxies (not speaking about quasars here). However, due to the multi-parameter nature of Ly$\alpha$ transfer, we do not have a full control of the galaxy populations detected in Ly$\alpha$, and complementary methods are necessary for characterizing galaxy evolution at high redshift. A good lesson was provided by the extremely low-metallicity, dust-poor local galaxy IZw18, where bright Ly$\alpha$ emission was expected and where the *HST* revealed a deep absorption instead [39]. We understand now from the UV metal lines that the neutral gas is static in IZw18, which disables the Ly$\alpha$ escape. We have to take into account these effects when interpreting high-$z$ samples, especially in experiments where only emission is targeted.

At present, Ly$\alpha$ imaging and spectroscopy are among the main objectives of all currently developed facilities such as the James Webb Space Telescope (*JWST*) and the generation of extremely large telescopes (ELTs). The Subaru telescope has its powerful Hyper-Suprime Cam, the ESO's Very Large Telescope (VLT) recently obtained the highly performant integral-field spectrograph MUSE optimized for Ly$\alpha$, and its successor HARMONI is in development for the ESO's ELT. The Hobby Eberly Dark Energy Experiment (HETDEX) is expected to detect millions of Ly$\alpha$ galaxies. Ly$\alpha$ will therefore stay a substantial cosmological tool in the years to come, despite all its limitations.

## *2.4 Optical Emission Lines*

Starburst galaxies are characterized by bright optical emission lines that are formed in the ionized interstellar gas (see Fig. 5). Part of the lines arise from recombination, and represent thus reprocessed ionizing radiation of O and B stars. This is the case of the hydrogen Balmer series, for instance. The H$\alpha$ line can reach equivalent widths over thousand Å in starburst galaxies. The H$\alpha$ luminosity is also one of the favourite tools for SFR determination [1].

Forbidden lines represent another type of transitions, excited by collisions and de-excited radiatively. Their luminosities are not directly coupled to the ionizing luminosity, but they are bright in starburst galaxies thanks to the high temperatures and the energy pumped into the ISM by star formation. Forbidden lines of $S^+$, $N^+$, $O^+$, $O^{++}$ are the most prominent ones. Their formation is sensitive to metallicity, electron temperature and density, therefore the ratios of the line fluxes are useful for measuring the fundamental gas properties. In addition, their combinations allow classifying the galaxies to AGN and star-forming galaxies. Statistical classification trends emerged from the pioneering work of Baldwin, Phillips and Terlevich (their so called BPT diagram [40]) and similar. Building on these early works, the big data brought about by the Sloan Digital Sky Survey (SDSS) at the turn of this century led to the discovery of a clear separation between star-forming galaxies and AGN into two narrow sequences in the BPT diagram [41]. Theoretical works such as [42–44] provided interpretation and prescription for the observational patterns. The oxygen [O III]$\lambda$5007 line is among the brightest optical lines under certain conditions.

Sensitive to the temperature, the line is bright in low-metallicity environments (around 10% solar, e.g. in dwarf starburst galaxies) and in high-excitation regions (in AGN). A second line, such as [N II] $\lambda$6583 in the BPT diagram, is necessary for discriminating between the low-metallicity starburst and the AGN. In extreme starbursts, such as those at high redshift, the division between SF and AGN becomes more delicate: the extreme ISM conditions produce emission line fluxes originally believed to exist only in AGN. Recent detections of powerful starbursts thus make us revise existing models and our understanding of star formation [44].

The optical band also hosts spectral lines that are related to specific astrophysical questions connected to the starburst. One of the examples is the lines produced by Wolf-Rayet (WR) stars, the evolved descendants of O stars. A blue bump around 4600 Å and a red bump around 5700 Å, both formed by a gathering of several emission lines, are characteristic WR features. He II lines such as 4686 Å can also be of WR origin. The WR evolution is strongly affected by metallicity: at low metallicities, only the most massive stars become Wolf-Rayet, probing thus the high-mass end of the initial mass function (IMF). For more details on WR observations, we refer to [45]. Another example is the use of optical lines for probing the primordial helium abundance. While the oxygen abundance (O/H) varies among galaxies, the helium abundance (Y) stays nearly constant. By extrapolating the measured trend to zero O/H, one should retrieve the primordial Y. Observations of low-metallicity star-forming dwarfs play a decisive role in this problem [45].

Detection of the brightest rest-frame optical lines (H$\alpha$, H$\beta$, [O III], [O II]) at high redshift has been achieved in the last decade using the 8–10 m telescopes such as the Very Large Telescope (VLT), the Keck Telescope or Subaru. Future larger samples and a more complete coverage of lines will provide important clues to the conditions in the distant galaxies.

## 2.5 Infrared Emission

Star formation takes place in dense molecular and dusty clouds. It has been approximated that dust absorbs half of the stellar light produced in the universe and re-emits it in the infrared. As the absorption is particularly efficient in the UV, the IR band is an important complement to the UV SFR estimators. The IR alone is not sufficient to correctly retrieve the SFR, but the combined UV and IR measurements proved to be an efficient method. The IR correction can be essentially negligible in dust-poor dwarf galaxies and metal-poor regions, but its importance increases with growing metallicity and dust content. On the other hand, the IR emission can significantly overestimate the SFR in the cases where evolved stars contribute to dust heating. The conversion factor between dust emission and SFR must therefore be a function of the stellar population [3].

The far-IR emission is from 99% composed of continuum radiation produced by dust grains. The remainder is line emission from atomic and molecular transitions in the ISM gas, concentrated in a small, sub-kpc region. The IR continuum depends on the dust content and composition, but also on the spectrum of the starlight: luminous young stars will heat dust to higher temperatures than old stars and their thermal IR will thus peak at shorter wavelengths ($\lesssim$60 μm). The complexity of dust emission features corresponding to small and large dust grains at different temperatures have been summarized in excellent review papers by Calzetti [17] and Kennicutt [3], together with the IR SFR calibrators. For the purposes of this text, we only need the qualitative statement that the IR emission provides an important window for detecting obscured starbursts at both low and high redshift (Sects. 3.5 and 4.3).

## *2.6 X-rays*

X-rays trace the star-forming activity through its final products—X-ray binaries—and through accompanying phenomena such as heating of the surrounding ISM by stellar feedback and supernovae. Binary systems composed of a main-sequence star and a compact object—black hole or neutron star—emit X-rays by accretion of matter, transferred from the donor star onto the compact object (see review [46]). The systems are classified as high-mass or low-mass X-ray binaries according to the mass of the donor star. Starbursts are characterized by prominent X-ray emission that is dominantly provided by the rapidly evolving high-mass X-ray binaries (HMXBs). Additional X-ray luminosity is provided by interstellar hot gas peaking at $\sim$1 keV and a few additional sources such as low-mass X-ray binaries or cataclysmic variables. In addition, more exotic sources may contribute decisively: ultra-luminous X-ray sources (ULX) whose nature is debated (extreme X-ray binaries, anisotropic emission, neutron stars) [47], and intermediate-mass black holes (IMBH). Such objects seem to be more probable at low metallicity and at high SFR [48].

X-ray emission has been mapped in a number of local dwarf galaxies including the low-metallicity IZw18, SBS0335-052 [49], Haro 11 [50] and other starbursts [51–55]. Their X-ray luminosity is generally dominated by several extremely bright, point-like sources of debatable origin—usually too bright for "classical" X-ray binaries, they are suspected to be ULXs, IMBHs or even AGN (though often without optical counterparts). Empirically, the total X-ray luminosity of star-forming galaxies is proportional to the SFR [56] and inversely correlates with metallicity [53, 57, 58]. The theoretical basis for starburst X-ray emission was provided by numerical simulations of X-ray binaries and of hot gas emission [59–61]. X-ray observations remain challenging at high redshift, where they are restricted to images of stacked samples [53]. Together with sensitive, detailed, resolved mapping of nearby sources, the high-$z$ observations will be a task for the new generation of X-ray satellites.

# 3 Local Starburst Galaxies

We will focus on global starbursts, rather than individual starbursting regions of massive galaxies (such as the central region of M82, or 30 Doradus in the Large Magellanic Cloud—Fig. 1). The latter are frequent in the local universe, the former are rare. Global starbursts are interesting from the cosmological point of view, as they are traceable out to the earliest epochs of the universe, where they were bright and numerous. The focus on global starbursts here predefines the focus on dwarf galaxies and on extremely dusty IR galaxies.

While dwarfs constitute the most common galaxy type in the nearby universe (>70%), only a few percent of them are starbursting. They have attracted attention for many decades [62, 63]. Their subsets appeared under various names in the past, such as H II galaxies [64] or blue amorphous galaxies [65], reflecting the techniques of their discovery and the focus on a particular feature at a time. Some of the starbursting dwarf classes will therefore partially or totally overlap. Our intention here is to mention those that have been extensively studied in recent past with the motivation to search for analogies with high redshift.

## *3.1 Blue Compact Dwarf Galaxies*

Blue galaxies, associated with the bright continuum of young stars, attracted attention since the 1960s. The colour selection alone resulted in a mixed bag of galaxy morphologies and luminosities. A subset that appeared barely resolved on optical plates of that time got the name blue compact galaxies (BCGs). It is still a heterogeneous class which some authors, with the presently improved knowledge, further sub-divide by compactness, luminosity or morphology. Not all of the BCGs would comply with all the definitions of a starburst. Their common properties are relatively small mass ($<10^9\,M_\odot$), low metallicity, bright optical emission lines, and sizes within a few kpc. Galaxies discovered later (for instance by emission lines) and having similar properties have been added to the class. The most centrally peaked ones (<2 kpc) are referred to as blue compact dwarfs.

BCGs include some of the most metal-poor galaxies (e.g. IZw 18, see [66]). This led to the thinking that they formed only recently (which would have profound implications for galaxy formation). However, this was disproved by the discovery of older stellar populations (see review [45]). Instead, their star formation has been inefficient for most of the history, interleaved with time-limited bursts. BCGs are gas-rich ($\gtrsim$50% of mass is in gas) and dust-poor. Their morphologies range from H II-region-like (Pox 186) to irregular and clumpy (see Haro 11 in Fig. 4) to more symmetric ones. Their morphologies and velocity fields often bear signs of a recent interaction, despite the relative isolation of many of them [14, 15, 45]. The presence of low surface brightness neighbours cannot be excluded in such cases. Interaction with neighbours may, in turn, be responsible for the starburst activity in BCGs. This

**Fig. 4** A multi-color *HST* image of Haro 11, one of the nearest BCGs (and Lyman break analogues). Its irregular structure, of total diameter ∼2 kpc, is probably the result of a galaxy merger, which also induced the SFR $\sim 20\,M_{\odot}\,year^{-1}$. The galaxy is a weak Lyman continuum leaker (Sect. 5.2). It is a Ly$\alpha$ emitter in some of the star-forming knots and a Ly$\alpha$ absorber elsewhere. Credit: ESO/ESA/Hubble and NASA (https://www.eso.org/public/images/)

comes with a limitation that it is uncertain if interaction of two dwarfs is sufficient for triggering SF as in massive galaxies—in dwarfs, the gravitational effects are equally strong as the turbulent velocities caused by stellar winds. Starbursts in BCGs challenge the models of star formation anyhow: dwarfs are generally inefficient in converting gas to stars due to the weak gravitational bounding. In addition, the existence of molecular gas in BCGs is difficult to prove due to their low metallicity. Nevertheless, starbursts in BCGs do exist.

There have been long-standing debates about the evolutionary relation between BCGs and other dwarf galaxy classes [14]. Answering these questions is obviously interesting with respect to the high-$z$ universe, where galaxies of similar masses were prevailing. For the same reason, BCG observations in the UV and Ly$\alpha$ are important. For their low dust amount, a bright Ly$\alpha$ emission was naively expected in all BCGs. This hypothesis was disproved by observing a similar incidence of emission and absorption Ly$\alpha$ profiles [38]. The extremely metal-poor IZw 18 is one of the surprising examples of a deep absorption. This finding illustrates the importance of H I kinematics for the Ly$\alpha$ escape [67], which we described in Sect. 2.3. The absence of Ly$\alpha$ emission in a subset of BCGs is an important realization for high-$z$ surveys, where the Ly$\alpha$ technique will certainly be missing such galaxy populations.

## 3.2 Lyman-Alpha Reference Sample

Efforts to understand high-redshift galaxies have led to a definition of the local Lyman-alpha reference sample (LARS) [68]. The Ly$\alpha$ line becoming one of the dominant tools for galaxy detection at high redshift, the need was to understand the mechanisms of its formation and transfer through the ISM. This is achievable in the local universe, where spatial resolution and multi-wavelength data are available.

The LARS sample targeted star-forming galaxies (requiring H$\alpha$ equivalent widths above 100 Å) with FUV continuum luminosities similar to the $z \sim 3$ LAEs and LBGs. The sample comprises over forty galaxies at redshifts $z = 0.03$–$0.2$, observed with multiple *HST* broad-band and narrow-band filters, complemented with *HST* COS spectra, and ground-based data including optical integral-field spectroscopy and H I radio interferometry. To obtain *HST* Ly$\alpha$ images [35], a dedicated method of broad-band filter subtraction was developed, due to the lack of existing appropriate narrow-band Ly$\alpha$ filters.

The UV-based selection produced a sample composed of irregular galaxies bearing signs of interactions, and disk/spiral galaxies. Their Ly$\alpha$ shows a variety of properties, from absorption to bright emission, including extended diffuse Ly$\alpha$ halos that reach beyond the stellar or H$\alpha$ emission [35]. The high-resolution images comparing the stellar light, the optical gas emission and Ly$\alpha$ thus show how Ly$\alpha$ travels from the productions sites (which should be identical to H$\alpha$) to regions where it can easier escape. In complement, the UV and optical spectroscopy probe the effects of gas kinematics on the Ly$\alpha$ escape [69, 70], while the radio 21cm data demonstrate the importance of the H I mass [71]. Such detailed observations are by far not available for high-$z$ galaxies, and therefore LARS, as its name indicates, provides a local reference sample for aiding the interpretation of high-$z$ observations.

## 3.3 Lyman-Break Analogues

As soon as the Lyman-break technique proved efficient for detecting galaxies at high redshift (Sect. 4.1), the question arose what type of galaxies were selected by this approach, especially in the situation where the FUV was the only available signal. The rest-frame UV properties were better mapped in high redshift than locally, and therefore it was impossible to draw analogies with nearby galaxies. Launch of the GALEX mission remedied this situation in the early 2000s and finally enabled the construction of statistical samples of low-$z$ UV galaxies.

Lyman-break analogues (LBAs; formerly called UV-luminous galaxies, UVLGs) were selected from the GALEX archive as galaxies at $z < 0.2$ with FUV luminosities similar to those observed by the Lyman-break technique at high redshift [72]. Such galaxies are rare in the local universe; the FUV selection resulted in ~70 targets. The GALEX and SDSS images reveal that their morphologies range from compact systems with radii of 1 kpc to large, late-type, 10 kpc-size and $10^{11}\,M_{\odot}$ mass galaxies. The compact targets share many properties with the high-$z$ samples: they have stellar masses of $\sim 10^{9.5-10.7}\,M_{\odot}$, SFR ~5–25 $M_{\odot}$ year$^{-1}$, and metallicities 12+log(O/H) ~8.2–8.7. Lyman-break galaxies at high $z$, on the other hand, can reach significantly higher SFRs (few hundred $M_{\odot}$ year$^{-1}$) and they span a significantly larger metallicity range (12+log(O/H)$\sim 7.7 - 8.8$). Also, the local LBAs do not reach the maximum FUV luminosities of high-$z$ targets, the overlap between the FUV luminosities is achieved only close to the lower edge of the high-$z$ FUV luminosity interval.

Today, spectra and images at several wave bands are available for a subset of LBAs, including optical and UV data obtained with the *HST*, and X-ray data from *Chandra*. Thirty percent of the large LBAs and 15% of the compact LBAs were found to be type-2 AGN [72] (while type-1 AGN were removed from the sample upon selection). Among the star-forming LBAs, a subset have been studied in more detail, revealing starburst character of their spectra, including strong optical emission lines, emission in the Ly$\alpha$ line [19], and in X-rays [58, 73]. Their X-ray emission is among the brightest detected so far, with luminosity similar to high-redshift Lyman-break galaxies, and with resolved ULX sources or AGN candidates. LBAs are also possible Lyman continuum leakers thanks to their fast stellar winds [19]. The Lyman continuum escape has indeed been confirmed for several LBAs [74].

## 3.4 Green Peas and Luminous Compact Galaxies

Green Peas were identified in the SDSS by the citizen science Galaxy Zoo project [75]. The targets that appear green and point-like, i.e. unresolved by the SDSS, turned out to be compact, highly star-forming galaxies at redshift $z \sim 0.2$. Their green colour is caused by bright optical emission lines. In particular, the [O III] $\lambda$5007 line reaches equivalent widths above 1000 Å (Fig. 5). It was shown later that the original eighty Green Peas are part of a more general population extending over a large interval of redshifts (at least $z \sim 0.02 - 0.6$, probed by the SDSS), while they change colour accordingly with the emission line shift to other filters. More than 800 such objects were identified in the SDSS and were named Luminous Compact Galaxies in [76].

GP stellar masses are in the range $10^{8}$–$10^{10}\,M_{\odot}$, their SFR $\sim 1 - 60\,M_{\odot}$ year$^{-1}$, and oxygen abundances $12 + \log(O/H) \sim$ 8.05–8.15 [76]. For those where UV data are available (~40 targets [77–80]), the Ly$\alpha$ line is observed in emission and many of the GPs would pass the selection criteria of high-$z$ Lyman-alpha emitters

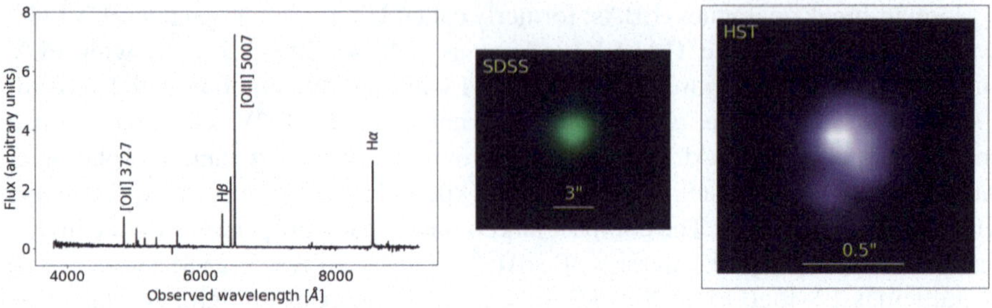


**Fig. 5** Green Pea optical SDSS spectrum (left); optical SDSS image—unresolved (middle); and near-UV *HST* image—obtained for Cosmic Spectrograph spectral acquisition (right). All three panels are for the galaxy SDSS J092532.36+140313.0 at $z \sim 0.3$. The diameter 0.5″ corresponds to ~2 kpc. Credit: SDSS and *HST* archives

(see Sect. 4.2). Their resolved *HST* images typically show a knotty structure (Fig. 5). Some of targets appear compact and structureless even with the *HST* resolution [81], which can however be a question of data depth.

Do the Peas represent a new class or are they related to the other dwarf populations, such as the BCGs? An evolutionary link is improbable, as the average GP masses and metallicities are larger (at an earlier cosmic time) than those of the average BCG. The continuity of the dwarf populations across redshifts is still being debated. Nevertheless, both GPs and BCGs represent the rare types of nearby dwarf starburst galaxies that are more typical of high-$z$ universe, with low masses and metallicities, irregular morphologies, and low dust amounts. In contrast, there is a clear overlap between Green Peas and LBAs. While LBAs were UV-selected, the GPs were identified in the optical, but their follow-up UV observations show that they are conform with the LBA selection criteria.

Green Peas (or Luminous Compact Galaxies) were among the first targets where a large escape of ionizing UV continuum was discovered [81, 82], as we describe in Sect. 5.2. This strengthens their role of local laboratories for the reionization era galaxies ($z > 6$). GPs are bright in X-rays [55], consistently with their high SFR and low metallicity. They are among the brightest star-forming galaxies known to date, exceeding the empirical calibrations such as [56, 58]. This implies the presence of powerful X-ray sources, most likely the end products of massive stars, or potentially low-mass AGN [83]. Nevertheless, this X-ray enhancement is not present in all GPs, some are X-ray sub-luminous [55]. This diversity in X-ray properties probably requires the dominance of extreme, short-lived sources such as ULXs.

### 3.5 Dusty, Luminous Infrared Galaxies (LIRGs)

Galaxies with large amounts of dust are invisible or exceptionally faint in the UV and optical. Instead, their bolometric luminosities are dominated by the dust-reprocessed IR radiation. Galaxies bright in the far-IR wavelengths were first reported in the 1960s, but dramatic progress was achieved with the IR satellites—*IRAS* in the 1980s, Spitzer, Herschel and WISE after the year 2000—and the ground-based SCUBA array. The most luminous targets were classified (somewhat arbitrarily) into Luminous Infrared Galaxies (LIRGs) for luminosity $L_{\mathrm{IR}} \geq 10^{11}\,\mathrm{L}_{\odot}$, and Ultra-luminous Infrared Galaxies (ULIRGs) for $L_{\mathrm{IR}} \geq 10^{12}\,\mathrm{L}_{\odot}$. The dusty infrared galaxies are fascinating objects with colossal star-formation rates reaching $1000\,\mathrm{M}_{\odot}\,\mathrm{year}^{-1}$, which are among the largest known. Such a vigorous SF is necessarily short-lived because of the finite gas supply. The (U)LIRG may therefore represent an evolutionary phase, possibly coming after a violent event such as a gas-rich galaxy merger. The merger would trigger the SF, which is subsequently responsible for the dust and for the obscured (U)LIRG. The obscured phase can possibly become an optically luminous AGN by blowing away the dust. This scheme is still a matter of a lively debate, with uncertainties about its universality between different galaxy types, different redshifts and different environments. The literature about (U)LIRGs is vast, the puzzle of their origin and future as well. We here describe the basic properties and we refer to the detailed reviews provided by Blain et al. [84], Lonsdale et al. [85], and Casey et al. [86], which contain a plethora of additional papers to read.

LIRGs and ULIRGs are massive galaxies, their morphologies include ellipticals, lenticulars, and spirals. Their dust temperatures vary between different targets, may vary with redshift, and multiple dust components are often present in the same galaxy (mostly ranging from 20 to 100 K). It is interesting to note that Ly$\alpha$ emission has been observed in some (U)LIRGs despite the large amounts of dust [87]. Gas kinematics play a decisive role for the Ly$\alpha$ escape there—high-velocity outflows ($900\,\mathrm{km\,s}^{-1}$) were measured, using the UV absorption lines of metals. To account for the large IR luminosity, (U)LIRGs necessarily host a powerful UV source: an AGN, a starburst, or both. Optical studies have found AGN-like spectral lines in $>30\%$ ULIRGs, in parallel with evidence for a strong star formation in all targets. The incidence of AGN possibly increases with luminosity, some studies claim the AGN presence in more-or-less all ULIRGs. The difficulty of AGN confirmation lies in the ULIRG massive obscuration that makes even the X-ray detection uncertain. ULIRGs were originally claimed sub-luminous in X-rays, but sensitive observations and polarized light observations discovered previously unknown active nuclei. Some of the ULIRGs host two or more AGN, and thus provide direct evidence of galaxy mergers. Deep optical and NIR imaging demonstrated that a vast majority of ULIRGs are located in interacting systems, ranging from widely separated galaxies to advanced mergers. Evidence of interaction between more than two galaxies led to the hypothesis that ULIRGs trace the previous presence of compact galaxy groups. Nevertheless, they never reside in rich environments such as galaxy clusters.

The causal relation between (U)LIRGs and quasars remains an open question that is being studied across redshifts by evaluating the clustering properties of both populations as well as their internal structure. One hypothesis is that (U)LIRGs are obscured quasars seen edge-on, and can thus be part of the AGN unification scheme. A prevailing theory is the evolution of (U)LIRGs into unobscured quasars and then into elliptical galaxies [11, 85]. ULIRGs represent only <0.1% galaxies in the local universe, but were much more common in the past, with a distribution peak at redshifts $z = 1$–3. High redshift holds the key to understanding these galaxy populations that we will discuss more in Sect. 4.3.

# 4 Starburst Galaxies at High Redshift

Starburst was a popular mode of star formation in the early universe, unlike today where it is much rarer. High-$z$ starbursts formed a large fraction of today's stars and most probably provided the building blocks for galaxies as we know them today.

The knowledge attained to date about high-$z$ galaxies is less complete than that about the local universe. Until the 1990s, only AGN were observable at redshift $z > 1$ (i.e. universe younger than 7 Gyr). Since then, technological progress in telescopes and detectors has allowed probing galaxy populations out to $z \sim 10$ (i.e. 500 Myr after the Big Bang). Still, their low surface brightness and the small angular size only allow detection of their brightest features, mostly without angular resolution. It is therefore convenient to design observational methods so as to target the prominent starburst signatures that we described in Sect. 2. Essentially all galaxies detected so far at high $z$ are hence starbursts and the galaxies are classified by the detection method, instead of morphology or other physical properties. The samples of brightest high-$z$ galaxies are complemented with lensed galaxies, which provide more detailed, deeper and spatially resolved information, reaching surface brightness much fainter than would be detectable without the gravitational lens.

## *4.1 Lyman-Break Galaxies*

The Lyman break imaging method uses the fact that starbursts are UV-bright, with a sharp drop at the edge of the Lyman sequence, i.e. $\lambda < 912$ Å (Sect. 2.1). The Lyman break is conveniently shifted to optical wavelengths at $z > 2.5$, and to NIR at $z > 7$, reachable by ground-based telescopes. The sky is imaged in a set of optical and NIR filters that sample the galaxy spectral energy distribution (SED). Cross-correlation of the sky maps in different filters will show galaxies that are detected in longer wavelengths and undetected in bluer filters. The disappearance of a galaxy from the filters corresponds to its Lyman break. The dropout filter provides its approximate redshift, with precision $\delta z \sim 0.1 - 0.2$ (Fig. 6). Objects detected by this technique are referred to as Lyman-break galaxies (LBGs) or dropout galaxies. The advantage of the method is the use of broad-band filters, and thus an efficient retrieval of large galaxy samples. The precise redshifts are then secured with follow-up spectroscopy that requires significantly longer exposure times. As we approach the reionization era, the Ly$\alpha$ forest becomes so optically thick that it removes essentially all the flux at $\lambda < 1216$ Å (Sect. 5.1), and therefore the Lyman break technique actually

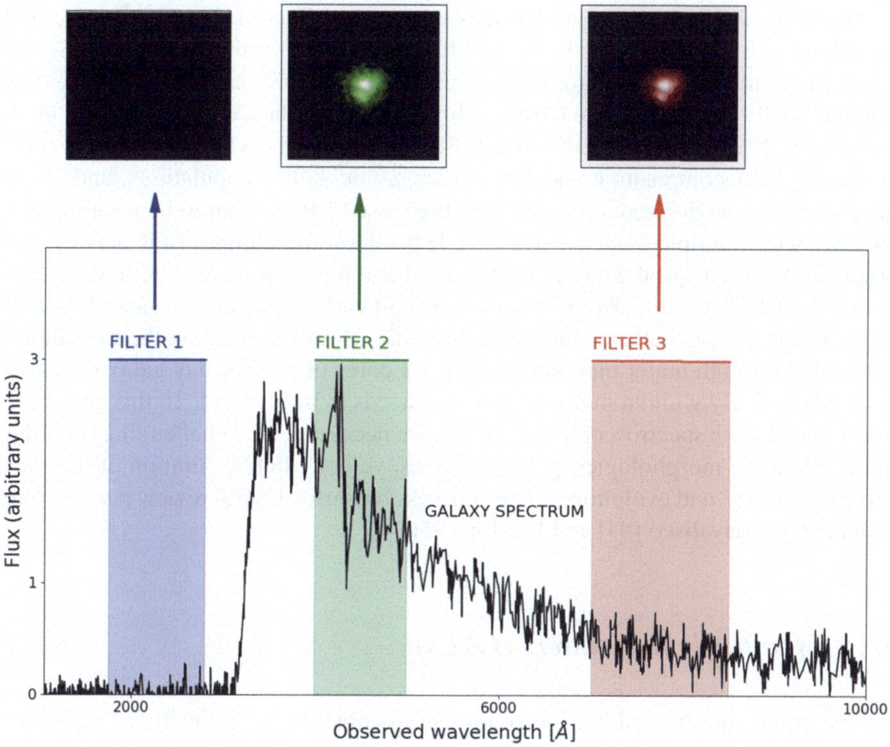


**Fig. 6** LBG detection technique: (mock) galaxy is detected in a set of filters (here green and red), while no detection is achieved in the blue filter. The blue filter thus corresponds to wavelengths shorter than the Lyman break. A typical galaxy spectrum (rest-frame UV) is shown with the black line, here at redshift $z = 2.5$

becomes the Gunn-Peterson trough technique, detecting the sharp drop blueward of Ly$\alpha$.

After the pioneering steps in the LBG technique, breakthrough discoveries were done by Steidel et al. [88] at $z \sim 3$. They started with 4-m telescopes such as the William Herschel Telescope in 1993 and then switched to the 10-m Keck Telescope and to the Hubble Space Telescope (Hubble Deep Field survey). Nowadays, the known LBG samples count tens of thousands of targets, 90% percent of which are situated in the redshift interval $z \sim 2.5 - 3.5$. Observations at $z > 5$ became possible later, with the development of more sensitive instruments, which was especially true for $z > 7$, where the Lyman break shifts to the NIR. The Hubble Deep Field, Hubble Ultra-Deep Field, and the GOODS survey have been prolific at identifying LBG populations out to redshifts $z = 7$–$10$ [89]. In addition, galaxies detected through the Ly$\alpha$ line (Sect. 2.3) may form a subset of the LBGs and provide thus complementary information.

The typical LBGs have stellar masses ranging from $10^9$ to $10^{11}\,M_\odot$, and metallicities from 10% solar to solar. More detailed information is available for a few LBGs that are in fortuitous alignment with a foreground galaxy and are thus gravitationally lensed: e.g. the Cosmic Horseshoe [90], the Cosmic Eye [91], the 8 o'clock arc [92], or the Sunburst Arc [93]. The lensing provides clues about their geometry, ISM composition and kinematics, about stellar populations, and about the escape of ionizing radiation. Most of the lensed LBGs seem to be rotating disk galaxies with multiple giant star-forming H II regions that have a SFR density two orders higher than spiral galaxies seen in the local universe. The comoving number density of LBGs is similar to present-day bright early-type galaxies, and therefore an evolutionary link between the two is possible. However, whether their evolution proceeded through major mergers, or they are cores of present-day galaxies which evolved passively without major events, still needs to be clarified. To this end, deep multi-wavelength spectroscopic follow-ups are necessary (and challenging) in order to constrain the morphologies, precise masses, velocity fields, duration of the star-forming activity, and evolution of LBGs across redshifts. Useful review papers were provided by Giavalisco [94] and Dunlop [95].

## *4.2 Lyman-Alpha Emitters (LAEs)*

Lyman-alpha emitting galaxies were important constituents of the high-$z$ universe, as they represent the majority of the known star-forming galaxies at redshifts close to reionization. Using Ly$\alpha$ as a detection tool was proposed as early as the 1960s [25]. However, the first Ly$\alpha$ detections were not available until the 1980s [28] when sizes and sensitivities of telescopes and detectors became sufficient. The Ly$\alpha$ fluxes were several factors lower than expected, as we explained in Sect. 2.3. Still today, predicting the Ly$\alpha$ luminosity for an individual galaxy is not straightforward due to the multi-parameter nature of the Ly$\alpha$ escape. The largest ground-based telescopes and space telescopes (*HST* and soon *JWST*) are used for observing high-$z$ LAEs today. The asymmetric Ly$\alpha$ spectral profile (Sect. 2.3, Fig. 3) presents an advantage for high-$z$ surveys, where it allows a unique identification of the line and hence of the galaxy redshift. Recently, LAE surveys have gained another dimension thanks to the use of integral-field spectrographs such as MUSE at the ESO VLT [37]. Simultaneous observation of spectra from different parts of the galaxy allows testing the theories of Ly$\alpha$ transfer from the production sites all the way to the outskirts.

Narrow-band Ly$\alpha$ imaging surveys pick galaxies where Ly$\alpha$ is produced in large quantities and is only weakly attenuated. This requires a powerful starburst activity together with low amounts of dust and neutral hydrogen, as well as the presence of outflows that shift Ly$\alpha$ out of resonance. These conditions are met in low-mass galaxies (typically $10^8$–$10^{10}\,M_\odot$), but not only. The low stellar mass facilitates the action of stellar feedback that can more easily outweigh the gravitational potential. However, the conditions that allow Ly$\alpha$ to escape span a certain range (Sect. 2.3), and therefore the possible LAE masses and dust amounts can be larger in special cases. The escape of Ly$\alpha$ photons may also be non-isotropic due to the disk galaxy morphology [34], due to gas clumpiness or preferential directions of escape [96, 97]. To improve our understanding of what galaxy population is selected by Ly$\alpha$, several studies have searched for the differences between LAEs and the continuum-selected LBG or the emission-line-selected (in the sense of rest-frame optical) galaxies on the same redshifts. No statistical differences were found between LAEs and optical emission-line galaxies in the morphology, inclination, sSFR, UV slope or the distance between neighbours [98], suggesting that the Ly$\alpha$ escape is driven by the detailed local physics rather than by global properties. Some LBGs are bright in Ly$\alpha$, therefore the LBG and LAE populations at least partially overlap. The mean LAE mass is usually found to be one or two orders of magnitude lower than the mean LBG mass. However, there are studies that claim no differences between the two populations (discussed in [98]). On the other hand, clustering properties of LAEs and LBGs seem to differ, suggesting their different evolutionary paths [95, 99].

The fraction of Ly$\alpha$ photons that escape from galaxies evolves with redshift [100]: the average escape fraction increases from $z = 0$ to $z = 6$ and declines at $z > 6.5$. The increase correlates with a decreasing dust amount in the cosmos, suggesting that dust could be the main driver of the *average* Ly$\alpha$ escape (modulated by other parameters [101]). The escape fraction drop at $z > 6.5$ can be attributed to the progressively more neutral IGM i.e. an effect of the environment rather than the galaxies themselves. The drop may also be an intrinsic feature of the reionization era galaxies, either due to a large fraction of static H I (as in the local IZw18) or the opposite, the Lyman continuum escape. At redshift $z > 6$, not only the escape fraction, but also the number of observed LAEs decreases, either due their intrinsically lower numbers or due to the neutral IGM. LAEs are currently observed out to redshifts $z > 10$, and play thus an important role as cosmological tools [99, 102–104]. Clustering properties map the dark matter structures and the evolution of LAEs through redshifts to the present day. The clustering suggests that they were the building blocks of the present-day MW-type galaxies [105]. The LAE clustering in the reionization era probably also maps the patchy IGM ionization. The importance of LAEs is reflected in the objectives of the new generation of telescopes, such as the *JWST* or the ELTs, with the ambition to use Ly$\alpha$ to detect the first galaxies in the universe.

### *4.3 Sub-millimeter Galaxies*

High-redshift counterparts of local ULIRGs, i.e. dust-obscured, massive starbursts (Sect. 3.5), are known under the name Sub-millimeter galaxies (SMGs). Their detection has been effectively possible since the operation of the SCUBA bolometer array on Mauna Kea in the 1990s (see review [84]). Nowadays, several sensitive mm/sub-mm telescopes are available, including IRAM (Spain), NOEMA (France), or ALMA (Chile). To facilitate the orientation in the vast literature on the subject, we refer preferentially to review papers in this text.

Typically scaled-up versions of ULIRGs, the SMGs are the most intense SF sites and the most bolometrically luminous galaxies in the universe ($L_{\rm bol} \sim 10^{13}\,{\rm L}_\odot$). Their molecular gas masses reach $10^{11} M_\odot$ and they fuel star-formation rates exceeding $1000\,{\rm M}_\odot\,{\rm year}^{-1}$. ULIRGs were once thought to be evolved versions of SMGs. However, today's view is that SMGs were more massive and cooler than ULIRGs and thus any evolutionary link between them is unlikely. While the number density of nearby ULIRGs is one per four square degrees, there were hundreds per one square degree at $z > 1$ [85]. Cosmologically, SMGs should be the tracers of massive dark matter halos and their evolution across cosmic history. Recent instrumental advances (especially ALMA) have permitted the first studies of SMG clustering properties (reviewed in [86]), which will allow testing the theories of the SMG origin and their fate. High-$z$ SMGs were predicted to reside in rich environments where they would accumulate their mass. According to today's knowledge, SMGs mostly appear to live in relative isolation (contrary to the predictions), even though they bear signs of galaxy mergers. Only the most luminous SMGs have been identified to reside in potential proto-clusters of galaxies [86].

SMGs show a surprisingly diverse range of optical properties, from undetected to bright sources. The optical spectra provide evidence for both starburst and AGN, residing in metal-rich gas. The bolometric luminosity is typically dominated by star formation. The lower limit for AGN incidence in SMGs is 30%, though in reality is probably significantly larger. The incidence seems to increase with luminosity and therefore with redshift. Intriguing results have been provided by radio observations showing that ~70% of SMGs are spatially extended on ~10 kpc scales, i.e. dramatically larger than the dust emission that tends to be concentrated in <1 kpc. Several interpretations of the radio emission exist, ranging from large-scale SF to jet-like structures similar to those in radio AGN [85]. SMGs are also known as sites of gamma-ray bursts (GRBs), ignited at the death of massive stars.

The present-day mainstream hypothesis is that SMGs evolved into quasars and eventually into early-type galaxies (nicely depicted in [11]). The SMG masses and SFRs indeed make them the ideal progenitors of elliptical galaxies. The existence of large numbers of massive galaxies such as SMGs out to high redshift represents a challenge to the simple hierarchical galaxy formation paradigm that predicted slow mass buildup through small galaxy mergers. On the other hand, it is not in contradiction with dark matter halo growth, which allows for rapid baryon accumulation in very massive halos [85]. The SMGs, their place in galaxy evolution, their progenitors and their successors thus play an important role in answering fundamental astrophysical questions such as the buildup of stellar mass in the universe, the origin of present-day galaxies, and the growth of supermassive black holes.

# 5 Role of Starburst Galaxies in Reionizing the Universe

## *5.1 Cosmic Reionization*

The present-day intergalactic medium is essentially fully ionized, but it was not so during all of the cosmic history. After the early evolutionary stages where matter and radiation were constantly interacting, the universe became cool enough for radiation to split from matter at $z \sim 1000$ (i.e. ~300,000 years after the Big Bang). The radiation then freely travelled through the universe without being absorbed by atoms. This was the start of Dark Ages, where the universe was neutral and without sources of radiation (apart from the H I line at 21 cm). The first stars are supposed to have formed at $z \sim 30$. Current observations have not yet reached such distant redshifts to capture their light. Nevertheless, an exceptional result was recently achieved: an absorption feature was observed in the H I 21 cm line profile, consistent with theoretical predictions of the effect of the first stars [106]. The observation thus proved that stars had been formed by $z \sim 20$. The phenomenon will further be explored by upcoming surveys such as LOFAR, HERA or SKA.

The birth of the first sources of light caused a phase change in the IGM: the IGM progressively evolved from neutral to a fully ionized state. The era between the appearance of the first stars and the redshift $z \sim 6$ where the IGM became essentially completely ionized is called the Cosmic Reionization. The end of reionization is well demonstrated by the Gunn-Peterson effect in quasars [107]: the FUV light of a distant quasar (or another bright source) encounters hydrogen clouds along the line of sight and the H I scatters away the light that gets in resonance with Ly$\alpha$ at the cloud's redshift. At different redshifts, different parts of the spectrum "become" Ly$\alpha$, and, as a result, the original spectrum adopts numerous absorption features, known as the "Ly$\alpha$ forest". At $z > 6$, the discrete features turn into a continuous Gunn-Peterson trough that removes all the FUV at wavelengths shorter than the quasar's rest-frame Ly$\alpha$ [108]. Additional constraints for reionization come from the cosmic microwave background radiation measured with the *WMAP* and Planck satellites [109, 110].

## *5.2 Sources of Reionization: Starburst Galaxies?*

Details of the cosmic reionization process depend on the density and distribution of gas, and, most importantly, on the available ionization sources, their density, distribution, evolution and energy spectrum (see review [111]). The first galaxies provided ionizing FUV radiation from hot, massive stars, and are thus natural candidates for reionization. Furthermore, quasars set in at some point, providing copious amounts of energetic radiation. Additional sources such as X-ray binaries or cosmic rays are being considered as possible ionization contributors (see references in [55]).

The distribution of the first galaxies was associated with dark matter structures, which are possible to model using cosmological simulations. Their ionizing photon production rate depends on the nature of the first stars and on the star formation rate density, for both of which we need an observational input in order to achieve realistic predictions [111]. However, yet another parameter controls the ionizing radiation (Lyman continuum) flux to the IGM: its escape fraction, reflecting the galaxy's ISM structure (total H I amount, dust, asymmetry, clumpiness). Numerical simulations predict that despite a strong variability of the escape fraction in time and in angular direction, an average value of $\sim$20% per galaxy is necessary to achieve the reionization solely by star-forming galaxies [112]. On the observational side, direct detection of the Lyman continuum from the reionization era galaxies is unfeasible: any ionizing photon that escapes from the galaxy will be absorbed by the neutral IGM and will not reach our telescopes. Only indirect signatures can be explored at the reionization era, while direct studies of the Lyman continuum leakage are restricted to lower redshifts, in practice $z \lesssim 3$, where the IGM H I amount is sufficiently low.

Searches for escaping Lyman continuum at any redshift were fruitless during two decades of observational efforts, except for a few sources with low leakage fractions [74, 113]. The situation has dramatically changed in the past few years. In 2016–2019, Lyman continuum was spectroscopically confirmed in approximately fourty $z \sim 3$ galaxies [114–118]. In parallel, a significant leakage was detected with the *HST* in eleven starburst galaxies of Green-Pea type at $z \sim 0.3$ (summarized in [81]). The Lyman continuum escape fraction reaches 70% in some of these targets, while their mean is $\sim$20%. Interestingly, this percentage is consistent with the numerical predictions mentioned above. Despite their too low redshift, the leaking galaxies support the possibility of cosmic reionization by their $z > 6$ analogues.

The low-$z$ samples serve as laboratories for understanding the physical conditions that lead to a low optical depth at Lyman continuum wavelengths. From the current observations, we may speculate about the role of the galaxy compactness and low mass, which favour the escape of ISM gas from a shallow potential well and thus decrease the H I column. Second, low metallicity modifies stellar evolution, leading to a larger production of ionizing photons. Third, feedback from starburst-related processes (stellar winds, energetic photons and jets/outflows from compact binary systems) deposits radiative and mechanical energy into the ISM, which results in gas ionization and/or its removal along certain paths. Indirect indicators are being explored that would help us preselecting galaxies with Lyman continuum escape: various works probe the ISM ionization [119, 120], ISM velocities [19], saturation of the ISM spectral lines [121], X-ray emission [55], and the Ly$\alpha$ line profile [97, 122]. So far, the Ly$\alpha$ line was proven, both theoretically and observationally, to be the best indicator of the Lyman continuum escape, thanks to its sensitivity to the ISM properties [81, 97, 122].

A natural question is whether quasars, efficient producers of ionizing photons, could be the true sources of cosmic reionization instead of the star-forming galaxies. The question remains open so far. The first quasars formed in the reionization era, but it seems that they were not abundant enough at $z > 6$ to dominate the reionization. However, with the scarcity of data, any new detection may change the picture [123–125]. Nevertheless, the Lyman continuum of starburst galaxies appears to be the most viable mechanism for reionization, winning over quasars or any other sources.

## 6 Conclusions and Future Prospects

It has been estimated in the literature that equal amounts of stars in the universe were formed in the following four types of environments: optically visible regions, dust-enshrouded regions of optical galaxies, heavily obscured galaxies with $L_{\rm IR} < 10^{12}\,{\rm L}_{\odot}$, and ULIRGs with $L_{\rm IR} > 10^{12}\,{\rm L}_{\odot}$ [126]. Dwarf galaxies dominated the star formation among the optically visible ones, while massive starburst galaxies are dusty and IR-bright. We have reviewed the different types of starburst galaxies both at low and high redshift. Without being complete, this overview illustrated the various possible manifestations of vigorous star formation, the corresponding detection techniques, the properties of various starburst galaxy populations, and their role in cosmic evolution. We demonstrated the search for analogies between low and high-redshift galaxies. Analogies allow complementary views on star formation and galaxy evolution, and on the origin and the future of starburst galaxy populations. In light of the analogies with high redshift, local galaxies have seen an increased interest in their classification by parameters such as the SFR, metallicity, FUV luminosity or compactness. While the overlap between high-$z$ and low-$z$ galaxy parameters cannot be perfect, the nearby galaxies serve as local laboratories that provide essential clues for the interpretation of processes in the distant universe where the details are unattainable.

Research on starburst galaxies has good prospects for future, and in particular galaxies at the cosmic dawn are among the main drivers for the new generation of astronomical instruments. MUSE, a sensitive optical integral-field spectrograph is already operating at ESO VLT. Its ability to simultaneously obtain hundreds of spectra per galaxy combines the advantages of classical imaging and spectroscopy. In combination with adaptive optics, MUSE reaches an angular resolution comparable to that of the *HST* and it can outperform the *HST* in detecting distant galaxies [127, 128]. Spatially resolved spectra of LAEs observed at $z \in (3, 6)$ with MUSE have brought unprecedented details on the structure of these distant starbursts and resolved individually their diffuse Ly$\alpha$ halos [37, 129]. ALMA, a recently built international facility in Chile, is an interferometer composed of 66 antennas observing at mm and sub-mm wavelengths. It is well suited for observing dusty starbursts and SMGs at $z > 2$, their distribution across redshifts, and possibly their evolution into massive early-type galaxies [130, 131]. ALMA is also efficient at mapping the distribution of dust in low-mass predecessors of MW-type galaxies, such as the UV-selected LBGs and LAEs [132] or the Chandra-detected X-ray galaxies [130]. ALMA has a sufficient sensitivity to resolve high-$z$ infrared atomic and molecular lines that probe galaxy metallicities, ionizations and kinematics, and determine their precise redshifts [133, 134]. In the local universe, ALMA is mapping hundreds of thousands of stellar nurseries in nearby galaxies with the goal to shed light on the star-formation mechanisms and the conditions under which the formation proceeds by intense bursts.

In the years to come, several major instruments will become active. HARMONI is a visible/NIR integral-field spectrograph in construction for the E-ELT's first light in the mid-2020s. Assisted with adaptive optics, the HARMONI angular resolution will be comparable to that achieved with space-borne instruments. The optical/NIR wavelengths will detect rest-frame UV (Ly$\alpha$, continuum, lines) in the reionization era. One of the goals is the detection of Population-III stars, i.e. primordial stars with no heavy elements, possibly reaching hundreds of solar masses. Their detection through the ionized helium lines has been claimed possible with HARMONI out to $z > 10$. HARMONI will thus play an important role in probing the high-mass end of the IMF. In the local universe, HARMONI will resolve individual stars in nearby galaxies and supernovae out to $z \sim 3 - 4$, testing our models of star-formation and galaxy assembly histories. HARMONI already has a dedicated simulation tool to predict the future observational outcomes, using mock data from cosmological simulations [135]. The tool may also be an essential preparatory step for observers once the instrument is operational.

Space missions play an ever larger role in cosmic studies. The *JWST* will operate in the IR domain and will detect the rest-frame UV of reionization era galaxies. The telescope will possess the first space-borne integral-field spectrograph, NIRSPEC. Compared to the ground-based instruments, NIRSPEC will have much more severe limitations on the number of spectra that it can cover simultaneously (100 as opposed to 31,000 in HARMONI). On the other hand, it will have the advantages of the space mission, unaffected by the Earth's atmosphere. Among other future missions, ESA's *EUCLID* will map the large cosmic structure and its evolution. It will produce deep optical and NIR images and spectra of galaxies out to $z \sim 2$ across approximately half of the sky. The unprecedented extent of this catalogue will provide the basis for statistical studies mapping the galaxy assembly in the past ten billion years. In X-rays, the newly planned mission *ATHENA* will have a high sensitivity and high angular resolution. It will complete the census of evolved binary sources in starburst regions down to faint X-ray fluxes. It will resolve the spatial structure of X-ray emitting regions in galaxies and will constrain the dominant sources, discriminating between X-ray binaries, AGN, hot gas and other. Eventually, ATHENA will open the X-ray window for high-redshift galaxies, too faint to be reachable individually by current instruments.

Complementary to direct detections of starburst signatures, radio telescopes such as SKA, HERA or LOFAR will map the atomic hydrogen distribution in the universe at its different epochs, including the reionization era. Using arrays of hundreds of radio antennas, these interferometers will resolve how the neutral IGM structures evolved, in what conditions formed the first stars and galaxies and how they continue to form today.

Astronomy is experiencing a golden age and the quest for understanding how stars formed and how galaxies assembled through the cosmic history is one of its flagships. The existing and upcoming facilities, using multiple spectral windows, take us progressively to the very first structures that formed in the universe. Supported by numerical simulations and by detailed, resolved local studies, they will complete our picture of the conditions for star formation at different epochs, the origin of galaxies as we know them today, the existence of extreme objects, and the mechanisms driving galaxy evolution.

**Acknowledgements** The author is grateful to the referee for their thoughtful comments that made this review more complete. The author was supported by the Czech Science Foundation project 17-06217Y while working on the manuscript.

## References

1. R.C. Kennicutt Jr., Astrophys. J. **498**, 541 (1998). https://doi.org/10.1086/305588
2. F. Bigiel, A. Leroy, F. Walter, E. Brinks, W.J.G. de Blok, B. Madore, M.D. Thornley, Astron. J. **136**(6), 2846 (2008). https://doi.org/10.1088/0004-6256/136/6/2846
3. R.C. Kennicutt, N.J. Evans, Annu. Rev. Astron. Astrophys. **50**, 531 (2012). https://doi.org/10.1146/annurev-astro-081811-125610
4. R.B. Larson, B.M. Tinsley, Astrophys. J. **219**, 46 (1978). https://doi.org/10.1086/155753
5. P.A. Duc, I.F. Mirabel, Astron. Astrophys. **289**, 83 (1994)
6. S.L. Ellison, D.R. Patton, L. Simard, A.W. McConnachie, Astron. J. **135**(5), 1877 (2008). https://doi.org/10.1088/0004-6256/135/5/1877
7. P. Di Matteo, F. Combes, A.L. Melchior, B. Semelin, Astron. Astrophys. **468**(1), 61 (2007). https://doi.org/10.1051/0004-6361:20066959
8. P.F. Hopkins, E. Quataert, Mon. Not. R. Astron. Soc. **407**(3), 1529 (2010). https://doi.org/10.1111/j.1365-2966.2010.17064.x
9. B.G. Elmegreen, in *The Spectral Energy Distribution of Galaxies - SED 2011*, ed. by R.J. Tuffs, C.C. Popescu. IAU Symposium, vol. 284 (2012), pp. 317–329
10. B.G. Elmegreen, *Formation of Stars and Clusters over Cosmological Time* (2015), p. 477
11. D.M. Alexander, R.C. Hickox, New Astron. Rev. **56**(4), 93 (2012). https://doi.org/10.1016/j.newar.2011.11.003
12. H. Gerola, P.E. Seiden, Astrophys. J. **223**, 129 (1978). https://doi.org/10.1086/156243
13. T.M. Heckman, in *Starbursts: From 30 Doradus to Lyman Break Galaxies*, ed. by R. de Grijs, R.M. González Delgado. Astrophysics and Space Science Library, vol. 329 (2005), p. 3
14. N. Bergvall, Astrophys. Space Sci. Proc. **28**, 175 (2012)
15. G. Östlin, P. Amram, N. Bergvall, J. Masegosa, J. Boulesteix, I. Márquez, Astron. Astrophys. **374**, 800 (2001). https://doi.org/10.1051/0004-6361:20010832
16. N. Bergvall, T. Marquart, M.J. Way, A. Blomqvist, E. Holst, G. Östlin, E. Zackrisson, Astron. Astrophys. **587**, A72 (2016). https://doi.org/10.1051/0004-6361/201525692
17. D. Calzetti, *Star Formation Rate Indicators* (2013), p. 419
18. N.R. Walborn, J. Nichols-Bohlin, R.J. Panek, NASA Reference Publication 1155 (1985)
19. T.M. Heckman, S. Borthakur, R. Overzier, G. Kauffmann, A. Basu-Zych, C. Leitherer, K. Sembach, D.C. Martin, R.M. Rich, D. Schiminovich, M. Seibert, Astrophys. J. **730**, 5 (2011). https://doi.org/10.1088/0004-637X/730/1/5
20. J. Chisholm, C.A. Tremonti, C. Leitherer, Y. Chen, A. Wofford, B. Lundgren, Astrophys. J. **811**, 149 (2015). https://doi.org/10.1088/0004-637X/811/2/149
21. A.E. Shapley, C.C. Steidel, M. Pettini, K.L. Adelberger, Astrophys. J. **588**, 65 (2003). https://doi.org/10.1086/373922
22. B.D. Savage, K.R. Sembach, Astrophys. J. **379**, 245 (1991). https://doi.org/10.1086/170498

23. J. Chisholm, I. Orlitová, D. Schaerer, A. Verhamme, G. Worseck, Y.I. Izotov, T.X. Thuan, N.G. Guseva, Do galaxies that leak ionizing photons have extreme outflows? Astron. Astrophys. **605**, A67 (2017). https://doi.org/10.1051/0004-6361/201730610
24. S. Gazagnes, J. Chisholm, D. Schaerer, A. Verhamme, J.R. Rigby, M. Bayliss, Astron. Astrophys. **616**, A29 (2018). https://doi.org/10.1051/0004-6361/201832759
25. R.B. Partridge, P.J.E. Peebles, Astrophys. J. **147**, 868 (1967). https://doi.org/10.1086/149079
26. M. Dijkstra, Publ. Astron. Soc. Aust. **31**, e040 (2014). https://doi.org/10.1017/pasa.2014.33
27. D.L. Meier, R. Terlevich, Astrophys. J. Lett. **246**, L109 (1981). https://doi.org/10.1086/183565
28. S. Djorgovski, H. Spinrad, P. McCarthy, M.A. Strauss, Astrophys. J. Lett. **299**, L1 (1985). https://doi.org/10.1086/184569
29. D. Kunth, C. Leitherer, J.M. Mas-Hesse, G. Östlin, A. Petrosian, Astrophys. J. **597**(1), 263 (2003). https://doi.org/10.1086/378396
30. D.A. Neufeld, Astrophys. J. **350**, 216 (1990). https://doi.org/10.1086/168375
31. M. Dijkstra, Z. Haiman, M. Spaans, Astrophys. J. **649**, 14 (2006). https://doi.org/10.1086/506243
32. A. Verhamme, D. Schaerer, A. Maselli, Astron. Astrophys. **460**, 397 (2006). https://doi.org/10.1051/0004-6361:20065554
33. M. Gronke, M. Dijkstra, M. McCourt, S. Peng Oh, Resonant line transfer in a fog: using Lyman-alpha to probe tiny structures in atomic gas. Astron. Astrophys. **607**, A71 (2017). https://doi.org/10.1051/0004-6361/201731013
34. A. Verhamme, Y. Dubois, J. Blaizot, T. Garel, R. Bacon, J. Devriendt, B. Guiderdoni, A. Slyz, Astron. Astrophys. **546**, A111 (2012). https://doi.org/10.1051/0004-6361/201218783
35. M. Hayes, G. Östlin, D. Schaerer, A. Verhamme, J.M. Mas-Hesse, A. Adamo, H. Atek, J.M. Cannon, F. Duval, L. Guaita, E.C. Herenz, D. Kunth, P. Laursen, J. Melinder, I. Orlitová, H. Otí-Floranes, A. Sandberg, Astrophys. J. Lett. **765**, L27 (2013). https://doi.org/10.1088/2041-8205/765/2/L27
36. R. Momose, M. Ouchi, K. Nakajima, Y. Ono, T. Shibuya, K. Shimasaku, S. Yuma, M. Mori, M. Umemura, Mon. Not. R. Astron. Soc. **442**, 110 (2014). https://doi.org/10.1093/mnras/stu825
37. L. Wisotzki, R. Bacon, J. Blaizot, J. Brinchmann, E.C. Herenz, J. Schaye, N. Bouché, S. Cantalupo, T. Contini, C.M. Carollo, J. Caruana, J.B. Courbot, E. Emsellem, S. Kamann, J. Kerutt, F. Leclercq, S.J. Lilly, V. Patrício, C. Sandin, M. Steinmetz, L.A. Straka, T. Urrutia, A. Verhamme, P.M. Weilbacher, M. Wendt, Astron. Astrophys. **587**, A98 (2016). https://doi.org/10.1051/0004-6361/201527384
38. D. Kunth, J.M. Mas-Hesse, E. Terlevich, R. Terlevich, J. Lequeux, S.M. Fall, Astron. Astrophys. **334**, 11 (1998)
39. D. Kunth, J. Lequeux, W.L.W. Sargent, F. Viallefond, Astron. Astrophys. **282**, 709 (1994)
40. J.A. Baldwin, M.M. Phillips, R. Terlevich, Publ. Astron. Soc. Pac. **93**, 5 (1981). https://doi.org/10.1086/130766
41. G. Kauffmann, T.M. Heckman, C. Tremonti, J. Brinchmann, S. Charlot, S.D.M. White, S.E. Ridgway, J. Brinkmann, M. Fukugita, P.B. Hall, Ž. Ivezić, G.T. Richards, D.P. Schneider, Mon. Not. R. Astron. Soc. **346**(4), 1055 (2003). https://doi.org/10.1111/j.1365-2966.2003.07154.x
42. L.J. Kewley, M.A. Dopita, R.S. Sutherland, C.A. Heisler, J. Trevena, Astrophys. J. **556**, 121 (2001). https://doi.org/10.1086/321545
43. G. Stasińska, R. Cid Fernandes, A. Mateus, L. Sodré, N.V. Asari, Mon. Not. R. Astron. Soc. **371**(2), 972 (2006). https://doi.org/10.1111/j.1365-2966.2006.10732.x
44. L.J. Kewley, M.A. Dopita, C. Leitherer, R. Davé, T. Yuan, M. Allen, B. Groves, R. Sutherland, Astrophys. J. **774**, 100 (2013). https://doi.org/10.1088/0004-637X/774/2/100
45. D. Kunth, G. Östlin, Astron. Astrophys. Rev. **10**, 1 (2000). https://doi.org/10.1007/s001590000005
46. R.A. Remillard, J.E. McClintock, Annu. Rev. Astron. Astrophys. **44**(1), 49 (2006). https://doi.org/10.1146/annurev.astro.44.051905.092532

47. P. Kaaret, H. Feng, T.P. Roberts, Annu. Rev. Astron. Astrophys. **55**(1), 303 (2017). https://doi.org/10.1146/annurev-astro-091916-055259
48. J.E. Greene, J. Strader, L.C. Ho (2019). e-Prints. arXiv:1911.09678
49. T.X. Thuan, F.E. Bauer, P. Papaderos, Y.I. Izotov, Astrophys. J. **606**(1), 213 (2004). https://doi.org/10.1086/382949
50. J.P. Grimes, T. Heckman, D. Strickland, W.V. Dixon, K. Sembach, R. Overzier, C. Hoopes, A. Aloisi, A. Ptak, Feedback in the local Lyman-break Galaxy Analog Haro 11 as probed by far-ultraviolet and X-ray observations. Astrophys. J. **668**(2), 891–905 (2007). https://doi.org/10.1086/521353
51. H. Otí-Floranes, J.M. Mas-Hesse, E. Jiménez-Bailón, D. Schaerer, M. Hayes, G. Östlin, H. Atek, D. Kunth, Astron. Astrophys. **546**, A65 (2012). https://doi.org/10.1051/0004-6361/201219318
52. H. Otí-Floranes, J.M. Mas-Hesse, E. Jiménez-Bailón, D. Schaerer, M. Hayes, G. Östlin, H. Atek, D. Kunth, Astron. Astrophys. **566**, A38 (2014). https://doi.org/10.1051/0004-6361/201323069
53. A.R. Basu-Zych, B.D. Lehmer, A.E. Hornschemeier, T.S. Gonçalves, T. Fragos, T.M. Heckman, R.A. Overzier, A.F. Ptak, D. Schiminovich, Astrophys. J. **774**, 152 (2013). https://doi.org/10.1088/0004-637X/774/2/152
54. M. Brorby, P. Kaaret, Mon. Not. R. Astron. Soc. **470**(1), 606 (2017). https://doi.org/10.1093/mnras/stx1286
55. J. Svoboda, V. Douna, I. Orlitová, M. Ehle, Astrophys. J. **880**(2), 144 (2019). https://doi.org/10.3847/1538-4357/ab2b39
56. P. Ranalli, A. Comastri, G. Setti, Astron. Astrophys. **399**, 39 (2003). https://doi.org/10.1051/0004-6361:20021600
57. V.M. Douna, L.J. Pellizza, I.F. Mirabel, S.E. Pedrosa, Astron. Astrophys. **579**, A44 (2015). https://doi.org/10.1051/0004-6361/201525617
58. M. Brorby, P. Kaaret, A. Prestwich, I.F. Mirabel, Mon. Not. R. Astron. Soc. **457**(4), 4081 (2016). https://doi.org/10.1093/mnras/stw284
59. M. Cerviño, J.M. Mas-Hesse, D. Kunth, Astron. Astrophys. **392**, 19 (2002). https://doi.org/10.1051/0004-6361:20020785
60. J.M. Mas-Hesse, H. Otí-Floranes, M. Cerviño, Astron. Astrophys. **483**(1), 71 (2008). https://doi.org/10.1051/0004-6361:20078398
61. S. Mineo, M. Gilfanov, B.D. Lehmer, G.E. Morrison, R. Sunyaev, Mon. Not. R. Astron. Soc. **437**, 1698 (2014). https://doi.org/10.1093/mnras/stt1999
62. F. Zwicky, Astrophys. J. **142**, 1293 (1965). https://doi.org/10.1086/148411
63. W.L.W. Sargent, Astrophys. J. **160**, 405 (1970). https://doi.org/10.1086/150443
64. R. Terlevich, J. Melnick, J. Masegosa, M. Moles, M.V.F. Copetti, Astron. Astrophys. Suppl. Ser. **91**, 285 (1991)
65. I. Gallagher, S. John, D.A. Hunter, Astron. J. **94**, 43 (1987). https://doi.org/10.1086/114445
66. L. Searle, W.L.W. Sargent, Astrophys. J. **173**, 25 (1972). https://doi.org/10.1086/151398
67. M. Hayes, Publ. Astron. Soc. Aust. **32**, e027 (2015). https://doi.org/10.1017/pasa.2015.25
68. G. Östlin, M. Hayes, F. Duval, A. Sandberg, T. Rivera-Thorsen, T. Marquart, I. Orlitová, A. Adamo, J. Melinder, L. Guaita, H. Atek, J.M. Cannon, P. Gruyters, E.C. Herenz, D. Kunth, P. Laursen, J.M. Mas-Hesse, G. Micheva, H. Otí-Floranes, S.A. Pardy, M.M. Roth, D. Schaerer, A. Verhamme, Astrophys. J. **797**, 11 (2014). https://doi.org/10.1088/0004-637X/797/1/11
69. T.E. Rivera-Thorsen, M. Hayes, G. Östlin, F. Duval, I. Orlitová, A. Verhamme, J.M. Mas-Hesse, D. Schaerer, J.M. Cannon, H. Otí-Floranes, A. Sandberg, L. Guaita, A. Adamo, H. Atek, E.C. Herenz, D. Kunth, P. Laursen, J. Melinder, Astrophys. J. **805**, 14 (2015). https://doi.org/10.1088/0004-637X/805/1/14
70. E.C. Herenz, P. Gruyters, I. Orlitova, M. Hayes, G. Östlin, J.M. Cannon, M.M. Roth, A. Bik, S. Pardy, H. Otí-Floranes, J.M. Mas-Hesse, A. Adamo, H. Atek, F. Duval, L. Guaita, D. Kunth, P. Laursen, J. Melinder, J. Puschnig, T.E. Rivera-Thorsen, D. Schaerer, A. Verhamme, Astron. Astrophys. **587**, A78 (2016). https://doi.org/10.1051/0004-6361/201527373

71. S.A. Pardy, J.M. Cannon, G. Östlin, M. Hayes, T. Rivera-Thorsen, A. Sandberg, A. Adamo, E. Freeland, E.C. Herenz, L. Guaita, D. Kunth, P. Laursen, J.M. Mas-Hesse, J. Melinder, I. Orlitová, H. Otí-Floranes, J. Puschnig, D. Schaerer, A. Verhamme, Astrophys. J. **794**, 101 (2014). https://doi.org/10.1088/0004-637X/794/2/101
72. T.M. Heckman, C.G. Hoopes, M. Seibert, D.C. Martin, S. Salim, R.M. Rich, G. Kauffmann, S. Charlot, T.A. Barlow, L. Bianchi, Y.I. Byun, J. Donas, K. Forster, P.G. Friedman, P.N. Jelinsky, Y.W. Lee, B.F. Madore, R.F. Malina, B. Milliard, P.F. Morrissey, S.G. Neff, D. Schiminovich, O.H.W. Siegmund, T. Small, A.S. Szalay, B.Y. Welsh, T.K. Wyder, Astrophys. J. Lett. **619**(1), L35 (2005). https://doi.org/10.1086/425979
73. A.R. Basu-Zych, B. Lehmer, T. Fragos, A. Hornschemeier, M. Yukita, A. Zezas, A. Ptak, Astrophys. J. **818**(2), 140 (2016). https://doi.org/10.3847/0004-637X/818/2/140
74. E. Leitet, N. Bergvall, M. Hayes, S. Linné, E. Zackrisson, Astron. Astrophys. **553**, A106 (2013). https://doi.org/10.1051/0004-6361/201118370
75. C. Cardamone, K. Schawinski, M. Sarzi, S.P. Bamford, N. Bennert, C.M. Urry, C. Lintott, W.C. Keel, J. Parejko, R.C. Nichol, D. Thomas, D. Andreescu, P. Murray, M.J. Raddick, A. Slosar, A. Szalay, J. Vandenberg, Mon. Not. R. Astron. Soc. **399**, 1191 (2009). https://doi.org/10.1111/j.1365-2966.2009.15383.x
76. Y.I. Izotov, N.G. Guseva, T.X. Thuan, Astrophys. J. **728**, 161 (2011). https://doi.org/10.1088/0004-637X/728/2/161
77. A.E. Jaskot, M.S. Oey, Astrophys. J. **766**, 91 (2013). https://doi.org/10.1088/0004-637X/766/2/91
78. A. Henry, C. Scarlata, C.L. Martin, D. Erb, Astrophys. J. **809**, 19 (2015). https://doi.org/10.1088/0004-637X/809/1/19
79. H. Yang, S. Malhotra, M. Gronke, J.E. Rhoads, C. Leitherer, A. Wofford, T. Jiang, M. Dijkstra, V. Tilvi, J. Wang, Astrophys. J. **844**(2), 171 (2017). https://doi.org/10.3847/1538-4357/aa7d4d
80. I. Orlitová, A. Verhamme, A. Henry, C. Scarlata, A. Jaskot, M.S. Oey, D. Schaerer, Astron. Astrophys. **616**, A60 (2018). https://doi.org/10.1051/0004-6361/201732478
81. Y.I. Izotov, G. Worseck, D. Schaerer, N.G. Guseva, T.X. Thuan, Fricke, A. Verhamme, I. Orlitová, Mon. Not. R. Astron. Soc. **478**(4), 4851 (2018). https://doi.org/10.1093/mnras/sty1378
82. Y.I. Izotov, I. Orlitová, D. Schaerer, T.X. Thuan, A. Verhamme, N.G. Guseva, G. Worseck, Nature **529**, 178 (2016). https://doi.org/10.1038/nature16456
83. T. Kawamuro, Y. Ueda, K. Ichikawa, M. Imanishi, T. Izumi, A. Tanimoto, K. Matsuoka, Astrophys. J. **881**(1), 48 (2019). https://doi.org/10.3847/1538-4357/ab2bf6
84. A.W. Blain, I. Smail, R.J. Ivison, J.P. Kneib, D.T. Frayer, Phys. Rep. **369**(2), 111 (2002). https://doi.org/10.1016/S0370-1573(02)00134-5
85. C.J. Lonsdale, D. Farrah, H.E. Smith, in *Ultraluminous Infrared Galaxies* (2006), p. 285
86. C.M. Casey, D. Narayanan, A. Cooray, Phys. Rep. **541**(2), 45 (2014). https://doi.org/10.1016/j.physrep.2014.02.009
87. C. Leitherer, R. Chandar, C.A. Tremonti, A. Wofford, D. Schaerer, Astrophys. J. **772**, 120 (2013). https://doi.org/10.1088/0004-637X/772/2/120
88. C.C. Steidel, M. Giavalisco, M. Dickinson, K.L. Adelberger, Astron. J. **112**, 352 (1996). https://doi.org/10.1086/118019
89. B. Vulcani, M. Trenti, V. Calvi, R. Bouwens, P. Oesch, M. Stiavelli, M. Franx, Characterization and modeling of contamination for Lyman break galaxy samples at high redshift. Astrophys. J. **836**(2), 239 (2017). https://doi.org/10.3847/1538-4357/aa5caf
90. A.M. Quider, M. Pettini, A.E. Shapley, C.C. Steidel, Mon. Not. R. Astron. Soc. **398**, 1263 (2009). https://doi.org/10.1111/j.1365-2966.2009.15234.x
91. A.M. Quider, A.E. Shapley, M. Pettini, C.C. Steidel, D.P. Stark, Mon. Not. R. Astron. Soc. **402**, 1467 (2010). https://doi.org/10.1111/j.1365-2966.2009.16005.x
92. M. Dessauges-Zavadsky, S. D'Odorico, D. Schaerer, A. Modigliani, C. Tapken, J. Vernet, Astron. Astrophys. **510**, A26 (2010). https://doi.org/10.1051/0004-6361/200913337

93. T.E. Rivera-Thorsen, H. Dahle, M. Gronke, M. Bayliss, J.R. Rigby, R. Simcoe, R. Bordoloi, M. Turner, G. Furesz, Astron. Astrophys. **608**, L4 (2017). https://doi.org/10.1051/0004-6361/201732173
94. M. Giavalisco, Annu. Rev. Astron. Astrophys. **40**, 579 (2002). https://doi.org/10.1146/annurev.astro.40.121301.111837
95. J.S. Dunlop, in *Observing the First Galaxies*. Astrophysics and Space Science Library, vol. 396 (2013), p. 223
96. C. Behrens, M. Dijkstra, J.C. Niemeyer, Astron. Astrophys. **563**, A77 (2014). https://doi.org/10.1051/0004-6361/201322949
97. A. Verhamme, I. Orlitová, D. Schaerer, M. Hayes, Astron. Astrophys. **578**, A7 (2015). https://doi.org/10.1051/0004-6361/201423978
98. A. Hagen, G.R. Zeimann, C. Behrens, R. Ciardullo, H.S. Grasshorn Gebhardt, C. Gronwall, J.S. Bridge, D.B. Fox, D.P. Schneider, J.R. Trump, G.A. Blanc, Y.K. Chiang, T.S. Chonis, S.L. Finkelstein, G.J. Hill, S. Jogee, E. Gawiser, Astrophys. J. **817**, 79 (2016). https://doi.org/10.3847/0004-637X/817/1/79
99. M. Ouchi, B. Mobasher, K. Shimasaku, H.C. Ferguson, S.M. Fall, Y. Ono, N. Kashikawa, T. Morokuma, K. Nakajima, S. Okamura, M. Dickinson, M. Giavalisco, K. Ohta, Astrophys. J. **706**, 1136 (2009). https://doi.org/10.1088/0004-637X/706/2/1136
100. M. Hayes, D. Schaerer, G. Östlin, J.M. Mas-Hesse, H. Atek, D. Kunth, Astrophys. J. **730**(1), 8 (2011). https://doi.org/10.1088/0004-637X/730/1/8
101. H. Atek, D. Kunth, D. Schaerer, J.M. Mas-Hesse, M. Hayes, G. Östlin, J.P. Kneib, Astron. Astrophys. **561**, A89 (2014). https://doi.org/10.1051/0004-6361/201321519
102. D. Sobral, J. Matthee, B. Darvish, D. Schaerer, B. Mobasher, H.J.A. Röttgering, S. Santos, S. Hemmati, Astrophys. J. **808**, 139 (2015). https://doi.org/10.1088/0004-637X/808/2/139
103. A. Zitrin, I. Labbé, S. Belli, R. Bouwens, R.S. Ellis, G. Roberts-Borsani, D.P. Stark, P.A. Oesch, R. Smit, Astrophys. J. Lett. **810**, L12 (2015). https://doi.org/10.1088/2041-8205/810/1/L12
104. P.A. Oesch, G. Brammer, P.G. van Dokkum, G.D. Illingworth, R.J. Bouwens, I. Labbé, M. Franx, I. Momcheva, M.L.N. Ashby, G.G. Fazio, V. Gonzalez, B. Holden, D. Magee, R.E. Skelton, R. Smit, L.R. Spitler, M. Trenti, S.P. Willner, Astrophys. J. **819**, 129 (2016). https://doi.org/10.3847/0004-637X/819/2/129
105. L. Guaita, E. Gawiser, N. Padilla, H. Francke, N.A. Bond, C. Gronwall, R. Ciardullo, J.J. Feldmeier, S. Sinawa, G.A. Blanc, S. Virani, Astrophys. J. **714**, 255 (2010). https://doi.org/10.1088/0004-637X/714/1/255
106. J.D. Bowman, A.E.E. Rogers, R.A. Monsalve, T.J. Mozdzen, N. Mahesh, Nature **555**(7694), 67 (2018). https://doi.org/10.1038/nature25792
107. J.E. Gunn, B.A. Peterson, Astrophys. J. **142**, 1633 (1965). https://doi.org/10.1086/148444
108. G.D. Becker, J.S. Bolton, A. Lidz, Publ. Astron. Soc. Aust. **32**, e045 (2015). https://doi.org/10.1017/pasa.2015.45
109. B.E. Robertson, S.R. Furlanetto, E. Schneider, S. Charlot, R.S. Ellis, D.P. Stark, R.J. McLure, J.S. Dunlop, A. Koekemoer, M.A. Schenker, M. Ouchi, Y. Ono, E. Curtis-Lake, A.B. Rogers, R.A.A. Bowler, M. Cirasuolo, Astrophys. J. **768**(1), 71 (2013). https://doi.org/10.1088/0004-637X/768/1/71
110. B.E. Robertson, R.S. Ellis, S.R. Furlanetto, J.S. Dunlop, Astrophys. J. Lett. **802**(2), L19 (2015). https://doi.org/10.1088/2041-8205/802/2/L19
111. B.E. Robertson, R.S. Ellis, J.S. Dunlop, R.J. McLure, D.P. Stark, Nature **468**(7320), 49 (2010). https://doi.org/10.1038/nature09527
112. J.P. Paardekooper, S. Khochfar, C. Dalla Vecchia, Mon. Not. R. Astron. Soc. **451**, 2544 (2015)
113. S. Borthakur, T.M. Heckman, C. Leitherer, R.A. Overzier, Science **346**, 216 (2014). https://doi.org/10.1126/science.1254214
114. E. Vanzella, S. de Barros, K. Vasei, A. Alavi, M. Giavalisco, B. Siana, A. Grazian, G. Hasinger, H. Suh, N. Cappelluti, F. Vito, R. Amorin, I. Balestra, M. Brusa, F. Calura, M. Castellano, A. Comastri, A. Fontana, R. Gilli, M. Mignoli, L. Pentericci, C. Vignali, G. Zamorani, Astrophys. J. **825**, 41 (2016). https://doi.org/10.3847/0004-637X/825/1/41

115. A.E. Shapley, C.C. Steidel, A.L. Strom, M. Bogosavljević, N.A. Reddy, B. Siana, R.E. Mostardi, G.C. Rudie, Q1549-C25: a clean source of Lyman-continuum emission at z = 3.15. Astrophys. J. **826**(2), L24 (2016). https://doi.org/10.3847/2041-8205/826/2/L24
116. F. Bian, X. Fan, I. McGreer, Z. Cai, L. Jiang, Astrophys. J. Lett. **837**, L12 (2017). https://doi.org/10.3847/2041-8213/aa5ff7
117. C.C. Steidel, M. Bogosavljević, A.E. Shapley, N.A. Reddy, G.C. Rudie, M. Pettini, R.F. Trainor, A.L. Strom, Astrophys. J. **869**(2), 123 (2018). https://doi.org/10.3847/1538-4357/aaed28
118. T.J. Fletcher, M. Tang, B.E. Robertson, K. Nakajima, R.S. Ellis, D.P. Stark, A. Inoue, Astrophys. J. **878**(2), 87 (2019). https://doi.org/10.3847/1538-4357/ab2045
119. K. Nakajima, M. Ouchi, Mon. Not. R. Astron. Soc. **442**, 900 (2014). https://doi.org/10.1093/mnras/stu902
120. G. Stasińska, Y. Izotov, C. Morisset, N. Guseva, Astron. Astrophys. **576**, A83 (2015). https://doi.org/10.1051/0004-6361/201425389
121. J. Chisholm, S. Gazagnes, D. Schaerer, A. Verhamme, J.R. Rigby, M. Bayliss, K. Sharon, M. Gladders, H. Dahle, Astron. Astrophys. **616**, A30 (2018). https://doi.org/10.1051/0004-6361/201832758
122. A. Verhamme, I. Orlitová, D. Schaerer, Y. Izotov, G. Worseck, T.X. Thuan, N. Guseva, Astron. Astrophys. **597**, A13 (2017). https://doi.org/10.1051/0004-6361/201629264
123. E. Giallongo, A. Grazian, F. Fiore, A. Fontana, L. Pentericci, E. Vanzella, M. Dickinson, D. Kocevski, M. Castellano, S. Cristiani, H. Ferguson, S. Finkelstein, N. Grogin, N. Hathi, A.M. Koekemoer, J.A. Newman, M. Salvato, Astron. Astrophys. **578**, A83 (2015). https://doi.org/10.1051/0004-6361/201425334
124. P. Madau, F. Haardt, Astrophys. J. Lett. **813**(1), L8 (2015). https://doi.org/10.1088/2041-8205/813/1/L8
125. S. Mitra, T.R. Choudhury, A. Ferrara, Mon. Not. R. Astron. Soc. **473**(1), 1416 (2018). https://doi.org/10.1093/mnras/stx2443
126. N. Trentham, in *Starbursts: From 30 Doradus to Lyman Break Galaxies*, ed. by R. de Grijs, R.M. González Delgado. Astrophysics and Space Science Library, vol. 329 (2005), p. 115
127. R. Bacon, J. Brinchmann, J. Richard, T. Contini, A. Drake, M. Franx, S. Tacchella, J. Vernet, L. Wisotzki, J. Blaizot, N. Bouché, R. Bouwens, S. Cantalupo, C.M. Carollo, D. Carton, J. Caruana, B. Clément, S. Dreizler, B. Epinat, B. Guiderdoni, C. Herenz, T.O. Husser, S. Kamann, J. Kerutt, W. Kollatschny, D. Krajnovic, S. Lilly, T. Martinsson, L. Michel-Dansac, V. Patricio, J. Schaye, M. Shirazi, K. Soto, G. Soucail, M. Steinmetz, T. Urrutia, P. Weilbacher, T. de Zeeuw, Astron. Astrophys. **575**, A75 (2015). https://doi.org/10.1051/0004-6361/201425419
128. R. Bacon, S. Conseil, D. Mary, J. Brinchmann, M. Shepherd, M. Akhlaghi, P.M. Weilbacher, L. Piqueras, L. Wisotzki, D. Lagattuta, B. Epinat, A. Guerou, H. Inami, S. Cantalupo, J.B. Courbot, T. Contini, J. Richard, M. Maseda, R. Bouwens, N. Bouché, W. Kollatschny, J. Schaye, R.A. Marino, R. Pello, C. Herenz, B. Guiderdoni, M. Carollo, Astron. Astrophys. **608**, A1 (2017). https://doi.org/10.1051/0004-6361/201730833
129. L. Wisotzki, R. Bacon, J. Brinchmann, S. Cantalupo, P. Richter, J. Schaye, K.B. Schmidt, T. Urrutia, P.M. Weilbacher, M. Akhlaghi, N. Bouché, T. Contini, B. Guiderdoni, E.C. Herenz, H. Inami, J. Kerutt, F. Leclercq, R.A. Marino, M. Maseda, A. Monreal-Ibero, T. Nanayakkara, J. Richard, R. Saust, M. Steinmetz, M. Wendt, Nature **562**(7726), 229 (2018). https://doi.org/10.1038/s41586-018-0564-6
130. A. Karim, A.M. Swinbank, J.A. Hodge, I.R. Smail, F. Walter, A.D. Biggs, J.M. Simpson, A.L.R. Danielson, D.M. Alexander, F. Bertoldi, C. de Breuck, S.C. Chapman, K.E.K. Coppin, H. Dannerbauer, A.C. Edge, T.R. Greve, R.J. Ivison, K.K. Knudsen, K.M. Menten, E. Schinnerer, J.L. Wardlow, A. Weiß, P. van der Werf, Mon. Not. R. Astron. Soc. **432**(1), 2 (2013). https://doi.org/10.1093/mnras/stt196
131. J.S. Dunlop, R.J. McLure, A.D. Biggs, J.E. Geach, M.J. Michałowski, R.J. Ivison, W. Rujopakarn, E. van Kampen, A. Kirkpatrick, A. Pope, D. Scott, A.M. Swinbank, T.A. Targett, I. Aretxaga, J.E. Austermann, P.N. Best, V.A. Bruce, E.L. Chapin, S. Charlot,

M. Cirasuolo, K. Coppin, R.S. Ellis, S.L. Finkelstein, C.C. Hayward, D.H. Hughes, E. Ibar, P. Jagannathan, S. Khochfar, M.P. Koprowski, D. Narayanan, K. Nyland, C. Papovich, J.A. Peacock, G.H. Rieke, B. Robertson, T. Vernstrom, P.P.v.d. Werf, G.W. Wilson, M. Yun, Mon. Not. R. Astron. Soc. **466**(1), 861 (2017). https://doi.org/10.1093/mnras/stw3088

132. D.A. Riechers, A. Pope, E. Daddi, L. Armus, C.L. Carilli, F. Walter, J. Hodge, R.R. Chary, G.E. Morrison, M. Dickinson, H. Dannerbauer, D. Elbaz, Astrophys. J. **786**(1), 31 (2014). https://doi.org/10.1088/0004-637X/786/1/31
133. J.D. Vieira, D.P. Marrone, S.C. Chapman, C. De Breuck, Y.D. Hezaveh, A. Wei$\beta$, J.E. Aguirre, K.A. Aird, M. Aravena, M.L.N. Ashby, M. Bayliss, B.A. Benson, A.D. Biggs, L.E. Bleem, J.J. Bock, M. Bothwell, C.M. Bradford, M. Brodwin, J.E. Carlstrom, C.L. Chang, T.M. Crawford, A.T. Crites, T. de Haan, M.A. Dobbs, E.B. Fomalont, C.D. Fassnacht, E.M. George, M.D. Gladders, A.H. Gonzalez, T.R. Greve, B. Gullberg, N.W. Halverson, F.W. High, G.P. Holder, W.L. Holzapfel, S. Hoover, J.D. Hrubes, T.R. Hunter, R. Keisler, A.T. Lee, E.M. Leitch, M. Lueker, D. Luong-van, M. Malkan, V. McIntyre, J.J. McMahon, J. Mehl, K.M. Menten, S.S. Meyer, L.M. Mocanu, E.J. Murphy, T. Natoli, S. Padin, T. Plagge, C.L. Reichardt, A. Rest, J. Ruel, J.E. Ruhl, K. Sharon, K.K. Schaffer, L. Shaw, E. Shirokoff, J.S. Spilker, B. Stalder, Z. Staniszewski, A.A. Stark, K. Story, K. Vanderlinde, N. Welikala, R. Williamson, Nature **495**, 344 (2013). https://doi.org/10.1038/nature12001
134. T. Hashimoto, N. Laporte, K. Mawatari, R.S. Ellis, A.K. Inoue, E. Zackrisson, G. Roberts-Borsani, W. Zheng, Y. Tamura, F.E. Bauer, T. Fletcher, Y. Harikane, B. Hatsukade, N.H. Hayatsu, Y. Matsuda, H. Matsuo, T. Okamoto, M. Ouchi, R. Pelló, C.E. Rydberg, I. Shimizu, Y. Taniguchi, H. Umehata, N. Yoshida, Nature **557**, 392 (2018). https://doi.org/10.1038/s41586-018-0117-z
135. R. Augustin, S. Quiret, B. Milliard, C. Péroux, D. Vibert, J. Blaizot, Y. Rasera, R. Teyssier, S. Frank, J.M. Deharveng, V. Picouet, D.C. Martin, E.T. Hamden, N. Thatte, M. Pereira Santaella, L. Routledge, S. Zieleniewski, Mon. Not. R. Astron. Soc. **489**(2), 2417 (2019). https://doi.org/10.1093/mnras/stz2238

The manufacturer's authorised representative in the EU is Springer Nature Customer Service Centre GmbH, Europaplatz 3, 69115 Heidelberg, Germany. If you have any concerns regarding our products, please contact ProductSafety@springernature.com

Printed and bound by CPI Group (UK) Ltd, Croydon, CR0 4YY
05/10/2026
02228124-0017